NUTRITIONAL COSMETICS: BEAUTY FROM WITHIN

PERSONAL CARE AND COSMETIC TECHNOLOGY

Series Editor: Meyer Rosen

President, Interative Consulting, Inc., NY, USA

9780815515043 Delivery System Handbook for Personal Care and Cosmetic Products: Technology, Applications and Formulations (2005) Edited by Meyer Rosen

9780815515678 Global Regulatory Issues for the Cosmetics Industry, volume 1 (2007) Edited by C. I. Betton

9780815515692 Global Regulatory Issues for the Cosmetics Industry, volume 2 (2009) Edited by Karl Lintner

9780815515722 Cosmetic Applications of Laser and Light-Based Systems (2009) Edited by Gurpreet S. Ahluwalia

9780815515845 Skin Aging Handbook: An Integrated Approach to Biochemistry and Product Development (2008) Edited by Nava Dayan

9780815520290 Nutritional Cosmetics: Beauty from Within (2009) Edited by Aaron Tabor and Robert M. Blair

NUTRITIONAL COSMETICS

Beauty from Within

Aaron Tabor

MD, CEO and Medical Research Director of Physicians Laboratories, North Carolina, USA

Robert M. Blair

Research Manager, Physicians Laboratories, North Carolina, USA

William Andrew is an imprint of Elsevier
Linacre House, Jordan Hill, Oxford OX2 8DP, UK
30 Corporate Drive, Suite 400, Burlington, MA 01803, USA

First edition 2009

British Library Cataloguing in Publication Data
A catalogue record for this book is available from the British Library

Library of Congress Cataloging-in-Publication Data
A catalog record for this book is available from the Library of Congress

ISBN: 978-0-8155-2029-0

For information on all William Andrew publications
visit our website at elsevierdirect.com

Printed and bound in United States of America

09 10 11 12 11 10 9 8 7 6 5 4 3 2 1

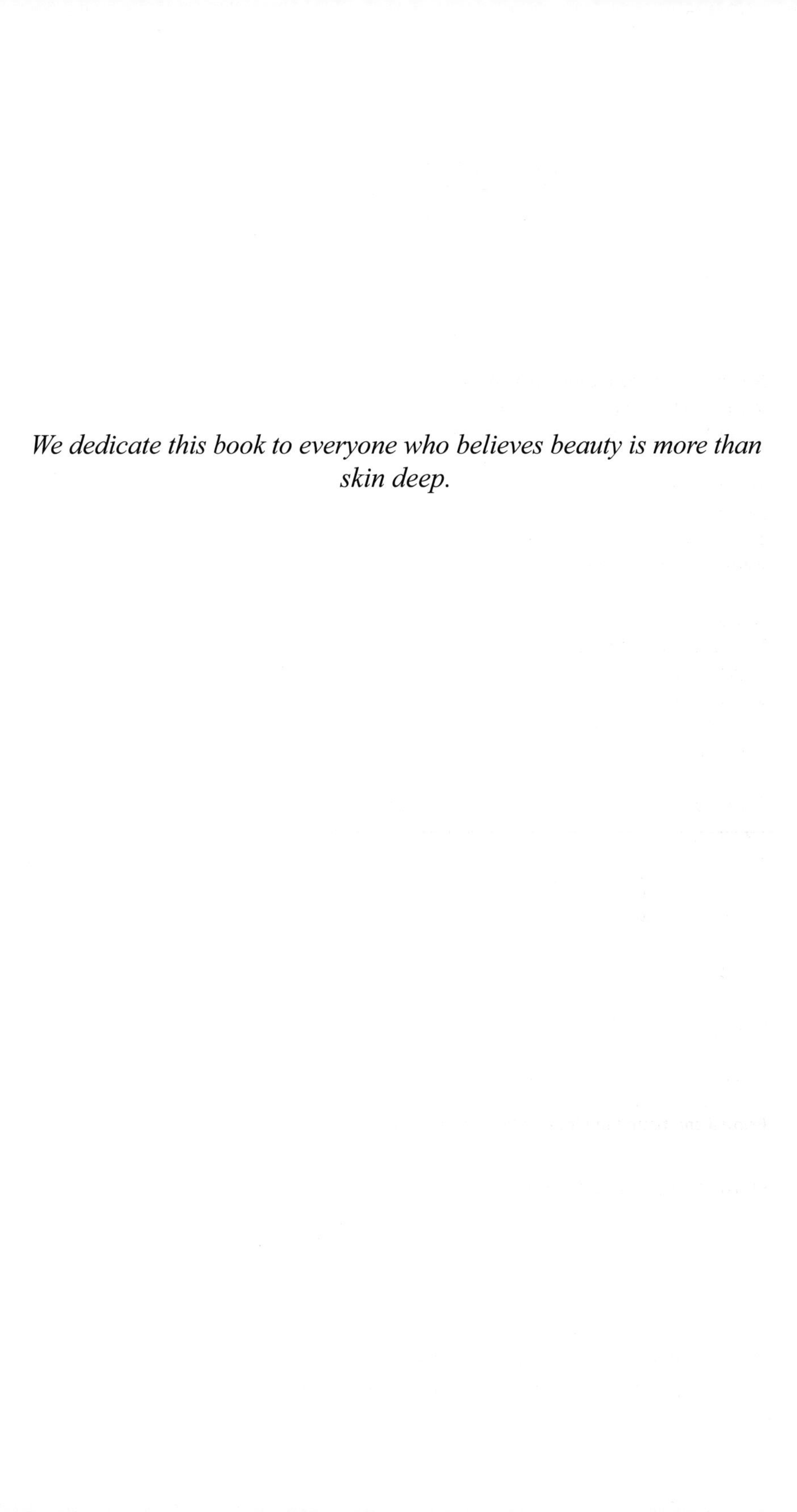

We dedicate this book to everyone who believes beauty is more than skin deep.

Contents

Contributors

Farrukh Afaq, PhD
Associate Scientist
Department of Dermatology
University of Wisconsin
4385 Medical Sciences Center
1300 University Avenue
Madison, WI 53706
Phone: +1 608 263 5519
Email: fafaq@dermatology.wisc.edu

Rajesh Agarwal, PhD
Professor
UCD School of Pharmacy
4200 East 9th Avenue, C-238
Denver, CO 80262
Phone: +1 303 315 1381
Fax: +1 303 315 6281
Email: Rajesh.Agarwal@uchsc.edu

Christian Artaria, BSc
Indena S.p.A.
Viale Ortles, 12
20139 Milan
Italy

Yutaka Ashida, PhD
Shiseido Co., Ltd.
Frontier Science Labs
2-12-1 Fukuura Kanazawa-Ku
Yokohama 236-8643
Japan
Email: yutaka.ashida@to.shiseido.co.jp

Debasis Bagchi, PhD, FACN, CNS
Department of Pharmacological & Pharmaceutical Sciences
College of Pharmacy
University of Houston
Houston, TX
Email: dbagchi@interhealthusa.com

Manashi Bagchi, PhD, FACN
Interhealth Research Center
Benicia, CA
Email: mbagchi@interhealthusa.com

Jalil Benyacoub, PhD
Nestle Research Center
Vers-Chez-Les-Blancs
P.O. Box 44
1000 Lausanne 26
Switzerland

Bruno Berra, PhD
Professor
Institute of General Physiology and Biochemistry, "G. Esposito"
Via Trentacoste 2
20134 Milan
Italy
Phone: +39 02 50315777
Fax: +39 02 50315775
Email: bruno.berra@unimi.it

Robert M. Blair, PhD
Physicians Pharmaceuticals, Inc.
1031 E. Mountain Street, Building 302
Kernersville, NC 27284
Phone: +1 336 722 2337 Ext. 1734
Fax: +1 336 722 7712
Email: drblair@revivalsoy.com

Stephanie Blum, PhD
Nestle Research Center
Vers-Chez-Les-Blancs
P.O. Box 44
1000 Lausanne 26
Switzerland

Lionel Breton, PhD, DrSc
L'Oreal
Charles Zviak Center
90 Rue du Général Roguet
92 583 Clichy Cedex
France
Phone: +33 1 47 56 7887
Fax: +33 1 47 56 40 07

Petra Caessens, PhD
DMV International
40196 State Highway 10
Delhi, NY 13753

Isabelle Castiel, PharmD, PhD
Scientific Coordinator Food Supplement Research
Life Science Research
L'Oreal Research and Development
L'Oreal
Centre C. Zviak
90 Rue du Général Roguet
92 583 Clichy Cedex
France
Phone: +33 4 93 83 09 65
Fax: +33 1 47 56 40 07
Email: icastiel@rd.loreal.com

Dr. Leonardo Celleno
Department of Dermatology
Catholic University of Sacred Heart
Largo A. Gemelli 8
00168 Rome
Italy
Phone: +39 06 3503591
Fax: +39 06 3503547
Email: lcelleno@libero.it

Meera Chandramouli
PL Thomas & Co., Inc., NJ, USA
Phone: +1 973 984 0900 Ext. 215
Email: PLT@PLThomas.com

Aldo Cristoni, PhD
Indena S.p.A.
Viale Ortles, 12
20139 Milan
Italy
Email: aldo472006@gmail.com

Gagan Deep, PhD
Research Associate
UCD School of Pharmacy
4200 East 9th Avenue, C-238
Denver, CO 80262
Phone: +1 303 315 1228
Fax: +1 303 315 6281
Email: Gagan.Deep@uchsc.edu

Rick de Waard, PhD
R&D Director, Nutritionals
DMV International
Innovium
Nieuwe Kanaal 7r
NL-6709 PA Wageningen
The Netherlands
Phone: +31 0 6 10377844

Dushka Dimitriejvic, MSc
Vitiva d.o.o.
Nova vas pri Markovcih 98
2281 Markovci
Slovenia
Phone: +386 2 788 87 38
Fax: +386 2 7888 745
Email: dushka@vitiva.si

David Djerassi
President
Intrachem Technologies
New York, NY
Email: daviddjerassi@msn.com

Zoe Diana Draelos, MD
Dermatology Consulting Services, Inc.
2444 N. Main Street
High Point, NC 27262-7833
Email: zdraelos@northstate.net

Craig A. Elmets, MD
Department of Dermatology
University of Alabama at Birmingham
Birmingham, AL 35294

Andrea Giori, PhD
Indena S.p.A.
Viale Ortles, 12
20139 Milan
Italy
Email: andrea.giori@indena.com

Audrey Gueniche, PharmD, PhD
Project Director
Oral Route Research—Skin and Hair
L'Oreal
Charles Zviak Center
90 Rue du Général Roguet
92 583 Clichy Cedex
France
Phone: +33 1 47 56 40 15 or +33 6 60 99 24 67
Fax: +33 1 47 56 78 88
Email: agueniche@rd.loreal.com

Rebat M. Halder, MD
Professor and Chairman
Department of Dermatology
Howard University College of Medicine
2041 Georgia Avenue, N.W.
Washington, DC 20060
Phone: +1 202 865 6725
Fax: +1 202 865 1757
Email: halderderm@rcn.com

Takeshi Ikemoto, PhD
Kanebo Cosmetics
Basic Research Laboratory
3-28, 5-Chome, Kotobuki-Cho, Odawara-Shi
Kanagawa-Ken, 250-0002, Japan

Eunsun Jung
Biospectrum Life Science Institute
SK Ventium 101-701
#552 Dangjung Dong
Gunpo City
Gyunggi Do, 435-776
Republic of Korea
Phone: +82 31 436 2090
Fax: +82 31 436 0605

Santosh K. Katiyar, PhD
Associate Professor of Dermatology
Department of Dermatology
1670 University Boulevard
Volker Hall 557, Box 202
University of Alabama at Birmingham
Birmingham, AL 35294
Phone: +1 205 975 2608
Fax: +1 205 934 5745
Email: skatiyar@uab.edu

Suchitra Katiyar, BS, MPH
Department of Dermatology
University of Alabama at Birmingham
Birmingham, AL 35294

Manjinder Kaur, PhD
Research Associate
UCD School of Pharmacy
4200 East 9th Avenue, C-238
Denver, CO 80262
Phone: +1 303 315 2187
Fax: +1 303 315 6281
Email: Manjinder.Kaur@uchsc.edu

Saebom Kim, PhD
Biospectrum Life Science Institute
SK Ventium 101-701
#552 Dangjung Dong
Gunpo City
Gyunggi Do, 435-776
Republic of Korea
Phone: +82 31 436 2090
Fax: +82 31 436 0605

Su Na Kim
Natural Products Research Institute
College of Pharmacy
Seoul National University
28 Yeonkun-Dong, Jongno-Ku
Seoul 110-460
Korea

Yeong Shik Kim, PhD
Professor
Natural Products Research Institute
College of Pharmacy
Seoul National University
28 Yeonkun-Dong, Jongno-Ku
Seoul 110-460
Korea
Phone: +82 2 740 8929
Fax: +82 2 765 4768
Email: kims@snu.ac.kr

Chesahna Kindred, MD, MBA
Department of Dermatology
Howard University College of Medicine
2041 Georgia Avenue, N.W.
Washington, DC 20060

Majda Hadolin Kolar, PhD
Vitiva d.d.
Nova vas 98
2281 Markovci
Slovenia
Phone: +386 2 7888 736
Fax: +386 2 7888 745
Email: majda@vitiva.si

Dr. Jean Krutmann, MD
Professor of Dermatology & Environmental Medicine
Director
Institut für Umweltmedizinische Forschung (IUF)
at the Heinrich-Heine-University Düsseldorf gGmbH
Auf'm Hennekamp 50
D-40225 Düsseldorf
Germany
Phone: +49 211 3389 224
Fax: +49 211 3389 226
Email: krutmann@rz.uni-duesseldorf.de

Francis C. Lau, PhD, FACN
Interhealth Research Center
Benicia, CA
Email: flau@interhealthusa.com

Jongsung Lee
Natural Products Research Institute
College of Pharmacy
Seoul National University
28 Yeonkun-Dong, Jongno-Ku
Seoul 110-460
Korea

Juhyeon Lee
Natural Products Research Institute
College of Pharmacy
Seoul National University
28 Yeonkun-Dong, Jongno-Ku
Seoul 110-460
Korea

Howard I. Maibach, MD
Department of Dermatology
University of California
90 Medical Center Way
Surge 110, Box 0989
San Francisco, CA 94143-0989
Phone: +1 949 533 6892
Fax: +1 415 753 5304
E-mail: MaibachH@derm.ucsf.edu

Giada Maramaldi, BSc
Indena S.p.A.
Viale Ortles, 12
20139 Milan
Italy
Email: giada.maramaldi@indena.com

Pierfrancesco Morganti, PhD
R&D Directdor
Mavi Sud. S.r.1.
Via dell'Industria 1
04011 Aprilia (LT)
Italy
Email: morganti@mavicosmetics.it

Hasan Mukhtar, PhD
Helfaer Professor of Cancer Research
Director and Vice Chair of Research
Department of Dermatology
University of Wisconsin
4385 Medical Sciences Center
1300 University Avenue
Madison, WI 53706
Phone: +1 608 263 3927
Email: hmukhtar@wisc.edu

Ayako Noguchi, MSc
Kyowa Hakko U.S.A., Inc.
767 Third Avenue, 19th Floor
New York, NY 10017
Phone: +1 212.319.5353
Email: info@kyowa-usa.com

Christian O. Oresajo, PhD
Department of Dermatology
Howard University College of Medicine
2041 Georgia Avenue, N.W.
Washington, DC 20060

Deokhoon Park, PhD
Biospectrum Life Science Institute
SK Ventium 101-701
#552 Dangjung Dong
Gunpo City
Gyunggi Do, 435-776
Republic of Korea
Phone: +82 31 436 2090
Fax: +82 31 436 0605

Nadine Pomarede, MD
Isocell, Paris, France
and
PL Thomas & Co., Inc., NJ, USA
Phone: +1 973 984 0900 Ext. 215
Email: PLT@PLThomas.com

Myriam Richelle, PhD
Nestle Research Center
Nestec Ltd
P.O. Box 44
CH-1000
Lausanne 26
Switzerland
Phone: +41 21 785 84 07
Fax: +41 21 785 85 44
Email: myriam.richelle@rdls.nestle.com

Angela Maria Rizzo, PhD
Professor
Institute of General Physiology and Biochemistry, "G. Esposito"
Via Trentacoste 2
20134 Milan
Italy
Phone: +39 02 50315789
Fax: +39 02 50315775
Email: angelamaria.rizzo@unimi.it

Hiroshi Shimoda, PhD
Oryza Oil & Fat Chemical Co., Ltd.
1 Numata, Kitagata-cho
Ichinomiya-city, Aichi
493-8001, Japan
Phone: +81 586 86 5141
Fax: +81 586 86 6191
Email: kaihatsu@mri.biglobe.ne.jp

Heike Steiling, PhD
Nestle Research Center
Nestec Ltd
P.O. Box 44
CH-1000
Lausanne 26
Switzerland

Deeba N. Syed, MBBS
Research Specialist
Department of Dermatology
University of Wisconsin
4385 Medical Sciences Center
1300 University Avenue
Madison, WI 53706
Phone: +1 608 263 5519
Email: dsyed@dermatology.wisc.edu

Aaron Tabor, MD
Physicians Pharmaceuticals, Inc.
1031 E. Mountain Street, Building 302
Kernersville, NC 27284
Phone: +1 336 722 2337
Fax: +1 336 722 7712
Email: drtabor@revivalsoy.com

Dr. Federica Tamburi
Department of Dermatology
Catholic University of Sacred Heart
Largo A. Gemelli 8
00168 Rome
Italy
Phone: +39 06 3503591
Fax: +39 06 3503547

Simona Urbancic
Vitiva d.d.
Nova vas 98
2281 Markovci
Slovenia
Phone: +386 2 7888 736
Fax: +386 2 7888 745
Email: service@vitiva.si

James Varani, PhD
Department of Pathology
The University of Michigan
1301 Catherine Road/Box 0602
Ann Arbor, MI 48109
Phone: +1 734 615 0298
Fax: +1 734 763 6476
Email: varani@umich.edu

Jeanette Waller, MD
Department of Dermatology
University of California
90 Medical Center Way
Surge 110, Box 0989
San Francisco, CA 94143-0989
Email: WallerJ@derm.ucsf.edu

Angela Walter
Market Development Manager
DMV International
40196 State Highway 10
Delhi, NY 13753
Phone: +1 607 746 0206
Email: angela.walter@dmv-ny.com

Wendeline Wouters, PhD
Nutrition Science & Technical Sales Support Manager
DMV International
40196 State Highway 10
Delhi, NY 13753
Phone: +1 607 746 0126
Fax: +1 607 746 0183
Email: wendeline.wouters@dmv-ny.com

Shirley Zafra-Stone, BS
Interhealth Research Center
Benicia, CA
Email: sstone@interhealthusa.com

Danny Zaghi
1299 Bedford Ct.
Sunnyvale, CA 94087
Email: dannyzaghi@yahoo.com

Mohammad Abu Zaid, PhD
Assistant Researcher
Department of Dermatology
University of Wisconsin
4385 Medical Sciences Center
1300 University Avenue
Madison, WI 53706
Phone: +1 608 263 5519
Email: azaid@dermatology.wisc.edu

Foreword

We will all live decades longer than our grandparents and parents and will have better health in our later years. With this enhanced longevity, we all want to look as young as we feel. Especially because we are inundated with images of youthful beauty on television, in films, and in glossy fashion magazines, we feel social pressure to improve our appearance. Aspiring to maintain "eternal youth" and to look good is not new and is not just superficial "vanity." Our human nature dictates that we take care of ourselves and enhance our appearance. Throughout history, women from Cleopatra to Marie Antoinette to today's movie stars have used cosmetics and nutrients to beautify and rejuvenate. Indeed recent psychological studies have proven that the better we look, the better we feel, and the healthier, happier, and more productive we become.

Today in the United States the number of over-45-year-olds is growing at three times the rate of the general population. To meet this demand, the cosmetic, nutrition, and health care industries have created an enormous variety of anti-aging nutritional products, supplements, topical treatments, and medical and cosmetic procedures. Information and misinformation abound. Sales representatives and retail clerks, popular press editorials and advertisements, television commercials and infomercials, news and beauty segments, home shopping networks, and the internet—all present nutritional supplements and skin care products promising beautification with reversal of the appearance of aging. How can we navigate our way through this labyrinth of claims and counterclaims?

Over 25 years ago, the prominent dermatologist Dr. Albert Kligman defined the category "cosmeceuticals" to describe topical formulations that improve the appearance of the skin by actually altering the function of the skin in a scientifically measurable way. More recently the terms "nutricosmetics" or "nutraceuticals" have defined foods and dietary supplements that benefit the health and beauty of the skin by directly affecting mechanisms and metabolism. Such cosmeceuticals and nutricosmetics are not subject to the stringent criteria and costly studies required for "drugs" in order to provide scientific and medical proof of safety and efficacy for approval by the U.S. Food and Drug Administration. Therefore, while savvy marketers promote products based loosely on science, consumers and even physicians cannot truly judge the merits of most claims.

This text, *Nutritional Cosmetics: Beauty from Within*, edited by Aaron Tabor, MD, and Robert M. Blair, PhD, presents for the first time a compendium of the science behind dietary ingredients with the potential to benefit the health and therefore the appearance of the skin. The authors are medical doctors and scientists from academia and industry who recognize that careful, controlled studies are required to demonstrate efficacy of specific ingredients in humans, even after mechanisms of action are demonstrated in *in vitro* and in animal models. Furthermore, substantiation of each specific formulation of each product applied topically or taken orally is of utmost importance. The molecular form and the purity of the active component, the concentration, the vehicle, the pH, and the coating all affect the absorption and activity and therefore the efficacy.

This text opens with chapters describing in detail the structure and the physiology of the skin as well as the ethnic variations in certain properties and disorders of the skin and differences in cultural practices. In the following chapters, we realize that, amazingly, even several decades ago, we did not understand that environmental pollutants affect the skin, particularly sun exposure and smoking. These chapters summarize current research on mechanisms and consequences of these external onslaughts, particularly describing recent discoveries regarding the distinction between intrinsic, natural aging and extrinsic, premature aging, elucidating distinct clinical manifestations and mechanisms of each. With this review of skin physiology, how nutricosmetic ingredients can be studied to prove efficacy can be understood.

The rest of this book compiles evidence and verification as previously published in the scientific and medical literature as well as internal studies by industry about dietary ingredients with potential for skin health—including those already being applied topically. With this documentation, the reader will recognize that there are indeed nutritional supplements that have scientific substantiation of preventing and reversing aging, of protecting, of moisturizing, and of treating specific problems of the skin.

The most comprehensively researched nutricosmetics are antioxidants (vitamins C and E, carotenoids, coenzyme Q10, as well as botanicals such as the polyphenols [catechins] in green tea, the tannins and flavanoids in pomegranate, resveratrol in grapes, anthocyanins in berries, silibinin in milk thistle, carnosic and rosmarinic acids in rosemary, and genistein in soy), as well as antioxidant enzymes and trace mineral cofactors for these enzymes (such as selenium and zinc). Studies on each of these are excellently and comprehensively reviewed.

Certain nutricosmetics affect epidermal moisturization and barrier function and/or the extracellular matrix. Some ingredients can increase synthesis of collagen and prevent the UV-induced degradation of collagen and elastic tissue by matrix metalloproteinases; others increase fibroblast proliferation and cell turnover and may even induce signaling for cellular differentiation or apoptosis. A review of evidence that amino acids, ceramides, and other plant extracts can influence epidermal moisturization and dermal mechanisms is presented.

One important chapter summarizes the benefits to the skin of probiotics (living microorganisms), which influence the composition and/or metabolism of the endogenous gut and skin microbiota. By modulating immune function, probiotics are prophylactic and therapeutic for atopic dermatitis, skin sensitivity, and food allergies. *Lactobacillus johnsonii* protects Langerhans cells from depletion after UV radiation, whereas *Lactobacillus pesodoris* inhibits odor-producing bacteria of the gut and armpits.

Numerous epidemiological studies link the abundance of particular nutrients with disease prevention and improved health. Armed with these correlations, doctors and scientists have incorporated certain of these nutrients into topical and oral formulations to improve the health and appearance of the skin. This text comprehensively reviews research on nutrients for which there is evidence of benefit to the skin "from within." Future skin care will undoubtedly focus on these cosmeceuticals and nutraceuticals. In confirming that we can indeed achieve "beauty through science," this book is a springboard to stimulate new ideas and future research.

Karen E. Burke, MD, PhD
Department of Dermatology
Mount Sinai Medical Center
New York, NY, USA

Preface

Nutritional cosmetics, more commonly referred to as nutricosmetics, embraces the idea that beauty can be enhanced through the consumption of functional dietary products that may support healthier and thus more beautiful skin. The term nutricosmetics appears to borrow from the terms nutraceuticals and cosmeceuticals to reflect the goal of these products, that is, to provide health and beauty benefits to the skin via nutritional products consumed on a regular basis.

The idea that food or other dietary ingredients can support healthy skin and beauty has existed for ages. Some of the earliest dietary products marketed for skin health and beauty included Merz Spezial-Dragees in 1964 and Oenobiol's Solaire in 1989. Despite these early forays into the nutrition and beauty arena, the market for nutricosmetic products has only recently begun to take off with any real force. A recent market research report by Kline & Company indicates that the global nutricosmetic market is valued at $1.5 billion with the vast majority of that due to the markets in Japan and Europe. According to Euromonitor International, the market for nutricosmetics was $2.1 billion, which was only 3% of the overall skincare market, suggesting that nutriticosmetics is still very much an emerging market. In fact, various market analyses suggest that the market for nutricosmetic products will continue to increase at a substantial rate.

At the moment it appears that the marketing of nutricosmetics is ahead of the science for these products and ingredients in general, though that is not to say that quality research has not been done on some nutricosmetic ingredients. To date, there are a multitude of ingredients now being marketed for their purported skin health and beauty benefits and their utility for incorporation into nutricosmetic products. These ingredients often are marketed based on their antioxidant capacity and the resulting inferred skin benefits. However, in many cases no research on their actual dermatological benefits has been published or conducted. Nonetheless, many of these ingredients are already being applied topically with positive results, so it would not be totally surprising if oral consumption produced similar effects.

A look into the peer-reviewed scientific literature reveals an emerging body of evidence in support of the potential benefits of nutritional products for skin health. The studies are wide ranging and explore the effects of ingredients on both mechanistic endpoints (antioxidant capacity, anti-inflammatory properties, modulation of enzymes involved in extracellular matrix restructuring, etc.) and functional endpoints (anti-wrinkling, reduction of erythema, skin hydration, etc.). While the science in this area is quickly emerging, it is far from extensive enough to provide definitive conclusions at this time. In addition to the growing body of evidence published in the scientific literature, a plethora of information has been presented only in industry white papers or marketing materials. These materials show the promise of many of the marketed ingredients and also provide a great starting point for more definitive studies.

The purpose of this book was to compile the scientific evidence showing the potential benefits of at least some of the better-studied nutricosmetic ingredients. We started this project by searching the peer-reviewed literature for nutritional ingredients with evidence of skin health benefits. We additionally contacted several industry leaders currently marketing nutricosmetic products backed by scientific research. The response from the thought leaders in this field was overwhelming and the result is this compilation of excellent contributions that delve into the science behind dietary ingredients for improved skin health.

It is apparent that consumers today are very well informed and take their personal care very seriously. More and more consumers are looking for products that are both environmentally friendly and provide specific health benefits that meet their specific needs. With this interest in the potential health benefits of the products they consume, the importance of the science behind the products will continue to grow. While many functional ingredients will initially do well in the market based on their novelty, we believe that ingredients and products with sound science behind them will lead the way in the field of nutricosmetics.

Aaron T. Tabor, MD, and Robert M. Blair, PhD
June 2009

Introduction

Aaron Tabor, MD, and Robert M. Blair, PhD
Physicians Pharmaceuticals Inc., Kernersville, NC, USA

I.1 What is "Nutritional Cosmetics"?

Nutritional cosmetics, which is probably better known in the industry as nutricosmetics, encompasses the concept that orally ingestible dietary products may support healthier and thus more beautiful skin. This is not totally unlike the term nutraceutical; however, this latter term typically refers to foods and dietary supplements that support better overall health. Similarly, the term cosmeceutical refers to products generally designed for topical application and which contain active ingredients with benefits for improved skin health.

The term nutricosmetics appears to borrow from the terms nutraceuticals and cosmeceuticals to reflect the goal of these products, that is to provide beauty and health benefits to the skin via nutritional products consumed on a regular basis. This concept encompasses a unique amalgamation of the nutrition and personal care industries.

More and more nutricosmetic products are reaching the retail shelves and can be found as functional beverages (e.g., NutriSoda from Andrea Beverage Co., Skin Balance Water from Borba), dietary supplements (e.g., Murad's Firm & Tone dietary supplement, Perricone's Skin & Total Body dietary supplement), and functional foods (e.g., Danone's Essensis beauty yogurt, Ecco Bella's Chocolate Instant Bliss Beauty Bar).

I.2 The Nutricosmetic Market

While the idea that dietary ingredients can support healthy skin and beauty has been around for some time, it seems that it is only now receiving much attention. As such, the market for nutricosmetic products has only recently begun to take off with any real force. According to the 2006 Datamonitor report "Seeking Beauty Through Nutrition" (from Seaton T. Wellness trends in personal care. *Beauty from the Inside & Out* 2006; 1:16–22), sales of oral beauty supplements in 2005 was $742 million in the United States, $224 million in France, $162 million in Germany, and $95 million in the United Kingdom. Expected growth of oral beauty supplement sales through the year 2010 ranges from approximately 7% to 12%.

According to this article, the Datamonitor report further indicated that half of the survey participants considered themselves conscious of skin nutrition, while nearly half of the respondents believed that what they ate could have just as much effect on skin health as topical products. With this belief and consumers' ever-growing interest in the science behind the products they choose, we believe that ingredients and products with sound science behind them will lead the way in the field of nutricosmetics.

I.3 "Nutritional Cosmetics: Beauty from Within"—An Overview

There are a multitude of ingredients now being marketed for their purported skin health and beauty benefits and their utility for incorporation into nutricosmetic products. These ingredients often are marketed based on their antioxidant capacity and inferred skin benefits, though little or no research on their dermatological benefits has been published. However, many of these ingredients are already being applied topically with a reduction in wrinkles, so it is not that much of a surprise that oral consumption may produce similar effects.

Nonetheless, there are quite a few dietary ingredients with scientific backing for their potential skin health benefits. Though some of this evidence has been published in the scientific literature, a plethora of information has been presented only in industry white papers or marketing materials. The purpose of this book is to compile the scientific evidence showing the potential benefits of at least some of these nutricosmetic ingredients. By including the efforts of both academic and industry investigators, we believe that the presentation of the known scientific data with new information brings the state of nutricosmetic science up to the present and provides a foundation from which to generate new ideas and information. Where possible, information specifically about the benefits of ingredients consumed orally for skin health is presented.

Part 1 of this book consists of four chapters that provide an excellent overview of skin biology, including an in-depth look at the structure and function of skin and its components by Dr. Leonardo Celleno, an examination of ethnic skin by Dr. Chesahna Kindred, and overviews of both natural (Mr. Danny Zaghi) and premature (Dr. Jean Krutmann) aging of the skin.

In Chapter 5, Dr. Pierfrancesco Morganti discusses the concept of "beauty from within" and the integration of functional foods with cosmeceuticals for total body beauty from the inside and the outside. Vitamins and minerals have been reported to have healthful benefits for the skin. These are explored in Chapters 6 (Dr. Myriam Richelle) and 7 (Dr. Bruno Berra).

One of the most popular categories of nutricosmetic ingredients is the antioxidants. These ingredients have been proposed to support skin health and beauty through their ability to suppress and/or reverse oxidative damage to the skin caused by such stressors as UV light from the sun. Some of the more popular and efficacious ingredients are examined in Part 4 of the book, which starts off with an excellent overview of the potential benefits of botanical antioxidants by Dr. Mohammad Abu Zaid, which briefly touches upon the role of several specific ingredients. This part of the book also includes chapters on carotenoids (Dr. Pierfrancesco Morganti), coenzyme Q10 (Dr. Yutaka Ashida), healthy fruits (Dr. Francis Lau), olive fruit (Dr. Aldo Cristoni), and the skin's natural antioxidant enzymes (Dr. Nadine Pomarede).

One of the more important aspects of healthy looking skin is a smooth appearance. However, a smooth appearance can be difficult to maintain without a firm foundation and proper hydration. Dr. James Varani and Dr. Zoe Draelos discuss the importance of these aspects. Natural dietary ingredients that may support a firm skin foundation and proper skin hydration are discussed in Parts 5 and 6 of the book. A firm, skin-supporting foundation may be boosted with ingredients that are precursors to the skin's extracellular matrix like amino acids. Additionally, ingredients that stimulate collagen synthesis and inhibit enzymes responsible for the breakdown of the extracellular matrix may support the skin's foundation. Proper hydration is important for both the skin's barrier function and for a smooth appearance. Such ingredients as rice ceramides and tocotrienols are discussed in this regard. Additionally, ingredients that support fibroblast proliferation and cell turnover may lead to smoother, healthier-looking skin.

In addition to dietary ingredients that may help support a firm foundation and good skin hydration, a number of other ingredients have been shown to have potential benefits for overall skin health and appearance through a variety of possible mechanisms. Dr. Audrey Gueniche discusses the potential skin health benefits of probiotics, which may function by supporting a healthy immune system and having beneficial effects on skin reactivity. Additionally, the potential benefits of whey protein (Dr. Petra Caessens), rosemary (Dr. Majda Hadolin Kolar), and soy (Dr. Robert Blair) are discussed.

According to the American Cancer Society, most of the non-melanoma skin cancer cases in the United States are considered to be sun related. The risk of skin cancer can be reduced by following a number of sun-safety rules, including avoiding the sun during the middle part of the day, covering your skin with light clothing, wearing a hat, and using sunscreen. There is increasing evidence that dietary ingredients may also have some beneficial effects related to photocarcinogenesis. In the final part of this book, Dr. Suchitra Katiyar and Dr. Manjinder Kaur discuss the potential benefits and mechanisms of action of green tea and milk thistle, respectively, for the protection against sun damage and photocarcinogenesis.

The combined efforts of these exceptional authors have allowed us to present much of the current scientific evidence surrounding the potential benefits of some of the most-researched nutricosmetic ingredients. We believe that as the market for nutricosmetics grows, the importance of the scientific support behind these and new ingredients will continue to be critical.

We hope that you find this book a solid addition to your library.

PART 1
THE BIOLOGY OF HEALTHY AND AGING SKIN

1

Structure and Function of the Skin

Leonardo Celleno, MD, and Federica Tamburi, MD

Department of Dermatology,
Catholic University of Sacred Heart, Rome, Italy

Aaron Tabor and Robert M. Blair (eds.), *Nutritional Cosmetics: Beauty from Within*, 1–45, © 2009 William Andrew Inc.

1.1 Introduction

"The skin draws the line between the end of the organism and the beginning of the world outside. Internally, the skin shelters and protects all the physiochemical phenomenon necessary for life, externally it is a barrier against mechanical forces, both physical and chemical, which can be hostile to life. The most important role of the skin both for man and for every other organism, vertebrate or invertebrate, unicellular or multicellular, is to create an obstacle for all those things outside the organism: the rest of the world.

"If life were hermetically sealed, like a pod, survival would be impossible for a being that depended on the outside world. An organism must, therefore, develop in relation to the environment in which it must live and with which it must communicate. So, the barrier which protects the organism from the outside, must at the same time inform the interior of all that is occurring outside itself. It is the perfect balance of these two barrier functions that determines survival." Furthermore, human skin acts as an organ of attraction between individuals. The appearance of the skin and hair is the "first image" that others have of us. Personal expression changes with variations in the condition of our hair and skin, whose appearance is derived from their intrinsic well-being. Modern cosmetology has the task of interacting with physiology in maintaining its "good condition."

1.2 The Structure of the Skin

1.2.1 Macroscopic Characteristics

The skin is the largest, most extensive organ of our body. In fact, the average adult has about 170–200 cm^2 of skin with a weight that varies between 15 kg and 17 kg (obviously varying according to the subject's height and dimensions).

The thickness of the epidermis, the outermost layer of the skin, can be from 0.5 mm in the thinnest areas (the eyelids, for example) to 4–6 mm at its thickest points (as on the palm of the hand and the sole of the foot). This thickness parameter becomes especially important when a substance is applied to the skin, be it a pharmaceutical or cosmetic product. In fact, once in contact with the skin any substance can penetrate the cutaneous barrier in a way directly proportional to the skin's thickness at that point.

The skin, even if it appears smooth and compact to our eyes, is in reality marked over its entire surface by grooves, some shallow, others deeper, which by their layout mark many small polygons (Fig. 1.1). On the palm of the hand and the sole of the foot these grooves (dermatoglyphics) have become so evident as to characterize each individual and so unique that they are a distinct identification for each person. These apparently unimportant grooves are, however, necessary to accomplish an essential function—that of permitting the skin to stretch; if the skin were completely smooth, many of our movements would be impossible (Fig. 1.1).

The skin tissue houses within its structure other important constituents: hairs, nails, etc. (the skin's annexes). Even with the naked eye one can

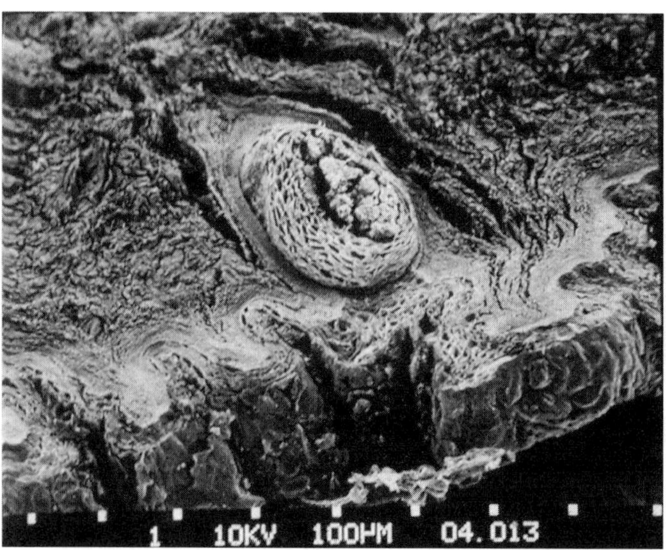

Figure 1.1 Human skin as seen with the scanning electron microscope.

see that (with the exception of the palm and sole) the whole of the skin is covered with hairs. In some areas the hairs are more developed and more coloured, as on the scalp, in the pubic region, and in the armpit. In other areas they are finer and much paler. These characteristics vary above all according to sex but also with individual biology and in the presence of certain pathologies.

Furthermore, tiny, invisible openings are found over the entire skin surface. These are the outlets of the eccrine sudoriparous glands, which, together with the apocrine sudoriparous glands and the sebaceous glands, will be handled in more detail later.

1.2.2 Microscopic Characteristics

When we consider the skin in its entirety, three different superimposed tissues can be identified (Fig. 1.2):

1. epidermis: the most external layer in contact with the environment
2. dermis: below the epidermis it is the structural component of the skin and the underlying organs

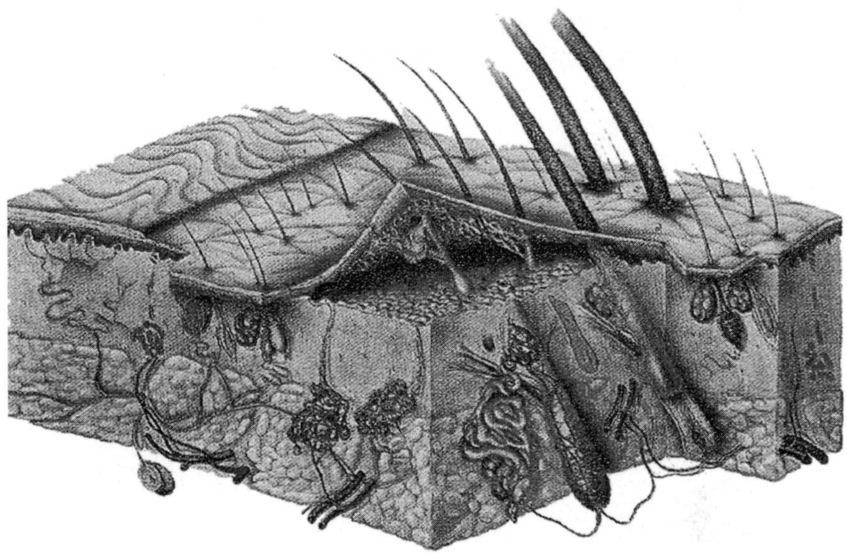

Figure 1.2 Diagram of a skin section (from W. Montagna).

3. hypodermis: immediately below the dermis, composed of a layer of adipose cells and representing a "cushion" of fat between the skin and the organs underneath

The boundary between epidermis and dermis, the "dermo-epidermal junction" (Fig. 1.2), is an "undulating" area resulting from many introflexions of the dermis and extroversions of the epidermis, dermal papillae, and epidermic crests, respectively. Along the entire dermo-epidermal junction there is a thin membrane, the "basal membrane." The interconnection of the dermic papillae and epidermic crests is made functional by the presence of the basal membrane; this junction is a true structure fundamental for the relationship that exists between exchange and semipermeable barrier between the epidermis and the dermis and, consequently, also between the external environment and the internal organs. This boundary changes from zone to zone: it is flatter in the area of the forehead and highly accentuated on the back and on the soles of the feet.

The basal membrane (dermo-epidermal junction) is really a complex structure formed of many components: the cytoplasmic membrane of the basal keratinocytes, two thin layers (the lucid layer and basal lamina), and finally a fibrous structure in contact with the dermis. Toward the innermost side of the basal membrane there are special structures called "emidesmosomes," which have an anchoring function. The fibrous structures below the basal lamina are of dermal origin and ensure the correct adhesion between epidermis and dermis.

We will now examine the epidermis, dermis, and hypodermis individually as each one contributes to the physiology of the cutaneous organ through its specific functions.

1.2.3 The Epidermis

The epidermis is composed of different types of cells that overlap, not randomly but in a well-defined manner. There are four different cell types:

1. keratinocytes
2. melanocytes
3. Langerhans cells
4. Merkel cells

1.2.3.1 The Keratinocytes

The "keratinocytes" are the predominant cell type and owe their name to the characteristic protein they produce in the course of their life, "keratin." This protein is responsible for specific, important skin functions.

Keratinocytes are formed, grow, and die "rising" toward the surface of the epidermis. Gradually as they mature they gain particular morphological characteristics. This mechanism is defined as "epidermal cell turnover" and is the basis of the continuous and incessant renewal of the epidermis.

Under the microscope the epidermis is an obvious superimposition of cell layers, each clearly different from the others, these being the maturing phases of the keratinocytes (Fig. 1.3).

The basal layer is composed of a single line of more or less cylindrical cells that are densely packed and adherent to the basal membrane. These cells show intense metabolic activity due to their rapid division. They are in fact the "parent" cells of all the epidermal keratinocytes and thanks to

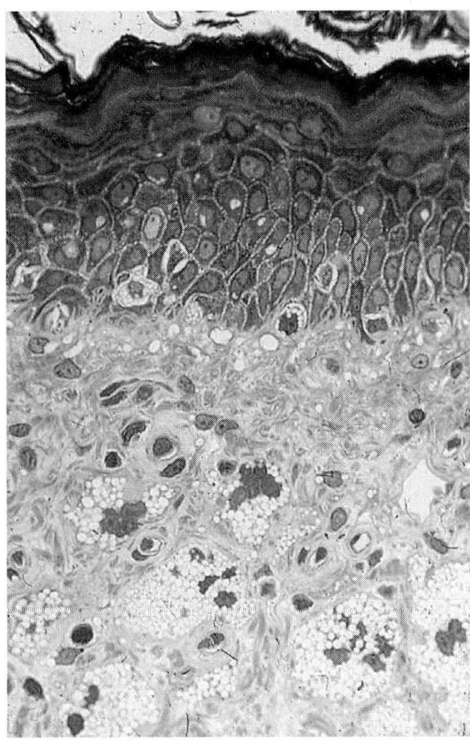

Figure 1.3 Image of a skin section seen under the light microscope, from the dermis to the upper epidermis (semi-thin section). The different layers of the epidermis are well visible. In the dermis istyocites and between the dermal fibres are other dermal cells.

their continuous division they are able to replace all the surface cells, which are continuously lost by exfoliation. Situated above the basal layer is the *spinous layer* (or *Malpighian layer*) formed by many layers of polygonal cells (moving upwards, the keratinocytes tend to flatten) that have already begun to produce keratin. Keratin, as previously mentioned, is the structural protein specific to the epidermis. The chemical structure of this protein makes possible certain fundamental functions of the skin: resistance against environmental attack and its impermeability to substances with which it comes into contact.

Under the microscope the keratinocytes of the spinous layer are seen to bear thin "spines" that protrude from the cell membrane. It is from this characteristic that the spinous layer derives its name. The cellular spines are really firm points of contact between one cell and another.

Proceeding to the surface one finds the *granular layer*, where the various lines of largely polygonal, but by now flattened, keratinocytes contain "granules" in their cytoplasm. The granules are aggregates of keratin, produced in abundant quantities by the keratinocytic cells of the granular layer.

From this point on the keratinocytes mature, changing their function and also their morphology: from polygons they tend to become squashed and to lose their regular shape, and they become less active and accumulate intracellularly large quantities of keratin. So in the next layer, the *lucid layer*, is formed of one or two strata of keratinocytes. The cells are flat, without either a nucleus or other cytoplasmic organelles, and contain a characteristic homogenous substance called "eleidin." This layer is found only in certain areas of the skin, namely the palms of the hand and soles of the feet.

This last formation of keratinocytes constitutes the *horny layer*. In this area the keratinocytes take the name of "corneocytes" and possess particular characteristics. They are completely without metabolic activity (due to the loss of the nucleus and cytoplasmic organelles), contain large amounts of keratin, and possess a peculiar cell structure, resembling "leaves" tiled one on top of the other.

In the most external layers the corneocytes gradually lose their cohesion and are desquamated, to then be replaced by keratinocytes pushing up from the underlying layers.

1.2.3.2 The Melanocytes

Interposed between the keratinocytes of the basal layer there are "melanocytes" (Fig. 1.4) that synthesize "melanin," the pigment responsible for skin colour. The number of melanocytes can vary according to body area and are usually present as a ratio of the number of keratinocytes: it has been calculated that one melanocyte is present for every 5–10 keratinocytes.

The embryologic origin of the melanocytes is different from that of the keratinocytes, melanocytes being derived from the "neural crest," the structure

(a)

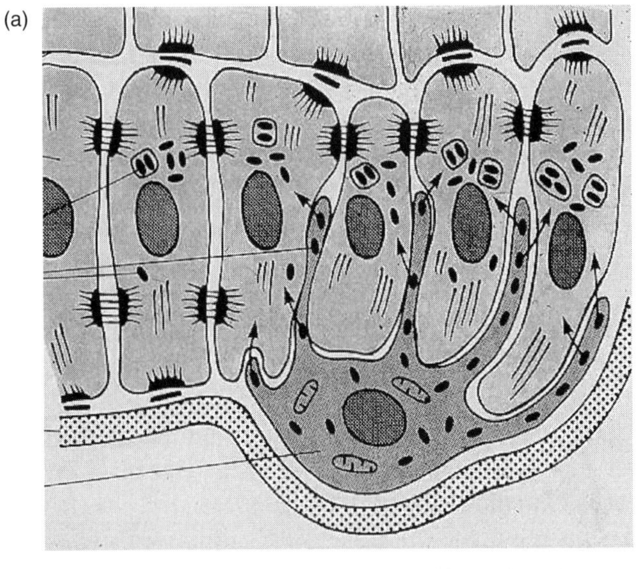

(b)

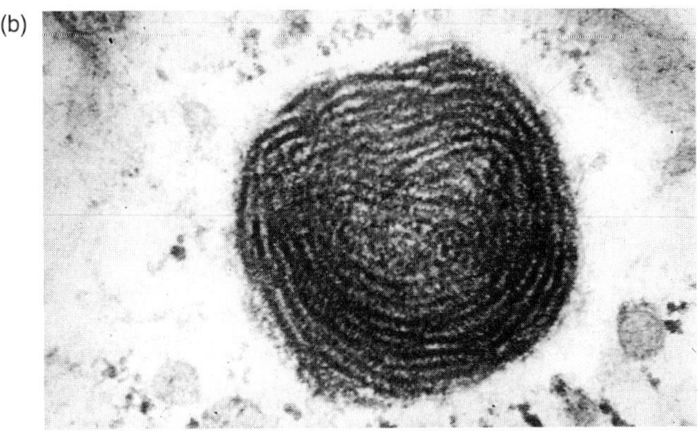

Figure 1.4 (a) Diagram of a melanocyte. (b) Electron microscope image of a melanosome in the cytoplasm.

from which the nervous system will develop. During growth of the embryo the melanocytes migrate from the neural crest toward the skin surface.

In order to fulfil their function the melanocytes have a very special cell structure. They possess long, thin extensions called "dendrites" that begin at the cytoplasmic membrane and infiltrate between the keratinocytes of the basal and spinous layers.

On stimulation by the sun's ultraviolet rays, the melanocytes begin to produce melanin, which is absorbed into special cytoplasmic bodies called "melanosomes." At this point the melanocytes distribute the melanosomes to all neighbouring keratinocytes using the cellular dendrites as a means of transport. The melanin pigment enclosed in the melanosomes passes out through the melanocyte cell membrane and enters the adjacent keratinocyte cells. In the keratinocytes the melanosomes position themselves around the nucleus, almost as an "umbrella," forming a defence against ultraviolet rays. The distribution of melanin from the melanosomes to the keratinocytes is responsible for the phenomenon of tanning on exposure to the sun.

The differences in skin colouring between individuals and between one race and another does not depend on the number of melanocytes but simply on a higher synthesis of melanosomes within the cell. Therefore, an individual of negroid race has the same number of melanocytes as an individual of the caucasian race, but possess melanocytes that synthesize a greater number of melanosomes and, consequently, a larger quantity of melanin.

1.2.3.2.1 Biochemistry of Melanin

Melanin is the pigment contained in the structures called melanosomes produced by the melanocytes. It is transferred to the surrounding epidermal keratinocytes, which maintain functional contact forming an epidermal melanin unit. Each melanocyte provides melanosomes to a group of about 36 keratinocytes, melanocytes being responsible for the skin's colouration.

There are different types of melanin classifiable into two main groups:

1. black and brown eumelanins (insoluble in all solvents)
2. red-brown pheomelanins (soluble in alkaline solutions)

There is also a third group of melanins called trichochromes, which are intermediate between the other two groups and are found in brown-haired individuals.

Eumelanins are pigments containing nitrogen groups, whereas pheomelanins, in addition to having nitrogen groups, also contain sulphydril

Table 1.1 Summary Table of Skin "Phototypes"

Phototype	UV Sensitivity	Sun Exposure Response
I	Elevated	Always burns, never tans
II	Elevated	Burns easily, tans minimally
III	Medium	Burns moderately, tans gradually to light brown
IV	Low	Burns minimally, always tans well to moderately brown
V	Very low	Rarely burns, tans profusely to dark
VI	None	Never burns, deeply pigmented

groups. The two groups are derived from tyrosine through processes whose initial steps are common to both. Tyrosine is oxidated to 3,4-dihydroxyphenylalanine (DOPA) and subsequently to dopaquinone by the same tyrosinase. The eumelanins are formed by the transformation of dopaquinone to cyclodopa to dopachrome to 5,6-dihydroxyindol through oxidative polymerization processes or to 5,6-dihydroxyindol 2 carboxylic acid.

The pheomelanins' structure is composed of benzothiazolic groups and tetrahydro-isoquinoline groups. The first two reactions in their formation are the same as those for the eumelanins. Dopaquinone is then transformed into 5 cysteinyldopa, which is transformed in its turn to cis-dopaquinone to cyclocisdopaquinonimin, benzothiazinylalanine, and so to pheomelanin.

It is interesting to note that cysteinyldopa can be found in the plasma and urine in patients with melanoma though its significance remains unknown and controversial.

The presence of different types of melanin determines an individual's pigmentation and, consequently, their response to light. As a result subjects can be divided into different "phototypes" (Table 1.1).

1.2.3.3 Langerhans Cells

These cells are situated above the basal layer and, like the melanocytes, have a dendritic appearance. Unlike the other two cell types in the epidermis, the Langerhans cells are visible under the microscope only when stained. Their function is one of "immunocompetence," that is, they belong to the organic defence system that unleashes a rapid response against "attack" on the human organism. The Langerhans cells unite against exogenous antigens and "present" them to both skin and lymph-node T lymphocytes. They are also involved in immune surveillance against viral and tumour antigens. They are thought to be involved in the genesis of skin neoplasia caused by the action of ultraviolet rays, which damage the cells and inhibit their immune surveillance functions.

1.2.3.4 Merkel Cells

These cells are found mainly in certain areas of the body: the fingertips, the oral mucosa, the lips, and the hair follicles. They are easily visible under the electron microscope and are always associated with a nerve fibre. For this reason Merkel cells are considered as "tactile receptors," that is they are the structures responsible for our sense of touch. However, in man this ability has not yet been clearly proven.

1.2.4 The Dermis

The dermis (Fig. 1.3) is positioned below the epidermis and is the tissue that supports the skin and its annexes (hair, nails, etc.). Its thickness varies from area to area, being thinnest on the eyelids and thickest on the back. The dermis tends to become progressively thinner with age.

This layer is formed by cells, fibres, and ground substance, and, unlike the epidermis, is richly innervated and vascularized. The most abundant cells are the *fibroblasts*. These cells are the production site of the other dermal components: both the fibres of the dermis and the ground substance are synthesized within the fibroblasts. In addition to fibroblasts, *mastocytes* (mast cells), *lymphocytes*, and *histiocytes* are present, which are mainly concerned with immunocompetence.

The fibres produced by the fibroblasts are of different types, according to the function that they perform:

- *The collagen fibres* are the most abundant fibres and are collected together into variably orientated bundles. Under the electron microscope these fibres show characteristic transverse striations owing to their peculiar structure. The

Figure 1.5 Collagen fibres in the dermis.

collagen fibres are composed of specific amino acids like proline, hydroxyproline, and glycine, all positioned to form a fibrous structure. The principle function of the collagen fibres is to support the internal structure of the skin (Fig. 1.5).

- *The elastic fibres* form a loose mesh that is also only visible with specific staining. The main component is a protein called "elastin." The principle function of this type of fibre is to provide the skin with the elasticity fundamental for all our movements. Changes in or overextensions of these fibres are the main cause of the phenomenon of "striae distensae," stretchmarks (Fig. 1.6).
- *The ground substance* of the dermis is the "cementing" component, a group of different constituents combining to compress the dermis structure. It is formed by substances such as mucopolysaccharide acids (glycosaminoglycans, chemically classifiable as complex sugars), hyaluronic acid, and chondroitin sulphate. Glycoproteins, dermatan sulphate, and keratin sulphate, for example, are also present. Glycosaminoglycans and other specific proteins form large molecular aggregates termed "proteoglycans." Their main characteristic is the ability to bind numerous water molecules, so producing an amorphous gel that performs two essential functions: allowing nutrients and oxygen down into the tissues and protecting the dermal structure.

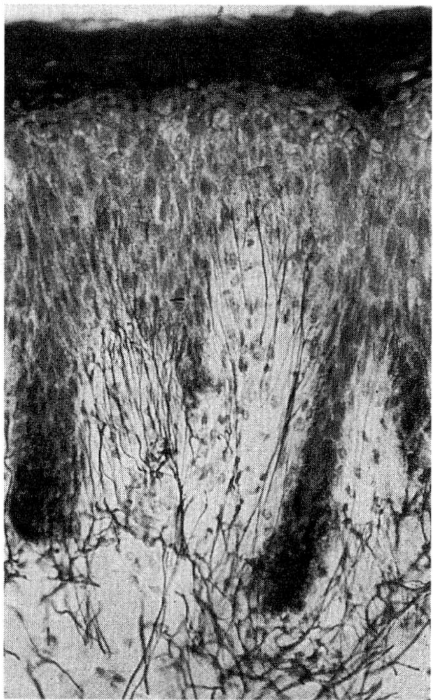

Figure 1.6 Elastic fibres seen under the light microscope.

The morphological differences within the dermis allow it to be schematically subdivided into two zones: the papillary dermis and the reticular dermis.

The papillary dermis is closest to the skin surface and, therefore, situated just below the epidermis. This layer houses a rich capillary network responsible for the "nutrition" of the overlying epidermis and annexes. Nerve endings are also present in the papillary dermis as both free, that is to say "opened out" into many ramifying fibres, and corpusculated terminals. Fibres are orientated in an irregular manner; the ground substance is abundant and capillaries and lymphocytes are numerous.

The reticular dermis is situated deeper and is composed of large bundles of collagen fibres arranged parallel to the epidermal surface. The cells and ground substance are present in smaller quantities than in the papillary dermis and also vessels, especially arterioles and venules, are less numerous. The main function of the reticular dermis is to give greater support to the skin structure.

1.2.5 The Hypodermis

The hypodermis (Fig. 1.7) is situated below the dermis and is composed of "adipocytes" grouped into "lobes" separated by an area of connective tissue. Arterioles and venules infiltrate the connective tissue and are responsible for the nutrition of the hypodermis tissue.

Adipose cells are characterized by the remarkable quantity of lipids they contain. These lipids are grouped into a single globule in the cytoplasm, so large that the nucleus is forced to occupy a peripheral position.

The thickness of the hypodermis varies considerably according to the body site, nutritional state, and sex of the individual. In fact, the distribution of subcutaneous fat is a strictly hormone-dependant secondary sexual characteristic. The number of adipose cells changes in the first phases of child development, at first increasing and then remaining stable. During puberty in women there is an increase in subcutaneous fat predominantly in the

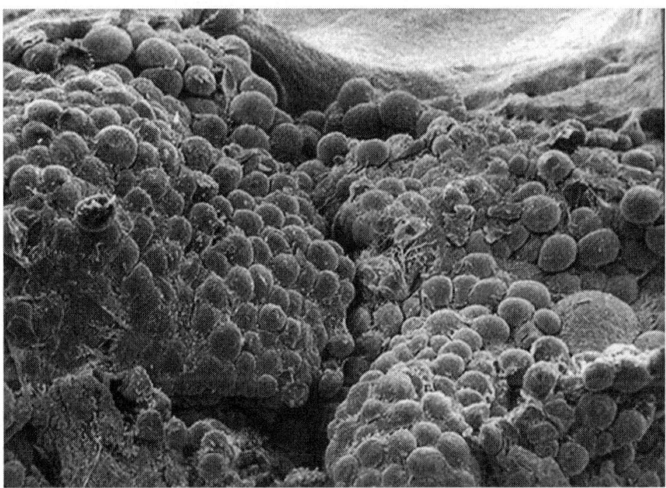

Figure 1.7 Adipocytes seen with the scanning electron microscope.

area of the buttocks, hips, thighs and breasts. In men, on the other hand, the increase in lipid content is in the area of the torso and abdomen.

Weight loss or increase is owed, not to an alteration in the number of adipocytes, but to a change in the lipid content of the adipose cells (Fig. 1.7).

1.2.6 Vascularization

Blood vessels have the task of distributing throughout the body the blood, which reaches the cardiac muscle from the lungs. The heart allows the distribution of oxygen-rich blood, necessary for the survival of all cells, to all parts of the body.

Blood vessels can be subdivided into *arteries, capillaries*, and *veins*. Arteries have the widest gauge and transport oxygenated blood from the heart to the peripheral organs. The capillaries facilitate the exchange of oxygen and nutrients from the blood to the tissues. They can be subdivided in two different portions: the first, the arterioles, transport oxygenated blood and nutrients to the tissues, and the second, the venules, collect liquid and metabolic by-products. Finally, the veins carry blood, poor in oxygen, to the lungs for re-oxygenation.

Together with the blood vessels there are the "lymphatic vessels," which reabsorb intracellular fluid and drain by-products and cellular waste. Lymph nodes are found along the lymphatic vessels, where immune cells, dedicated to the body's defence, ensure that the fluid transported by the lymphatic system is free from bacteria, germs, and harmful substances.

The skin has a vast vascular network subdivided into "plexuses" composed of arteries, capillaries, and veins running parallel to the skin surface. The dermis and hypodermis are vascularized, whereas the epidermis is completely devoid of such vessels. The principle plexuses are the *subepidermal plexus* between the dermal papillae and the *deep plexus* positioned between the dermis and epidermis. Between the two parallel vascular networks there are many linking vessels that allow the exchange of blood from one plexus to the other. The capillaries from the subepidermal plexus are close to the basal membrane and provide the epidermis with nutrients by allowing the diffusion of oxygen and principle nutrients into the epidermal cells. The arterial and venous sections of the capillaries are directly linked by interconnecting vessels that provide "anastomosis." The opening of these structures allows blood to pass directly from one section to the other, so excluding the capillary circulation. This complex vascular architecture is necessary for the skin's essential function of thermoregulation, which will be dealt with in more detail later.

1.2.7 Innervation

The nervous system is composed of a vast network of fibres that originate from a central base, the brain, and go to all peripheral areas of the body before returning to their source. The brain continuously receives and sends stimuli throughout the body, so completely controlling all our bodily functions.

Nerve fibres can be classified into three categories: motor fibres, sensory fibres, and vegetative fibres. Motor neurons facilitate movement through the stimulation of the muscles by the nervous system. Sensory neurons collect tactile, thermal, and pain stimuli, etc., and transmit them back to the brain for processing. Finally, the vegetative neurons control the activity of the organs independently of our will (e.g., the intestine).

The skin is considered a "sensory organ" in that it allows us to relate to our external environment and is, therefore, supplied with a rich system of sensory neurons and obviously devoid of motor neurons. In the skin the sensory neurons can be divided into two categories: free nerve endings and corpusculated nerve endings. The former are found deep in the dermis and end just below the epidermis. They are formed of "openings" of the sensory nerve fibres and their function is to collect pain sensations. Corpusculated nerve endings are formed by nerve fibre endings around which other cells are arranged, so forming the so-called corpuscle structures. In the skin the corpusculated nerve endings are the Pacini (which gather tactile stimuli) and Meissner corpuscles.

Vegetative fibres, which are beyond the conscious control, innervate the erector muscles of the hairs and eccrine and apocrine sweat glands. In fact, the phenomena of "horripilation" (raising of the hair by contraction of the erector muscles) and sweating are responses to temperature and emotional stimuli that we cannot control.

1.2.8 Skin Annexes

The term *skin annexes* describes all those structures present in the skin that are embryonically derived from the epidermis. The hair system, eccrine and apocrine sweat glands, sebaceous sweat glands, and nails are all annexes of the skin.

1.2.8.1 The Hair System

The entire surface of the skin is supplied with hairs, but on the head the hairs are longer and more pigmented. The structure of both head and body hair is practically identical yet they differ in their physiological responses, especially to hormones. Only two areas of the body, the palm of the hand and the sole of the foot, are devoid of hairs.

Men and women have the same number of hair follicles, the organs that produce the hairs. In fact, the "number" of hair follicles is all but equal in all individuals, even the bald. In these subjects hairs are not completely absent; it is only that they have become so small as to be invisible.

There are three types of hair:

1. vellus hair
2. terminal hair
3. lanugo hair

The lanugo is the thin and unpigmented hair present in the foetus before and just after birth, which is subsequently transformed into terminal hair (like that of an adult). Vellus hair is that present in women, adolescents, and also on the scalp of bald individuals. All three hair types have the same structure and even the same histological profile.

1.2.8.2 Anatomy and Histology of Hair

Hairs develop from the hair follicles (Fig. 1.8(a)), which are true organs formed by the invagination of the epidermis during the foetal development. These organs (Fig. 1.8(b)) are divided into three main parts:

1. *The infundibulum* is the portion from the opening on the skin surface to the mouth of the sebaceous gland duct.
2. *The isthmus* is the section between the sebaceous gland duct and the point of intersection with the hair erector muscle.
3. *The inferior portion* extends from the erector muscle to the bulb of the hair follicle.

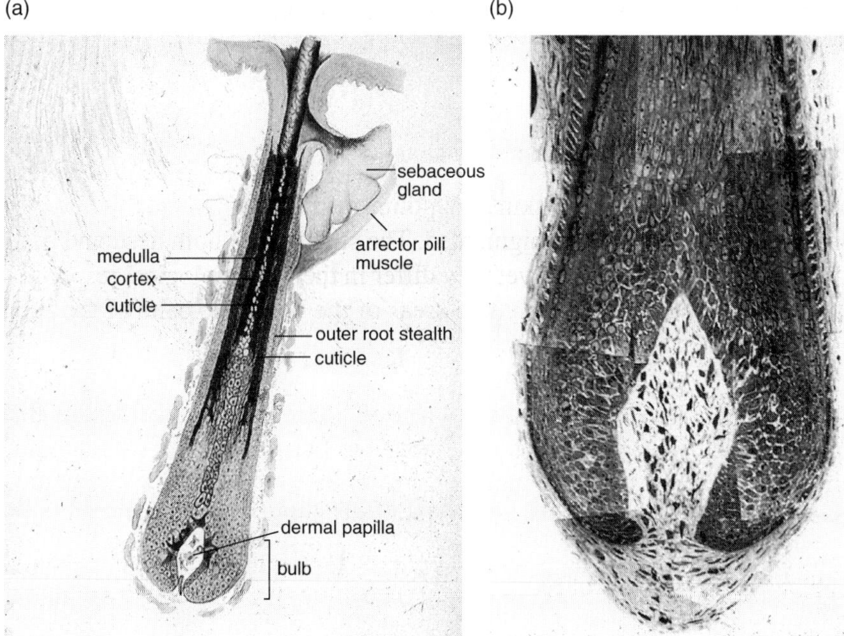

Figure 1.8 (a) Diagram of a hair follicle (from W. Montagna). (b) Semi-thin section of a hair bulb.

The *hair root* is a generic term that includes all of the inferior portion and the isthmus (Fig. 1.8(a)). The *bulb* is the enlarged terminal part of the hair follicle in the root (Fig. 1.8(a) and (b)). The root produces the hair *shaft*, which is defined as the hair structure visible above the skin surface. Below the bulb the dermis re-enters the hair follicle, producing the *dermal papillae*, as the hair follicle is formed by introflexions of the epidermis; this re-establishes the junction with the dermis. In this way the hair follicle is surrounded by the basal membrane. In this region it is particularly thick and assumes the name *vitreous membrane*, because together with the basal membrane it forms a kind of connective sheath that surrounds the whole follicle.

The part on the hair follicle between the bulb and the isthmus is known as the *keratogenous zone*. The maturation of the cells that produce the hair shaft occurs in this area.

The most important part of the hair follicle is the bulb. In the basal layer, immediately above the papillae, cells continuously divide and begin their ascent. Slowly as this process proceeds the cells subdivide in the layers that make up the follicle. In fact, if a hair is cut perpendicularly to the follicle one can easily see how, independently of the zones and structure described earlier, each of these is formed by the union of layers and from inwards towards the outside of the follicle we can recognize (Fig. 1.9(a) and (b)) the following layers: the *medulla, cortex*, and *cuticle*. These three layers form the hair shaft.

The cuticle is followed by the *cuticle of the internal epithelial sheath, Huxley's layer*, and *Henle's layer*. Beyond these layers the *external epithelial sheath* begins and this, too, is composed of cell layers becoming increasingly thinner from the epidermal surface to the bulb.

In the germinative layer, or *hair matrix*, the keratinocytes are in contact with melanocytes, which are present only in this region. Here the melanocytes produce the pigment that will give rise to our hair colouring.

The connective sheath also envelops the sebaceous gland and the hair erector muscle. The latter structure, important in furry animals in controlling the angle of the hair and so regulating heat exchange, is all but useless in man—nevertheless, this thin muscle is able to raise the hair. The follicles are set obliquely to the skin's surface (Fig. 1.8(a)) and the erector muscle, by contracting and nipping both the follicle and the epidermis, is able to raise the hair. This is the phenomenon of *horripilation*, commonly called goose flesh, which occurs during shivering due to cold or fear.

1.2.8.3 The Hair Life Cycle

Hairs live according to particular cycles. They do not all grow together and consequently their life and loss are staggered. In furry animals "hair" growth and loss are synchronized, allowing these animals to change their fur entirely with the seasons. In man, on the other hand, hair lives in a way that is defined "mosaic."

There are approximately 180,000 hairs on the scalp. The cycle of each hair is staggered with respect to the others and thus each is at a different phase of growth. Three distinct phases can be identified in the hair life cycle (Fig. 1.10):

1. *anagen* or growing phase
2. *catagen* or quiescent phase
3. *telogen* or loss phase

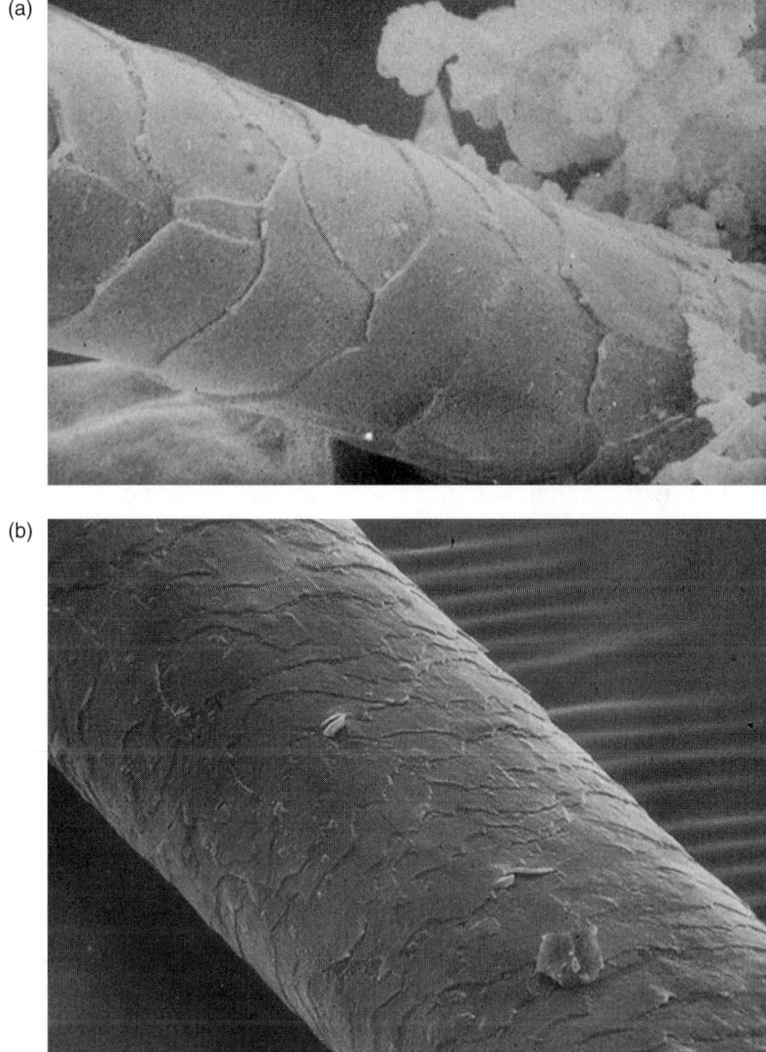

Figure 1.9 Scanning electron microscope image of the hair shaft showing the keratin "tiles": (a) hair shaft at the emerging point; (b) hair shaft.

The growing phase (anagen) lasts about three years and involves 90–95% of the hair at any one time. During this period the cells of the matrix divide continuously and the hair shaft grows in length. After this phase the hair follicle enters a resting state that will normally last about three

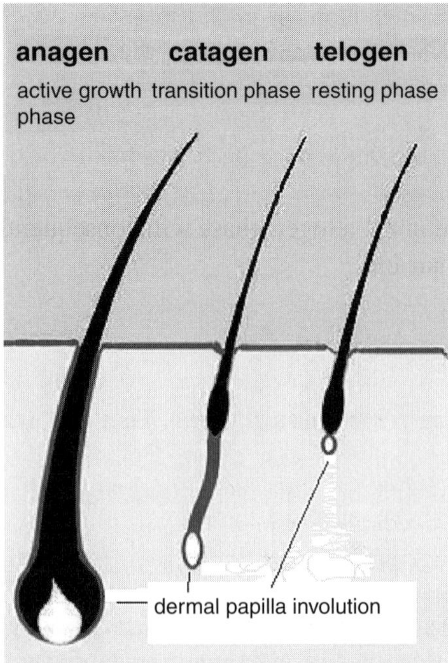

Figure 1.10 Diagram of the hair life cycle.

months. During this period, which will daily involve approximately 13% of the hair, the follicle undergoes profound changes and the bulb becomes thinner and loses its connections with the papilla. Later in catagen the follicle can regain the growing phase or enter the final stage (telogen). If the latter occurs a new hair follicle will develop at the bottom of the old one after approximately three weeks. This new growth pushes the old hair out of the follicle, which is subsequently lost. One percent of the hairs in the telogen phase are lost daily and therefore, even in healthy subjects, 100–180 hairs can be lost in a day. The important thing is that these hairs get replaced by new ones.

1.2.8.4 The Tricogram

Knowing the percentage of hairs normally found in the three growth phases, it is possible to establish whether an individual has some illness that has brought about an increase in hair loss. With this method, known as a *tricogram*, it can be determined whether the three stages are regular or, in the event of a problem, which is over- or under-represented.

To perform the test a sample of at least 100 hairs are taken and observed under the microscope. The number of hairs in each phase is then determined. An increase in the telogen phase will consequently be indicative of a real increase in hair loss.

1.2.8.5 Alopecias

Hair loss or thinning is known as alopecia. There are three types:

1. noncicatricial or dysfunctional alopecia
2. cicatricial or destructive alopecia
3. hair breakage alopecia

Dysfunctional alopecias are the most frequent forms, with both common baldness (*androgenic alopecia*) and temporary hair loss (*telogen effluvium*) belonging to this category. The latter condition very often occurs after giving birth. Destructive alopecias are those conditions of the scalp that may be either congenital or acquired. This group also includes all those alopecias caused by any "trauma" leading to scarring of the scalp.

Hair breakage alopecias can also be either congenital or acquired. Acquired forms include the *tricoressi nodosa* that can follow aggressive cosmetic treatments like, for example, excessive heat or over-energetic brushing.

1.2.8.6 Cutaneous Glands

The skin is supplied with three types of glands, all of which are formed from the epidermis during foetal development.

1.2.8.7 Sebaceous Glands

These are defined as *holocrine* glands due to the secretions they produce. The secretions are particularly thick and rich in organic substances and their production brings about the destruction of the gland cells themselves. With the exception of the palms and the soles of the feet, sebaceous glands are present all over the skin. They are always connected to a hair follicle, but in some areas like those around the anus, genitals, eyelids, and lips, they are directly connected with the skin surface because in these regions an adequate protection and lubrication must be guaranteed. This applies to man as much as to other mammals.

The sebaceous glands are not uniformly distributed over the body, some areas—the face, the back, and the torso—are particularly richly endowed, so much so that their density can reach over 200 glands per cm^2. In other areas, such as the front of the leg, they are at their least dense and are found at less than 40 glands per cm^2. This non-uniform distribution of the sebaceous glands is the main difference responsible for both the different skin types and so-called combination skin.

1.2.8.7.1 Anatomical and Histological Features

The sebaceous glands are defined as being *ramified-alveolar* because their structure resembles a bunch of grapes, each grape (alveolus) connected by a stem to the main branch of the bunch (Fig. 1.11(a) and (b)). The entire gland, like the hair follicle, is surrounded by a connective sheath that invaginates to surround even the individual alveoli. A few alveoli become surrounded by the connective sheath and form a *lobe*, as if several alveoli were held together by a thin membrane. The alveolus is the secretory unit of the gland. The cells that constitute the outer layer of this region and are in contact with the connective sheath divide and multiply, progressing toward the centre of the alveolus and beginning sebum production.

Each alveolus ends in a small excretory duct that collects the sebum produced by the rupture of the mature cells. The alveolar ducts unite at the excretory duct of the different lobes, the lobe ducts coming together to form the *common excretory duct*, which opens into the hair follicle at the base of the infundubulum (Fig. 1.11(a) and (b)). This way hairs are already covered with sebum before they emerge from the skin surface.

In areas where the sebaceous glands are not associated with a hair follicle the excretory duct opens directly onto the skin surface. In these areas the sebaceous glands are larger and at times modified to such an extent that they constitute different entities. This is true in the case of the *Meibomian*

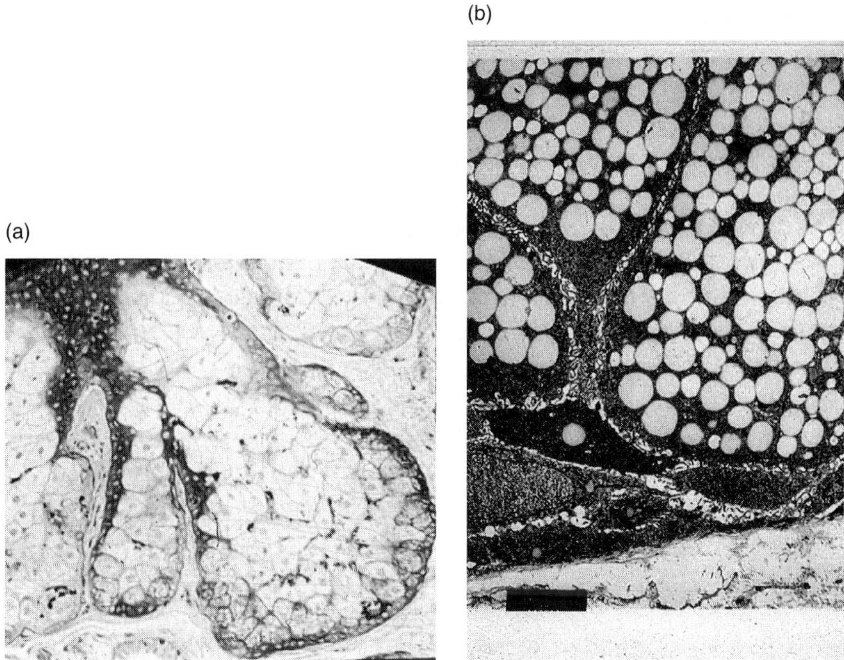

Figure 1.11 (a) Light microscope image of a sebaceous gland showing the ramified-alveolar structure. (b) Electron microscope image of sebaceous gland containing lipid droplets.

glands of the eyelids, which when inflamed produce a hard filtrate in the shape of an ear of corn so giving rise to styes.

1.2.8.7.2 Sebum

Sebum is a product of the sebaceous glands. Production of sebum is not constant, and it changes with different stages of life.

In the foetus the sebaceous glands begin secreting sebum during the fourth month. This occurs to such an extent that at birth a newborn baby is often covered by a *vernix caseosa*, which is in fact a layer of sebum that protects the foetus from the amniotic fluid. Later when the baby is no longer being breastfed and, therefore, no longer receiving the maternal hormones that stimulate the glands, the sebaceous glands enter a state of quiescence which only ends when the child reaches puberty and the sexual glands have developed and begun to secrete sex hormones—predominantly androgens in the male and estrogens in the female. In both sexes it is the androgenic hormones that stimulate both the sebaceous glands and hair growth.

The production of sebum, beginning again at puberty and continuing until adulthood, is always higher in males than in females. In females there is a reduction in sebum production after the menopause caused by the progressive inactivation of the ovaries, which consequently stop synthesizing androgens. In later years both men and women produce less sebum even though the sebaceous glands enlarge (become hypertrophic) and their openings often dilate.

1.2.8.7.3 Composition of Sebum

Sebum represents about 95% of the lipids in our skin. On synthesis it has a different composition from that when it reaches the skin surface. The hair follicle contains many bacteria that possess a variety of enzymatic activities (particularly lipases) able to hydrolyse and, therefore alter, the complex molecules contained in the newly synthesized sebum.

The composition of sebum on the skin surface:

- free fatty acids: 20%
- triglycerides: 40%
- squalene: 15%
- wax: 20%
- sterols and glycophopholipids: 5%

1.2.8.7.4 Functions of the Sebum

Sebum is fundamental in all mammals because it represents the best protection for their fur. Man has lost most of his covering of hair and so sebum, though still important, is less so than for other mammals. However, even in man sebum continues its role of protecting the skin. By emulsifying itself, that is, mixing with the water derived from sweating or transpiration, sebum forms a film over the skin that is able within certain limits to protect it from harmful chemical substances and from the action of pathogenic microorganisms.

Sebum contributes to the skin's emollience (softness) by retaining water. The hairs are also protected by sebum without which they would appear opaque and dull. The presence of sebum around the anus, genitals, eyes, and lips has protective, emolient, and antimicrobic functions.

1.2.8.8 Sweat Glands

These are divided into *eccrine* and *apocrine sweat glands*. The first, the common sweat glands, are the more numerous and are found all over the skin. They are defined "eccrine" due to their mechanism of secretion that produces an almost aqueous liquid.

The apocrine sweat glands, on the other hand, are present only in certain regions of the body. Examples are the armpits, around the genitals, on eyelids, and in the external part of the ear canal, etc. The apocrine glands differ from the eccrine glands in various ways including: their secretion product, their distribution, their being limited to certain areas of the body, their localization deeper in the skin, their having their opening in the hair sheath, and their strict linkage of function with the sexual phases of life.

Eccrine sweat glands are innervated by sympathetic nerves, and sweating does not occur in the denervated areas of the skin. In rare cases, individuals with congenital defects may lack eccrine sweat glands and in hot climates must take special precautions to avoid excessive increases in their body temperature.

The apocrine sweat glands are under double nerve control but their activity is not notably influenced by variations in temperature. Therefore, strictly speaking, they should not be classified as sweat glands.

1.2.8.8.1 Eccrine Sweat Glands

Eccrine sweat glands are "merocrinous" glands in that they produce an aqueous solution of low density that does not appreciably cause disintegration of the epithelial cells. Their principle function is that of thermoregulation: sweating to allow the body to rid itself of excessive heat. These glands begin to function only at temperatures above 18–20°C, when normal mechanisms of thermoregulation become insufficient. The secretory capacity of these glands means that several litres of water can be lost from the body hourly, taking with it toxins and other substances. It is the evaporation of this water that brings about cooling of the body.

Another essential point is that, in combination with transpiration (*perspiration insensibilis*), sweat accounts for a large proportion of the water that remains on the skin, forming part of the hydrolipid film.

Eccrine sweat has an odour between that of musk and urine and contains numerous amino acids and organic substances that, together with secreted salts, constitute the *buffering system* of the skin. It is, therefore, essential because sweat and sebum together constitute the hydrolipid film on the skin's surface, which is the indispensable protective covering keeping the skin in good condition and, consequently, allowing it to perform its many essential functions.

1.2.8.8.2 Anatomical and Histological Aspects of Eccrine Glands

Eccrine glands are defined as *simple tubular*, being composed of a thin canal connected directly to the surface and extending into the dermis almost until the hypodermis, where it winds around itself (Fig. 1.2). There are two recognisable parts to the gland: the first, the *secretory segment*, is the "ball" of the gland situated in the mid-dermis. Here the cells absorb liquid derived from the surrounding capillaries and deliver it into the lumen, the gland interior in the duct that leads to the *excretory portion*. In this part of the gland the composition of the liquid is modified due to reabsorption of mineral salts, potassium, and some of the sodium. Sweat is, in fact, salty, even though the body has at least in part attempted to extract the precious mineral salts from it.

The sweat glands are, as mentioned earlier, richly innervated by the sympathetic nerve system. This explains why intense emotions can be accompanied by excessive sweating.

1.2.8.8.3 Apocrine Sweat Glands

These glands have little physiological importance. Their importance is derived more from the negative effects of their product, from which unpleasant odours arise. The product of these glands is a secretion rich in organic substances including large amounts of protein and steroids. This is due to their mechanism of secretion that results in the loss of the upper part of the sudoral cells. The secretion is extremely dense and provides an excellent broth for bacteria, which break down the organic compounds so producing volatile pungent substances. These glands are present only in certain regions of the body: the armpits, the groin, and around the genitals.

In other mammals these glands are important because they produce "pheromones," substances that are at the base of olfactory communication between animals both in sexual attraction and in the demarcation of territory (Fig. 1.2).

1.2.8.8.4 Anatomical and Histological Aspects of Apocrine Glands

In their structure the apocrine glands are similar to the eccrine sweat glands in that they too are *simple tubular-alveolar*. The apocrine glands are also composed of a *secretory segment* in the form of a "ball" and an *excretory duct*.

The apocrine sweat glands, like the sebaceous glands, are always associated with a hair follicle, although these glands open directly onto the skin surface in the same region as that described for the sebaceous glands (Fig. 1.11(a) and (b)). In the outer ear, in particular, these glands have become the "ceruminous" glands characterized by a dense yellowish secretion.

1.2.8.9 The Nails

In addition to the *nail lamina*, the hard part generally identified as nail, nails are also composed of surrounding tissues. Anatomically, the nail is a complex of the skin composed of the following tissues:

- perionychium
- eponychium
- hyponychium
- nail bed
- nail matrix

The formation of the nail lamina begins in the matrix. Here the cells multiply and become keratinized in a manner analogous to that in the hair. Though keratinization is similar in the epidermis, the hair, and the nail, in each case it is stabilized by a different pattern of linking disulphide bonds, which determine the anatomical and structural differences between the skin and its annexes.

The hyponychium is a short portion of normal epidermis just below the lamina that defines the end of the fingertip and the beginning of the nail bed. The latter is below the nail lamina and is covered by a particular epidermis. The peronychium is the epidermis that surrounds the entire nail, bounding it with a deep furrow. The characteristic little membranes, or *cuticles*, that surround the nail continue below it in the form of the eponychium, which cannot be seen. Finally, the *lunula* is the portion of the matrix seen at the base of the nail; its whitish appearance is due to its distance from the nail bed.

The nails, though in man less important than the claws of animals, still play a fundamental role both in protection and in the numerous functions fulfilled by the fingers. For such reasons the nails are protected and looked after. Their growth is stimulated most by the activity of the fingers themselves. In other words, the more the nails are used, the faster they grow. The surrounding tissues are fundamental to the health of the nails and, therefore, must be protected and not removed. As the nail lamina is composed of hard keratin it is therefore impermeable and can be covered with enamel and other substances which can at times be useful in increasing the nail's strength.

1.3 Skin Functions and Physiology

1.3.1 The Physiology of the Skin

Physiology is the study of the mechanisms associated with the life of cells and the entire organism. In this chapter we will deal with the principle aspects of the basic mechanisms controlling the life of our skin. The understanding of these mechanisms will allow us to intervene with them, even with cosmetics, to maintain the skin in better health and as a result give it a nicer appearance.

1.3.1.1 Keratinization

Keratinization refers to the life of the epidermal cells, from the moment they are formed in the basal layer to their loss from the horny layer. In this period of time, which lasts about 28 days, the cells synthesize and accumulate keratin so that the protein fills the cells of the epidermal horny layer, the hair cortex, and the nail lamina. However, keratinization also refers to other fundamental processes that occur during this period, most important the production of the epidermal lipids, which are indispensable for the maintenance of epidermal integrity (Fig. 1.3).

1.3.1.2 Cellular Dynamics of Keratinization

Basal layer cells divide, giving rise to another layer of cells, which take their place on the basal layer. The cell that divided commences its ascent toward the upper layers and begins its metabolic activity, putting aside an energy reserve composed mainly of *glycogen* (a complex sugar). In the spinous layer the cell re-establishes its intercellular contacts, lost at the moment of division, and immediately develops the cell organelles necessary for the synthesis of various substances. In this way the cell begins to synthesize a keratin "precursor" protein, *keratohyalin*, which increasingly collects into intracellular bundles until the upper portion of the granular layer, where the keratinohyalin is finally converted to keratin by specific enzymes.

In the upper section of the granular layer the cell nucleus is expelled and the cell, by now no longer alive, is transformed into an envelope containing mostly keratin and other residual cell substances. The walls of the envelope are all that remains of the cell membrane and are covered with lipids. In the horny layer, composed of several layers of cells laid one on top of the other, the cells are reduced to lifeless, loosely connected sacks. Driven by the new cells forming in the basal layer, these cells, or rather what remains of the original cells, are exfoliated either singularly or in small groups.

This mechanism is of fundamental importance in guaranteeing that there are always new cells at the skin surface, which is daily exposed to abrasions and environmental attack. Moreover, the skin is continuously subjected to small injuries. The keratinization mechanism is able to respond to the demands imposed by such phenomena, to repair its surface, and, as a direct consequence to an injury, to set off processes (in addition to the usual keratinization) that repair the damage. The latter is achieved by the immediate division of cells in the damaged region or adjacent areas.

The process of keratinization and the consequent exfoliation occurs over the entire skin surface as a continuous yet staggered process; otherwise, the skin cells would all be lost together, as in those animals which, like the reptiles, are able to shed their skins seasonally. Common dandruff that falls from the scalp is, therefore, a normal consequence of this continual change of the skin surface; however, if some changes occur in the mechanism and the balance is upset, dandruff may become excessive. Other areas can also be affected by disturbances in keratinization; in some cases these can even be serious illnesses, psoriasis, for example.

1.3.1.3 Epidermal Lipids

Although they constitute only 5% of all the lipids in the skin, the epidermal lipids are extremely important. In fact, whereas sebum protects the surface of the skin, the lipids produced by the epidermal cells allow the cohesion of the epidermis, retaining water within its structure and allowing the horny layer to act as a barrier.

1.3.1.4 Keratinosomes (Odland Bodies)

Keratinocytes produce a specific organelle containing lipid compounds stored in the form of "bistratified discs" stacked one on top of the other. These organelles were first identified by Odland, from whom they take their name. In the granular layer the contents of these organelles are poured into the intercellular spaces, where they form the abundant leaves of bistratified lipoid material found between the cells of the horny layer.

These substances have been biochemically characterized and are composed mainly of *ceramides*. Though other substances such as glucosylceramide, sphingosine, cholesterol esters, and fatty acids are also present, it has been clearly demonstrated that the ceramides play an essential role. In fact, these substances together with other natural hydration factors play an indispensable role both in the retention of water in the epidermis and in the construction of a real barrier to the entry and exit of substances into and out of the epidermis. It has been calculated that the removal of the intercellular lipids from the horny layer would provoke a much larger loss of water than would the removal of the cells themselves. Moreover, without the ceramides interposed between the horny tiles, substances applied to the skin (sometimes even dangerous ones) would have free access into the body.

The possibility of artificially reproducing the major constituents of the epidermal lipids, the *ceramides*, today allows us to physiologically supplement the skin's functional barrier and to significantly improve both the hydration and normal structure of the epidermis.

1.3.1.5 The Keratins

Keratin filaments are grouped into bundles by the action of a basic protein containing histidine called filagrine. The physiological process implies a cellular stratification and a regulation of the differentiation of specific epithelial gene products. The keratinized cells move from the epithelial basal layer to the surface of the tissue, undergoing during their migration a series of morphological changes that reflect differential epithelial gene expression. The result of the differentiation is the formation of the horny layer due to the processing of a series of proteins and the degradation of the cell organelles.

The keratins, protagonists of the keratinization process, are a vast family of related proteins. They are insoluble, contain cysteine, and have molecular weights ranging from 40 kDa to 67 kDa. They are present in the cells of the epidermis, hair, and nails and form the characteristic cytoskeleton of epithelial cells, including those found in the gastrointestinal and urothelial areas and the orogenital mucosa. They consist of intermediate filaments 7–10 nm in diameter. There are six types of intermediate filaments: type 1 (hard keratin, basic epithelial keratin), type 2 (hard keratin, acidic epithelial keratin), type 3 (GFAP, desmin, vimentin, peripherin), type 4 (neurofilaments, NF-L, NF-M, NF-H, alpha-internexin), type 5 (nuclear laminins A,B,C), and type 6 (nestin). Of these intermediate filaments the most important examples are vimentin, present in the mesenchymal cells, GFAP acidic protein, which composes the glial filaments of the glial cells, the neurofilaments present in the neurones, desmin of the muscle cells, and the proteins of and the nuclear matrix, nuclear laminins A,B,C.

The polypeptide structures of all the intermediate filaments have a similar skeletal part composed of structural blocks of polypeptide subunits. The number of subunits varies between 1 and 30. The proteins are products of separate genes that, by cross-hybridization, fall into two gene families: type 1 (basic) and type 2 (acidic). The epithelial keratins are classified numerically and are, therefore, divided into two groups: basic keratins (1–8) and acidic keratins (9–19).

The fundamental polypeptide chain of keratin is a classical alpha helix with repeated groups of seven. It is composed of four separate helical zones linked by interhelical sequences and is preceded and followed by nonhelical carboxyterminal and aminoterminal sequences. During keratinization two polypeptide chains are assembled as a heterodimer called a protofilament. Two protofilaments then assemble into larger protofilaments, so forming a protofibril, the basic element of the keratin fibre heterodimer. It is possible that type 2 keratins are present in the cell before type 1 keratins and is thought likely that the former induce the synthesis of the latter. In epithelial cells keratin filaments radiate from the perinuclear region to the internal face of the plasma membrane. Bundles of these filaments are associated with specialized structures of the plasma membrane called desmosomes. Looking at their chemical structure it is evident that keratin filaments are heterogenous. This heterogeneity explains how the filaments perform diverse functions in different cells. Many keratins are produced in fixed associations as well as in particular cell types.

The most important keratins of the basal layer are keratins 5 and 14. These are synthesized only in the basal cells and are transported in the suprabasal compartment. In epibasal cells the keratinocytes synthesize a new pair of keratins, keratins 1 and 10, which are characteristic of epidermal differentiation. Recent studies have considered the Fos family of proteins, transcriptional activators that are rapidly induced by extracellular stimuli. The Fos proteins are expressed immediately before cell death and the formation of the horny layer in keratinized tissue. It is supposed that they play an important role in the transcriptional activation of the genes mediating the formation of the horny layer. The keratinized epithelial cells that express the Fos proteins must be the target of a differentiation signal that is as yet unknown. A small number of basal cells will determine the differentiation of the keratin leaving the basal layer. After minimal post-transcriptional modification they are transported in the cell layers of granules, where after the release of filagrin by keratohyalin bodies the keratins are processed to allow their alignment into macrofibres. Filagrin is the substance produced by keratohyalin granules that link the tonofilaments.

During the passage of the cells toward the horny layer profilagrin is cleaved by a specific phosphatase to produce the filagrin, which interacts with keratin filaments so promoting their aggregation and forming the interfilamentous matrix in the corneocytes. The degradation of filagrin produces amino acids that help the horny layer to retain water. The assembly of protofilaments occurs along the vertical axis of the helical regions, the nonhelical parts probably playing a role in their stabilization. To allow this alignment some regions are eliminated, meaning the loss of keratin antigens and a reduction in molecular weight. It has been observed *in vitro* that retinoids can stimulate the production of epithelial keratin n 19, which implies that *in vivo* they would be able to influence the expression of keratin. Currently studies involving keratins 6 and 16 are particularly interesting; they hypothesize a possible role for these proteins in cellular migration and mitosis. Their mRNAs are found in normal skin and could be responsible for the rapid cellular synthesis, though this suggests the presence of post-transcriptional regulation.

It has further been proposed that keratin 19 may play the part of activator, stabilizing the type 2 keratins until type 1 keratins are synthesized. It has in fact been found in large quantities in squamous epithelium, in the basal cells of the mucosa and hair follicles, junction regions, areas of fast turnover, and close to staminal cells.

The role of extracellular calcium has been controversial since it was demonstrated that a low level of calcium *in vitro* causes the loss of stratification and formations of desmosomes with production of stratified cultures.

1.3.1.6 The Hydrolipid Film

We have already touched upon the subject of how the hydrolipid film is produced by the sebaceous glands, sweat, and transpiration and have further stressed its importance in the maintenance of the skin's emollience and defence against pathogenic bacteria.

Though once produced the film can flow into the grooves from one region to neighbouring ones, differences in the distribution density of sweat and sebaceous glands mean that not all areas of the skin are equally covered by the hydrolipid film.

Despite its physiological functions the hydrolipid film is considered in a negative sense due to the problems that it can give our skin. For example, regions that are poorly supplied with the hydrolipid film will have less natural emollience and, retaining less water, will have a higher tendency of desquamation. Furthermore, in some individuals the sebaceous glands produce a hydrolipid film that is qualitatively altered and tends to irritate the skin itself.

The hydrolipid film is essentially formed of:

- sebum components
- cellular waste (keratinocyte residues)
- bacterial substances (derived from bacteria normally present on the skin)

- water (sweating and transpiration)
- exogenous substances (cosmetics, dirt)

An important point to remember regarding the hydrolipid film is that it constitutes part of the *normal dirt* that we remove daily from our skin by cleansing. Cleansing is not only to remove exogenous dirt but also the undesirable residues formed by the degredative action of the skin's bacteria on the organic components of the hydrolipid film. However, the excessive removal of the hydrolipid film strips the epidermis of its protection and exposes the skin to environmental attack and dehydration.

1.3.1.7 Skin Hydration

As water is the life for the entire body, so it is for the skin. Hydration represents the most important parameter in the health of our skin. Numerous factors determine the water content of the skin, although, overall it is directly related to the ambient humidity and the skin can only retain adequate concentrations of water at a relative humidity of 60%.

The skin's factors that govern the maintenance of its hydration are:

- epidermal lipids
- surface hydrolipid film
- natural moisturizing factors (NMF)
- the horny layer
- organic substances (salts, amino acids, hyaluronic acid)

Epidermal lipids have a fundamental role both in binding water and in occluding the intercellular spaces. Their action is expressed in the correct structure of the lamellae in the horny layer that otherwise would not be able to retain water.

The hydrolipid film covers the epidermis with a thin protective layer that softens it, slows down its desquamation, and defends it against both chemical and bacterio-micotic agents. Removal of the film by excessive cleansing, for example, leads to damage of the horny layer, which desquamates and loses intercellular lipids, so opening the door for the entry of germs and harmful substances.

The NMF are a series of substances produced by the epidermal cells and by sweating that function to bind water both intracellularly and extracellularly together with the intercellular lipids. Of the NMFs the most important are PCA (pirrolidon carboxylic acid) and urea, but many other organic substances and mineral salts are part of the group.

Below the epidermis, in the dermis, there is a larger quantity of water. It is this water of the dermis that hydrates the epidermis by two principal mechanisms:

1. eccrine sweating
2. transpiration (*perspiratio insensibilis*)

The latter is essential for life: the body must continuously exchange the heat produced by the energy reactions that occur uninterruptedly in all its cells.

In the dermis the majority of water is bound to *glycosaminoglycans* and *proteoglycans*. The first are formed by sugars complexed with organic acids and the second by the union of these with protein. Of particular interest is *hyaluronic acid*, which is able to bind many molecules of water. All substances that bind water are defined as *hygroscopic*, whereas the property of being soluble in water is defined as *hydrosolubility*. Hyaluronic acid is also present in the epidermis, where, in addition to hydration, it is also involved in other activities connected with the life of keratinocytes.

1.3.1.8 Skin pH

pH can be expressed, in simple terms, as the level of acidity of alkalinity of a solution or defined system. The pH scale is from 0 to 14. On this scale 7 represents neutrality. Above 7 is alkaline (or basic); values below 7 are acidic.

Under normal physiological conditions our skin has an acidic pH that varies depending on the region between 4.5 and 6.5. This value is determined by substances present in the hydrolipid film, where fatty acids are responsible for the slight but essential acidity of the skin. If the pH of the skin is altered, becoming more alkaline, this aids not only the proliferation of undesirable germs but also the denaturation of many of the substances present in the hydrolipid film itself. Therefore, the acidity of the skin must be maintained in order to protect the skin's physiology. Factors largely responsible for the modification of the skin's pH include cleansing, environmental attack, and anything that can provoke "stress" of the skin.

In the case of changes in skin pH, it is eccrine sweat, with its content of salts and amino acids, that is the skin's principle *buffering system*, ensuring a rapid restoration of the optimal acidic pH.

1.3.1.9 Skin Flora

Many bacteria and fungi naturally live on our skin. This population, which is normally present but not producing illness, is called the skin's *resident flora* and its presence is necessary to hinder the development of harmful germs.

The flora's composition changes from region to region and also with the skin's pH. The flora on the epidermal surface is different from that inside the hair follicle. Whereas *Staphylococcus epidermis* is common on the skin's surface, bacteria such as the *Propionibacterium acnes* are found in the hair follicle. Occasionally some potentially pathogenic organisms such as *Candida*, normally kept under control by the resident flora, manage to proliferate and to cause illnesses by altering thc normal physiological condition of the skin. Excessive sweating or aggressive cleansing (particularly with alkaline detergents) can easily facilitate the prevalence of these germs, especially around the genitals and armpits, which can lead to real illness.

A rich micotic and microbiotic flora exists on the scalp, thanks to the abundant sebaceous secretions in this region. A major component of this flora is *Pityrosporum orbicolare*. This fungus causes versicolor pityriasis, the common mycoses with clear marks that usually appear after the exposure to sun at the sea and is also mainly responsible for dandruff and maybe also for "acne rosacea."

1.3.2 Functions of the Skin

The structure and physiology of the skin are obviously much more complex than we have been able to describe in the present work. Nevertheless, by knowing them even in part it is easier to identify the functions (Table 1.2) that our skin fulfills:

- barrier
- protection
- immunological

Table 1.2 Summary Table of the Function of the Skin

Function	Defence Against	
Protection	Chemical agents (mechanical, chemical, biochemical) Electromagnetic rays	
Sensitivity	Perception of the outside world through the rich nerve network	
Thermoregulation	*Heat* Increase in blood flow in the surface layers Sweating	*Cold* Decrease in blood flow in the surface layers
Elimination	*Of substances* Natural substances Drugs Toxins	*Through* Sweating Sebaceous secretion Cellular desquamation Transpiration
Synthesis	*Of* Vitamin D Other functional substances	
Secretion	*Of substances* Antimicotic Antibacterial Hydrating	*Through* Sebaceous glands Apocrine sweat glands Eccrine sweat glands
Barrier immunity	Allows the entry and exit of substances Recognition by the Langerhan cells of foreign agents (haptens, allergens, antigens) that penetrate the horny layer	
Communication	Through the appearance of the skin, hair, nails, etc.	

- secretion
- thermoregulation
- sensitivity
- absorption

Some aspects of each function are reported.

1.3.2.1 Barrier Function

The function of the skin's barrier is to protect the skin and, therefore, the body from the entry of chemical substances and also preventing the loss of bodily substances. This function is valid from the outside to the inside and vice versa. Only lipid-soluble substances can penetrate it, provided that their molecular weight is not excessive. This selective permeable barrier is mainly due to the basal membrane, horny layer, and intercellular lipids. Thanks to this functional barrier, water, the essential source of life, does not escape from our bodies.

1.3.2.2 Protective Function

The skin performs a protective function against biological (bacteria, viruses, and mycetes), physical, and chemical agents. An alkaline substance placed on the skin is neutralized by the hydrolipid film and the horny layer before it can damage the organs below. In the same way the sun's radiation is neutralized, at least in part, by melanin or by the horny layer. Finally, the skin plays the essential role of mechanical protection that we all appreciate every day when large or small mechanical traumas are cushioned by our skin.

1.3.2.3 Immunological Function

The first site of entry for foreign substances and bacteria is the skin. With the Langerhan cells the skin is able to identify these and to prepare a defence. Sometimes, as in the case of contact dermatitis, the defences themselves do us harm, resulting in inflammation that is normally the essential response marking the invasion of a foreign agent.

1.3.2.4 Secretory Function

The skin's secretory functions are carried out both by the cutaneous glands and the epidermis itself. Sebum, sweat, and epidermal lipids are products that perform functions for the skin (protecting it) and for the whole body. In fact sweating, like keratinization, is one of the means by which drugs and harmful substances are removed from the body. The apocrine sweat glands participate in this function, as becomes clear, for example, when these expel dietary herbs and spices such as garlic.

1.3.2.5 Thermoregulatory Function

The mechanisms used by our skin in thermoregulation are insensible perspiration, eccrine sweating, and changes in cutaneous vascularization.

By these mechanisms the skin is able to adapt our body temperature as a function of the ambient temperature.

When the individual is hot it means that they have not been able to exchange the heat produced by the body with the environment. If this exchange does not occur, then life may be in danger because an increase in body temperature, as with fever, can block cellular reactions and, therefore, cause their death. This is why, when one is hot, the sweat glands become active and the blood vessels dilate to allow more blood to flow just below the skin, producing heat loss both in physical ways (irradiation, convection, and conduction) and by sweating. Perspiration, on the other hand, remains quite stable and changes occur only under conditions of intense heat.

The opposite occurs when one is cold: there is a reduction in sweating and a restriction of the blood vessels with consequent lower blood flow and less heat under the skin. In other words, the ability to increase or decrease the quantity of blood flowing under the skin provides the means of dissipating more or less heat into the environment.

1.3.2.6 Sensitivity Functions

The chance of survival in an environment is linked to the capacity of the individual to be in contact with it. Together with the senses of sight, hearing, and smell, the skin's sensitivity provides the individual not only with the sense of touch, but also allows us to recognize our position and its variation in space. In addition the skin is able to sense pain due its nerve receptors and through other receptors present in the encephalon to "read" the blood, so facilitating the identification of hot and cold. The skin also detects itching, which, together with pain, heat, and cold, is vital for the survival of the individual, as these sensations warn us of danger or injury. Without the skin and its sensitivity we could burn ourselves without being aware of it or we could freeze without knowing.

1.3.2.7 Absorption Function

Strictly linked to the skin's barrier function, absorption allows substances applied to the skin to be conveyed into the blood system. This important function is being ever more exploited by medicine to avoid the damage that can be caused by gastrolesive drugs when administered orally, or to favour a continuous slow drug absorption.

This function obviously varies according to the area and thickness of the epidermis. Eyelids and other analogous regions characterized by a thin layer of skin will absorb more of the substances applied to them than do other areas. Other regions, like the scalp and the armpits, which are rich in sweat glands and hair follicles, will absorb through these structures more than in other regions. This function can also have negative aspects in that it can also cause absorption of dangerous and even lethal substances.

Logically, any change in the horny layer will favour absorption by the skin. Absorption is larger for lipid-soluble substances and is divided into three stages:

1. penetration or passage of a substance from the hydrolipid film into the epidermis
2. permeation, which is the diffusion through the cells of the epidermis and dermis
3. reabsorption, the penetration of local blood vessels by a compound that has reached the dermis

Though with limitations, especially for large molecules such as collagen and other proteins, the epidermis is easily permeable to many substances, particularly if they are appropriately carried.

Liposomes, resembling the constitution and structure of membranes and intercellular lipids, penetrate the skin better than other substances and can carry many active principles. For these characteristics, liposomes, like other special structures (some even smaller than liposomes), are particularly well used in cosmetology and medicine.

Two other functions are also worth remembering—the skin communicates our image to others and has a fundamental role in the production of vitamin D, the compound necessary for normal bone development. The skin receives ultraviolet rays and utilizes them in the production of vitamin D. The discovery of this role is quite recent. In fact, at the beginning of the industrial age, when the main fuel was coal, the skies over the cities of Northern Europe were darkened by pollution, which, in combination with climate, resulted in those populations receiving little sunlight. Children, therefore, grew up affected by rickets. Exposure to sunlight was found to improve the condition, leading to the discovery that the skin is the organ that synthesizes the active anti-ricket compound, vitamin D.

References

Ackerman AB. Skin: structure and function. In: Histologic Diagnosis of Inflammatory Skin Diseases. Lea & Febiger-Lippincott, Philadelphia-Toronto, 1978, pp. 3–88.

Breathnach AS. Aspects of epidermal ultrastructure. J Invest Dermat 1975; 65: 2–15.

Caputo R, Peluchetti D. The junctions of normal human epidermis. A freeze-fracture study. J Ultrastruct Res 1977; 61:44–61.

Champion RH, Burton JL, Ebling FJG (eds.). Rook/Wilkinson/Ebling Textbook of Dermatology, 5th Edition, Blackwell Scientific Publications, London, 1992.

Giannetti A. Trattato di dermatologia (founded by Serri F), Edizioni Piccin Padova, 2002.

Hashimoto K. Fibroblast, collagen and elastin. In: Zelickson AS (ed.) Ultrastructure of Normal and Abnormal Skin. Lea and Febiger, Philadelphia, 1967, pp. 228–260.

Hashimoto K. Normal and abnormal connective tissue of the human skin I. Fibroblast and collagen. Int J Dermatol 1978; 17:459–471.

Montagna W. In: Serri F. Trattato di Dermatologia – Edizioni Piccin Padova, 1987.

Montagna W, Lobitz WC (eds.). The epidermis, Academic Press, New York, 1964.

Montagna W, Ellis RA, Silver AF (eds.). Advances in Biology of Skin. Vol. IV. The Sebaceous Glands. Proceedings of the Brown University Symposium on the Biology of Skin, 1962. Pergamon Press, Oxford, 1963.

Serri F, Montagna W. The structure and function of the epidermis. Pediatr Clin North Am 1961; 8:917–941.

2

Overview of the Structure and Function of Ethnic Skin

Chesahna Kindred[1], MD, MBA, Christian O. Oresajo[2], PhD, and Rebat M. Halder[1], MD

[1]Department of Dermatology, Howard University College of Medicine, Washington, DC, USA
[2]L'Oreal USA, Clark, NJ

2.1 Introduction

As the population of people with skin of color (Africans, African-Americans, Afro-Caribbean, Asians, Hispanics, and Native Americans) increases,

Aaron Tabor and Robert M. Blair (eds.), *Nutritional Cosmetics: Beauty from Within*, 47–62, © 2009 William Andrew Inc.

the importance of understanding the ethnic differences in the properties of skin increases. Not only do the differences in skin properties influence the presentation of skin disorders, but cultural practices alter presentation as well. Moreover, there has been a recent surge of information available about the properties of ethnic skin. This chapter covers the differences in the structure and function of ethnic skin.

2.2 Epidermis

The epidermal layer of skin is made up of five different layers: stratum basale, stratum spinosum, stratum granulosum, stratum lucidum, and stratum corneum. The stratum basale (also called the basal layer) is the germinative layer of the epidermis. The time required for a cell to transition from the basal layer through the other epidermal layers to the stratum corneum is 24–40 days. The morphology and structure of the epidermis is very similar among different races, though a few differences do exist. Additional differences are explained below and listed in Table 2.1 [1].

The stratum corneum, the most superficial layer, is the layer responsible for preventing water loss and providing mechanical protection. The cells of the stratum corneum, the corneocytes, are flat cells measuring 50 μm across and 1 μm thick. The corneocytes are arranged in layers; the number of layers varies with anatomic site and race. There is no difference between races in corneocyte surface area, which has a mean size of 900 μm^2 [2]. The stratum corneum of black skin is more compact than that of white skin. Though the mean thickness of the stratum corneum is the same among blacks and whites, black skin contains 20 cell layers whereas white skin contains 16. This may reflect greater intercellular adhesion in black skin [3].

Various studies with contradictory results evaluated spontaneous desquamation of corneocytes in various races. Spontaneous desquamation can reflect stratum corneum cohesiveness. In one study, the investigators evaluated spontaneous desquamation on the upper outer arm in African-Americans, Caucasians, and Chinese-Americans [2]. Spontaneous desquamation, measured as corneocyte count, was 2.5 times higher in blacks than in whites and Asians. These authors acknowledged that these findings were not consistent with previous studies that concluded that black skin had increased cellular adhesion and transepidermal water loss. That is, increased desquamation should correlate with decreased cellular adhesion.

Table 2.1 Differences in Structural and Functional Characteristics of Ethnic Skin

	Blacks	Whites	Asians	Hispanics
Corneocyte surface area	900 μm^2	900 μm^2	NA	NA
Corneocyte thickness	20 cell layers	16 cell layers	NA	NA
Spontaneous desquamation	?	?	?	NA
Transepidermal water loss	?	?	?	?
Conductance	Higher	Lower	NA	Higher
Lipid content	Higher	NA	NA	NA
Dryness	Higher	NA	NA	NA
Mast cells	Larger granules 15% more parallel-linear striations	30% more curved lamellae Tryptase reactivity confined to peripheral area of granules	NA	NA
Melanocyte	>200 melanosomes	<20 melanosomes	NA	NA
Amount of eumelanin	Highest	Lowest	Intermediate (East Indians)	
Gene expression of tyrosinase	–	–	NA	NA
Melanosomes	Located mostly in basal layer 0.5–0.8 mm in diameter No limiting membrane Stuck closely together Individually distributed in epidermis	Located mostly in stratum corneum 0.3–0.8 mm in diameter Limited membrane Distributed in clusters with spaces between them Degraded faster	NA	NA

(*Continued*)

Table 2.1 Differences in Structural and Functional Characteristics of Ethnic Skin (*Continued*)

	Blacks	Whites	Asians	Hispanics
Dermis	Thicker More compact Loosely stacked smaller collagen fiber bundles	More distinct papillary and reticular layers Larger collagen fiber bundles	NA	NA
Melanophages	Larger	NA	NA	NA
Fibroblasts	Larger More biosynthetic organelles More multinucleated	NA	NA	NA
Nerve fiber network	–	–	NA	NA
Elasticity	?	?	NA	?
Sebaceous glands	Larger	NA	NA	NA
Mixed apocrine-eccrine glands	More	NA	NA	NA
Eccrine glands	–	–	NA	NA

–: no difference; ?: data available, but inconclusive; NA: data not available.

Another study found spontaneous desquamation to be increased on the cheeks and foreheads of whites compared to blacks [4]. Finally, another study concluded that age and anatomic site but not race influenced spontaneous desquamation [5]. Therefore, the answer to whether or not there are racial differences in spontaneous desquamation is inconclusive.

Transepidermal water loss (TEWL) is the amount of water vapor loss from the skin, excluding sweat. TEWL increases with the temperature of the skin. Studies investigating differences in TEWL among races yield conflicting results as well. Some studies conclude there is no difference in TEWL among ethnic groups [4,6]. Others state that TEWL is greater in black skin [7]. One study found Asian skin to have the greatest TEWL among Asians, blacks, and whites [8], whereas another found Asian skin to have the least amount of TEWL [7]. This same study [7] found the skin of blacks to have the greatest TEWL, followed by that of whites, then Hispanics, and then Asians. Another study [6] found whites and Hispanics to have similar levels of TEWL. Finally, Reed et al. [9] compared TEWL of skin based on the degree of pigmentation, opposed to race alone. In a small sample size of seven, this study found that Fitzpatrick skin types V and VI had a more compact stratum corneum, increased intercellular adhesion, and increased barrier strength. Warrier et al. [4] compared the TEWL of skin at different anatomical sites suggesting that the anatomical sites used to measure TEWL in previous studies may have been a confounding factor. In this study, TEWL was lower on the cheeks and legs of black skin compared to white skin. In whites, posterior auricular and forehead stratum corneum had a higher degree of TEWL than that of the forearm, medial thigh, and back. In general, studies that measure TEWL of the forearm, medial thigh, and back concluded that black skin had a higher rate of TEWL than whites while another measuring TEWL of the checks and legs concluded black skin to have a lower rate of TEWL. In summary, concrete evidence regarding the difference in TEWL between different races has yet to be established.

Aside from TEWL, hydration is also a characteristic of skin. One of the ways to measure hydration, or water content, is conductance. Conductance, the opposite of resistance, is increased in hydrated skin because hydrated skin is more sensitive to the electrical field [10]. Skin conductance is higher in blacks and Hispanics than whites [10]. Lipid content in black skin is higher that that of white skin [11]. However, black skin is more prone to dryness, suggesting that a difference in lipid content plays a role. This includes the ratio of ceramide/cholesterol/fatty acids, the type of ceramides, and the type of sphingosine backbone. One study suggests that the degree of pigmentation influences lipid differences [9]. A more resistant barrier was demonstrated in darker skin.

2.3 Mast Cells

Sueki et al. [12] studied the mast cells of four African-American men and four white men (mean age 29) by evaluating the punch biopsy of the buttock with an electron microscope with the following results. The mast cells of black skin contained larger granules (the authors attributed this to the fusion of granules). Black skin also had 15% more parallel-linear striations and 30% less curved lamellae in mast cells. Tryptase reactivity was localized preferentially over the parallel-linear striations and partially over the dark amorphous subregions within granules of mast cells from black skin, whereas it was confined to the peripheral area of granules, including curved lamellae, in white skin. Cathepsin G reactivity was more intense over the electron-dense amorphous areas in both groups, while parallel-linear striations in black skin and curved lamellae in white skin were negative.

2.4 Melanocytes

Melanin is the major determinant of skin color. Melanin absorbs UV light and blocks free radical generation, thus protecting the skin from sun damage and aging. Melanocytes, the cells that produce melanin, synthesize melanin in special organelles called melanosomes. Melanin-filled melanosomes are transferred from one melanocyte to 30–35 adjacent keratinocytes in the basal layer [13]. Of note, the stratum lucidum in black skin is not altered by sunlight exposure [1].

There is more than one type of melanin. The first is eumelanin, a dark brown/black pigment. The second is pheomelanin, a yellow/reddish pigment. Pheomelanin has higher sulfur content than eumelanin because of the sulfur-containing amino acid cysteine. Pheomelanin is synthesized in spherical melanosomes and is associated with microvesicles [14]. Eumelanin, on the other hand, is deposited in ellipsoidal melanosomes that contain a fibrillar internal structure. Synthesis of eumelanin increases after UV exposure (tanning). Though not obvious to the naked eye, most melanin pigments of the hair, skin, and eyes are combinations of eumelanin and pheomelanin [15]. Thus, the eumelanin/pheomelanin ratio is an important determinant of skin color. It is generally believed that genetics determine constitutive levels of pheomelanin and eumelanin.

In a study by Wakamatsu et al. [15] the investigators studied melanocytes from 60 human melanocyte cultures. In this study, they found that eumelanin, and not pheomelanin, consistently increased with visual pigmentation. Lighter melanocytes had higher pheomelanin content than dark melanocytes. This indicates that eumelanin is more important in determining the degree of pigmentation. Whites had the least amount of eumelanin, Indians had more, and African-Americans had the highest. Of note, adult melanocytes contain significantly more pheomelanin than cultured neonatal melanocytes. Another study [1] found that the number of melanocytes also decreases with age.

The steps of melanogenesis are as follows: The enzyme tyrosinase hydroxylates tyrosine to dihydroxyphenylalanin (DOPA) and oxidizes DOPA to dopaquinone. Dopaquinone then proceeds down one of two pathways. If dopaquinone binds to cysteine, the oxidation of cysteinyldopa produces pheomelanin. In the absence of cysteine, dopaquinone spontaneously converts to dopachrome. Dopacrhome is then decarboxylated or tautomerized to eventually yield eumelanin. Melanosomal P-protein is involved in the acidification of the melanosome in melanogenesis [16]. Finally, the tyrosinase activity (not simply the amount of the tyrosinase protein) and cysteine concentration determine the eumelanin/pheomelanin content [15].

When α-melanocyte-stimulating hormone (α-MSH) or adrenocorticotropin binds to melanocortin-1 receptor (MC1R), a transmembrane receptor located on melanocytes, tyrosinase, and tyrosinase-related protein-1 and -2 (TRP-1 and -2) are upregulated [16–19]. In mice, loss-of-function mutations of MC1R promote pheomelanogenesis and yields the red hair phenotype [15]. In contrast, the loss-of-function mutations in humans does not alter the phenotype, but it does increase sensitivity to UV-induced DNA damage. Gene expression of tyrosinase is similar between blacks and whites, but other related genes are expressed differently. The MSH cell surface receptor gene for melanosomal P-protein is expressed differently between races. This gene may regulate tyrosinase, TRP-1, and TRP-2 [15].

As mentioned above, the melanosomes are transferred from melanocytes in the basal layer to adjacent keratinocytes that migrate to the stratum corneum. Regulatory factors in keratinocytes determine the distribution pattern of the recipient melanosome(s) [20]. The melanosomes also differ among different races. The melanosomes of blacks are mostly in the basal layer, but those of whites are mostly in the stratum corneum. This is evident in the site of UV filtration: the basal and spinous layers in blacks and the stratum corneum in whites. Of note, the epidermis of black skin rarely shows atrophied areas [1].

In black skin, the melanocytes contain more than 200 melanosomes. The melanosomes are 0.5–0.8 mm in diameter, do not have a limiting membrane, are stuck closely together, and are individually distributed throughout the epidermis. In white skin, the melanocytes contain less than 20 melanosomes. The melanosomes are 0.3–0.5 mm in diameter, associated with a limiting membrane, and distributed in clusters with spaces between them. The melanosomes of lighter skin degrade faster than that of darker skin. As a result, there is less melanin content in the upper layers of the stratum corneum. Thus, the melanocytes in black skin are more active in making melanin, and the melanosomes are packed, distributed, and broken down differently than in white skin.

There is also a difference in melanosomes between individuals within the same race but with varying degrees of pigmentation. Despite greater melanin content in darker skins, there is no evidence of major differences in the number of melanocytes [21]. Also, dark Caucasian skin resembles the melanosome distribution observed in black skin [22]. Blacks with dark skin have large, nonaggregated melanosomes and blacks with lighter skin have a combination of large nonaggregated and smaller aggregated melanosomes [23]. Whites with darker skin have nonaggregated melanosomes when exposed to sunlight, and whites with lighter skin have aggregated melanosomes when not exposed to sunlight [21,22,24].

2.5 Dermis

The dermis lies deep to the epidermis and is divided into two layers, the papillary and reticular dermis. The papillary dermis is tightly connected to the epidermis via the basement membrane at the dermo-epidermal junction. The papillary dermis extends into the epidermis with finger-like projections, hence the name "papillary." The reticular dermis is a relatively avascular, dense, collagenous structure that also contains elastic tissue and glycosaminoglycans. The dermis is made up of collagen fibers, elastic fibers, and an interfibrillar gel of glycosaminoglycans, salt, and water. Collagen makes up 77% of the fat-free dry weight of skin and provides tensile strength. Collagen types I, II, V, and VI are found in the dermis. The elastic fiber network is interwoven between the collagen bundles.

There are differences between the dermis of whites and blacks. The dermis of white skin is thinner and less compact compared to that of black skin [25]. In white skin, the papillary and reticular layers of the dermis are more distinct, they contain larger collagen fiber bundles, and the fiber fragments are sparse. The dermis of black skin contains closely stacked smaller collagen fiber bundles with a surrounding ground substance. The fiber fragments are more prominent in black skin than in white skin. Whereas the quantity is similar in both blacks and whites, the size of melanophages is larger in blacks. Also, the number of fibroblasts and lymphatic vessels are greater in black skin. The fibroblasts are larger, have more biosynthetic organelles, and are more multinucleated in black skin [1]. The lymphatic vessels are dilated and empty, with surrounding elastic fibers [25]. No racial differences in the epidermal nerve fiber network have been observed using laser-scanning confocal microscopy, suggesting that there is no difference in sensory perception between races, as suggested by capsaicin response to C-fiber activation [26].

Studies that investigated skin extensibility, elastic recovery, and skin elasticity have yielded conflicting results. Skin extensibility is how stretchable the skin is. Elastic recovery is the time required for the skin to return to its original state after releasing the stretched skin. Skin elasticity is elastic recovery divided by extensibility. Berardesca et al. concluded that differences exist between races in skin extensibility, elastic recovery, and skin elasticity and these differences are mainly related to the protective role of melanin present in races with darker skin [27]. The study found differences in sun-exposed (dorsal) and non-sun-exposed (volar) aspects of the forearms of 15 blacks, 12 whites, and 12 Hispanics. The dorsal forearm of whites had less skin extensibility than blacks and Hispanics. The dorsal and volar forearms of blacks had the same elastic recovery, but were different in whites and Hispanics. Elastic recovery was higher in black skin than Hispanic and white skin. The skin elasticity was the same on the volar forearm in all groups; it was lowest in blacks and highest on Hispanics on the dorsal forearm [27]. Warrier et al. arrived at contradictory conclusions [4]. Warrier et al. studied the difference in elastic recovery of the legs and cheeks of 30 black and 30 white women. There was no significant difference in the legs; the cheeks of black skin had a 50% greater elastic recovery [4]. Based on these conflicting results [4,27], it is likely that elastic recovery and extensibility vary by anatomic site, race, and age.

Herzberg and Dinehart [28] examined the features of sun-protected black skin from individuals 6 weeks to 75 years of age with light and electron microscopy. With age, the dermo-epidermal junction became flattened, with multiple zones of basal lamina and anchoring fibril reduplication. Microfibrils in the papillary dermis became somewhat more irregularly oriented. Compact elastic fibers showed cystic changes and separation of skeleton fibers with age. The area occupied by the superficial vascular plexus in specimens of equal epidermal surface length decreased from the infant to young adult (21–29 years old) to adult (39–52 years old) age groups, then increased in the aged adult (73–75 years old) age group. With the exception of the vascularity in the aged adult group, the above features are similar to those seen in aging white skin, and suggest that chronologic aging in white and black skin is similar. In photodamaged white skin, only elastic fibers in the dermis stain pink, whereas others stain lilac or deep blue [1]. In black skin, all dermal elastic fibers stain pink, similar to that of sun-protected white skin [1].

2.6 Sebaceous Glands and Sweat Glands

Sebaceous glands are small organs in the dermis that secrete sebum, which is a mixture of fat and debris, through a duct. Most ducts of sebaceous glands open into a hair follicle, but some open directly on the skin surface. There are three types of sweat glands: eccrine, apocrine, and apoeccrine. All three are innervated by post-ganglionic sympathetic fibers; thus acetylcholine is the neurotransmitter. Eccrine glands are sweat glands found all over the body; the highest density is on the palms and soles. Eccrine glands have no anatomic relationship to the hair follicle, secrete a clear odorless fluid, and are the primary means for thermoregulation via evaporative heat loss. Apocrine glands develop just before puberty and are localized to the axillae, anogenital region, mammary glands, external auditory canal, eyelid, face, and scalp. The duct opens into the pilosebaceous follicle at the level of the infundibulum and secretes a viscous fluid that, upon alteration by microbes, produces an odor. The function of the apocrine gland is unclear, but is a source of pheromones in animals. Apoeccrine glands, or "mixed sweat glands," have mixed characteristics of apocrine and eccrine glands. The apoeccrine gland is limited to the axillae, also develops at puberty, and is considered an eccrine gland that undergoes apocrinization. The gland plays a role in thermoregulation and axillary hyperhidrosis.

Several studies have investigated the difference in sebaceous glands and sebum production between races. Sebaceous glands of blacks are larger than that of whites [29]. One study of a small sample size of five blacks and five whites concluded that blacks produced more sebum than whites [29]. Other studies [30,31] with larger sample sizes concluded whites and blacks produce an equal amount of sebum. A study of 101 Japanese women revealed a positive correlation between the quantity of skin surface lipids and darker pigmentation [32]. Finally, Pochi and Strauss measured sebaceous gland secretion in 649 subjects, 10.3% of whom were black. Anatomically, the amount of sweat glands in black and white skins is identical and varies with climatic changes but not with racial factors. The authors concluded that, "No consistent difference in sebaceous gland activity was found between black and white skin. As sebum is an integral etiologic factor in acne, these findings are consistent with the clinical impression and with epidemiologic data, albeit scant, that the incidence of acne vulgaris in the black population differs little, if at all, from the incidence in the white population" [30]. McDonald's [33] conclusion coincides with the above. McDonald explained that a racial difference in sebum levels would lead to racial differences in the control of bacterial, viral, and other infections. Sebum and sweat contribute to the "acid mantle" of the skin and protect the skin from infection. However, blacks and whites have a similar rate of cutaneous infection, suggesting that there is not a racial difference in sebum and sweat secretion between blacks and whites.

The ratio of sebaceous glands to sweat glands is believed to be higher in blacks and the sweat glands in darker skin are believed to be larger, providing better tolerance to hot climates [34]. Black skin has many more mixed apocrine-eccrine sweat glands than does white skin [1]. There is no racial difference in eccrine glands between white and black skin [1]. Apocrine glands are found on all levels of the dermis in black skin. La Ruche and Cesarini concluded that, anatomically, the amount of sweat glands in black and white skin is identical and varies with climatic changes but not with racial factors [11].

2.7 Hair

There are two types of hair fibers, terminal and vellus. Terminal hair is found on the scalp and trunk. Vellus hair is fine and shorter and softer than terminal hair. The hair fiber grows from the epithelial follicle, which is an invagination of the epidermis from which the hair shaft develops via mitotic activity and into which sebaceous glands open. The hair follicle is one of the most proliferative cell types in the body and undergoes growth cycles. The cycles include anagen (active growth), catagen (regression), and telogen (rest). Each follicle follows a growth pattern independent of the rest. The hair follicle is lined by a cellular inner and outer root sheath of epidermal origin and is invested with a fibrous sheath derived from the dermis. Each hair fiber is made up of an outer cortex and a central medulla. Enclosing the hair shaft is a layer of overlapping keratinized scales, the hair cuticle that serves as protective layers.

Racial differences in hair include the hair type, shape, and bulb. There are four types of hair: helical, spiral, straight, and wavy. The vast majority of blacks have spiral hair [35]. The hair of blacks is naturally more brittle and more susceptible to breakage and spontaneous knotting than that of whites. The kinky form of black hair, the weak intercellular cohesion between cortical cells, and the specific hair grooming practices among black people account for these effects [35]. The shape of the hair is different between races: black hair has an elliptical shape, Asian hair has a round-shaped straight hair, and Caucasian hair is intermediate [36,37]. The bulb determines the shape of the hair shaft, indicating a genetic difference in hair follicle structure [2]. The cross-section of black hair has a longer major axis, a flattened elliptical shape, and curved follicles. Asian hair has the largest cross-sectional area and Western European hair has the smallest [36,37]. Similarities between white and black hair include: cuticle thickness, scale size and shape, and cortical cells [38].

The length and degree of curliness is genetically determined and varies by race. The curly nature of black hair is believed to result from the shape of the hair follicle [38]. By serial sectioning of hair follicles and computer-aided reconstruction from people of different races, Lindelof et al. [39] concluded that the black hair follicle has a helical form, whereas the Asian follicle is completely straight and the Caucasian hair form is intermediate. Taylor [21] reviewed and summarized articles about racial differences in hair. Blacks were found to have fewer elastic fibers anchoring the hair follicles to the dermis compared to white subjects. Melanosomes were in the outer root sheath and in the bulb of vellus hairs in blacks, but not in whites. Black hair also had more pigment and on microscopy had larger melanin granules compared to hair from light-skinned and Asian individuals.

There is no difference in keratin types between hair from different races and no difference in amino acid composition of hair from different races [40]. Among Caucasian, Asian, and African, no differences in the intimate structures of fibers were observed among these three types of hairs, whereas geometry, mechanical properties, and water swelling differed according to ethnic origin [41]. One study in 1941 did find variation in the levels of some amino acids between black and white hair [42]. Black subjects had significantly greater levels of tyrosine, phenylalanine, and ammonia in the hair, but were deficient in serine and threonine.

The morphologic features of African hair were examined using the transmission and scanning electron microscopic techniques in an unpublished study. The cuticle cells of African hair were compared with those of Caucasian hair. Two different electron density layers were shown. The denser exocuticle is derived from the aggregation of protein granules that first appear when the scale cells leave the bulb region. The endocuticle is derived from the zone that contains the nucleus and cellular organites. The cuticle of Caucasian hair is usually six to eight layers thick and constant in the hair perimeter, covering the entire length of each fiber. Black hair, in contrast, has variable thickness; the ends of the minor axis of fibers are six to eight layers thick, and the thickness diminishes to one or two layers at the ends of the major axis. The loss of several layers of cuticle cells revealing the cortex is usually seen on scanning electron microscopic micrographs of this area. The weakened endocuticle is subject to numerous fractures [43].

Black hair has a tight-curl pattern, which makes it particularly susceptible to breakage when manipulated mechanically [44]. In order to obtain the widest variety of styles, people with very curly hair, such as in blacks, often straighten the hair by either pressing (thermally straightening hair with a hot metal comb) or use of a chemical relaxer. This is among the many cultural practices that may impact the disorders of the skin and hair that are experienced by people of different ethnic groups.

2.8 Conclusion

Though more studies are needed, there are firm differences in the skin of various ethnic groups. The major differences are as follows: The stratum corneum of black skin is more compact and has greater intercellular adhesion than that of white skin. The epidermis of black skin rarely shows atrophied areas. The stratum lucidum in black skin is not altered by sunlight exposure. Lipid content in black skin is higher that that of white skin [34]. However, black skin is more prone to dryness, suggesting that a difference in lipid content plays a role. Whites have the least amount of eumelanin, East Indians have more, and African-Americans have the highest. The melanosomes of blacks are located mostly in the basal layer, are larger, and are distributed individually throughout the epidermis; melanosomes of whites are located mostly in the stratum corneum, are associated with a limiting membrane, and are distributed in clusters. Melanocytes in black skin contain more than 200 melanosomes, but those of white skin contain less than 20. The melanosomes of darker skin degrade more quickly than those of lighter skin. The dermis of white skin is thinner and less compact compared to that of black skin. Whereas the quantity is similar in both blacks and whites, the size of melanophages is larger in blacks. Also, the number of fibroblasts and lymphatic vessels are greater in black skin. The mast cells of black skin contain larger granules. Skin conductance is higher in blacks and Hispanics than in whites. In conclusion, despite the recent surge in information regarding the properties of ethnic skin, more studies are necessary for a complete understanding.

References

1. Montagna W, Carlisle K. The architecture of black and white facial skin. J Am Acad Dermatol 1991;24:929–937.
2. Courcuff P, Lotte C, Rougier A, Maibach HI. Racial differences in corneocytes—a comparison between black, white, and Oriental skin. Acta Dermatol Venereol 1991;71:146–148.
3. Weigand DA, Haygood C, Gaylor JR. Cell layers and density of Negro and Caucasian stratum corneum. J Invest Dermatol 1974;62:563–568.
4. Warrier AG, Kligman AM, Harper RA, Bowman J, Wickett RR. A comparison of black and white skin using noninvasive methods. J Soc Cosmet Chem 1996;47:229–240.
5. Manuskiatti W, Schwindt DA, Maibach HI. Influence of age, anatomic site and race on skin roughness and scaliness. Dermatology 1998;196:401–407.
6. Berardesca E, Maibach HI. Racial differences in sodium lauryl sulphate induced cutaneous irritation: black and white. Contact Dermatitis 1988;18: 65–70.
7. Sugino K, Imokawa G, Maibach HI. Ethnic difference of stratum corneum lipid in relation to stratum corneum function. J Invest Dermatol 1993; 100:587.
8. Kompaore F, Marty JP, Dupont C. Skin permeability modifications in vivo in men after application of surfactants. Importance of experimental conditions in

the determination of transepidermal insensible water loss [Article in French]. Therapie 1991;46:79–82.

9. Reed JT, Ghadially R, Elias MM. Skin type but neither race nor gender, influence epidermal permeability barrier function. Arch Dermatol 1995;131: 1134–1138.
10. Triebskorn A, Gloor M. Noninvasive methods for the determination of skin hydration. In: Forsch PJ, Kligman AM (eds.). Noninvasive Methods for the Quantification of Skin Functions. Springer-Verlag, Berlin, 1993:42–55.
11. La Ruche G, Cesarini JP. Histology and physiology of black skin. Ann Dermatovenereologica 1992;119:567–574.
12. Sueki H, Whitaker-Menezes D, Kligman AM. Structural diversity of mast cell granules in black and white skin. Br J Dermatol 2001;144:85–93.
13. Fitzpatrick TB, Szabo G. The melanocytes: cytology and cytochemistry. J Invest Dermatol 1959;32:197–209.
14. Jimbow K, Oikawa O, Sugiyama S, Takeuchi T. Comparison of eumelanogenesis and pheomelanogenesis in retinal and follicular melanocytes: role of vesiculo-globular bodies in melanosome differentiation. J Invest Dermatol 1979;73:278–284.
15. Wakamatsu K, Kavanagh R, Kadekaro AL, Terzieva S, Sturm RA, Leachman S, Abdel-Malek Zalfa, Ito S. Diversity of pigmentation in cultured human melanocytes is due to differences in the type as well as quantity of melanin. Pigment Cell Res 2006;19:154–162.
16. Abdel-Malek Z, Swope VB, Suzuki I, Akcali C, Harriger MD, Boyce ST, Urabe K, Hearing VJ. Mitogenic and melanogenic stimulation of normal human melanocytes by melanotropic peptides. Proc Natl Acad Sci USA 1995;92:1789–1793.
17. Sturm RA, Teasdale RD, Box NF. Human pigmentation genes: identification, structure and consequences of polymorphic variation. Gene 2001;277:49–62.
18. Rees JL. Genetics of hair and skin color. Annu Rev Genet 2003;37:67–90.
19. Suzuki I, Cone RD, Im S, Nordlund J, Abdel-Malek ZA. Binding of melanotropic hormones to the melanocortin receptor MC1R on human melanocytes stimulates proliferation and melanogenesis. Endocrinology 1996;137:1627–1633.
20. Minwalla L, Zhao Y, Le Poole IC, Wickett RR, Boissy RE. Keratinocytes play a role in regulating distribution patterns of recipient melanosomes in vitro. J Invest Dermatol 2001;117:341–347.
21. Taylor SC. Skin of color; biology, structure, function, implications for dermatologic disease. J Am Acad Dermatol 2002;46:S41–S62.
22. Toda K, Fatnak MK, Parrish A, Fitzpatrick TB. Alteration of racial differences in melanosome distribution in human epidermis after exposure to ultraviolet light. Nat New Biol 1972;236:143–144.
23. Olson RL, Gaylor J, Everett MA. Skin color, melanin, and erythema. Arch Dermatol 1973;108:541–544.
24. Jimbow M, Jimbow K. Pigmentary disorders in Oriental skin. Clin Dermatol 1989;7:11–27.
25. Montagna W, Giusseppe P, Kenney JA. The structure of black skin. In: Montagna W, Giusseppe P, Kenney JA (eds.). Black Skin Structure and Function. Academic Press, 1993;37–49.

26. Reilly DM, Ferdinando D, Johnston C, Shaw C, Buchanan KD, Green MR. The epidermal nerve fiber network: characterization of nerve fibers in human skin by confocal microscopy and assessment of racial variations. Br J Dermatol 1997;137:163–170.
27. Berardesca E, Rigal J, Leveque JL, Maibach HI. In vivo biophysical characterization of skin physiological differences in races. Dermatologica 1991;182:89–93.
28. Herzberg AJ, Dinehart SM. Chronologic aging in black skin. Am J Dermatopathol 1989;11:319–328.
29. Kligman AM, Shelley WB. An investigation of the biology of the human sebaceous gland. J Invest Dermatol 1958;30:99–125.
30. Pochi PE, Strauss JS. Sebaceous gland activity in black skin. Dermatol Clin 1988;6:349–351.
31. Grimes P, Edison BL, Green BA, Wildnauer RH. Evaluation of inherent differences between African American and white skin surface properties using subjective and objective measures. Cutis 2004;73:392–396.
32. Abe T, Arai S, Mimura K, Hayakawa R. Studies of physiological factors affecting skin susceptibility to ultraviolet light irradiation and irritants. J Dermatol 1983;10:531–537.
33. McDonald CJ. Structure and function of the skin. Are there differences between black and white skin? Dermatol Clin 1988;6:343–347.
34. Nicolaides N, Rothman S. Studies on chemical composition of human hair fat: the overall composition with regard to age, sex, and race. J Invest Dermatol 1952;21:90.
35. Halder RM. Hair and scalp disorders in blacks. Cutis. 1983;32:378–80.
36. Bernard BA. Hair shape of curly hair. J Am Acad Dermatol 2003; 48(6 Suppl):S120–S126.
37. Vernall DG. A study of the size and shape of hair from four races of men. Am J Phys Anthropol 1961;19:345–350.
38. Brooks O, Lewis A. Treatment regimens for "styled" black hair. Cosmetics Toiletries 1983;98:59–68.
39. Lindelof B, Froslind B, Hedblad MA, Kaveus U. Human hair form: morphology revealed by light and scanning electron microscopy and computer aided three dimensional reconstruction. Arch Dermatol 1988;124:1359–1363.
40. Gold RJ, Scriver CG. The amino acid composition of hair from different racial origins. Clin Chim Acta 1971;33:465–466.
41. Franbourg A, Hallegot P, Baltenneck F, Toutain C, and Leroy F. Current research on ethnic hair. J Am Acad Dermatol 2003;48(6 Suppl):S115–S119.
42. Menkart J, Wolfram L, Mao I. Causacian hair, Negro hair and wool: similarities and differences. J Soc Cosmet Chem 1966:17:769–787.
43. Handjur C, Fiat, Huart M, Tang D, Leory F. Morphology of the cuticle of African hair. Unpublished data.
44. Syed AN. Ethnic hair care: history, trends and formulation. Cosmetics Toiletries 1993;108:99–107.

3

The Effects of Aging on Skin

Danny Zaghi[1], *Jeanette M. Waller*[2], *MD, and Howard I. Maibach*[2], *MD*

[1]*Albert Einstein College of Medicine, Bronx, NY, USA*
[2]*University of California, San Francisco, CA, USA*

Aaron Tabor and Robert M. Blair (eds.), *Nutritional Cosmetics: Beauty from Within*, 63–77, © 2009 William Andrew Inc.

3.1 Introduction

The aging process has been studied with fervor recently, given our shifting demographics. Since age's effects are so manifest in skin's appearance, structure, mechanics, and barrier function, it is not surprising that much effort has been placed in research to better understand them. Quantitative measurements permitted by bioengineering have allowed us to objectively and precisely study aging skin.

The anatomic facets of skin are infinite, making a complete review of age-related changes in skin structure problematic. Therefore, we focus on certain readily quantifiable aspects of skin: blood perfusion, pH, thickness of strata, protein, glycosaminoglycan, water, and lipid content.

3.2 Cutaneous Blood Perfusion

A Laser Doppler Flowmetry study of 201 people ages 10–89 revealed that in areas with high blood flow, such as the lip, cushion of the third finger, nasal tip, and forehead, blood flow decreased with age [1]. However, in areas where the cutaneous blood perfusion was initially lower, such as the trunk, no clear variation occurred with age [1]. This study also measured cutaneous blood flow and surface temperature in finger cushions for 20 minutes after a 10-minute immersion in 10°C water. The decrease in blood flow after immersion was greater in people over age 50, and the restorative ability poorer in those over age 70, compared to subjects younger than age 50 [1]. A Laser Doppler Velocimetry study found that skin's vasodilation response to heat stress and vasoconstriction in response to cold challenge appears delayed in elderly subjects (ages 70–83), indicating a possibly reduced vessel density in aged skin [2]. This study only included 9–10 people from young and old age groups, so it may not provide the most conclusive answer to the question of the effect of age on cutaneous perfusion.

Intravital capillaroscopy measurements of 26 subjects using fluorescin angiography and native microscopy suggest a decrease in dermal papillary loops and little change in horizontal vessels (post-capillary venules, ascending arterioles, and part of subpapillary plexus) with increasing age [3]. An immunohistochemical study of 19 individuals ranging from age 20 to 84 revealed little effect of intrinsic aging of buttock skin on blood perfusion, but progressive and marked decrease in cutaneous perfusion in the photoaged eye corners [4]. A photoplethysmographic study including 69 individuals ages 3–99 revealed significantly decreased capillary circulation in forehead skin with advancing age. Specific numbers were not given, however.

A review of this data makes a clear conclusion difficult. There may be a trend toward decreased cutaneous perfusion in older individuals, most notably in photoaged areas; however, future studies with more subjects will be helpful in further clarification.

3.3 pH

Skin's acidity, an important part of the skin surface ecosystem and a contributor to defense against microbiological or chemical insults, also plays a role in skin barrier homeostasis and stratum corneum desquamation [5].

pH appears relatively constant, at least from childhood through age 70 [6–8]. However, skin surface pH rises significantly in adults near or older than 70 years. Some have noticed that this effect is especially marked in lower limbs, and therefore have attributed the increased pH to stasis and reduced oxygen supply frequently observed on the lower limbs in older individuals [9].

We expect that even this well-studied parameter may yield new insights when the stratum corneum is examined in a microdissected manner [10]. Furthermore, pH measurements taken from the skin surface should be interpreted cautiously as "apparent pH"; we are measuring extracted material of the stratum corneum (SC) diffusing into water applied at the stratum corneum surface, where hydrogen ions are not in solution [11].

3.4 Skin Thickness

The effect of age on the thickness of the skin strata is one of the more controversial topics among dermatological researchers. Comparing measures of skin layer thickness between individuals (and between studies) is especially challenging due to significant variation in measurements between individuals and between sites within each individual.

Qualitative histologic data from young and old skin provide a foundation for discussing skin strata thickness quantitatively. Microscopic appearance of aged skin reveals a thinner epidermis than young skin, although the intrafollicular epidermis maintains a constant thickness. The epidermal thinning is primarily due to a retraction of the rete pegs, resulting in a flattened interface between the epidermis and dermis. One consequence is that aged epidermis becomes less resistant to shearing forces and more easily torn after trauma [4,12].

The flattening of the interface is also observed histologically at the ultrastructural level. In young skin, basal cells display numerous villous cytoplasmic projections into the dermis, resulting in a highly convoluted dermal/epidermal interface. In contrast, basal cells from aged skin lack these serrations, and the dermal/epidermal junction is flattened. Hull and Warfel used scanning electron microscopy to determine that the corrugated papillary interface between the dermis and epidermis is visible up through the sixth decade; flattening occurs in the sixth through tenth decades [12]. This flattening may also be associated with a decreased proliferative potential of aged epidermis and could also affect absorption [12,13].

Aside from increased basal cell atypia, keratinization does not appear markedly different in aged epidermis. Keratin filaments, lamellar bodies, and keratohyalin granules are histologically present in usual amounts. The appearance and number of horny cells does not appear to change or diminish with age, so the stratum corneum retains its normal thickness of approximately 14–17 layers [12].

Ultrasound and microscopy studies focusing specifically on changes in SC thickness with age show little or no appreciable change, even up to the age of 97 [14,15]. However, note that there is significant variation in SC thickness between individuals and between sites within each individual.

Studies of epidermal thickness agree less. A confocal microscopy study of 34 women ages 18–69 found that the living epidermis thins on the back of the arm with increased age [15]. A morphometric analysis of histological sections from 64 individuals between 20 and 80 years old found progressive and significant (6.4% per decade) thinning of the epidermis, beginning in subjects as young as 30 years [16]. However, a microscopic study of punch biopsies from the dorsal forearm, buttock, and shoulder included 71 people ages 20–68 and found no significant change in epidermal thickness with age [17]. It is difficult to explain these differences between studies, other than to tentatively attribute them to differences in site. Branchet et al.

[17] analyzed the upper inner arm, whereas the other studies looked at different areas. To reach consensus, it may be important for future studies to focus on standardized skin sites so that their results can be more easily compared. Analyses regarding dermal thickness reveal similar results: no change on the back of the arm [17] and progressive thinning on the upper inner arm [16]. Again, skin site may be the main explanatory factor for these contradictory results.

With studies conflicting on whether changes occur in epidermal and dermal thickness, clearly studies of whole skin thickness are just as challenging to interpret. An ultrasound study analyzed the effects of aging on two anatomically similar areas of neck, one exposed to the sun throughout life, and the other covered. This experiment included 30 women age 81 ± 6 and found thinning of the skin (~ 0.1 mm difference) in the photoaged region compared to the covered area [18]. A larger study of 170 females ages 17–76 used A-mode ultrasound to find significant thinning with age on sun-protected skin, and thickening with age on sun-exposed skin [19]. The differences between these studies might be accounted for by the older population in the study by Richard et al. [18], or by the use of different modes of ultrasound. Richard et al. suggested that perhaps solar elastosis causes thickening in photoaged skin of younger individuals, but that this ceases to occur in older people [18]. Other studies found an increase in skin thickness over the first 20 years of life, then a period of either constant skin thickness or progressive thinning, followed by more marked thinning in older individuals, such that a diagram of skin thickness versus age might look like a bell curve [20–23]. However, using B-mode ultrasonography, Pellacani and Seidenari found an increase in facial skin thickness with age in 40 people between 25 and 90 years old [24]. Yet other ultrasound studies showed thinning of forehead skin with age [25,26]. One B-mode ultrasound study of 61 women ages 18–94 found that skin thickness increased on the forehead and buttock but decreased in extremity skin with increasing age [27].

Despite extensive data, it is difficult to define the effects of aging on whole skin thickness. Like the skin strata, individual and regional variation likely plays a large role in the answer to this question. Also, elastoic effects of chronic sun exposure are probably involved. Furthermore, hormonal differences between individuals and throughout the aging process may confound studies of skin thickness. The effects of estrogen on skin aging are reviewed by Shah and Maibach [28]. More concordant future results relating skin thickness and age might be obtained by greater standardization of body site, patient population, and ultrasound method, such that different laboratories use the same frequency and gain curves.

3.5 Proteins

3.5.1 Collagen

Collagen, which comprises approximately 70–80% of the dry weight of the dermis, is primarily responsible for skin's tensile strength. Each collagen molecule consists of three polypeptide chains, each containing about 1,000 amino acids in their primary sequence. In the collagen molecule the alpha-chains are wrapped around each other to make a triple helical conformation [29]. In chronologically aged skin, the rate of collagen synthesis, activity of enzymes that act in the post-translational modification, collagen solubility, and thickness of collagen fiber bundles in the skin all decrease [27,30]. Also, the ratio of type III to type I collagen increases with increasing age [27,31]. In photoaged skin, however, collagen fibers are fragmented, thickened, and more soluble [27]. It is plausible that reduced collagen deposition in elderly skin could explain development of dermal atrophy and might relate to poor wound healing in the elderly [30].

Histological data, though not quantitative, reveals important information about orientation and arrangement of collagen fibers in skin. Lavker et al. compared skin from the upper inner arm of old (ages 70–85) and young (ages 19–25) individuals using light, transmission electron, and scanning electron microscopy [32]. Interestingly, they suggested that the upper inner arm might be an optimal site for analyzing sun-protected skin, because it is not exposed to the pressure deformations and reformations occurring in the buttock. They found that in young adults, collagen in the papillary dermis forms a meshwork of randomly oriented thin fibers and small bundles. The reticular dermis consists of loosely interwoven, large, wavy, randomly oriented collagen bundles. However, the collagen within each bundle is packed together closely [32]. In aged skin the density of the collagen network appears to increase. This likely reflects a decrease in ground substance that would otherwise form spaces between the collagen [32]. Also, rather than appearing in discrete ropelike bundles of tightly packed fibers, collagen forms aggregates of loosely woven, mostly straight fibers. As fibers become straighter in aged skin, there is less room for the skin to be stretched, so tensile strength increases [32]. Using immuno-electron microscopy, Vitellaro-Zuccarello et al. found similar age-related trends in skin collagen. They also noted greater intensity of collagen III staining in subjects over 70 years of age [33]. Hence histological and more recent methods are in agreement, revealing that increased age is associated with decreased collagen content and straightening of collagen fibers, forming looser bundles, an increased type III:type I collagen ratio, and decreased ground substance.

3.5.2 Elastin

The skin's intact elastic fiber network, which occupies approximately 2–4% of the dermis by volume, provides resilience and suppleness. This network shows definite changes associated with aging, especially between the ages of 30 and 70. In sun-exposed skin, an excessive accumulation of elastoic material occurs. Accumulation of new elastin in response to photoaging is also apparent from upregulation of the elastin promoter activity and increased abundance of elastin mRNA [34,35]. Bernstein et al. compared photoaged to intrinsically aged skin, and found a 2.6-fold increase in elastin mRNA, a 5.3-fold increase in elastin expression, and a 5-fold increase in elastin promoter activity in photodamaged skin [34]. However, these apparent increases in elastin synthesis do not account for the massive accumulation of elastoic material seen histologically in photoaged skin [34]. Some attribute this to elastin degradation being slower than synthesis, leading to an accumulation of partially degraded fibers. In purified skin elastin, the amount of racemized aspartic acid increases rapidly and is highly correlated with age ($r = 0.98$) [35]. This indicates that skin's elastin, like elastin in the aorta and lung, is long-lived and accumulates damage over time [36,37].

In innate aging, fragmentation of elastic fibers results in their decreased number and diameter. Computerized image analysis of elastin-stained skin biopsies from the buttock and upper inner arm reveal an age-related increase in mean elastin fiber length and percentage surface area coverage in the dermis, but these fibers are thought to be abnormally enriched in polar amino acids, carbohydrates, lipids, and calcium [36]. Through different mechanisms, photoaging and intrinsic aging ultimately result in a deficiency of functional, structurally intact elastic fibers [30]. The finer oxytalan fibers in the papillary dermis are depleted or lost altogether; eluanic and elastic fibers become progressively abnormal. These alterations largely account for the widely recognized decrease in skin's physiological elasticity with increased age [36].

Examination of intrinsically aged skin elastin and fibrillin with immunohistochemical staining revealed that elastin was located in the papillary dermis just below the basement membrane, as small fibers mostly oriented perpendicular to the epidermis. In the deeper dermis, fibers were thicker and oriented differently. Areas surrounding adnexal structures and larger vessels in the deep dermis were also intensely stained [34]. Photoaged skin demonstrated similar small-diameter fibers just below the basement membrane within a zone lacking excessive staining, which was of variable thickness [34]. This may correspond to the subepidermal low echogenic band seen in ultrasound imaging. Beneath this area of relatively sparse staining was a region of poorly formed, clumped, thick fibers. This staining pattern occupied the superficial to mid-dermis, below which staining again resumed its well-defined pattern as seen in sun protected skin [34].

Elastin, therefore, exhibits numerous age-related changes, including slow degradation and accumulation of damage in existing elastin with intrinisic aging, increased synthesis of apparently abnormal elastin in photoexposed areas, and abnormal localization of elastin in the upper dermis of photodamaged skin. These factors lead to the histologically evident elastoic accumulation and contribute to characteristic changes in ultrasound images of aged skin.

3.5.3 General Protein Structure

Through Raman spectroscopy, little difference is seen between photoexposed and protected areas in young individuals; the majority of proteins in young skin are in helical conformation. Intrinsically aged skin shows slightly altered protein structure, and photoaged skin reveals markedly altered protein conformation, with increased folding and less exposure of aliphatic residues to water [38,39]. Amino acid composition of proteins and free amino acids in aged skin also differ significantly from that of young skin, including an increase in overall hydrophobicity of amino acid fractions from the elderly [40]. Because free amino acids are believed to play a key role in stratum corneum water binding, this shift in their composition, combined with the evidence of altered tertiary protein structure, provides insight into the increased incidence of xerosis in aged individuals.

3.6 Glycosaminoglycans

Glycosaminoglycans (GAGs) are composed of specific repeating disaccharide units. Those attached to a core protein are referred to as proteoglycans and are found widely distributed throughout the skin. GAGs most often present in human skin are hyaluronic acid (not attached to a protein core) and the proteoglycan family of chondroitin sulphates, including dermatan sulphate [41]. GAGs are especially important in skin because they bind up to 1,000 times their volume in water. Therefore, skin hydration is highly related to the content and distribution of dermal GAGs, especially hyaluronic acid [41].

GAGs increase in photoaged skin compared to young or intrinsically aged skin [38,41]. This seems paradoxical, because photoaged skin appears leathery and dry, unlike newborn skin, which also contains high levels of GAGs. Confocal laser scanning microscopy reveals that GAGs in photodamaged skin are abnormally deposited on elastoic material, rather than diffusely scattered as in young skin [41]. This aberrant localization may interfere with normal water binding by GAGs, despite their increased number.

3.7 Water

In young skin, most water is bound to proteins and, appropriately, is called bound water [39]. This is important for the structure and mechanical properties of many proteins and their mutual interactions. Water molecules not bound to proteins bind to each other, and are called tetrahedron or bulk water [39].

Intrinsic aging does not appear to alter water structure significantly [38]. However, in photoaged skin, Raman spectroscopy reveals an increase in total water content. Again, this seems paradoxical, because aged skin is often dry and weathered. However, structurally, significantly more of the water in aged skin is in tetrahedron form. Thus, because proteins are more hydrophobic and folded, and GAGs are clumped on elastoic material, they interact less with water, and water in aged skin binds to itself instead. This lack of interaction between water and surrounding molecules in photoaged skin likely contribute to its characteristically dry and wrinkled appearance.

3.8 Lipids

The "brick and mortar" model is often employed to describe the stratum corneum's protein-rich corneocytes embedded in a matrix of ceramides, cholesterol, and fatty acids, and smaller amounts of cholesterol sulphate, glucosylceramides, and phospholipids. These lipids form multilamellar sheets amidst the intercellular spaces of the stratum corneum, and are critical to the SC's mechanical and cohesive properties, enabling it to function as an effective water barrier [42]. Changes in SC lipid content have been linked to skin conditions such as xerosis and possibly atopic dermatitis [42].

Many authors agree that overall lipid content of human skin decreases with age [42–44]. Using high performance thin layer chromatography (HPTLC), Rogers et al. found a 30% decrease in the face, hand, and leg in older subjects, but the older group only extended to age 50. No significant change was seen in proportional composition of lipid classes or ceramide species [42]. Schreiner et al. used small angle X-ray diffraction to compare lipid composition in subjects aged 23–27 years to subjects aged 63–69 years [45]. They did not see any overall difference in lipid quantity or composition between the groups. However, the aged group consisted of only four subjects, and again, included a narrow age range. Saint Leger et al. studied the lower legs of 50 subjects and found that the lipid profile was constant from age 50 upward; overall, aging was associated with a slight decrease in sterol esters and triglycerides [44]. Cua et al. noted significant regional variation within individuals as they studied 11 sites on 29 people, comparing individuals in their third decade of life to those in their eighth. Interestingly, they, too, found little relation between skin surface lipid content and age, except on the ankle, where the elderly demonstrated decreased lipid content [46]. From these conflicting studies, it is difficult to conclude with certainty whether lipid content decreases with age. Many confounding factors may hinder such studies, including seasonal and diurnal variation, general interindividual variation, and the use of several different methodologies by different researchers.

3.9 Conclusions

Despite numerous impressive technological methodologies, data regarding age-related changes in skin structure and function are often conflicting and difficult to interpret. Multiple studies of blood flow present multiple conclusions. There may be a trend toward decreased cutaneous perfusion in older individuals, most notably in photoaged areas. pH measurements are more concordant; it appears that skin pH is constant throughout adulthood until approximately 70 years of age, after which it rises significantly. This may be in part due to stasis of blood in elderly individuals, especially in lower limbs [9].

We have great difficulty in reaching consensus regarding age-related changes in skin strata thickness. Though it is accepted that aging presents retraction of rete pegs and flattening of the epidermal-dermal junction, and though most agree that the stratum corneum maintains relatively constant thickness throughout life, quantitative measurements of epidermal, dermal, and whole skin thickness versus age provide no single answer. It appears that the epidermis thins with age on some body sites, such as the upper inner arm [17] and back of the upper arm [14], but remains constant on others, such as the buttock, dorsal forearm, and shoulder [15]; this variation is clearly not accounted for by sun or environmental exposure alone. Dermal thickness presents similar variation.

Whole skin thickness has been seen by some to increase in youth, remain constant during adulthood, and decrease in the elderly [20–23]. However, others maintain that photoexposed areas thicken with age, whereas protected areas become thinner [19]. Some suggest that changes with age are more related to skin's location on the extremities or axially than to sun exposure [40]. Differences in study method, population, and body site likely account for these markedly different results and tend to obscure a reasonable conclusion.

Collagen becomes less soluble, thinner, and sparser in intrinsically aged skin, but is thickened, fragmented, and more soluble with photoaging [38]. The ratio of type III to type I collagen is reported to increase with age [31,33,38]. Histologically, young collagen is randomly organized into a meshwork of loosely interwoven bundles. Age leads to a loosening within these bundles and straightening of collagen fibers, increasing skin's tensile strength [24].

Elastin is a long-lived protein in human skin; it appears to accumulate damage with age and sun-exposure. New elastin is synthesized in greater quantities in aged skin, but it is thought that this synthesis results in abnormally structured elastin [34,47]. Also, elastin degradation does not appear to keep pace with new synthesis in aged skin. This results in massive accumulations of elastoic material, especially in photoaged skin. The abnormal structure of this elastin prevents it from functioning as it does in young skin.

Studies of primary and tertiary skin protein structure in aged skin reveal an environment unfriendly to water, with an overall increase in hydrophobic amino acids and greater folding such that aliphatic residues are more hidden from water [38,40]. Also, although total amounts of glycosaminoglycans appear to be increased in aged skin, these are abnormally localized on the elastoic material in the superficial dermis; thus, they are unable to bind water as well as if they were scattered appropriately throughout the whole dermis [41]. Hence, it is not surprising that, although aged skin contains increased amounts of water, most of this water is bound to itself in tetrahedral form, rather than being bound to proteins and GAGs as it is in young skin [38]. These factors together likely contribute to increased xerosis and withered appearance of aged skin.

Though it tends to be an accepted assumption that lipid content decreases with age, quantitative studies are conflicting. Some indicate a marked age-related decrease in skin lipids, at least up to age 50 [42], whereas others indicate little or no relationship [45,46].

References

1. Ishihara M, Itoh M, Ohsawa K, Kinoshita M, Satoh Y. Blood Flow. In: Cutaneous Aging, Kligman AM, Takase Y (eds.), University of Tokyo Press, Tokyo, 1988, pp. 167–181.
2. Tolino MA, Wilkin JK. Aging and cutaneous vascular thermoregulation responses. J Invest Dermatol 1988; 90(4):613.
3. Kelly RI, Pearse R, Bull R, Leveque JL, de Rigal J, Mortimer P. The effects of aging on cutaneous microvasculature. J Am Acad Dermatol 1995; 33:749–756.
4. Chung JH, Yano K, Lee MK, Youn CS, Seo JY, Kim KH, Cho KH, Eun HC, Detmar M. Differential effects of photoaging vs. intrinsic aging on the vascularization of human skin. Arch Dermatol 2002; 138:1437–1442.
5. Rippke F, Schreiner V, Doering T, Maibach HI. Stratum corneum pH in atopic dermatitis: impact on skin barrier function and colonization with staphylococcus aureus. Am J Clin Dermatol 2004; 5(4):217–223.
6. Fluhr JW, Pfistener S, Gloor M. Direct comparison of skin physiology in children and adults with bioengineering methods. Pediatric Dermatology 2000; 17(6):436–439.
7. Diktein S, Hartzshtark A, Bercovici P. The dependence of low pressure indentation, slackness, and surface pH on age in forehead skin of women. J Soc Cosmet Chem, 1984; 35:221–228.
8. Shoen A. The Skin Surface pH as a Function of Age. M Pharm. Thesis, Hebrew University, 1982.
9. Wilhelm KP, Cua AB, Maibach HI. Skin aging: effect on transepidermal water loss, stratum corneum hydration, skin surface pH, and casual sebum content. Arch Dermatol 1991; 127:1806–1809.

10. Fluhr JW, Elias PM. Stratum corneum pH: Formation and function of the 'acid mantle'. Exogenous Dermatology 2002; 1(4):163–175.
11. Escoffier C, de Rigal J, Rochefort A, Vasselet R, Leveque JL, Agache PG. Age related mechanical properties of human skin: an in vivo study. J Invest Dermatol 1989; 93(3):353–357.
12. Hull MT, Warfel KA. Age-related changes in the cutaneous basal lamina: scanning electron microscopic study. J Invest Dermatol 1983; 81:378–380.
13. Ya-Xian Z, Suetake T, Tagami H. Number of cell layers of the stratum corneum in normal skin: relationship to the anatomical location on the body, age, sex, and physical parameters. Arch Dermatol Res 1999; 291: 555–559.
14. Batisse D, Bazin R, Baldeqeck T, Querleux B, Leveque JL. Influence of age on the wrinkling capacities of skin. Skin Research and Technology 2002; 8:148–154.
15. Sandby-Moller J, Poulsen T, Wulf HC. Epidermal thickness at different body sites: relationship to age, gender, pigmentation, blood content, skin type, and smoking habits. Acta Derm Venereol 2003; 83:410–413.
16. Shuster S, Black M, McVitie E. The influence of age and sex on skin thickness, skin collagen, and density. Br J Dermatol 1975; 93:639.
17. Branchet MC, Boisnic S, Frances C, Robert AM. Skin thickness changes in normal aging skin. Gerontology 1990; 36:28–35.
18. Richard S, de Rigal J, de Lacharriere O, Berardesca E, Leveque JL. Noninvasive measurement of the effect of lifetime exposure to the sun on the aged skin. Photodermatol Photoimmunol Photomed 1994; 10:164–169.
19. Takema Y, Yorimoto Y, Kawai M, Imokawa G. Age-related changes in the elastic properties and thickness of human facial skin. Br J Dermatol 1994; 141:641–648.
20. Agache P. Metrology of the stratum corneum. In: Measuring the Skin, Agache P, Humbert P (eds.), Springer-Verlag, Berlin, 2004, pp. 101–111.
21. Seidenari S, Pagnoni A, Nardo AD, Giannetti A. Echographic evaluation with image analysis of normal skin: Variation according to age and sex. Skin Pharmacol 1994; 7:201–209.
22. Dykes PJ, Marks R. Measurement of skin thickness: A comparison of two in vivo techniques with a conventional histometric method. J Invest Dermatol 1976; 21:418–429.
23. Denda M, Takahasi M. Measurement of facial skin thickness by ultrasound method. J Soc Cosmet Chem Japan 1990; 23:316–319.
24. Pellacani G, Seidenari S. Variations in facial skin thickness and echogenicity with site and age. Acta Derm Venereol 1999; 79:366–369.
25. Nishimura M, Tuji T. Measurement of skin elasticity with a new suction device. Jpn J Dermatol 1992; 102:1111–1117.
26. Gniadecka M, Jemec GBE. Quantitative evaluation of chronological aging and photoaging in vivo: Studies on skin echogenicity and thickness. Br J Dermatol 1998; 139:815–821.
27. Gniadecka M, Gniadecki R, Serup J, Sondergaard J. Ultrasound structure and digital image analysis of the subepidermal low echogenic band in aged human skin: diurnal changes and interindividual variability. J Invest Dermatol 1994; 102(3):362–365.

28. Shah MG, Maibach HI. Estrogen and Skin. An Overview. Am J Clin Dermatol 2001; 2(3):143–150.
29. Oikaren A. Aging of the skin connective tissue: how to measure the biochemical and mechanical properties of aging dermis. Photodermatol Photoimnunol Photomed 1994; 10:47–52.
30. Uitto J. Connective tissue biochemistry of the aging dermis. Clin Geriatric Medicine 1989; 5(1):127–147.
31. Lovell CR, Smolenski KA, Duance VC, Light ND, Young S, Dyson M. Type I and III collagen content and fibre distribution in normal human skin during aging. Br J Dermatol 1987; 117:419–428.
32. Lavker RM, Zheng P, Dong G. Aged skin: a study by light, transmission electron microscopy, and scanning electron microscopy. J Invest Dermatol 1987; 88:44s–53s.
33. Vitellaro-Zuccarello L, Garbelli R, Rossi VD. Immunocytochemical localization of collagen types I, II, IV, and fibronectin in the human dermis: modifications with aging. Cell Tissue Res 1992; 268:505–511.
34. Bernstein EF, Chen YQ, Tamai K, Shepley KJ, Resnik KS, Zhang H, Tuan R, Mauviel A, Uitto J. Enhanced elastin and fibrillin gene expression in chronically photodamaged skin. J Invest Dermatol 1994; 103:182–186.
35. Ritz-Timme S, Laumier I, Collins MJ. Aspartic acid racemization—evidence for marked longevity of elastin in human skin. Br J Dermatol 2003; 149:951–959.
36. Robert C, Lesty C, Robert AM. Ageing of the skin: study of elastic fiber network modifications by computerized image analysis. Gerontology 1988; 34:91–96.
37. Powell JT, Vine N, Crossman M. On the accumulation of D-aspartate in elastin and other proteins of the aging aorta. Atherosclerosis 1992; 97(2–3): 201–208.
38. Gniadecka M, Nielsen OF, Wessel S, Heidenhcim M, Christensen DH, Wulf HC. Water and protein structure in photoaged and chronically aged skin. J Invest Dermatol 1998; 11:1129–1133.
39. Gniadecka M, Niclsen OF, Christensen DH, Wulf HC. Structure of water, proteins, and lipids in intact human skin, hair, and nail. J Invest Dermatol 1998; 110(4):393–398.
40. Jacobson T, Yuksel Y, Geesin JC, Gordon JS, Lane AT, Gracy RW. Effects of aging and xerosis on the amino acid composition of human skin. J Invest Dermatol 1990; 95(3):296–300.
41. Bernstein EF, Underhill CB, Hahn PJ, Brown DB, Uitto J. Chronic sun exposure alters both the content and distribution of dermal glycosaminoglycans. Br J Dermatol 1996; 135:255–262.
42. Rogers J, Harding C, Mayo A, Banks J, Rawlings A. Stratum corneum lipids: the effects of aging and the seasons. Arch Dermatol Res 1996; 288: 765–770.
43. Roskos KV, Maibach HI, Guy RH. The effect of aging on percutaneous absorption in man. J Pharmacokinet Biopharm 1989; 17(6):617–630.
44. Saint Leger D, Francois AM, Leveque JL, Stoudemayer TJ, Grove GL, Kligman AM. Age-associated changes in the stratum corneum lipids and their relation to dryness. Dermatologica 1988; 177:159–164.

45. Schreiner V, Gooris G, Pfeiffer S, Lanzendorfer G, Wenck H, Diembeck W, Proksch E, Bouwstra J. Barrier characteristics of different human skin types investigated with x-ray diffraction, lipid analysis, and electron microscopy imaging. J Invest Dermatol 2000; 114:654–660.
46. Cua AB, Wilhelm KP, Maibach HI. Skin surface lipid and skin friction. Relation to age, sex, and anatomical region. Skin Pharmacology 1995; 8(5): 246–251.
47. Ritz-Timme S, Collins MJ. Racemization of aspartic acid in human proteins. Age Res Rev 2002; 1:43–59.

4

Premature Aging of Skin from Environmental Assaults

Jean Krutmann, MD

Professor of Dermatology & Environmental Medicine and Director Institut für Umweltmedizinische Forschung (IUF) at the Heinrich-Heine-University Düsseldorf gGmbH, Düsseldorf, Germany

Aaron Tabor and Robert M. Blair (eds.), *Nutritional Cosmetics: Beauty from Within*, 79–92, © 2009 William Andrew Inc.

4.1 Introduction

Among all environmental factors, solar ultraviolet (UV) radiation is most important for premature skin aging, a process accordingly also termed photoaging [1]. Other factors include exposure to near-infrared radiation (IRA; 760–1440 nm) from sunlight or artificial IRA radiation devices, tobacco smoke, and particulate matter from airborne traffic pollution [2,3].

Within recent years substantial progress has been made in elucidating the molecular mechanisms underlying extrinsic skin aging. These studies have initially focused on photoaging of the skin. From this large body of research it is now clear that both UVB (290–320 nm) and UVA (320–400 nm) radiation contribute to photoaging. UV-induced alterations at the level of the dermis are best studied and appear to be largely responsible for the phenotype of photoaged skin. It is also generally agreed that UVB acts preferentially on the epidermis, where it not only damages DNA in keratinocytes and melanocytes but also causes the production of soluble factors including proteolytic enzymes, which then in a second step affect the dermis. In contrast, UVA radiation penetrates far more deeply on average and hence exerts direct effects on both the epidermal and the dermal compartment. UVA is also 10–100 times more abundant in sunlight than UVB, depending on the season and time of day. It has therefore been proposed that, although UVA photons are individually far less biologically active than UVB photons, UVA radiation may be at least as important as UVB radiation for the pathogenesis of photoaging [4].

The exact mechanisms by which UV radiation causes premature skin aging are not yet clear, but a number of molecular pathways have been described to explain one or more of the key features of photoaged skin. Some of these models are based on irradiation protocols, which use single or few UV exposures, whereas others take into account the fact that photoaging results from chronic UV damage and as a consequence employ chronic repetitive irradiation protocols. Still others rely on largely theoretic constructs rather than on experimental observations.

It should be noted that many if not most of the detrimental effects that are induced by UVB and UVA radiation also occur upon exposure of the skin to other environmental assaults, but that the molecular mechanisms and signaling pathways involved may differ substantially. As an example, UV radiation as well as IRA radiation cause collagen degradation in the skin (and thereby wrinkle formation) by upregulating the expression of collagen degrading proteases such as matrix metalloproteinase-1 (MMP-1) in dermal fibroblasts and in both instances this effect is causally related to the intracellular generation of oxidative stress [1,5–7]. In the case of IRA radiation, the oxidative stress is being generated within mitochondria (reactive species leak out of the mitochondrial respiratory chain as a consequence of IRA irradiation) and this intramitochondrial production of reactive oxygen species initiates a retrograde signaling response, which is directed from the mitochondria toward the cell nucleus [7]. In marked contrast, UV radiation–induced MMP-1 induction is not mediated by a retrograde mitochondrial signaling pathway, but involves a singlet oxygen-triggered disturbance of cell membrane microdomains (in the case of UVA-1), or activation of intracytoplasmatic transcription factors such as the arylhydrocarbon receptor pathway (in the case of UVB) or the generation of cyclobutane pyrimidine dimers in nuclear DNA (in the case of UVB) [8,9]. In other words, all these environmental assaults may cause similar clinical, histological, and even molecular problems, but they are initiated and mediated through distinct intracellular pathways and thus are differentially susceptible to prevention strategies. Indeed, effective prevention of UV versus IRA radiation–induced premature skin aging requires distinct antioxidants: in the case of IRA, antioxidants that preferentially localize within mitochondria work best, whereas for UV, antioxidants that accumulate in the cytoplasm or enzymes that are capable of repairing nuclear DNA will be more effective [7,10].

In this chapter we will focus on the most popular pathogenic concepts relevant for UV-induced skin aging, because photoaging is by far the best studied and probably also most relevant example of extrinsic skin aging. For more detailed information on IRA radiation and tobacco smoke–induced skin aging, interested readers should refer to a recent review [2].

4.2 Mitochondrial DNA Mutations and Photoaging

Mitochondria are organelles whose main function is to generate energy for the cell. This is achieved by a multistep process called oxidative phosphorylation or electron-transport-chain. Located at the inner mitochondrial membrane are five multiprotein complexes that generate an electrochemical proton gradient used in the last step of the process to turn Adenosine diphosphate (ADP) and organophosphate into Adenosine triphosphate (ATP). This process is not completely error free and ultimately this leads to the generation of reactive oxygen species (ROS), making the mitochondrion the site of the highest ROS turnover in the cell. In close proximity to this site lies the mitochondria's own genomic material, the mitochondrial (mt) DNA. The human mtDNA is a 16,559-bp-long, circular, and double-stranded molecule of which 4 to 10 copies exist per cell. Mitochondria do not contain any repair mechanism to remove bulky DNA lesions. Although they do contain base excision repair mechanisms and repair mechanisms against oxidative damage, the mutation frequency of mtDNA is approximately 50-fold higher than that of nuclear DNA. Mutations of mtDNA have been found to play a causative role in degenerative diseases such as Alzheimer's disease, chronic progressive external ophthalmoplegia, and Kearns-Sayre syndrome [11]. In addition to degenerative diseases, mutations of mtDNA may play a causative role in the normal aging process, with an accumulation of mtDNA mutations accompanied by a decline of mitochondrial functions. Recent evidence indicates that mtDNA mutations are also involved in the process of photoaging [4].

Photoaged skin is characterized by increased mutations of the mitochondrial genome [1,12,13]. Intraindividual comparison studies have revealed that the so-called common deletion, a 4,977 base pair deletion of mtDNA, is increased up to 10-fold in photoaged skin, as compared with sun-protected skin of the same individuals. The amount of the common deletion in human skin does not correlate with chronological aging [14], and it has therefore been proposed that mtDNA mutations such as the common deletion represent molecular markers for photoaging. In support of this concept it was shown that repetitive, sublethal exposure to UVA radiation at doses acquired during a regular summer holiday induces mutations of mtDNA in cultured primary human dermal fibroblasts in a singlet oxygen-dependent fashion [15]. Even more importantly, *in vivo* studies have revealed that repetitive three-times daily exposure of previously unirradiated buttock skin for a total of two weeks to physiological doses of UVA radiation leads to an approximately 40% increase in the levels of the common deletion in the dermal, but not epidermal, compartment of irradiated skin [16]. Also, use of sunbeds for a period of only 3 months increased mtDNA mutagenesis *in vivo* in human skin [17]. Furthermore, it was shown that, once induced, these mutations persist for at least 16 months in UV-exposed skin [16]. Interestingly, in a number of individuals, the levels of the common deletion in irradiated skin continued to increase with a magnitude up to 32-fold.

It has been postulated for the normal aging process as well as for photoaging that the induction of ROS generates mtDNA mutations, in turn leading to a defective respiratory chain and, in a vicious cycle, inducing even more

ROS and subsequently allowing mtDNA mutagenesis independent of the inducing agent [16]. It is the characteristic of vicious cycles that they evolve at ever increasing speeds. Thus, the increase of the common deletion up to levels of 32-fold, independent of UV exposure, may represent the first *in vivo* evidence for the presence of such a vicious cycle in general and in human skin in particular.

The mechanisms by which generation of mtDNA mutations by UVA exposure translates into the morphologic alterations observed in photoaging of human skin are currently being unraveled. In general, a cause-effect relationship between premature aging and mtDNA mutagenesis is strongly suggested by studies employing homozygous knock-in mice that express a proofreading-deficient version of PolgA, the nucleus-encoded subunit of mtDNA polymerase [18]. As expected, these mice develop an mtDNA mutator phenotype with increased amounts of deleted mtDNA. This increase in somatic mtDNA mutations was found to be associated with reduced lifespan and premature onset of aging-related phenoytypes such as weight loss, reduced subcutaneous fat, alopecia, kyphosis, osteoporosis, anemia, reduced fertility, and heart enlargement.

In addition, recent studies demonstrate that UVA radiation-induced mtDNA mutagenesis is of functional relevance in primary human dermal fibroblasts and apparently has molecular consequences suggestive of a causative role of mtDNA mutations in photoaging of human skin as well [19]. Accordingly, induction of the common deletion in human skin fibroblasts is paralelled by a measurable decrease of oxygen consumption, mitochondrial membrane potential, and ATP content as well as an increase of MMP-1, whereas TIMP remains unaltered, an inbalance that is known to be involved in photoaging of human skin (see below). These observations suggest a link not only between mutations of mtDNA and cellular energy metabolism, but also between mtDNA mutagenesis, energy metabolism, and a fibroblast gene expression profile that would functionally correlate with increased matrix degradation and thus premature skin aging. In order to provide further evidence for a role of the energy metabolism in mtDNA mutagenesis and the development of this "photoaging phenotype," the effect of creatine was studied in these cells. This applied the hypothesis that generation of phosphocreatine, and consequently ATP, is facilitated if creatine is abundant in cells. This would allow easier binding of existing energy-rich phosphates to the energy precursor creatine. Indeed, experimental supplementation of normal human fibroblasts with creatine normalized mitochondrial mutagenesis as well as the functional parameters of oxygen consumption and MMP-1, whereas an inhibitor of creatine uptake abrogated this effect [19].

In another experimental approach, partial depletion of mtDNA from dermal fibroblasts caused a gene expression profile in these cells reminiscent of photoaging [20]. Specifically, there was significant upregulation of the expression of genes involved in collagen degradation, but expression of genes relevant for collagen *de novo* synthesis was decreased. Taken together, these studies strongly indicate that UV-induced mtDNA mutagenesis leads to photoaging of human skin.

4.3 Connective-Tissue Alterations in Photoaging: The Role of Matrix Metalloproteinases and Collagen Synthesis

Photoaged skin is characterized by alterations of the dermal connective tissue. The extracellular matrix in the dermis mainly consists of type I and type III collagen, elastin, proteoglycans, and fibronectin. In particular, collagen fibrils are important for the strength and resiliency of skin, and alterations in their number and structure are thought to be responsible for wrinkle formation. In photoaged skin, collagen fibrils are disorganized and abnormal elastin-containing material accumulates [21]. Biochemical studies have revealed that in photoaged skin levels of types I and III collagen precursors and cross-links are reduced, whereas elastin levels are increased [22,23].

How does UV radiation cause these alterations? In principle it is conceivable to assume that UV radiation leads to an enhanced and accelerated degradation and/or a decreased synthesis of collagen fibers and our current knowledge indicates that both mechanisms may be involved.

A large number of studies unambiguously demonstrate that the induction of matrix metalloproteinases (MMPs) plays a major role in the pathogenesis of photoaging. As indicated by their name, these zinc-dependent endopeptidases show proteolytic activity to degrade matrix proteins such as collagen and elastin. Each MMP degrades different dermal matrix proteins, for example MMP-1 cleaves collagen type I, II, and III, whereas MMP-9, which is also called gelatinase, degrades collagen types IV and V, and gelatin. Under basal conditions, MMPs are part of a coordinate network and are precisely regulated by their endogenous inhibitors, i.e., tissue-specific inhibitors of MMPs (TIMPs), which specifically inactivate certain MMPs. An inbalance between activation of MMPs and their respective TIMPs could lead to excessive proteolysis.

It is now very well established that UV radiation induces MMPs without affecting the expression or activity of TIMPs [24,25]. These MMPs can be induced by both UVB and UVA radiation, but the underlying photobiological and molecular mechanisms differ depending on the type of irradiation. In a very simplified scheme, UVA radiation would mostly act indirectly through the generation of reactive oxygen species, in particular singlet oxygen, which subsequently can exert a multitude of effects such as lipid peroxidation, activation of transcription factors, and generation of DNA-strand breaks [24]. While UVB radiation–induced MMP induction has been shown to involve the generation of ROS as well [26], the main mechanism of action of UVB is the direct interaction with DNA via the induction of DNA damage. Recent studies have indeed provided evidence that enhanced repair of UVB-induced cyclobutane pyrimidine dimers in the DNA of epidermal keratinocytes through topical application of liposomally encapsulated DNA repair enzymes on UVB-irradiated human skin prevents UVB radiation–induced epidermal MMP expression [10].

The activity of MMPs is tightly regulated by transcriptional regulation, and elegant *in vivo* studies by Fisher et al. have demonstrated that exposure of human skin to UVB radiation leads to the activation of the respective transcription factors [27]. Accordingly, UV exposure of human skin not only leads to the induction of MMPs within hours after irradiation, but already within minutes, transcription factors AP-1 and NFkB, which are known stimulatory factors of MMP genes, are induced. These effects can be observed at low UVB dose levels, because transcription factor activation and MMP-1 induction could be achieved by exposing human skin to one-tenth of the dose necessary for skin reddening (0.1 minimal erythema dose). Subsequent work by the same group clarified the major components of the molecular pathway by which UVB exposure leads to the degradation of matrix proteins in human skin. Low-dose UVB irradiation induced a signaling cascade that involves upregulation of epidermal growth factor receptors, the GTP-binding regulatory protein p21Ras, extracellular signal-regulated kinase, c-jun amino terminal kinase, and p38. Elevated c-jun together with constitutively expressed c-fos increased activation of AP-1. Identification of this UVB-induced signaling pathway does not only unravel the complexity of the molecular basis that underlies UVB radiation–induced gene expression in human skin, but also provides a rationale for the efficacy of tretinoin (all-trans-retinoic acid) in the treatment of photoaged skin. Accordingly, topical pretreatment with tretinoin inhibited the induction and activity of MMPs in UVB-irradiated skin through prevention of AP-1 activation.

In addition to destruction of existing collagen through activation of MMPs, failure to replace damaged collagen is thought to contribute to photoaging as well. Accordingly, in chronically photodamaged skin, collagen synthesis is downregulated as compared to sun-protected skin [28]. The mechanism by which UV radiation interferes with collagen synthesis is not yet known, but in a recent study evidence has been provided that fibroblasts in severely (photo)damaged skin have less interaction with intact collagen and are thus exposed to less mechanical tension, and it has been proposed that this situation might lead to decreased collagen synthesis [29].

4.4 UV-Induced Modulation of Vascularization

There is increasing evidence that cutaneous blood vessels may play a role in the pathogenesis of photoaging. Photoaged skin shows vascular damage that is absent from intrinsically aged skin. In mildly photodamaged skin, there is venular wall thickening, whereas in severely damaged skin the vessel walls are thinned and supporting perivascular veil cells are reduced in number [30]. The number of vascular cross-sections is reduced [31], and there are local dilations, corresponding to clinical telangiectases. Overall, there is a marked change in the horizontal vascularization pattern, with dilated and distorted vessels. Studies in humans as well as in the hairless Skh-1 mouse model for skin aging have demonstrated that acute and chronic UVB irradiation greatly increases skin vascularization [32,33].

The formation of blood vessels from pre-existing vessels is tightly controlled by a number of angiogenic factors as well as factors that inhibit angiogenesis. These growth factors include basic fibroblast growth factor, interleukin-8, tumor growth factor-beta, platelet-derived growth factor, and vascular endothelial growth factor (VEGF). VEGF appears to be involved in chronic UVB damage because UVB radiation–induced dermal angiogenesis in Skh-1 mice is associated with increased VEGF expression in the hyperplastic epidermis of these animals [33]. Even more important, targeted overexpression of the angiogenesis inhibitor Thrombospondin-1 does not only prevent UVB radiation–induced skin vascularization and endothelial cell proliferation, but significantly reduces dermal photodamage and wrinkle formation. These studies suggest that UVB radiation–induced angiogenesis plays a direct biological role in photoaging.

4.5 Photoaging as a Chronic Inflammatory Process

In contrast to intrinsically aged skin, which shows an overall reduction in cell numbers, photoaged skin is characterized by an increase in the number of dermal fibroblasts, which appear hyperplastic, but also by increased numbers of mast cells, histiocytes, and mononuclear cells. The presence of such a dermal infiltrate indicates the possibility that a chronic inflammatory process takes place in photoaged skin, and in order to describe this situation the terms heliodermatitis and dermatoheliosis have been coined [34]. More recent studies have shown that increased numbers of CD4+ T-cells are present in the dermis, whereas intraepidermally, infiltrates of indeterminate cells and a concomitant reduction in the number of epidermal Langerhans cells have been described [35,36]. It is currently not known whether the presence of inflammatory cells represents an epiphenomenon or these cells play a causative role in the pathogenesis of photoaging, for example, through the production of soluble mediators that could affect the production and/or degradation of extracellular matrix proteins.

4.6 Protein Oxidation and Photoaging

The aging process is accompanied by enhanced oxidative damage. All cellular components including proteins are affected by oxidation [37]. Protein carbonyls may be formed either by oxidative cleavage of proteins or by direct oxidation of lysine, arginine, proline, and threonine residues. In addition, carbonyl groups may be introduced into proteins by reactions with aldehydes produced during lipid peroxidation or with reactive carbonyl derivatives generated as a consequence of the reaction with reducing sugars or their oxidation products with lysine residues of proteins.

Within the cell, the proteasome is responsible for the degradation of oxidized proteins. During the aging process this function of the proteasome is diminished and oxidized proteins accumulate. In addition, lipofuscin, a highly cross-linked and modified protein aggregate, is formed. This aggregate accumulates within cells and is able to inhibit the proteasome. These alterations mainly occur within the cytoplasm, and lipofuscin does not accumulate in the nucleus.

In biopsies from individuals with histologically confirmed solar elastosis, an accumulation of oxidatively modified proteins was found specifically within the upper dermis [38]. Protein oxidation in photoaged skin was most likely due to UV irradiation, because repetitive exposure of human buttock skin for 10 days to increasing UV doses as well as *in vitro* irradiation of cultured dermal fibroblasts to UVB or UVA radiation caused protein oxidation. The functional relevance of increased protein oxidation in UV-irradiated dermal fibroblasts, in particular with regard to the pathogenesis of photoaging, is currently not known. Very recent studies, however, indicate that increased protein oxidation, which may result from a single exposure of cultured human fibroblasts to UVA radiation, inhibits proteasomal functions and thereby affects intracellular signaling pathways that are involved in MMP-1 expression.

4.7 Conclusion

From the above it is evident that major progress has been made recently in identifying molecular mechanisms involved in photoaging. In this regard, skin has proven to serve as an excellent model organ to understand basic mechanisms relevant for extrinsic aging.

Despite all this progress, however, a general, unifying concept linking the different mechanisms and molecular targets described in the previous paragraphs is still missing. In other words, the critical question to answer is: How do mitochondrial DNA mutagenesis, telomere shortening, neovascularization, protein oxidation, downregulation of collagen synthesis, and increased expression of matrix metalloproteinases together cause photoaging of human skin? Which of these mechanisms are of primary importance and responsible for inducing others? Are some or all of the above-mentioned characteristics of photoaged skin merely epiphenomena and, if so, to what extent causally related to premature skin aging? The current state of knowledge does not allow us to answer these questions in a definitive manner.

In this regard we have recently proposed a hypothesis that tries to reconcile most of the research discussed above in one model [1,20]. We envision photoaging of human skin to include UV radiation–induced mitochondrial DNA mutagenesis in the dermis of human skin. We believe that the persistence of UV radiation–induced mitochondrial DNA mutations and the resulting vicious cycle with further increases in mitochondrial

DNA mutations leads to a situation that can best be described as a "defective powerhouse" where inadequate energy production leads to chronic oxidative stress. In the dermis, functional consequences of direct DNA damage and aberrant ROS production in human dermal fibroblasts could be an altered gene expression pattern that would affect neovascularization and collagen metabolism and possibly also the generation of an inflammatory infiltrate, as well as the oxidation of intracellular proteins and inhibition of the proteasome.

Evidence supporting this model has recently been generated in studies employing human skin equivalent models. If these models were engineered by using skin fibroblasts from patients with Kearns-Sayre syndrome, which constitutively show a high frequency of UV-inducible large-scale mtDNA deletions, then signs of photoaging were found to develop over time at a molecular and histological level. In order to corroborate these results, development of appropriate mouse models characterized by different susceptibilities towards UV radiation–induced mitochondrial DNA mutagenesis is of utmost importance [39].

References

1. Krutmann, J. and Gilchrest, B.A. (2006). Photoaging of skin. In: Skin Aging. Gilchrest B, Krutmann J (eds.), Springer-Verlag, Berlin/Heidelberg, pp. 33–42.
2. Schroeder, P., Schieke, S.M., and Morita, A. (2006) Premature skin aging by infrared radiation, tobacco smoke and ozone. In: Skin Aging. Gilchrest B, Krutmann J (eds.), Springer-Verlag, Berlin/Heidelberg, pp. 45–51.
3. Schofer, A., Krämer, U., Ranft, U., Sugiri, D., Momna, T., Kaneko, N., Yamamoto, A., Morita, A., and Krutmann, J. (2008) Determinants of spot and wrinkle formation in German and Japanese women. J Invest Dermatol, in press.
4. Berneburg, M., Plettenberg, H., and Krutmann, J. (2000) Photoaging of human skin. Photodermatol. Photoimmunol Photomed 16:239–244.
5. Schieke, S.M., Stege, H., Kürten, V., Grether-Beck, S., Sies, H., and Krutmann, J. (2002) Infrared A radiation-induced matrixmetalloproteinase-1 expression is mediated through ERK1/2 activation in human dermal fibroblasts. J Invest Dermatol 119:1323–1329.
6. Schroeder, P., Lademann, J., Darvin, M., Stege, H., Marks, C., Bruhnke, S., and Krutmann, J. (2008) Analysis of the in vivo relevance of infrared radiation induced signaling in human skin: evidence for skin damage and implications for photoprotection. J Invest Dermatol, in press.
7. Schroeder, P., Pohl, C., Calles, C., Marks, C., Wild, S., and Krutmann, J. (2007) Cellular response to infrared radiation involves retrograde mitochondrial signalling. Free Radical Bio Med 43:128–135.

8. Grether-Beck, S., Salashour-Fard, M., Timmer, A., Brenden, H., Felsner, I., Walli, R., Fuellekrug, J., and Krutmann, J. (2008) Ceramide and raft signaling are linked with each other in UVA radiation–induced gene expression. Oncogene, in press.
9. Fritsche, E., Schäfer, C., Bernsmann, T., Calles, C., Wurm, M., Hübenthal, U., Cline, J.E., Schroeder, P., Rannug A., Klotz, LO., Fürst, P., Hanenberg, H., Abel, J., and Krutmann, J. (2007) Lightening up the UV response by identification of the arylhydrocarbon receptor as a cytoplasmatic target for ultraviolet B radiation. Proc Natl Acad Sci 104:8851–8856.
10. Damaghi, N., Dong, K., Picart, S., Markova, N., Masaki, H., Grether-Beck, S., Krutmann, J., Smiles, K., and Yarosh, D. (2008) UV-induced DNA damage initiates release of MMP-1 in human skin. Photoderm Photoimmun and Photomed, in press.
11. DiMauro, S. and Schon, E.A. (2003) Mitochondrial respiratory-chain diseases. N Engl J Med 348:2656–2668.
12. Berneburg, M., Gattermann, N., Stege, H., Grewe, M., Vogelsang, K., Ruzicka, T., and Krutmann, J. (1997) Chronically ultraviolet-exposed human skin shows a higher mutation frequency of mitochondrial DNA as compared to unexposed skin and the hematopoietic system. Photochem Photobiol 66: 271–275.
13. Yang, J.H., Lee, H.C., and Wei, Y.H. (1995) Photoageing-associated mitochondrial DNA length mutations in human skin. Arch Dermatol Res 287: 641–648.
14. Koch, H., Wittern, K.P., and Bergemann, J. (2001) In human keratinocytes the common deletion reflects donor variabilities rather than chronologic aging and can be induced by ultraviolet A irradiation. J Invest Dermatol 117: 892–897.
15. Berneburg, M., Grether-Beck, S., Kurten, V., Ruzicka, T., Briviba, K., Sies, H., and Krutmann, J. (1999) Singlet oxygen mediates the UVA-induced generation of the photoaging-associated mitochondrial common deletion. J Biol Chem 274:15345–15349.
16. Berneburg, M., Plettenberg, H., Medve-Konig, K., Pfahlberg, A., Gers-Barlag, H., Gefeller, O., and Krutmann, J. (2004) Induction of the photoaging-associated mitochondrial common deletion in vivo in normal human skin. J Invest Dermatol 122:1277–1283.
17. Reimann, V., Krämer, U., Sugiri, D., Schröder, P., Hoffmann, B., Medve-Koenigs, K., Jöckel, K.H., Ranft, U., and Krutmann J. (2007) Sunbed use induces the photoaging–associated mitochondrial common deletion in human skin. J Invest Dermatol, published online, 8 November.
18. Trifunovic, A., Wredenberg, A., Falkenberg, M., Spelbrink, J.N., Rovio, A.T., Bruder, E., Bohlooly, Y.M., Gidlof S., Oldfors, A., Wibom, R., Tornell, J., Jacvobs, H.T., and Larsson, N.G. (2004) Premature ageing in mice expressing defective mitochondrial DNA polymerase. Nature 429:417–423.
19. Berneburg, M., Gremmel, T., Kurten, V., Schroeder, P., Hertel, I., Mikecz, A.V., Wild, S., Chen, M., Declercq, L., Matsui, M., Ruzicka, T., and Krutmann, J. (2005) Creatine supplementation normalizes mutagenesis of mitochondrial DNA as well as functional consequences. J Invest Dermatol 125:213–220.

20. Schroeder, P., Gremmel, T., Berneburg, M., and Krutmann, J. (2008) Partial depletion of mitochondrial DNA from human skin fibroblasts induces a gene expression profile reminiscent of photoaged skin. J Invest Dermatol, published online, 13 March.
21. Smith, J.G., Davidson, E.A., Sams, W.M., and Clark, R.D. (1962) Alterations in human dermal connective tissue with age and chronic sun damage. J Invest Dermatol 39:347–350.
22. Braverman, I.M. and Fonferko, E. (1982) Studies in cutaneous aging: I. The elastic fibre network. J Invest Dermatol 78:434–443.
23. Talwar, H.S., Griffiths, C.E.M., Fisher, G.J., Hamilton, T.A., and Voorhees, J.J. (1995) Reduced type I and type III procollagens in photodamaged adult human skin. J Invest Dermatol 105:285–290.
24. Scharffetter-Kochanek, K., Brenneisen, P., Wenk, J., Herrmann, G., Ma, W., Kuhr, L., Meewes, C., and Wlaschek, W. (2000) Photoaging of the skin: From phenotype to mechanisms. Exp Gerontol 35:307–331.
25. Fisher, G.J., Talwar, H.S., Lin, J., Lin, P., McPhillips, F., Wang, Z., Li, X., Wan, Y., Kang, S., and Voorhees, J.J. (1998) Retinoic acid inhibits induction of c-jun protein by ultraviolet radiation that occurs subsequent to activation of mitogen-activated protein kinase pathways in human skin in vivo. J Clin Invest 101:1432–1440.
26. Wenk, J., Brenneisen, P., Meewes, C., Wlaschek, M., Peters, T., Blaudschun, R., Ma, W., Kuhr, L., Schneider, L., and Scharffetter-Kochanek, K. (2001) UV-induced oxidative stress and photoaging. Curr Probl Dermatol 29:83–94.
27. Fisher, G.J., Wang, Z.Q., Datta, S.C., Varani, J., Kang, S., and Voorhees, J.J. (1997) Pathophysiology of premature skin aging induced by ultraviolet light. N Engl J Med 337:1419–1428.
28. Fisher, G., Datta, S., Wang, Z., Li, X., Quan, T., Chung, J., Kang, S., and Voorhees, J. (2000) c-Jun dependent inhibition of cutaneous procollagen transcription following ultraviolet irradiation is reversed by all-trans retinoid acid. J Clin Invest 106:661–668.
29. Varani, J., Schuger, L., Dame, M.K., Leonhard, Ch., Fligiel, S.E.G., Kang, S., Fisher, G.J., and Vorhees, J.J. (2004) Reduced fibroblast interaction with intact collagen as a mechanism for depressed collagen synthesis in photodamaged skin. J Invest Dermatol 122:1471–1479.
30. Braverman, I.M. and Fonferko, E. (1982) Studies in cutaneous aging: II. The microvasculature. J Invest Dermatol 73:59–66.
31. Kligman, A.M. (1979) Perspectives and problems in cutaneous gerontology. J Invest Dermatol 73:39–46.
32. Bielenberg, D.R., Bucana, C.D., Sanchez, R., Donawho, C.K., Kripke, M.L., and Fidler, I.J. (1998) Molecular regulation of UVB-induced angiogenesis. J Invest Dermatol 111:864–872.
33. Yano, K., Ouira, H., and Detmar, M. (2002) Targeted over expression of the angiogenesis inhibitor thrombospondin-1 in the epidermis of transgenic mice prevents ultraviolet-B-induced angiogenesis and cutaneous photodamage. J Invest Dermatol 118:800–805.
34. Lavker, R.M. and Kligman, A. (1988) Chronic heliodermatitis: a morphologic evaluation of chronic actinic dermal damage with emphasis on the role of mast cells. J Invest Dermatol 90:325–330.

35. DeLeo, V.A., Dawes, L., and Jackson, R. (1981) Density of Langerhans cells (LC) in normal versus chronic actinically damaged skin (CADS) of humans. J Invest Dermatol 76:330–334.
36. Gilchrest, B.A., Murphy, G.F., and Soter, N.A. (1982) Effects of chronologic aging and ultraviolet irradiation on Langerhans cells in human skin. J Invest Dermatol 79:85–88.
37. Levine, R.L. and Stadtman, E.R. (2001) Oxidative modification of protein during ageing. Exp Gerontol 36:1495–1502.
38. Sander, C.S., Chang, H., Salzmann, S., Muller, C.S., Ekanayake-Mudiyanselage, S., Elsner, P., and Thiele, J.J. (2002) Photoaging is associated with protein oxidation in human skin in vivo. J Invest Dermatol 118: 618–625.
39. Schroeder, P., Maresch, T.D., Schneider, M., Bernerd, F., and Krutmann, J. (2007) The photoaging-associated common deletion of the mitochondrial genome increases oxidative stress and alters cellular functions in human dermal fibroblasts in a 3D environment. J Invest Dermatol 127:41.

PART 2
BEAUTY FROM THE INSIDE AND THE OUTSIDE

5

Natural Products Work in Multiple Ways

Pierfrancesco Morganti, PhD

Professor of Applied Cosmetic Dermatology,
II University of Naples, Naples, Italy
Visiting Professor, China Medical University, Shenyang, China
R&D Director, Mavi Sud S.r.l., Aprilia (LT), Italy

Aaron Tabor and Robert M. Blair (eds.), *Nutritional Cosmetics: Beauty from Within*, 93–111, © 2009 William Andrew Inc.

5.1 Summary

Personal appearance is a good way to generate interest. Our self-esteem and physical confidence are largely influenced by how we perceive our body and skin, as well as by other people's opinions. The skin serves as an interface in our exchanges with others, filtering the stimuli that come from the outside and sending out messages. Indeed, it is one of the most intimate elements of our identity.

Through the process of aging and exposure to the sun and to environmental pollutants the skin loses moisture, tension, and elasticity, resulting in wrinkles and spots. Today there is a tendency to try to make up for nutritional deficiencies by resorting to dietary supplements and topical cosmetics. Products that claim to produce skin health benefits from the outside as well as the inside are also beginning to appear on the market. Moreover, innovative cosmetics and functional foods strive to stimulate the imagination and produce emotions by combining exciting images, sensual fragrances, and the feeling of a caress on the skin to convey the sensation of a total beauty effect.

The combination of effective cosmetics and food supplements is the new multiple approach to providing an excellent skin care treatment that offers beauty, audacity, shyness, pleasure, and pride, to obtain the total wellness one desires.

5.2 Introductory Remarks

It is increasingly evident that free radicals play a key role in determining the overall well-being and general appearance of the skin. They are linked with decreased cell viability, contributing significantly to the skin aging process. It is also clear that a stressful environment can cause an overabundance of these reactive and potentially harmful free radicals and that to mitigate the effects of such an excess, the natural defenses of the skin may not suffice and may have to be supplemented [1,2]. This includes topical application of cosmetics and use of dietary supplements enriched with chelating agents, excited state quenchers, antioxidants, free radical scavengers, etc.

Consumers therefore remain extremely interested in products, especially natural ones that may prevent the onset of disease and preserve quality of life. Diet, nutrition, and a growing list of nontraditional sources are some of the key factors involved in achieving these goals, together with multifunctional natural cosmetics [3].

5.2.1 What Are Free Radicals?

Free radicals are species capable of independent existence that contain one or more unpaired electrons, that is, electrons that are alone in their orbit. Free radicals and reactive oxygen and nitrogen species are either synthesized endogenously, for example, in energy metabolism and by the antimicrobial defense system of the body, or produced as reactions to exogenous exposure such as cigarette smoke, an imbalanced diet, exhaustive exercise, environmental pollutants, ultraviolet and blue light, and food contaminants (Fig. 5.1). They usually have a transitory existence, as a result of which their steady state concentration in the biological system is very low. Oxidative and other chemical stresses may modify not only polyunsatured lipids, but also carbohydrates, protcins, and complex macromolecules, forming atherogenic, carcinogenic, diabetogenic, and brain degenerating substances, depending on the target organ. Modified biomolecules also interfere with gene expression, causing metabolic disturbances and skin aging.

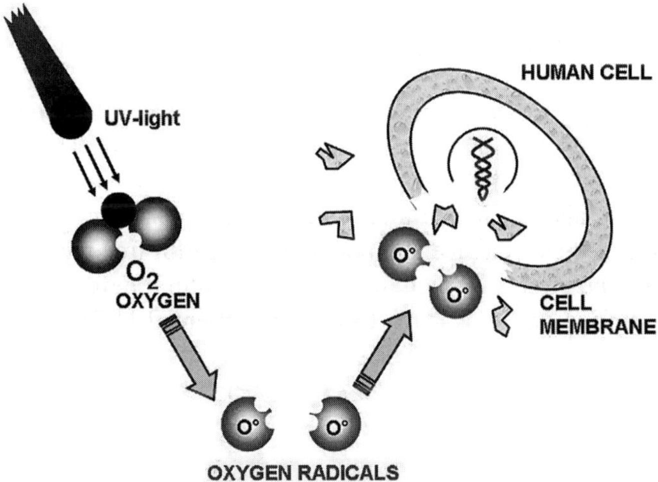

Figure 5.1 Schematic representation of free radicals and cell damage.

5.2.2 Oxidative Injuries

To counteract oxidative injury of structural lipids, carbohydrates, and proteins, human skin is equipped with a network of enzymatic and nonenzymatic antioxidant systems that are responsible for maintaining an equilibrium between pro- and antioxidant compounds [4]. Therefore, the use of natural antioxidants as topical and/or systemic agents, which reduce the onset of oxidative stress, may help to protect and increase the efficacy of the skin's biological system [5]. Usually, the use of cosmetics (beauty from the outside) associated with diet supplements (beauty from the inside) has a synergistic activity aimed to maintain the cells' antioxidant power [6]. The current tendency to apply and to ingest various kinds of vitamins and antioxidant compounds is notably on the rise.

However, major information is lacking on the mechanisms by which these compounds exert their antioxidative activity.

5.3 Beauty from the Inside

The concept of health and wellness is broad and segmented, offering a composite of several smaller specialty categories generally characterized by an emphasis on prevention and maintenance rather than on therapy [7] (Fig. 5.2). Perhaps this is due to an increased awareness of what appear to be the most frequently discussed diseases (i.e., heart disease, cancer, osteoporosis, arthritis, etc.) coupled with better information provided by the media. Every year more and more newspaper and magazine articles are dedicated to the relationship between diet, health, and beauty, and more specifically to nutraceutical and cosmeceutical concepts [8]. The fact that people tend to be more aware of health-related issues is also a result of the aging of the population.

5.3.1 The Market

The functional food market has risen from US \$8.9 billion in 1996 up to US \$16.1 billion in 2006. It is interesting to note that this market has grown at a steady rate of 10% a year and is expected to register a +56% increase worldwide by 2010 [9] (Fig. 5.3).

Interest in the role diet plays in optimizing personal health and well-being is stimulated by substantial work that is being carried out, and will continue to be carried out in the future, in six areas: (a) risk reducers, (b) life stage

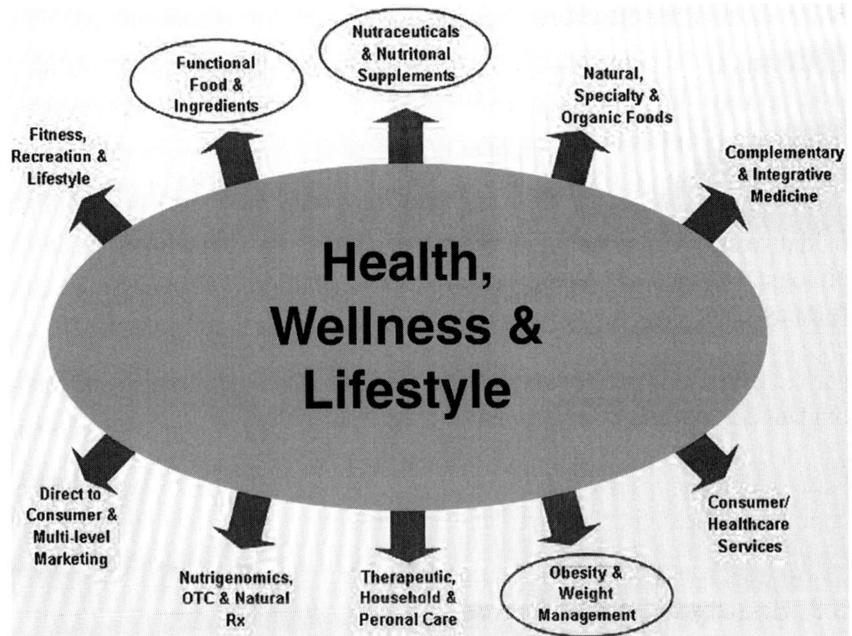

Figure 5.2 Categories with the health and wellness market. *Source*: David Thibodeau, Managing Director, Canaccord Adams, Vancouver, British Columbia, Canada, 2007.

Functional Foods Market	
Strict definition: +56% to USD 25.1bn	
➢Europe	+79%
➢US	+66%
➢Japan	+24%
➢Australia	+80%
Broad definition: +47% to USD 53.2bn	
➢Europe	+71%
➢US	+55%
➢Japan	+26%
➢Australia	+61%

Figure 5.3 Future prospects (strong growth to 2010) of the functional foods market. *Source*: Fiona Angus, Leatherhead Food International, 2007.

and gender nutrition, (c) getting through the day, (d) slimming solutions, (e) age of antioxidants, and (f) beauty from the inside. To achieve these objectives consumers are willing to pay a premium, purchasing so-called anti-aging nutraceutical and cosmeceutical products.

Thus, the growing link between diet and beauty/health, makes the consumer more open to the concept of beauty from within.

5.3.2 Consumer Expectations

One must also consider consumer expectations in terms of what is natural. Indeed, according to consumers, a natural product is:

- "a product containing ingredients that come from the earth rather than from a laboratory;
- "a product containing no man-made ingredients;
- "a product without added chemicals;
- "a product that is more expensive because chemical ingredients are cheaper than natural ones on account of the fact that they are mass produced [10]."

"What is the conclusion? By and large, according to consumers, when we say that a product is *natural*, we mean that the product's ingredients are derived from a plant or the earth and are not man-made or produced in a laboratory [10]." Moreover, "natural products are perceived as being less harmful, gentler on the skin, 100% safe, and containing no toxic agents". According to Michael Harmswoth from ESPA (UK) [10], "the key to benefiting from this demand lies in avoiding the pitfalls—this can be achieved through transparent communication and by continuously improving products and formulations as the development of natural ingredients improves." This applies to both oral and topical beauty products. However, regardless of whether or not they are natural, modern functional foods and cosmetics may represent a real alternative to minor dermatology for the prevention and treatment of imperfections caused primarily by skin aging.

5.3.3 The Impact of Functional Food on the Skin

What are functional foods? They have been defined as foods with ingredients (either natural or added) that provide a health benefit beyond their traditional value.

How do they impact the skin? Skin is more than beauty—it is a vital organ that protects us from the environment, maintains moisture and body temperature, and is part of the immune system. Therefore, nutraceuticals must renew the skin by rebuilding tissues, making skin more flexible, smoother, healthier, and younger looking. Moreover, they must help to prevent, reduce the risk of, or even treat minor diseases and, in some cases, do all of the above. Some specialized foods have certain weight loss, calming, anti-inflammatory, or immuno-stimulant properties. Triterpenes from shea nuts have displayed interesting anti-inflammatory properties; alpha acids, iso-alpha acids, and possibly beta-acids contained in hop extracts are tentatively identified as active pain relief compounds; and a soluble, digestible keratin protein extracted from sheep wool appears to improve joint health and mobility. Linoleic acid, an omega-6-fatty acid, is an important precursor of ceramides, which are essential for moisturizing the skin and preserving its barrier function; meanwhile, gamma-linolenic acid, as a precursor of anti-inflammatory prostaglandins, is capable of reducing inflammatory skin disorders and controlling skin moisturization and health.

5.3.4 Inflammation and the Immune System

What about inflammation? It is a function of immunity [11] and, as with other defense mechanisms, the immune system has a way of creating and curbing inflammation to suit immediate needs. Organized as an intricate network of cells, tissues, and organs, the immune system protects the body against pathogens. It produces and releases pro- and anti-inflammatory compounds in an intricate choreography. An improper immune response can lead to various illnesses. One of the main factors impacting immune function is single nutritional status. The impairment of immune function can be due to an insufficient intake of energy and macronutrients and/or to deficiencies in specific combinations. Among nutrients, vitamin C is well known for its antioxidant properties but may also benefit the functioning of specific immune cells.

Alpha lipoic acid [12] is a powerful antioxidant, with anti-inflammatory effects on diabetes, cataracts, and brain degeneration; vitamin E [13] has the ability to prevent intracellular signaling cascades in inflammatory cells. Lutein [14] modulates the skin's response to ultraviolet B, radiation-induced inflammation, and immuno-suppression, protecting also from UVA and blue light. Lutein is known to be effective on eye health, but it also appears to have anti-inflammatory benefits on the skin's health, especially in relation to sun exposure and aging [15]. Other essential nutrients with an antioxidant and immune function against inflammation are minerals such as selenium, chromium, magnesium, and zinc.

These and many other minerals, botanicals, and natural compounds are also used to formulate nutraceuticals and cosmeceuticals for consumers with inflammatory and other diseases connected with the aging process [16]. Nobody wants to age, and we all want to look younger.

Thus, consumers are looking for oral and topical foods capable of repairing permanent sun damage produced by actinic keratosis, which make the skin look healthier, livelier, tighter, firmer, more elastic, smoother, and with fewer fine lines and age spots. Everyone wants beauty from within, which can enhance collagen production and neutralize excess free radicals.

5.4 Beauty from the Outside

5.4.1 Skin Aging

The skin is probably the organ that is most susceptible to damage produced by free radicals because of its contact with oxygen and with other environmental stimuli. It reveals the first signs of physical aging at around age 30. But skin aging does not affect only the elderly; it affects everyone because people sunbathe too much. It is not only the passing of time that makes us age, but the accumulation of deleterious chemical events that deteriorate our bodies into the condition that we call aging. The aging process, which reduces the number of healthy cells, is soon visible on the skin. On facial skin, the cumulative and inexorable process of aging is there for all to see. The damage produced by free radicals impairs the cells' ability to transport nutrients, eliminate waste products, and reproduce adequately, and results in the accumulation of by-products, such as pigment lipofuscin (age spots), interfering with their normal functions. Moreover, too much sun exposure increases the production of free radicals, which, in turn, increases inflammation, and the degradation rates of collagen and elastin, thereby resulting in accelerated aging.

Thus, anti-aging products and global facial skincare remain the cosmetic category's growth driver (Fig. 5.4), by an increase of 7.6% in the U.S., 7%

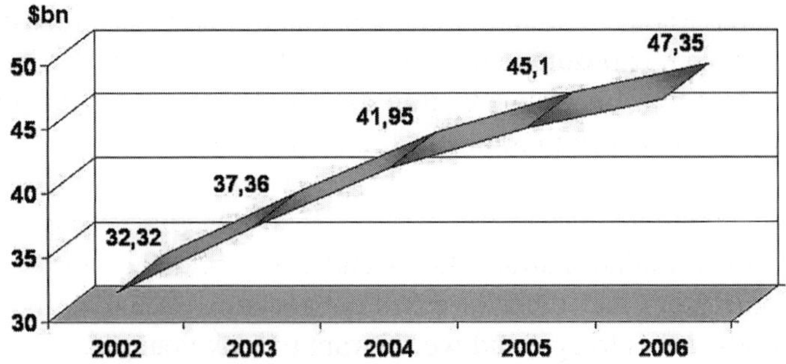

Figure 5.4 Global facial skincare sales (in billion dollars) from 2002 to 2005. *Source*: Euromonitor.

in Western Europe, 9% in the Asia Pacific region, 21.6% in Eastern Europe, 19.7% in Latin America, and 8.2% in Africa and the Middle East. So-called cosmeceuticals and doctor brands continue to be a major feature of the skincare segment [17].

5.4.2 The Cosmetic Challenge

In an attempt to defeat the undesirable effects of aging and photoaging, cosmetologists have developed many new cosmetic formulas, keeping abreast of what happens in the areas of biology and biochemistry and using increasingly sophisticated and active raw materials [18]. As a result, moisturizing and protective micro- and nano-emulsions containing vitamins A, E, B, C, collagen and its derivatives have been proposed (Fig. 5.5). Interesting studies have been conducted on peptides from gelatin enriched with glycine, which turned out to be very active, increasing moisturizing activity and reducing the inexorable process of skin aging (Fig. 5.6) [19,20]. A double-blind clinical study recently showed that gelatin-glycine may increase the moisturizing activity of lutein [21], which is currently pointed to as an active compound in skin aging [22].

Scientific evidence has been presented that shows the benefits of hydrolyzed collagen (gelatin), used both in cosmetics and in diet supplements, in preserving joint health. This is because both the topical application and the ingestion of hydrolyzed collagen may increase the pool of amino acids and peptides that are useful for collagen synthesis. Moreover, gelatin may also have a specific stimulating effect on the production of collagen by the

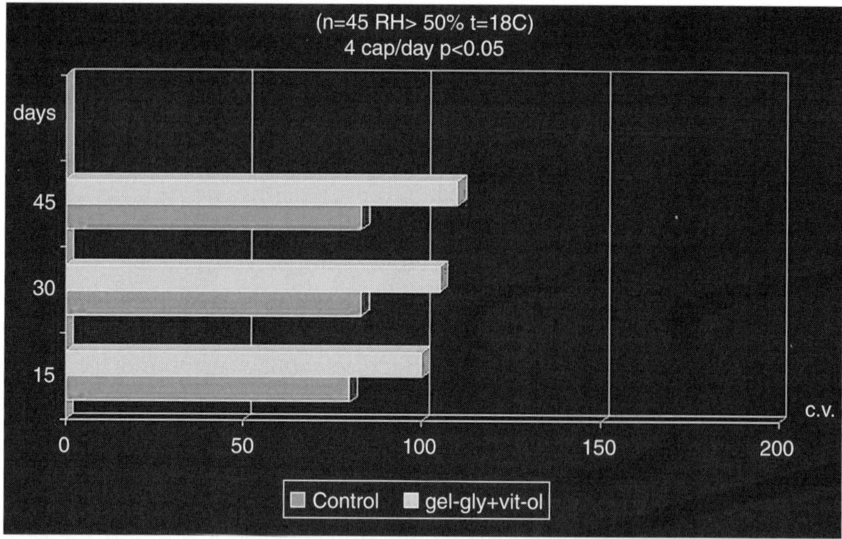

Figure 5.5 Moisturizing activity of gelatin-glycerine (gel-gly) plus vitamins and oligoelements (vit-ol).

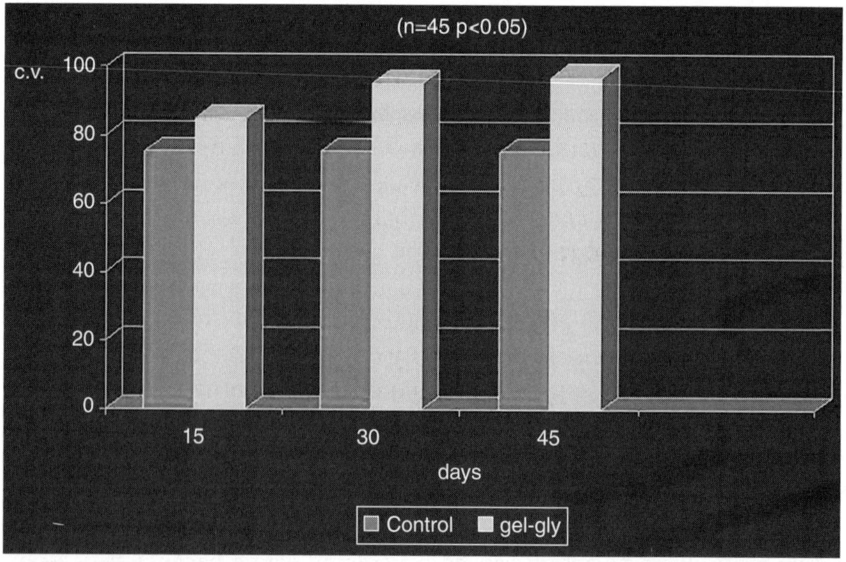

Figure 5.6 Moisturizing activity of gelatin-glycerine (gel-gly).

joint cells themselves [23]. Collagen is a structural protein of the body that ensures cohesion, elasticity, and regeneration of the skin, cartilage, and bone, which declines in terms of quality and quantity during the aging process.

Ceramide polymer is a new ingredient developed in Japan (Nof Co.) to help improve the elasticity and smoothness of skin. This polymer has a high affinity with the skin, and thanks to its similarity to natural ceramides, it can reinvigorate the skin and restore its softness and elasticity [24]. Goldschmidt (Germany) has proposed a hydrolyzed sodium hyaluronate obtained through the fermentation of *Bacillus sublitis* to improve the skin's elasticity and to reduce wrinkles. Vitamin B_6, an innovative liposoluble that serves as an enhancer of filagrin production, was presented this year by Jon Dekker (USA) at In Cosmetics. The hydrolyzed oat polypeptide obtained by enzymatic hydrolysis from Kelisema (Italy) seems to protect the stratum corneum against irritations and may be a solution for skin sensitivity.

Because consumers expect beauty and personal care products to provide them with a unique sensory experience before, during, and after application, Dow Corning (USA) has created specialized silicium polymers providing lubricity, emolliency, and moisturization both for skin and hair. Silicium plays an important role in connective tissues; the finding that it is a component of human glycosaminoglycans and their protein complexes suggests that this mineral plays a structural role. Indeed, silicium contributes to the structural framework of connective tissues by forming links or bridges within and between individual polysaccharide chains and perhaps by linking polysaccharide chains to proteins (Fig. 5.7) [25,26]. In this way, this mineral compound may help build the architecture of the fibrous elements of connective tissues and may contribute to their structural integrity by providing strength and resilience.

Chitin nanofibrils [27–30] are cosmetic components or pharmaceutical carriers that help to restore the integrity of the skin barrier and to increase the ability to link and retain water contained in the corneocytes (Fig. 5.8). They are used in cosmetic and/or pharmaceutical solutions or emulsions to stimulate the formation of a molecular film that slows down water evaporation and helps to keep the skin perfectly hydrated.

Their mechanism of action can be considered active because they repair the intercorneocitary cement that joins both ceramides and phospholipids

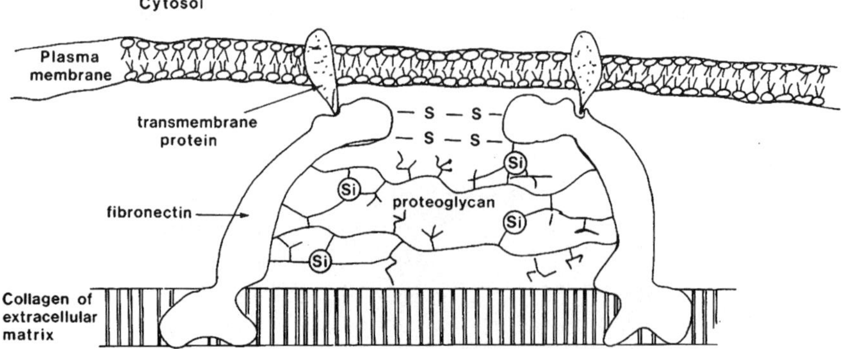

Figure 5.7 Model of the arrangement of fibronectin, collagen, and proteoglycans in the extracellular matrix showing the role of silicon.

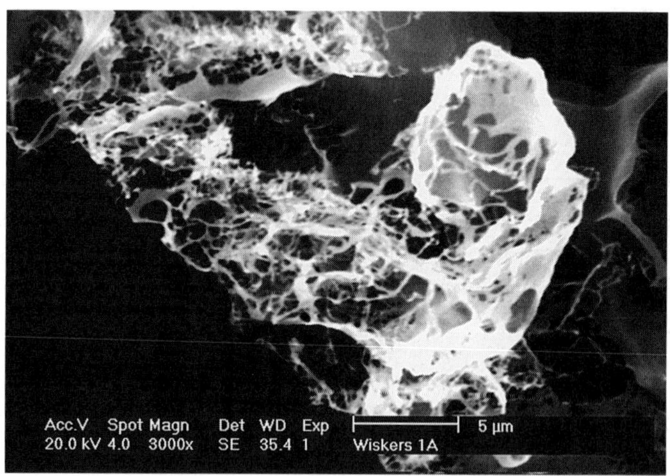

Figure 5.8 Photomicrograph of a chitin nanofibril at a magnification of 3000×.

to form the lipidic lamellae. Moreover, these nanofibrils are easily recognized by the cutaneous enzymes and hydrolyzed. In this way, the N-acetyl glucosamine content can regulate collagen synthesis in fibroblasts, also facilitating the granulation and repair of the altered skin tissue [31,32]. Thus, nanotechnology is now present also in consumer products (Fig. 5.9).

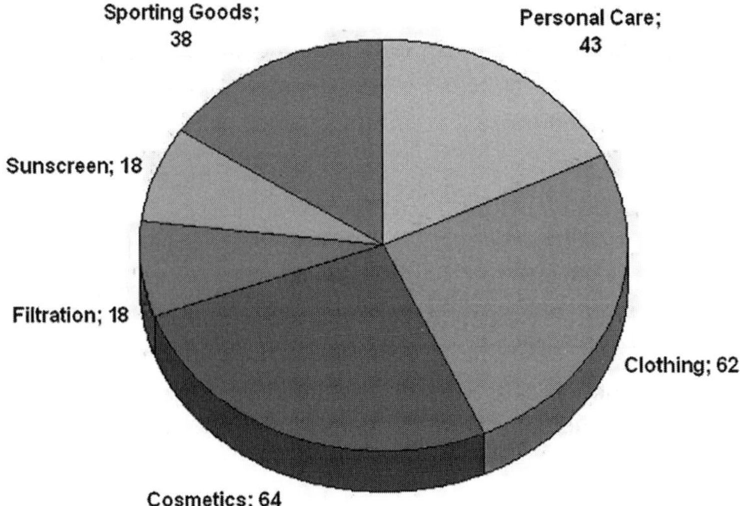

Figure 5.9 Presence of nanotechnology-enabled consumer products in the health and fitness arena. *Source*: Woodrow Wilson Center Nanotech Project Inventory.

Needless to say, thousands of active principles are used in cosmetic products. However, "natural and organic skin care is undergoing somewhat of a revolution, involving issues such as the environment. The integrity of ingredients and holistic well-being strike a chord with more and more consumers" [33].

5.5 Total Beauty Effects

Product efficacy seems insufficient to capture the essence of beauty and today "there is a prevailing trend in advertising that captures people in a narcissistic process [33]." Beauty, which defines performance and success, is driven by the concepts of perfection, harmony, and proportion. The chemistry of emotion is thus born that links the flavor of food with the flavor of cosmetics, partly due to the popularity of natural products. Givaudan has taken this concept to a new level by launching a new active ingredient based on yogurt powder. This new product is a combination of spray-dried yogurt powder and a natural vegetable-based probiotic ingredient that is said to enhance the growth of beneficial microflora on the skin. However, sensuality is an integral part of marketing, both in skin care and in nutraceuticals. Staying healthy is the new message driven by cosmetics and diet supplements. People are demanding products with a soul supplement: being beautiful is not enough; they want to feel beautiful and healthy in their body and in their mind. Therefore, "the food, nutraceutical, and cosmetic markets are hungry for new active ingredients with proven efficacy, be they from known food sources or from newly discovered ethnic foods or traditional herbal remedies" [34].

But in our era of cross-fertilization, boundaries are no longer that clear. The same actives are used in nutraceuticals and cosmetics; fragrances are becoming active and new foods are introducing sensual formulae in skin care. Beauty cannot be left just to skin doctors. Thus, a scientific approach to emotion and attraction has been developed by neurochemistry, which is capable of capturing this emotion through attitudes and sensual elements.

5.5.1 Beauty from Within Is the New Message

Therefore "many cosmetics that strongly appeal to the senses often evoke delicious foods. As if food were the most efficient emotional way to convey the sensual dimension of a formula. But evoking seasonal fruit may also reveal an irresistible urge to devour the beautiful! [34]" This is the new message that is sent out by different industries that produce raw materials both for cosmetics and food, such as Jan Dekker, Greentech, Solabia, Symrise, Cosmetochem, Croda, Cognis, etc., which has also ushered in the term *Feelosophy.*

Personal appearance is a good way to generate interest. Our self-esteem and physical confidence are largely influenced by how we perceive our body and skin, as well as by other people's opinions. The skin serves as an interface in our exchanges with others, filtering the stimuli that come from the outside and sending out messages. Indeed, it is one of the most intimate elements of our identity. Through the process of aging and exposure to the sun and to environmental pollutants, the skin loses moisture, tension, and elasticity, resulting in wrinkles and spots.

Today there is a tendency to try to make up for nutritional deficiencies by resorting to dietary supplements and topical cosmetics. Products that claim to produce skin health benefits from the outside as well as the inside are also beginning to appear on the market (Fig. 5.10). Moreover, innovative cosmetics and functional foods strive to stimulate the imagination and produce emotions by combining exciting images, sensual fragrances, and

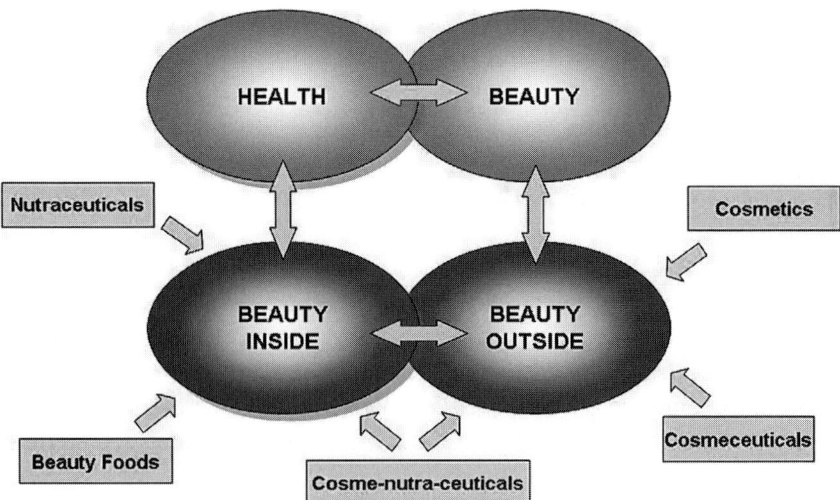

Figure 5.10 Diagrammatic representation of the overlap between nutraceuticals, beauty foods, cosmeceuticals, and cosmetics. *Source*: a&r (modified) 2007.

the feeling of a caress on the skin to convey the sensation of a total beauty effect. The consumer demands for more active and safety cosmeceuticals and nutraceuticals and the emerging technologies, create every day innovative skin-care beauty products to be used both from inside and outside.

This is why the existing direct relationship between Cosmetic and Beauty Food establishing more knowledge, new analytical toots and therapeutic modalities try to protect the skin from endogenous and exogenous stresses while preserving its healthy and young look for a longer period of time. Thus, the coming age of *biotech* cosmetics & foods together with the development of nanoscience and dermogenetics represent the future challenge to solve the new consumers needs of an evolving society [35].

References

1. Commoner B, Townsend J, Pake GW. (1954) Free radical in biological materials. *Nature* **174:** 689–697.
2. Halliwell B, Gutteridge JM. (1990) Role of free radicals and catalytic metal ions in human disease: an overview. *Methods Enzymol.* **186:** 1–85.
3. Salonen JT. (1999) Antioxidants, lipid peroxidation and cardiovascular diseases. In: Kumpulunen JT and Salonen JT, eds., *Natural Antioxidants and*

Anticarcinogenesis in Nutrition, Health and Disease. Cambridge, UK, RSC, pp. 3–8.

4. Thiele JJ, Dreher F, Packer L. (2000) Antioxidant defence systems in skin. In: Elsner P and Maibach HI, eds., *Drugs vs Cosmetics-Cosmeceuticals?* New York, Dekker, pp. 145–188.
5. Blatt T, Mundt C, Mummert C, Maksiuk T, Wolbert R, Keyhani R, Shreider V, Hoppe V, Schachtschabel DO, Stab F. (1999) Modulation of oxidative stress in human aging skin. Z. *Gerontol. Geriatr.* **32(2):** 83–88.
6. Passi S. (2003) The combined use of oral and topical lipophilic antioxidants increases their levels both in sebum and stratum corneum. *Biofactors* **18:** 1–4.
7. Wildman REC. (2001) Nutraceuticals: a brief review of historical and theological aspects. In: Wildman REC, ed., *Nutraceutical and Functional Food.* Boca Raton, CRC Press, pp. 1–12.
8. Fogg-Johnson N. (2007) Venture capital investment drives health & wellness innovation. *Nutraceuticals World* **10(4):** 76–86.
9. Angus F. (2007) Functional foods—market drivers and product innovation. In: *Vitafoods, Intern. Conferences*, 8–10 May, Geneva, CH, pp. 1–45, www.vitafoods.eu.com/conferences.
10. Harmsworth M. (2007) The consumer's expectation of natural-100%. *Natural Conference*, May 15, h.9.45, Daventry, UK.
11. Granato H. (2007) Enhancing immune function. *Natural Products* **13(4):** 16–28.
12. Podda A, Zoller TM, Grundmann-kolmann M, Thiele JJ, Packer L Kaufmann R. (2001) Activity of alpha-lipoic acid in the protection against oxidative stress in skin. *Curr. Problems. Dermatol.* **29:** 43–51.
13. Thiele JJ. (2001) Oxidative targets in the stratum corneum. A new basis for antioxidative strategies. *Skin Pharmacol. Appl. Skin Physiol.* **14(suppl 1):** 87–91.
14. Junghans A, Sies H, Stahl W. (2001) Macular pigments lutein and zeaxanthin as blue light filters studied in liposomes. *Arch. Biochem. Biophys.* **391(2):** 160–164.
15. Morganti P, Sousa M, Morganti G. (2008) Skin activity of lutein. *SÖFW Journal* **133(5):** 22–26.
16. Myers S. (2007) Burning man: chronic inflammation causes human degeneration. *Insider* **12(n4):** 40–50.
17. Demorest A. (2007) The cream of the crop. *ICN* **n397:** 34–35.
18. Morganti P. (1991) The future of cosmetic dermatology. Oral cosmesis: a new frontier. In: Morganti P and Ebling FJG, eds., *Every Day Problems in Dermatology: The Cosmetic Connection*, Vol. 2. Rome, Italy, International Ediemme, pp. 291–300.
19. Morganti P, Randazzo SD, Cardillo A. (1986) Role of insoluble and soluble collagen as skin moisturizer. *J. Appl. Cosmetol.* **4:** 141–152.
20. Morganti P, Randazzo SD. (1987) Enriched gelatin as skin hydration enhancer. *J. Applied Cosmetol.* **5:** 105–120.
21. Morganti P, Oshida K, Zhang F. (2008) Bleu light and skin's hydration. *Eurocosmetics* **16(6):** 24–28.
22. Palombo P, Fabrizi G, Ruocco V, Ruocco E, Flühr J, Roberts R, Morganti P. (2007) Beneficial long-term effects of combined oral/topical antioxidant

treatment with carotenoids lutein and zeaxanthin on human skin: a double-blinded, placebo-controlled study in humans. *Skin Pharmacology and Physiology* **20:** 199–210.

23. Oesser S, Seifest J. (2003) Stimulation of type II collagen biosynthesis and secretion in bovine chondrocytes cultured with degraded collagen. *Cell Tissue Res.* **311:** 393–399.
24. Henderson C. (2007). Multiple choice. *SPC* **80(6):** 41–48.
25. Berra B, Rapelli SC. (1986) Aspetti biochimico-funzionali degli oligoalimenti in traccia. *Rassegna chimico scientifica* **62:** 104–110.
26. Berra B, Zoppi S, Rapelli SC. (1991)Vitamins and minerals as skin nutrients. In: Morganti P and Ebling FJG, eds., *Every Day Problems in Dermatology: The Cosmetic Connection*, Vol. 2. Rome, Italy, International Ediemme, pp. 291–300.
27. Morganti P, Muzzarelli RAA, Muzzarelli C. (2006) Multifunctional use of innovative chitin nanofibrils for skin care. *J. Appl. Cosmetol.* **24:** 105–114.
28. Morganti P, Morganti G, Muzzarelli RAA, Muzzarelli C. (2007) Chitin nanofibrils: a natural compound for innovative cosmeceuticals. *Cosmetics & Toiletries* **122(4):** 81–88.
29. Biagini G, Zizzi A, Tucci G, Orlando F, Lucarini G, Buldreghini E, Mattioli-Belmonte E, Giantomassi F, Provinciali M, Carezzi F, Morganti P. (2007) Chitin nanofibrils linked to chitosan glycolate as spray, gel and gauze preparations for wound repair. *Journal Bioactive and Compatible Polymers* **22:** 525–538.
30. Muzzarelli RAA, Morganti P, Morganti G, Palombo P, Palombo M, Biagini G, Mattioli Belmonte M, Giantomassi F, Orlandi F, Muzzarelli C. (2007) Chitin nanofibrils/chitosan composites as wound medicaments. *Carbohydrate Polymers* **70:** 274–284.
31. Morganti P, Morganti G. (2007) Nanotechnology and wellness. *SÖFW Journal* **133(5):** 22–27.
32. Morganti P. (2007) Applied nanotechnology in cosmetic and functional food. *Eucocosmetics* **2:** 12–15.
33. Henderson C. (2007) Bridging the gulf. *SPC* **80(6):** 28–32.
34. Gruenwald J. (2007) Discovering active food and nutraceutical ingredients. *Nutraceuticals World* **10(4):** 30–31.
35. Morganti P, Chen HD, Gao XH, Li Y, Jacobson C, Arct J, Fabianowsky W. (2009) NANOSCIENCE the challenging cosmetics healthy food & biotextiles. *SÖFW Journal* **135(4):** 2–7.

PART 3
MICRONUTRIENT SUPPORT FOR BEAUTIFUL HAIR AND SKIN

6

Bioavailability and Skin Bioefficacy of Vitamin C and E

Myriam Richelle[1], PhD, Heike Steiling[1], PhD, and Isabelle Castiel[2], PharmD, PhD

[1]Nestlé Research Center, Nestec Ltd, Lausanne, Switzerland
[2]L'Oréal Recherche, Centre C. Zviak, Clichy Cedex, France

Aaron Tabor and Robert M. Blair (eds.), *Nutritional Cosmetics: Beauty from Within*, 113–138,

Abstract

The use of functional nutrients to alleviate skin changes, such as aging, photosensitivity, and dryness, is increasing in the over-the-counter market. The most popular nutrients in this context are vitamins, carotenoids, polyphenols, and minerals. For skin, the classical route of administration of active compounds is by topical application, and manufacturers have substantial experience in this field. However, another means to deliver bioactives to the skin is by using oral administration, and the use of oral supplements for improving the condition of skin is increasing. In this case, food bioactives have to cross several barriers before reaching the skin; they have to pass down the gastrointestinal tract, cross the intestinal barrier, reach the blood circulation, and then be distributed to the different tissues of the body including the skin. The advantages of this route of administration are that the food bioactives are metabolized and then presented to the entire tissue, potentially in an active form. Also, the blood continuously replenishes the skin with these bioactives, which can then be distributed to all skin compartments, that is, epidermis, dermis, subcutaneous fat, and also sebum. This chapter presents the mechanisms involved in the bioavailability of nutrients such as vitamins C and E at the intestinal as well as at the skin level but also their bioefficacies in skin. Vitamin E has been shown to present an antioxidant property, and to play a role in photoprotection and in the prevention of aging, alone or in association with vitamin C. Literature on vitamin C for skin benefits is more abundant, showing evidence for the beneficial effect of this ingredient on dermal matrix formation, or epidermal differentiation, against UV-induced skin damage, and oxidative stress, indicating that vitamin C supplementation may be of interest to target skin aging, photoprotection, and skin xerosis.

6.1 Introduction

Skin is constantly exposed to pro-oxidant environmental stresses from an array of sources, such as air pollutants, solar UV light, chemical oxidants, micro-organisms, cigarette smoke, and ozone (Cross et al., 1998; Thiele et al., 1997). Reactive oxygen species have been implicated in the etiology of several skin disorders including skin cancer and photoaging (Dalle & Pathak, 1992; Emerit, 1992; Guyton & Kensler, 1993; Perchellet & Perchellet, 1989). These reactive oxygen species are capable of oxidizing lipids, proteins, or DNA, leading to the formation of oxidized products such as lipid hydroperoxides, protein carbonyls, or 8-hydroxyguanosine, respectively (Beehler et al., 1992; Hu & Tappel, 1992; Podda et al., 1998). Reactive oxygen species constantly generated in skin are rapidly neutralized by nonenzymatic and enzymatic antioxidant substances, which prevent their harmful effects and maintain a pro-oxidant/antioxidant balance, resulting in cell and tissue stabilization. If the antioxidant defence is exhausted, cell damage can occur. Known nonenzymatic scavengers of free radicals in human skin are β-carotene, vitamin C, and vitamin E, and enzymatic scavengers are seleno-dependent glutathione peroxidases, copper-zinc superoxide dismutase, manganese superoxide dismutase, and catalase (Steenvoorden & van Henegouwen, 1997; Thiele et al., 2000). In recent years, particular antioxidants have gained considerable attention as a means to neutralize reactive oxygen species (Mukhtar & Ahmad, 1999). Green tea polyphenols (Katiyar & Mukhtar, 1997), resveratrol (Jang et al., 1997), curcumin (Stoner & Mukhtar, 1995), silymarin, ginger (Katiyar et al., 1996), and diallyl sulfide (Perchellet et al., 1990; Sadhana et al., 1988) afford protection against the development of skin cancer, both *in vitro* (in culture system) as well as *in vivo* (in animal models). Additionally, diets rich in bioactives such as vitamins C and E, β-carotene, lycopene, zinc, and selenium have also demonstrated a photoprotective effect against solar irradiation in humans (Cesarini et al., 2003; Fuchs, 1998; Fuchs & Kern, 1998; Greul et al., 2002; McKenzie, 2000; Rostan et al., 2002; Stahl et al., 2000, 2001; Stahl & Sies, 2002).

An increase in cellular antioxidants in skin can be obtained by the exogenous administration of antioxidant compounds. In the case of the skin, the classical route of antioxidant administration to skin is topical application. This approach allows delivery of antioxidants to the skin and at the same time avoids possible side effects of excess antioxidants to other organs. However, this topical route of administration can be efficiently achieved only if the particular antioxidant is stable in the preparation as well as on skin, is able to penetrate the skin, and is present in its active form, that is, a possible metabolite. Interestingly, a recent publication showed that keratinocytes also have potential to metabolize bioactives as demonstrated for phenolic acids (Poquet et al., 2008). In addition, penetration of antioxidants into the skin is influenced by environmental factors, such as temperature, hydration, and the presence of other chemicals. Another means to deliver antioxidants to the skin is using oral administration. In this case, antioxidants cross the intestinal barrier to reach the blood, from which they are distributed to different tissues, and specifically for skin to subcutaneous adipose tissue, dermis, epidermis, and sebum. The advantages of this oral administration are that antioxidants are metabolized and then presented to the entire skin potentially in their active forms. In addition, the blood continuously replenishes skin with these antioxidants, which are distributed to all skin compartments in which they could exert a biological activity.

In order to be active in skin, dietary bioactives must be able to cross the intestinal barrier and reach the blood circulation. This step could be a limiting factor of the efficacy of these dietary bioactives in skin. This chapter reviews current knowledge on the journey of dietary bioactives, that is, vitamin C and vitamin E from the mouth to skin, as well as their biological activities in skin.

6.2 Definition of Bioavailability

Bioavailability is defined by the relative amount of a dietary bioactive consumed that crosses the intestinal barrier, reaches the blood circulation, and is available for metabolic processes or storage in the body. Bioavailability comprises various steps summarized in the acronym LADME. L means liberation of the molecule from the dietary matrix (food or supplement); A is absorption, that is, transfer of the molecule from the gut lumen into the blood circulation; D represents distribution of molecules from the blood circulation in all body tissues; M is metabolism, consisting of the further processing of the molecule in the body either in the gastrointestinal tract or in various tissues; and E is elimination out of the body in urine, stools, sweat, tears, or expired air.

In the nutritional context, the commonly used definition of bioavailability refers to the proportion of dietary bioactive that crosses the intestinal barrier and reaches the blood circulation.

6.3 Mechanisms of Absorption in Gut

Dietary bioactives such as vitamin E and vitamin C belong to two main groups, that is, lipid-soluble and water-soluble dietary bioactives, respectively. Solubility of the dietary bioactive markedly affects mechanisms of their bioavailability.

In the mouth, the food matrix is disrupted by mastication, leading to a mash. α-Amylase starts the hydrolysis of carbohydrates, whereas, in infants, a lingual lipase also hydrolyses lipids. The efficiency of these enzymatic reactions is directly related to the duration of the mastication.

The mash is then transferred to the stomach, where mainly the gastric lipases begin hydrolysis of triacylglycerols, producing diacylglycerols and fatty acids. The chyme is mechanically mixed by forceful contractions of the antrum, in order to continue dietary matrix disruption. At this stage, the chyme starts to separate into two phases, with oil droplets containing mainly lipid-soluble dietary bioactives and with the bulk containing water-soluble dietary bioactives.

Lipid-soluble dietary bioactives such as vitamin E, present in oil droplets, enter the small intestine and are mixed with pancreatic juice and bile salts (Tyssandier et al., 2001). In consequence, bile salts cover the oil droplets, creating a negative surface charge and so permitting the binding of colipase to the lipid/aqueous interface. The colipase then binds to the pancreatic lipases, which hydrolyse the bond between dietary bioactives and fatty acids, liberating free dietary bioactives (Borel et al., 1996; Lombardo & Guy, 1980), for example, from esterified carotenoids. These hydrolytic processes drastically reduce the size of oil droplets, which exhibit a particle size about one million times smaller than oil droplets formed in the stomach (Borel et al., 1996). The micellar solubilization of these lipophilic dietary bioactives, that is, carotenoids and vitamin E, is mandatory for their absorption, and so if lipophilic nutrients escape this solubilization, they will continue their journey in the gastrointestinal tract and will be subjected to microflora metabolism in the colon or eliminated. The uptake of lipid-soluble dietary bioactives by enterocytes was long considered to be a passive mechanism (Hollander et al., 1975; Hollander & Ruble, 1978), but there is increasing evidence for facilitated, protein-carrier mediated transport (During et al., 2002; Reboul et al., 2005). Once inside the enterocyte, lining the small intestine, these lipid-soluble dietary bioactives are packed into new oil-droplet structures called pre-chylomicrons, which are expulsed by exocytosis into the extracellular space, entering the lymphatic system and then reaching the general blood circulation.

In the particular case of water-soluble dietary bioactive, such as vitamin C, it remains in the bulk of the meal and arrives in the small intestine (Rios et al., 2002). Ascorbic acid and dehydroascorbic acid are readily, rapidly, and efficiently absorbed from the upper part of the small intestine into the circulatory system (Moser & Bendich, 1991) via sodium-dependent active transport processes or by facilitated-diffusion glucose transporters (Liang et al., 2001). Above 400 mg daily intake, plasma vitamin C remains constant due to higher renal and faecal excretion.

6.4 Factors Affecting Gastrointestinal Bioavailability

Different parameters modulate the bioavailability of dietary bioactives and have been summarized by the acronym SLAMENGHI, which stands for: *s*pecies of active molecule; molecular *l*inking; *a*mount; *m*atrix in which it is entrapped; *e*ffect of modifiers; *n*utrient status; *g*enetic factors; *h*ost-related factors; and mathematical *i*nteractions (Castenmiller & West, 1998; West & Castenmiller, 1998). Modifying matrix, composition, processing, consumer habits, synergy or competition between different compounds, and physical-chemical properties of dietary bioactives modulates their absorption efficiency. Some of these factors are presented in the following paragraphs.

Class of molecules: The extent of bioavailability of dietary bioactives depends on class, with a high level for vitamins C and E and a lower one for carotenoids and polyphenols. However, the extent of bioavailability varies markedly between studies maybe due to the marked difference of experimental conditions used with respect to dose administered, standard meal, and formulation of species.

Consumer habits might markedly affect the bioavailability of dietary bioactives, that is, whether consumers eat a dietary bioactive in a fasting state or together with a meal. In the case of lipid-soluble dietary bioactives, the transfer of these dietary bioactives into lipid droplets of the meal is mandatory for their absorption (Iuliano et al., 2001). In addition, lipids present in the meal induce contraction of the gall bladder and bile secretion in the small intestine. Bile participates in the formation of mixed micelles. Therefore, dietary bioactives that do not undergo this transfer will be eliminated from the body without having the opportunity to exert a biological activity. Three grams of fat is enough to allow absorption of vitamin E, α-carotene, and β-carotene, whereas increasing the fat dose to 36 g is associated with increased lutein absorption (Roodenburg et al., 2000). Moreover, the type of lipid consumed also plays an important role; medium-chain triglycerides markedly reduce β-carotene bioavailability (Borel et al., 1998). Solubilization of vitamin E in long-chain triglycerides slows the absorption process compared to medium-chain triglycerides (Fukui et al., 1989a, 1989b; Gallo-Torres et al., 1978), which is attributed to an increased oxidation of tocopherol by polyunsaturated fatty acids during digestion.

The *dose* of dietary bioactives consumed markedly affects the efficiency of bioavailability. There is no evidence that vitamin E (Borel et al., 1997) bioavailability is affected by the dose consumed, at least in dietary doses, but further investigations need to clearly address this issue. In contrast, the percentage of vitamin C absorbed decreases with dosage exceeding the RDA (Levine et al., 1999; Padayatty et al., 2003, 2004). When low doses (4–64 mg) are ingested, the absorption efficiency is as high as 98%; that is, less than 2% of an ingested radioactive dose appears in the faeces (Baker et al., 1969). When larger doses (30–180 mg) are consumed, the absorption efficiency falls to 80–90% (Kallner et al., 1979), to 75% at 1 g, 50% at 1.5 g, 26% at 1.6 g, and 16% at 12 g (Kallner et al., 1977; Hornig & Moser 1981). Vitamin C plasma levels in individuals range from 10 μmol/L to 160 μmol/L. Brubacher et al. (2000) have carried out a meta-analysis on the relationship between vitamin C intake and plasma concentrations for different subgroups, that is, adult, elderly, smokers, and nonsmokers, to determine which intake is necessary to achieve a plasma concentration of 50 μmol/L. They estimate that 50% of the general population can achieve plasma concentrations of 50 μmol/L with a daily intake of 83.4 mg. The elderly and smokers, however, would need a higher intake, that is, 150.2 mg and 206.6 mg, respectively.

The structure of dietary bioactives also markedly affects their bioavailability, that is, degree of lipophilicity, crystal or solute form, presence and nature of conjugation, spatial configuration, etc. Natural vitamin E consists of RRR-α-tocopherol, whereas the synthetic form of vitamin E consists of a mixture of eight stereoisomers, namely all-rac-α-tocopherol. After consumption by humans, both natural and synthetic forms of vitamin E reach the blood after one day of supplementation (Vaule et al., 2004; Clifford et al., 2006), whereas its appearance in sebum takes some weeks (Vaule et al., 2004). The ester of vitamin E is often used in supplements because this form is more stable in the product. Natural and esterified forms of vitamin E (acetate or succinate) are similarly absorbed, because a hydrolytic process takes place very efficiently in the intestine releasing vitamin E, which is then taken up by enterocytes and reaches the blood circulation (Burton et al., 1988; Cheeseman et al., 1995). Recently, a phosphorylated form of α-tocopherol has also been reported to be present in a variety of foods (Gianello et al., 2005).

Competition or synergy between dietary bioactives: Dietary bioactives are rarely consumed alone and therefore synergy or competition processes with other food bioactives present in the meal or in the supplement could readily occur. Vitamin C increases the absorption of iron (Levine et al., 1999). Plant sterols also reduce β-carotene bioavailability by 50% and affecting by 20% that of vitamin E (Richelle et al., 2004). Phosphatidylcholine decreases absorption of carotenoids and vitamin E, whereas lysophosphatidylcholine enhances it (Koo & Noh, 2001; Sugawara et al., 2001). Moreover, excess α-tocopherol causes a reduction of γ-tocopherol concentration in plasma (Wolf, 2006). Vitamin C has no effect on selenate bioavailability, whereas a high vitamin C intake seems to increase selenite bioavailability, possibly by a protective effect of critical sulphydryl groups involved in gastrointestinal selenite uptake (Martin et al., 1989).

Formulation of dietary bioactives: Nanoformulation of vitamin E is able to increase by 10-fold its bioavailability in humans (Wajda et al., 2007), whereas polysaccharide formulation increases vitamin E bioavailability by about 2-fold in rabbits (Zimmer & Czarnecki, 2007). Another water-soluble formulation increases γ-tocopherol absorption in patients suffering from cystic fibrosis, a malabsorption disease due to pancreatic insufficiency and a diminished bile pool (Papas et al., 2007).

6.5 Distribution and Delivery to the Skin

When dietary bioactives arrive in the blood circulation, they are ready to be distributed to all body tissues, where they can exhibit a biological activity. Although certain dietary bioactives have been reported to exert a biological activity in skin such as photoprotection, collagen synthesis, and cancer prevention, information on their delivery mechanisms to skin is quite scarce.

6.5.1 Vitamin E

Vitamin E consists of a mixture of different molecules, that is, α-, β-, γ-, and δ-tocopherols and α-, β-, γ-, and δ-tocotrienols. Due to the presence of three chiral atoms, these molecules exhibit different stereoisomers ranging from RRR, RSR, etc., to SSS. In response to supplementation, the concentration of vitamin E increases immediately in plasma, whereas a rise in concentration is observed only after several days (7 d) in sebum (Vaule et al., 2004; Ekanayake-Mudiyanselage et al., 2004). The basis of this delay may be related to the process of sebum production, because it has been reported (Downing et al., 1981) to take approximately 8 d for newly synthesized lipids to be secreted in sebum. The delivery of vitamin E to skin seems to be very specific for certain isomeric forms. Indeed, following supplementation, both the natural form (RRR-α-tocopherol) and the synthetic form (all-rac-α-tocopherol) appear in the blood circulation, whereas only RRR-α-tocopherol appears in sebum (Vaule et al., 2004). This indicates that a specific protein could selectively transport this form of the vitamin into the sebum. A similar specificity has already been described in the liver (Hosomi et al., 1997). γ-Tocopherol is absorbed into plasma but it is still uncertain whether γ-tocopherol exerts a biological activity in man, because this form is mostly eliminated by the liver. However, γ-tocopherol is present in sebum and skin (Thiele et al., 1999; Vaule et al., 2004) and so it might exert an activity in human skin, although this needs further investigation. Skin vitamin E exhibits a gradient of concentration with a higher level in the dermis and a lower level in the stratum corneum (Shindo et al., 1994). In addition, there are regional variations of skin vitamin E; facial skin contains a several-fold higher level of vitamin E than unexposed skin sites such as skin from the upper arm (Thiele et al., 1998). This regional variability of vitamin E is supported by the fact that vitamin E is continuously delivered to sebum via the sebaceous glands (Thiele et al., 1999; Vaule et al., 2004). As a consequence, skin sites with high sebum production such as forehead skin exhibit higher vitamin E concentration (Lang et al., 1986).

6.5.2 Vitamin C

Vitamin C is an effective antioxidant and an essential cofactor in numerous enzymatic reactions. It comprises two major forms: L-ascorbic acid, the reduced form, and L-dehydroascorbic acid, the oxidized form. Man and other primates have lost the ability to synthesize vitamin C as a result of a mutation in the gene encoding for L-gulono-γ-lactone oxidase, an enzyme required for vitamin C biosynthesis. In man, plasma ascorbic acid concentrations are maintained between 10 mM and 160 mM (1–15 mg/ml) and any excess of the vitamin is excreted by the kidney (Fuchs & Podda, 1997). Analysis of tissue ascorbate levels in human subjects revealed highest amounts in adrenal glands (550 mg/kg), brain (140 mg/kg), and liver (125 mg/kg), followed by lungs (70 mg/kg), kidneys (55 mg/kg), heart (55 mg/kg), skeletal muscle (35 mg/kg), and skin (30 mg/kg), and low levels in adipose tissue (10 mg/kg) and blood (9 mg/kg; Brown & Jones, 1996;

Fuchs & Podda, 1997). Considering the different size and weight of the organs, it is evident that ascorbic acid is concentrated in specific tissues, with the highest concentrations in adrenal glands, brain, and liver, and the highest total amount present in skeletal muscle. In most of these tissues high ascorbate levels are probably important for maintaining structural integrity through collagen fibres, as well as for more specific functions, for example, hormone synthesis, immune response, and antioxidant protection. This concentration difference between different tissues clearly indicates that ascorbic acid uptake and distribution into tissues is mediated by an active transport mechanism. Indeed, two different ascorbic acid transporters, SVCT1 and SVCT2, have been identified by screening a rat kidney cDNA library (Tsukaguchi et al., 1999). SVCT1 is largely confined to epithelial surfaces involved in bulk transport, such as those of the intestine or kidney. In contrast, SVCT2 appears to account for tissue-specific uptake of vitamin C and is widely expressed, occurring in neurons, the endocrine system, and other tissues.

In skin, *in vitro* studies in HaCaT keratinocytes demonstrated the presence and functional activity of both transporters, as well as an efficient ascorbate recycling system in the epidermis (Savini et al., 1999, 2000, 2002; Liang et al., 2001). In the dermis, experiments with dermal fibroblasts cultured from SVCT2 knockout mice showed virtually abolished ascorbic acid uptake, indicating that SVCT2 is crucial for ascorbic acid delivery in fibroblasts (Sotiriou et al., 2002). A recent publication confirmed the presence of both SVCT isoforms in human skin biopsies and demonstrated that they fulfil specific functions. SVCT1 is responsible for epidermal ascorbic acid supply, whereas SVCT2 mainly facilitates ascorbic acid transport in the dermal compartment (Steiling et al., 2007).

Furthermore, the concentration and distribution of ascorbic acid and dehydroascorbic acid in the different skin compartments were studied in mice, a species in contrast to humans with functional ascorbic acid biosynthesis. Ascorbic acid concentration in the epidermis and dermis was 1.3 mmol/g (229 mg/g) and 1.0 mmol/g (176 mg/g), respectively. Similar concentrations were found for dehydroascorbate, with 1.3 mmol/g in epidermis and 0.9 mmol/g in dermis (Shindo et al., 1994). Within the murine stratum corneum, the outermost layer of the epidermis, ascorbic acid exhibits a gradient of concentration with low levels in the outer layers and a steep increase in the deeper parts (Weber et al., 1999). Although skin was always thought to be the most sensitive organ during deficiency status (e.g., scurvy), recent results indicate that skin can cope with marginal amounts of vitamin C whereas other organs like brain and lungs suffer much more. However, under certain conditions the skin vitamin C pool can be depleted selectively, for example, upon age, upon UV irradiation, in psoriatic skin, or in atopic dermatitic lesions (Shindo et al., 1994; Podda et al., 1998; Leveque et al., 2002, 2003, 2004; Hoppu et al., 2005b). Experiments in mice indicated that UVB-irradiation results in a significant decrease in SVCT1 mRNA expression, whereas SVCT2 mRNA levels are unchanged, suggesting that decreased levels might be a consequence of impaired SVCT1 transport (Steiling et al., 2007). This hypothesis was confirmed *in vitro*. However, these findings are in contrast to findings made by Kang and co-workers (2007). They reported that ascorbic acid uptake into keratinocytes is increased by UVB in a time- and dose-dependent manner through translocation of SVCT1 from the cytosol to the membrane (Kang et al., 2007). Discrepancies of results between the two studies might be due to differences in the UV protocol used.

In contrast to ascorbate, the intestinal uptake of dehydroascorbate is mediated by facilitated-diffusion glucose transporters GLUT1, GLUT3, and GLUT4 (Liang et al., 2001). Under physiological conditions however, the reduced form of vitamin C will predominate (95% in human plasma), and, thus, it is unlikely that GLUT-mediated dehydroascorbic acid uptake will be sufficient for the cellular demand of most cells. Furthermore, circulating levels of glucose are 1,000-fold higher than dehydroascorbic acid levels (2–5 mM) and marked competition by glucose of dehydroascorbic acid influx is most likely (Liang et al., 2001). Furthermore, dehydroascorbic acid is nearly undetectable in most tissues (Rumsey & Levine, 1998). However, higher concentrations may occur transiently during oxidative stress.

6.6 Bioefficacy in Skin

Vitamin E and vitamin C play a role in skin, and various biological effects are presented in the following paragraphs.

6.6.1 Vitamin E and Skin Photoprotection

In vitro, Jones et al. (1999) demonstrated that exposure of skin fibroblasts to ultraviolet radiation (UVR) leads to generation of reactive oxygen species as well as the oxidation of biomolecules and induction of adaptive responses. Interestingly, supplementation with Trolox, a vitamin E analog, suppressed UVR-induced oxidative stress, suggesting a photoprotective effect of vitamin E on skin fibroblasts. Similarly, Jin et al. (2007) reported that vitamin E was able to limit the generation of reactive oxygen species, reduce cell death, and increase endogenous antioxidant enzyme activity. Longas et al. (1993) reported in preclinical studies that dietary supplementation with vitamin E protected rat skin glycosaminoglycans against changes induced by UV radiation. In humans, vitamin E supplementation (at gram dose) failed to protect human skin against deleterious effects induced by UV exposure in two clinical studies (Werninghaus et al., 1991; Fuchs & Kern 1998), whereas another study exhibited a photoprotective effect (Mireles-Rocha et al., 2002).

6.6.2 Vitamin E and Skin Aging

Aging is associated with an increased protein kinase C activity, *in vitro* as a function of cell passage number and *in vivo* as a function of the donor's age. This increase of protein kinase C activity is associated with collagenase overexpression and activity, resulting in collagen degradation and skin aging. Ricciarelli et al. (1999) reported that vitamin E decreases collagen degradation *in vitro*. Although *in vitro* and animal studies have suggested that vitamin E supplementation alone could participate in skin photoprotection and in consequence may slow skin aging, only one human study showed a convincing effect (for reviews, see Boelsma et al., 2001 and Swindells & Rhodes, 2004). A major reason for this may be that the presence of other antioxidants, for example, vitamin C, are necessary for recycling UVR-induced α-tocopherol radicals (Kagan et al., 1992; Wefers & Sies, 1988). Others speculate that this lack of efficacy at physiological levels against oxidant insult rather supports the nonantioxidant molecular functions of α-tocopherol, for example, as ligands of proteins (receptors, transcription factors) capable of regulating signal transduction and gene expression (Azzi, 2007). The discovery of α-tocopheryl phosphate at low concentration in rat livers favours this hypothesis (Gianello et al., 2005).

In the elderly, vitamin E exhibited immuno-enhancing effect, showing an increased delayed-type hypersensitivity skin response, indicating that vitamin E seems to protect the skin's immune system from aging (Wu et al., 2006).

Moreover, vitamin E exhibits anti-carcinogenic (Heinen et al., 2007), photoprotective (Eberlein-König & Ring, 2005; Ahmed et al., 2006), as well as skin barrier–stabilizing properties when administrated orally or topically to the skin. All these properties in relation with skin benefits, *in vitro* or *in vivo*, have been reviewed recently (Thiele et al., 2005; Thiele & Ekanayake-Mudiyanselage, 2007). Serum tocopherol levels have been associated with a reduced atopic dermatitis condition and, in consequence, may play a role in the regulation of inflammatory diseases (Hoppu et al., 2005a).

6.6.3 Vitamin C and Antioxidant Properties

Vitamin C exhibited *in vitro* antioxidant activity. However, this activity was only reported in a few *in vivo* studies. In McArdle et al. (2002), a daily oral vitamin C supplementation (500 mg/day) for 8 weeks resulted in a reduction of skin malonaldehyde content, a biomarker of lipid oxidation, but also in a reduction of the skin content of total glutathione and protein thiols, which also are important antioxidant molecules. McCall and Frei (1999) reported that oral vitamin C supplementation induced a reduction of lipid oxidative damages, thought to be an important etiologic factor in skin aging, in nonsmokers as well as in smokers.

6.6.4 Vitamin C and Aging

In vitro, vitamin C enhances synthesis of collagens I and III in the dermis and collagens IV and VII, tenascin C, fibrillin, and versican at the basement membrane level, improving both skin dermal matrix quality and dermal-epidermal junction morphogenesis (Murad et al., 1981; Chan et al., 1990; Nusgens et al., 2001; Heber et al., 2006; Marionnet et al., 2006; Kim et al., 2006; Amano et al., 2007).

The role of vitamin C on collagen synthesis has been confirmed *in vivo* using preclinical models. Vitamin C deficiency induced a decrease of hydroxyproline, which is responsible for assembling triple-helical collagen molecules in the dermis (Bates & Tsuchiya, 2003). Vitamin C treatment caused a dose-dependent elevation in the wound contraction, indicating a protection of mice against radiation-induced damage as well as an improved healing of wounds after exposure to whole-body gamma radiation (Jagetia et al., 2004). A recent study reveals indeed that higher vitamin C intakes were associated with lower wrinkled appearance and senile dryness (Cosgrove et al., 2007).

6.6.5 Vitamin C and Barrier Function

In vitro, vitamin C plays a key role in epidermalization, keratinocyte differentiation, and formation of stratum corneum barrier lipids, and as a consequence improves skin barrier function (Ponec et al., 1997; Wha Kim et al., 2002; Savini et al., 2002). Vitamin C also appears crucial for dermal-epidermal junction, dermal matrix formation, epidermal turn-over, barrier function and wound repair, and thus may play a role in the prevention of skin aging and associated dry skin (Marionnet et al., 2006; Kim et al., 2006; Amano et al., 2007).

6.6.6 Vitamin C and Photoprotection

In vitro using a melanocyte-keratinocyte coculture model, vitamin C treatment induces modification of melanocyte dendricity (Regnier et al., 2005), which is known to participate in the beneficial effect of vitamin C for photoaging by preventing UV-induced hyperpigmentation and spots. In a human dermal fibroblast culture, vitamin C treatment counteracted matrix metalloproteinase 1 mRNA increase, which was induced 10- to 15-fold following UVA radiation exposure (Offord et al., 2002). Furthermore, vitamin C treatment of keratinocytes showed that vitamin C counteracted the increase of ROS in cells induced by acute UVB irradiation (Jin et al., 2007).

In vivo, using preclinical models, Kobayashi et al. (1996) reported that intraperitoneal (100 mg/kg body weight) administration of magnesium-L-ascorbyl-2-phosphate to mice induced a significantly decreased lipid peroxidation in skin, measured by TBARS assay (thiobarbituric acid reactive substance). Chandra Jagetia et al. (2003) reported that daily administration of 250 mg/kg body weight of vitamin C before exposure to gamma rays resulted in a significant increase in the activities of superoxide dismutase and glutathione peroxidase as well as glutathione content in the irradiated mouse skin. Moreover, the ascorbic acid pretreatment inhibited the radiation-induced increase in lipid peroxidation in skin. These results confirm the photoprotective effect of vitamin C in animal skin. Oral administration of vitamin C (60% in a mixture administered at 1,130 mg/kg/day) in combination with vitamin E, primrose oil, and pycnogenol was reported to significantly reduce UVB-induced wrinkle formation and epidermal thickness in a preclinical model (Cho et al., 2007). Topically applied vitamin C induced significant photoprotective effects at concentrations of at least 10% in animals and humans. This effect was also observed when vitamin C was applied in combination with vitamin E (Eberlein-König & Ring, 2005; Humbert et al., 2003). Furthermore, ferulic acid was shown to stabilize vitamins C and E when incorporated into a topical formulation and thereby doubled its photoprotection of skin as measured by erythema and sunburn cell formation (Lin et al., 2005). By contrast, oral vitamin C supplementation, even at high doses (gram), failed to confirm any photoprotection property (Fuchs & Kern, 1998; Mireles-Rocha et al., 2002; McArdle et al., 2002). Leveque et al. (2002) determined skin ascorbic acid concentration in face and abdomen skin samples of 39 volunteers, split in different groups according to their age (ranging from 15 years to greater than 50 years). The results showed that skin ascorbic acid significantly decreases with age and that ascorbic acid concentrations were lower in sun-exposed sites (face) versus sun-protected sites (abdomen). These results suggested that vitamin C supplementation can prevent deficiencies in skin, and thus skin photoaging, but more studies are needed to evaluate if the decrease of vitamin C is a consequence of impaired vitamin C bioavailability or simply a result of reduced food intake. With age the sensory capacity is often decreased, oral status is poor, and gastrointestinal changes are observed (e.g., *Helicobacter pylori* infection). Food intake–related lower vitamin C level with increasing age can be compensated for by oral supplementation. This might support the idea that oral vitamin C can counteract skin deficiencies and so prevent skin damage appearing with age.

6.7 Conclusions

Dietary bioactives such as vitamins E and C have demonstrated beneficial effects to maintain and improve skin integrity and physiology as well as to reduce deleterious effects induced by aging and environmental stresses, and more specifically UV-induced skin damages. Beneficial effects have been demonstrated with the use of topical application of these ingredients. More recently, oral supplements containing these ingredients have also been reported to be beneficial for skin. However, consumption of dietary bioactives does not guarantee obtaining a beneficial effect on human skin; indeed this ingredient must cross the intestinal barrier, and if necessary must be metabolized and distributed to the skin. Absorption of dietary bioactives in the gut and in the skin could be modulated based on understanding the key parameters involved in the absorption process, that is, role of transfer proteins, physical and chemical properties of the dietary bioactives, and competition and/or interaction with other dietary bioactives.

In conclusion, when dietary bioactives have been selected for their incorporation into an oral supplement, bioavailability of individual dietary bioactives present in the oral supplement as well their interaction with the different constituents of the supplement have to be carefully evaluated and, if necessary, improved. The importance of active versus inactive forms, appropriate concentrations, and product stability remain hurdles—and information that is missing in most of the published literature.

The administration of these bioactives by the oral route offers several advantages over their topical application: the intestines absorb bioactives, which are sometimes compromised in topical application due to their low stability or low skin penetration; and bioactives reach the entire skin of the body; bioactives are distributed to all the skin compartments, for example, epidermis, dermis, hypodermis, blood vessels, and sebum, allowing bioefficacy in all these compartments. The oral administration of these bioactives could be complementary to the topical application.

The skin health effects of aging will always remain an important issue for the population, and educating them about what has or has not been scientifically established is an important role for health care professionals. Additionally, interest and pursuit of further education on these ingredients, products, and oral skin health supplements in general may ultimately drive future research that will distinguish truly efficacious nutrients from misleading claims.

References

Ahmed RS, Suke SG, Seth V, Jain A, Bhattacharya SN, Banerjee BD (2006) Impact of oral vitamin E supplementation on oxidative stress & lipid peroxidation in patients with polymorphous light eruption. *Indian J Med Res* **123**, 781–787.

Amano S, Ogura Y, Akutsu N, Nishiyama T (2007) Quantitative analysis of the synthesis and secretion of type VII collagen in cultured human dermal fibroblasts with a sensitive sandwich enzyme-linked immunoassay. *Exp Dermatol* **16**, 151–155.

Azzi A (2007) Molecular mechanism of α-tocopherol action. *Free Radic Biol Med* **43**, 16–21.

Baker EM, Hodges RE, Hood J, Sauberlich H, March SC (1969) Metabolism of ascorbic-1-14C acid in experimental human scurvy. *Am J Clin Nutr* **22**, 549–558.

Bates CJ, Tsuchiya H (2003) Comparison of vitamin C deficiency with food restriction on collagen cross-link ratios in bone, urine and skin of weanling guinea-pigs. *Br J Nutr* **89**, 303–310.

Beehler BC, Przybyszewski J, Box HB, Kulesz-Martin MF (1992) Formation of 8-hydroxydeoxyguanosine within DNA of mouse keratinocytes exposed in culture to UVB and H_2O_2. *Carcinogenesis* **13**, 2003–2007.

Boelsma E, Hendriks HF, Roza L (2001) Nutritional skin care: health effects of micronutrients and fatty acids. *Am J Clin Nutr* **73**, 853–864.

Borel P, Grolier P, Armand M, Partier A, Lafont H, Lairon D, Azais-Braesco V (1996) Carotenoids in biological emulsions: solubility, surface-to-core distribution, and release from lipid droplets. *J Lipid Res* **37**, 250–261.

Borel P, Mekki N, Boirie Y, Partier A, Grolier P, Alexandre-Gouabau MC, Beaufrere B, Armand M, Lairon D, Azais-Braesco V (1997) Postprandial chylomicron and plasma vitamin E responses in healthy older subjects compared with younger ones. *Eur J Clin Invest* **27**, 812–821.

Borel P, Tyssandier V, Mekki N, Grolier P, Rochette Y, Alexandre-Gouabau MC, Lairon D, Azais-Braesco V (1998) Chylomicron β-carotene and retinyl palmitate responses are dramatically diminished when men ingest β-carotene with medium-chain rather than long-chain triglycerides. *J Nutr* **128**, 1361–1367.

Brown LAS, Jones DP (1996) The biology of ascorbic acid. In: *Handbook of Antioxidants*. Cadenas E, Packer L (eds.), Marcel Dekker Inc., New York, pp. 117–154.

Brubacher D, Moser U, Jordan P (2000). Vitamin C concentrations in plasma as a function of intake: a meta-analysis. *Int J Vit Nut Res* **70**, 226–237.

Burton GW, Ingold KU, Foster DO, Cheng SC, Webb A, Hughes L, Lusztyk E (1988) Comparison of free α-tocopherol and α-tocopheryl acctate as sources of vitamin E in rats and humans. *Lipids* **23**, 834–840.

Castenmiller JJ, West CE (1998) Bioavailability and bioconversion of carotenoids. *Annu Rev Nutr* **18**, 19–38.

Cesarini JP, Michel L, Maurette JM, Adhoute H, Bejot M (2003) Immediate effects of UV radiation on the skin: modification by an antioxidant complex containing carotenoids. *Photodermatol Photoimmunol Photomed* **19**, 182–189.

Chan D, Lamande SR, Cole WG, Bateman JF (1990) Regulation of procollagen synthesis and processing during ascorbate-induced extracellular matrix accumulation in vitro. *Biochem J.* **269**, 175–181.

Chandra Jagetia G, Rajanikant GK, Rao SK, Shrinath Baliga M (2003) Alteration in the glutathione, glutathione peroxidase, superoxide dismutase and lipid peroxidation by ascorbic acid in the skin of mice exposed to fractionated gamma radiation. *Clin Chim Acta* **332**, 111–121.

Cheeseman KH, Holley AE, Kelly FJ, Wasil M, Hughes L, Burton G (1995) Biokinetics in humans of RRR-α-tocopherol: the free phenol, acetate ester, and succinate ester forms of vitamin E. *Free Radic Biol Med* **19**, 591–598.

Cho HS, Lee MH, Lee JW, No KO, Park SK, Lee HS, Kang S, Cho WG, Park HJ, Oh KW, Hong JT (2007) Anti-wrinkling effects of the mixture of vitamin C, vitamin E, pycnogenol and evening primrose oil, and molecular mechanisms on hairless mouse skin caused by chronic ultraviolet B irradiation. *Photodermatol Photoimmunol Photomed* **23**, 155–162.

Clifford AJ, de Moura FF, Ho CC, Chuang JC, Follett J, Fadel JG, Novotny JA (2006) A feasibility study quantifying *in vivo* human α-tocopherol metabolism. *Am J Clin Nutr* **84**, 1430–1441.

Cosgrove MC, Franco OH, Granger SP, Murray PG, Mayes AE (2007) Dietary nutrient intakes and skin-aging appearance among middle-aged American women. *Am J Clin Nutr* **86**, 1225–1231.

Cross CE, Vandervliet A, Louie S, Thiele JJ, Halliwell B (1998) Oxidative stress and antioxidants at biosurfaces: Plants, skin, and respiratory tract surfaces. *Environ Health Perspect* **106**, 1241–1251.

Dalle CM, Pathak MA (1992) Skin photosensitizing agents and the role of reactive oxygen species in photoaging. *J Photochem Photobiol B* **14**, 105–124.

Downing DT, Stewart ME, Strauss JS (1981) Estimation of sebum production rates in man by measurement of the squalene content of skin biopsies. *J Invest Dermatol* **77**(4), 358–360.

During A, Hussain MM, Morel DW, Harrison EH (2002) Carotenoid uptake and secretion by CaCo-2 cells: β-carotene isomer selectivity and carotenoid interactions. *J Lipid Res* **43**, 1086–1095.

Eberlein-König B, Ring J (2005) Relevance of vitamins C and E in cutaneous photoprotection. *J Cosmet Dermatol* **4**, 4–9.

Ekanayake-Mudiyanselage S, Kraemer K, Thiele JJ (2004) Oral supplementation with all-Rac- and RRR-α-tocopherol increases vitamin E levels in human sebum after a latency period of 14–21 days. *Ann N Y Acad Sci* **1031**, 184–194.

Emerit I (1992) Free radicals and aging of the skin. *EXS* **62**, 328–341.

Fuchs J, Podda M (1997) Vitamin C in cutaneous biology. In: *Vitamin C in Health and Disease*. Packer L, Fuchs J (eds.), Marcel Dekker Inc., New York, pp. 333–340.

Fuchs J (1998) Potential and limitation of the natural antioxidants RRR alpha tocopherol, L-ascorbic acid and b-carotene in cutaneaous photoprotection. *Free Radic Biol Med* **25**, 848–873.

Fuchs J, Kern H (1998) Modulation of UV-light-induced skin inflammation by D-α-tocopherol and L-ascorbic acid: a clinical study using solar simulated radiation. *Free Radic Biol Med* **25**, 1006–1012.

Fukui E, Kurohara H, Kageyu A, Kurosaki Y, Nakayama T, Kimura T (1989a) Enhancing effect of medium-chain triglycerides on intestinal absorption of d-α-tocopherol acetate from lecithin-dispersed preparations in the rat. *J Pharmacobiodyn* **12**, 80–86.

Fukui E, Tabuchi H, Kurosaki Y, Nakayama T, Kimura T (1989b) Further investigations of enhancing effect of medium-chain triglycerides on d-α-tocopherol acetate absorption from lecithin-dispersed preparations in rat small intestine. *J Pharmacobiodyn* **12**, 754–761.

Gallo-Torres HE, Ludorf J, Brin M (1978) The effect of medium-chain triglycerides on the bioavailability of vitamin E. *Int J Vitam Nutr Res* **48**, 240–241.

Gianello R, Libinaki R, Azzi A, Gavin PD, Negis Y, Zingg J-M, Holt P, Keah H-H, Griffey A, Smallridge A, West SM, Ogru E (2005) α-Tocopheryl phosphate: a novel, natural form of vitamin E. *Free Radic Biol Med* **39**, 970–976.

Greul AK, Grundmann JU, Heinrich F, Pfitzner I, Bernhardt J, Ambach A, Biesalski HK, Gollnick H (2002) Photoprotection of UV-irradiated human skin: an antioxidative combination of vitamins E and C, carotenoids, selenium and proanthocyanidins. *Skin Pharmacol Appl Skin Physiol* **15**, 307–315.

Guyton KZ, Kensler TW (1993) Oxidative mechanisms in carcinogenesis. *Br Med Bull* **49**, 523–544.

Heber GK, Markovic B, Hayes A (2006) An immunohistological study of anhydrous topical ascorbic acid composition on ex vivo human skin. *J Cosmet Dermatol* **5**, 150–156.

Heinen MM, Hugues MC, Ibiele TI, Marks GC, Green AC, van der Pols JC (2007) Intake of antioxidant nutrients and the risk of skin cancer. *Eur J Cancer* **43**, 2707–2716.

Hollander D, Rim E, Muralidhara KS (1975) Mechanism and site of small intestinal absorption of α-tocopherol in the rat. *Gastroenterology* **68**, 1492–1499.

Hollander D, Ruble PE (1978) β-Carotene intestinal absorption: bile, fatty acid, pH, and flow rate effects on transport. *Am J Physiol* **235**, E686–E691.

Hoppu U, Salo-Väänänen P, Lampi AM, Isoleuri E (2005) Serum α- and γ-tocopherol levels in atopic mothers and their infants are correlated. *Biol Neonate* **88**, 24–26.

Hoppu U, Rinne M, Salo-Väänänen P, Lampi AM, Piironen V, Isoleuri E (2005) Vitamin C in breast milk may reduce the risk of atopy in the infant. *Eur J Nutr* **59**,123–128.

Hornig DH, Moser U (1981) The safety of high vitamin C intakes in man. In: *Vitamin C (Ascorbic Acid)*. Counsell JN, Hornig DH (eds.), Applied Science Publisher, New Jersey, pp. 225–248.

Hosomi A, Arita M, Sato Y, Kiyose C, Ueda T, Igarashi O, Arai H, Inoue K (1997) Affinity for α-tocopherol transfer protein as a determinant of the biological activities of vitamin E analogs. *FEBS Lett* **409**, 105–108.

Hu ML, Tappel AL (1992) Potentiation of oxidative damage to proteins by ultraviolet-A and protection by antioxidants. *Photochem Photobiol* **56**, 357–363.

Humbert PG, Haftek M, Creidi P, Lapière C, Nusgens B, Richard A, Schmitt D, Rougier A, Zahouani H (2003) Topical ascorbic acid on photoaged skin. Clinical topographical and ultrastructural evaluation: double-blind study vs. placebo. *Exp Dermatol* **12**(3), 237–244.

Iuliano L, Micheletta F, Maranghi M, Frati G, Diczfalusy U, Violi F (2001) Bioavailability of vitamin E as function of food intake in healthy subjects: effects on plasma peroxide-scavenging activity and cholesterol-oxidation products. *Arterioscler Thromb Vasc Biol* **21**, E34–E37.

Jagetia GC, Rajanikant GK, Baliga MS, Rao KV, Kumar P (2004) Augmentation of wound healing by ascorbic acid treatment in mice exposed to gamma-radiation. *Int J Radiat Biol* **80**(5), 347–354.

Jang M, Cai L, Udeani GO, Slowing KV, Thomas CF, Beecher CW, Fong HH, Farnsworth NR, Kinghorn AD, Mehta RG, Moon RC, Pezzuto JM (1997) Cancer chemopreventive activity of resveratrol, a natural product derived from grapes. *Science* **275**, 218–220.

Jin GH, Liy Y, Jin SZ, Liu XD, Liu SZ (2007) UVB induced oxidative stress in human keratinocytes and protective effect of antioxidant agents. *Radiat Environ Biophys* **46**(1), 61–68.

Jones SA, McArdle F, Jack CI, Jackson MJ (1999) Effect of antioxidant supplementation on the adaptive response of human skin fibroblasts to UV-induced oxidative stress. *Redox Rep* **4**(6), 291–299.

Kagan V, Witt E, Goldman R, Scita G, Packer L (1992) Ultraviolet light-induced generation of vitamin E radicals and their recycling. A possible photosensitizing effect of vitamin E in skin. *Free Radic Res Commun* **16**, 51–64.

Kallner A, Hartmann D, Hornig D (1977) On the absorption of ascorbic acid in man. *Int J Vit Nutr Res* **47**, 383–388.

Kallner A, Hartmann D, Hornig D (1979) Steady-state turnover and body pool of ascorbic acid in man. *Am J Clin Nutr* **32**, 530–539.

Kang JS, Kim HN, Jung DJ, Kim JE, Mun GH, Kim YS, Cho D, Shin DH, Hwang Y-I, Lee WJ (2007) Regulation of UVB-induced IL-8 and MCP-1 production in skin keratinocytes by increasing vitamin C uptake via the redistribution of SVCT-1 from cytosol to the membrane. *J Invest Dermatol* **127**, 698–706.

Katiyar SK, Agarwal R, Mukhtar H (1996) Inhibition of tumor promotion in SENCAR mouse skin by ethanol extract of *Zingiber officinale* rhizome. *Cancer Res* **56**, 1023–1030.

Katiyar SK, Mukhtar H (1997) Tea antioxidants in cancer chemoprevention. *J Cell Biochem* **27**, 59–67.

Kim SR, Cha SY, Kim MK, Kim JC, Sung YK (2006) Induction of versican by ascorbic acid 2-phosphate in dermal papilla cells. *J Dermatol Sci* **43**(1), 60–62.

Kobayashi S, Takehana M, Itoh S, Ogata E (1996) Protective effect of magnesium-L-ascorbyl-2 phosphate against skin damage induced by UVB irradiation. *Photochem Photobiol* **64**, 224–228.

Koo SI, Noh SK (2001) Phosphatidylcholine inhibits and lysophosphatidylcholine enhances the lymphatic absorption of α-tocopherol in adult rats. *J Nutr* **131**, 717–722.

Lang JK, Gohil K, Packer L (1986) Simultaneous determination of tocopherols, ubiquinols, and ubiquinones in blood, plasma, tissue homogenates, and subcellular fractions. *Anal Biochem* **157**, 106–116.

Leveque N, Muret P, Mary S, Makki S, Kantelip JP, Rougier A, Humbert P (2002) Decrease in skin ascorbic acid concentration with age. *Eur J Dermatol* **12**, 21–22.

Leveque N, Robin S, Makki S, Muret P, Rougier A, Humbert P (2003) Iron and ascorbic acid concentrations in human dermis with regard to age and body sites. *Gerontology* **49**, 117–122.

Leveque N, Robin S, Muret P, Mac-Mary S, Makki S, Berthelot A, Kantelip JP, Humbert P (2004) In vivo assessment of iron and ascorbic acid in psoriatic dermis. *Acta Derm Venereol* **84**, 2–5.

Levine M, Rumsey SC, Daruwala R, Park JB, Wang Y (1999) Criteria and recommendations for vitamin C intake. *JAMA* **281**, 1415–1423.

Liang WJ, Johnson D, Jarvis SM (2001) Vitamin C transport systems of mammalian cells. *Mol Membr Biol* **18**, 87–95.

Lin F-H, Lin J-Y, Gupta RD, Tournas JA, Burch JA, Selim MA, Monteiro-Riviere NA, Grichnik JM, Zielinski J, Pinell SP (2005) Ferulic acid stabilizes a solution of vitamins C and E and doubles its photoprotection of skin. *J Invest Dermatol* **125**, 826–832.

Lombardo D, Guy O (1980) Studies on the substrate specificity of a carboxyl ester hydrolase from human pancreatic juice. II. Action on cholesterol esters and lipid-soluble vitamin esters. *Biochim Biophys Acta* **611**, 147–155.

Longas MO, Bhuyan DK, Bhuyan KC, Gutsch CM, Breitweiser KO (1993) Dietary vitamin E reverses the effects of ultraviolet light irradiation on rat skin glycosaminoglycans. *Biochim Biophys Acta* **1156**(3), 239–244.

Marionnet C, Vioux-Chagnoleau C, Pierrard C, Sok J, Asselineau D, Bernerd F (2006) Morphogenesis of dermal-epidermal junction in a model of reconstructed skin: beneficial effects of vitamin C. *Exp Dermatol* **15**, 625–633.

Martin RF, Young VR, Blumberg J, Janghorbani M (1989) Ascorbic acid-selenite interactions in humans studied with an oral dose of 74SeO3(2-). *Am J Clin Nutr* **49**, 862–869.

McArdle F, Rhodes LE, Parslew R, Jack CI, Friedmann PS, Jackson MJ (2002) UVR-induced oxidative stress in human skin in vivo: effects of oral vitamin C supplementation. *Free Radic Biol Med* **33**, 1355–1362.

McCall MR, Frei B (1999) Can antioxidant vitamins materially reduce oxidative damage in humans? *Free Radic Biol Med* **26**, 1034–1053.

McKenzie RC (2000) Selenium, ultraviolet radiation and the skin. *Clin Exp Dermatol* **25**, 631–636.

Mireles-Rocha H, Galindo I, Huerta M, Trujillo-Hernández B, Elizalde A, Cortés-Franco R (2002) UVB photoprotection with antioxidants: effects of oral therapy with d-α-tocopherol and ascorbic acid on the minimal erythema dose. *Acta Derm Venereol* **82**, 21–24.

Moser U, Bendich A (1991) Vitamin C. In: *Handbook of Vitamins*. Machlin LJ (ed.), Marcel Dekker Inc., New York, p. 195.

Mukhtar H, Ahmad N (1999) Cancer chemoprevention: future holds in multiple agents. *Toxicol Appl Pharmacol* **158**, 207–210.

Murad S, Grove D, Lindberg KA, Reynolds G, Sivarajah A, Pinnell SR (1981) Regulation of collagen synthesis by ascorbic acid. *Proc Natl Acad Sci USA* **78**, 2879–2882.

Nusgens BV, Humbert P, Rougier A, Colige AC, Haftek M, Lambert CA, Richard A, Creidi P, Lapière CM (2001) Topically applied vitamin C enhances the mRNA level of collagens I and III, their processing enzymes and tissue inhibitor of matrix metalloproteinase 1 in the human dermis. *J Invest Dermatol* **116**, 853–859.

Offord EA, Gautier JC, Avanti O, Scaletta C, Runge F, Krämer K, Applegate LA (2002) Photoprotective potential of lycopene, β-carotene, vitamin E, vitamin C and carnosic acid in UVA-irradiated human skin fibroblasts. *Free Radic Biol Med* **32**, 1293–1303.

Padayatty SJ, Katz A, Wang Y, Eck P, Kwon O, Lee JH, Chen S, Corpe C, Dutta A, Dutta SK, Levine M (2003) Vitamin C as an antioxidant: evaluation of its role in disease prevention. *J Am Coll Nutr* **22**, 18–35.

Padayatty SJ, Sun H, Wang Y, Riordan HD, Hewitt SM, Katz A, Wesley RA, Levine M (2004) Vitamin C pharmacokinetics: implications for oral and intravenous use. *Ann Intern Med* **140**, 533–537.

Papas K, Kalbfleisch J, Mohon R (2007) Bioavailability of a novel, water-soluble vitamin E formulation in malabsorbing patients. *Dig Dis Sci* **52**, 347–352.

Perchellet JP, Perchellet EM (1989) Antioxidants and multistage carcinogenesis in mouse skin. *Free Radic Biol Med* **7**, 377–408.

Perchellet JP, Perchellet EM, Belman S (1990) Inhibition of DMBA-induced mouse skin tumorigenesis by garlic oil and inhibition of two tumor-promotion stages by garlic and onion oils. *Nutr Cancer* **14**, 183–193.

Podda M, Traber MG, Weber C, Yan LJ, Packer L (1998) UV-irradiation depletes antioxidants and causes oxidative damage in a model of human skin. *Free Radic Biol Med* **24**, 55–65.

Ponec M, Weerheim A, Kempenaar J, Mulder A, Gooris GS, Bouwstra J, Mommaas AM (1997) The formation of competent barrier lipids in reconstructed human epidermis requires the presence of vitamin C. *J Invest Dermatol* **109**, 348–355.

Poquet L, Clifford MN, Williamson G (2008) Effect of dihydrocaffeic acid on UV irradiation of human keratinocyte HaCaT cells. *Arch Biochem Biophys* **476**(2), 196–204.

Reboul E, Abou L, Mikail C, Ghiringhelli O, André M, Portugal H, Jourdheuil-Rahmani D, Amiot MJ, Lairon D, Borel P (2005) Lutein transport by Caco-2 TC-7 cells occurs partly by a facilitated process involving the scavenger receptor class B type I (SR-BI). *Biochem J* **387**(Pt 2), 455–461.

Regnier M, Tremblaye C, Schmidt R (2005) Vitamin C affects melanocyte dendricity via keratinocytes. *Pigment Cell Res* **18**, 389–390.

Ricciarelli R, Maroni P, Ozer N, Zingg JM, Azzi A (1999) Age-dependent increase of collagenase expression can be reduced by α-tocopherol via protein kinase C inhibition. *Free Radic Biol Med* **(7–8)**, 729–737.

Richelle M, Enslen M, Hager C, Groux M, Tavazzi I, Godin JP, Berger A, Metairon S, Quaile S, Piguet-Welsch C, Sagalowicz L, Green H, Fay LB (2004) Both free and esterified plant sterols reduce cholesterol absorption and the bioavailability of β-carotene and α-tocopherol in normocholesterolemic humans. *Am J Clin Nutr* **80**, 171–177.

Rios L, Bennett RN, Lazarus SA, Remesy C, Scalbert A, Williamson G (2002) Cocoa proanthocyanidins are stable during gastric transit in humans. *Am J Clin Nutr* **76**, 1106–1110.

Roodenburg AJ, Leenen R, van het Hof KH, Weststrate JA, Tijburg LB (2000) Amount of fat in the diet affects bioavailability of lutein esters but not of α-carotene, β-carotene, and vitamin E in humans. *Am J Clin Nutr* **71**, 1187–1193.

Rostan EF, DeBuys HV, Madey DL, Pinnell SR (2002) Evidence supporting zinc as an important antioxidant for skin. *Int J Dermatol* **41**, 606–611.

Rumsey SC, Levine M (1998) Absorption, transport, and disposition of ascorbic acid in humans. *J Nutr Biochem* **9**, 116–130.

Sadhana AS, Rao AR, Kucheria K, Bijani V (1988) Inhibitory action of garlic oil on the initiation of benzo[a]pyrene-induced skin carcinogenesis in mice. *Cancer Lett* **40**, 193–197.

Savini I, D'Angelo I, Ranalli M, Melino G, Avigliano L (1999) Ascorbic acid maintenance in HaCaT cells prevents radical formation and apoptosis by UV-B. *Free RadicBiolMed* **26**, 1172–1180.

Savini I, Duflot S, Avigliano L (2000) Dehydroascorbic acid uptake in a human keratinocyte cell line (HaCaT) is glutathione-independent. *Biochem J* **345**, 665–672.

Savini I, Catani MV, Rossi A, Duranti G, Melino G, Avigliano L (2002) Characterization of keratinocyte differentiation induced by ascorbic acid: protein kinase C involvement and vitamin C homeostasis. *J Invest Dermatol* **118**, 372–379.

Shindo Y, Witt E, Han D, Epstein W, Packer L (1994) Enzymic and non-enzymic antioxidants in epidermis and dermis of human skin. *J Invest Dermatol* **102**, 122–124.

Sotiriou S, Gispert S, Cheng J, Wang Y, Chen A, Hoogstraten-Miller S, Miller GF, Kwon O, Levine M, Guttentag SH, Nussbaum RL (2002) Ascorbic-acid transporter Slc23a1 is essential for vitamin C transport into the brain and for perinatal survival. *Nat Med* **8**, 514–517.

Stahl W, Heinrich U, Jungmann H, Sies H, Tronnier H (2000) Carotenoids and carotenoids plus vitamin E protect against ultraviolet light-induced erythema in humans. *Am J Clin Nutr* **71**, 795–798.

Stahl W, Heinrich U, Wiseman S, Eichler O, Sies H, Tronnier H (2001) Dietary tomato paste protects against ultraviolet light-induced erythema in humans. *J Nutr* **131**, 1449–1451.

Stahl W, Sies H (2002) Carotenoids and protection against solar UV radiation. *Skin Pharmacol Appl Skin Physiol* **15**, 291–296.

Steenvoorden DP, van Henegouwen GM (1997) The use of endogenous antioxidants to improve photoprotection. *J Photochem Photobiol B* **41**, 1–10.

Steiling H, Longet K, Moodycliffe A, Mansourian R, Bertschy E, Smola H, Mauch C, Williamson G (2007) Sodium-dependent vitamin C transporter isoforms in skin: Distribution, kinetics, and effect of UVB-induced oxidative stress. *Free Radic Biol Med* **43**, 752–762.

Stoner GD, Mukhtar H (1995) Polyphenols as cancer chemopreventive agents. *J Cell Biochem Suppl* **22**, 169–180.

Sugawara T, Kushiro M, Zhang H, Nara E, Ono H, Nagao A (2001) Lysophosphatidylcholine enhances carotenoid uptake from mixed micelles by Caco-2 human intestinal cells. *J Nutr* **131**, 2921–2927.

Swindells K, Rhodes LE (2004) Influence of oral antioxidants on ultraviolet radiation-induced skin damage in humans [Review]. *Photodermatol Photoimmunol Photomed* **20**, 297–304.

Thiele JJ, Podda M, Packer L (1997) Tropospheric ozone: an emerging environmental stress to skin [Review]. *Biol Chem* **378**(11), 1299–1305.

Thiele JJ, Traber MG, Packer L (1998) Depletion of human stratum corneum vitamin E: an early and sensitive in vivo marker of UV induced photo-oxidation. *J Invest Dermatol* **110**, 756–761.

Thiele JJ, Weber SU, Packer L (1999) Sebaceous gland secretion is a major physiologic route of vitamin E delivery to skin. *J Invest Dermatol* **113**, 1006–1010.

Thiele JJ, Dreher F, Packer L (2000) Cosmeceuticals. In: *Drugs vs Cosmetics*. Elsner P, Maibach H (eds.), Dekker, New York, pp. 145–188.

Thiele JJ, Hsieh SN, Ekanayake-Mudiyanselage S (2005) Vitamin E: critical review of its current use in cosmetic and clinical dermatology. *Dermatol Surg* **31**, 805–813.

Thiele JJ, Ekanayake-Mudiyanselage S (2007) Vitamin E in human skin: organ-specific physiology and considerations for its use in dermatology. *Mol Aspect Med* **28**, 646–667.

Tsukaguchi H, Tokui T, Mackenzie B, Berger UV, Chen XZ, Wang Y, Brubaker RF, Hediger MA (1999) A family of mammalian Na+-dependent L-ascorbic acid transporters. *Nature* **399**, 70–75.

Tyssandier V, Lyan B, Borel P (2001) Main factors governing the transfer of carotenoids from emulsion lipid droplets to micelles. *Biochim Biophys Acta* **1533**, 285–292.

Vaule H, Leonard SW, Traber MG (2004) Vitamin E delivery to human skin: studies using deuterated α-tocopherol measured by APCI LC-MS. *Free Radic Biol Med* **36**, 456–463.

Wajda R, Zirkel J, Schaffer T (2007) Increase of bioavailability of coenzyme Q(10) and vitamin E. *J Med Food* **10**, 731–734.

Weber SU, Thiele JJ, Cross CE, Packer L (1999) Vitamin C, uric acid, and glutathione gradients in murine stratum corneum and their susceptibility to ozone exposure. *J Invest Dermatol* **113**, 1128–1132.

Wefers H, Sies H (1988) The protection by ascorbate and glutathione against microsomal lipid peroxidation is dependent on vitamin E. *Eur J Biochem* **174**, 353–357.

Werninghaus K, Handjani RM, Gilchrest BA (1991) Protective effect of α-tocopherol in carrier liposomes on ultraviolet-mediated human epidermal cell damage in vitro. *Photodermatol Photoimmunol Photomed* **8**, 236–242.

West CE, Castenmiller JJ (1998) Quantification of the "SLAMENGHI" factors for carotenoid bioavailability and bioconversion. *Int J Vitam Nutr Res* **68**, 371–377.

Wha Kim S, Lee IW, Cho HJ, Cho KH, Han Kim K, Chung JH, Song PI, Chan Park K (2002) Fibroblasts and ascorbate regulate epidermalization in reconstructed epidermis. *J Dermatol Sci* **30**, 215–223.

Wolf G (2006) How an increased intake of α-tocopherol can suppress the bioavailability of γ-tocopherol. *Nutr Rev* **64**, 295–299.

Wu D, Han SN, Meydani M, Meydani SN (2006) Effect of concomittant consumption of fish oil and vitamin E and T cell mediated function in the elderly: a randomized double-blind trial. *J Am Coll Nutr* **25**, 300–306.

Zimmer Ł, Czarnecki W (2007) Polysaccharide formulation for improvement of racemic vitamin E bioavailability. *Acta Pol Pharm* **64**, 169–174.

7

Zinc, Selenium and Skin Health: Overview of Their Biochemical and Physiological Functions

Bruno Berra, PhD, and Angela Maria Rizzo, PhD

Department of Molecular Sciences applyed to Biosystems, University of Milan, Italy

Aaron Tabor and Robert M. Blair (eds.), *Nutritional Cosmetics: Beauty from Within*, 139–158, © 2009 William Andrew Inc.

7.1 Summary

Zinc is an essential element in human and animal nutrition with different and important roles. Zinc plays catalytic, structural, or regulatory functions in the more than 300 metalloenzymes that have been identified in biological systems in are involved in nucleic acid and protein metabolism and the production of energy. Zinc plays a structural role in the formation of the so-called zinc fingers that are exploited by transcription factors for interacting with DNA and regulating the activity of genes. Another structural role of zinc is in the maintenance of the integrity of biological membranes, resulting in their protection against oxidative injury.

Physiologically, zinc is vital for growth and development, sexual maturation and reproduction, dark vision adaptation, olfactory and gustatory activity, insulin storage and release, and a variety of host immune defences. Zinc deficiency can result in growth retardation, immune dysfunction, increased incidence of infections, hypogonadism, oligospermia, anorexia, diarrhea, weight loss, delayed wound healing, neural tube defects of the fetus, increased risk for abortion, alopecia, mental lethargy, and skin changes. Moderate to severe zinc deficiency is rare in industrialized countries. However, it is highly prevalent in developing countries. Zinc intake in many of the elderly may be suboptimal and, if compounded with certain drugs and diseases, can lead to mild or even moderate zinc deficiency.

Selenium, although toxic in large doses, is an essential micronutrient for animals. In plants, it occurs as a bystander mineral, sometimes in toxic proportions in forage (some plants may accumulate selenium as a defense against being eaten by animals, but other plants such as locoweed require selenium, and their growth indicates the presence of selenium in soil). It is a component of the unusual amino acids selenocysteine and selenomethionine. In humans, selenium is a trace element nutrient that functions as a cofactor for reduction of antioxidant enzymes such as glutathione peroxidases and certain forms of thioredoxin reductase found in animals and some plants (this enzyme occurs in all living organisms, but not all forms of it in plants require selenium).

Selenium deficiency is most commonly seen in parts of China where the selenium content in the soil, and therefore selenium intake, is very low. Selenium deficiency is linked to Keshan disease, whose signs are an enlarged heart and poor heart function. Keshan disease has been observed in low-selenium areas of China, where dietary intake (13–19 μg daily) is significantly lower than the current RDA.

Selenium deficiency results also in atrophy degeneration and necrosis of cartilage tissue. Because selenium is also necessary for the conversion of the thyroid hormone thyroxine into triiodothyronine, its deficiency can cause symptoms of hypothyroidism, including extreme fatigue, mental retardation, goitre, cretinism, and recurrent miscarriage.

7.2 Zinc and Zinc-Containing Biomolecules

The ubiquity of zinc in the human body (2.5 g) testifies the critical role it plays in normal health and development. Zinc is especially prevalent in the epidermis (71 μg/g dry weight), where it is needed owing to the highly proliferative nature of this tissue [1].

The importance of zinc to skin health is demonstrated by the prominent manifestation of dietary or genetically induced zinc deficiencies as dermatologic abnormalities. Low serum zinc levels generally accompany bullous pemphigoid and decubitus ulcers [2]. Moreover, the genetic inability to absorb zinc from the intestine results in acrodermatitis enteropathica, which manifests itself as severe dermatitis at orifices and acra [3]. Oral zinc treatment improves this condition [4].

Several thousand years ago, the ancient Egyptians were already exploiting the therapeutic benefits of topically applied zinc compounds for treating various skin conditions.

In the human genome, the prevalence of genes encoding for zinc proteins (over 3% of the 32,000 identified genes) is very large [5], as well as one of the largest classes of eukaryotic proteins, called zinc finger proteins [6]. They are a family of more than 2,000 transcription factors that bind specifically to DNA and activate transcription of growth factors [7–9]; moreover, they are regulators of adult hematopoietic stem cells [10]. Two of the most important zinc finger proteins relevant to skin are DNA and RNA polymerases, which catalyze the replication of DNA and its transcription [11].

Zinc biomolecules include over 300 characterized enzymes. The catalysts that have specific relevance to skin are: (1) matrix metalloproteinases (MMPs), which hydrolyze nearly all of the structural proteins of the extracellular matrix (ECM) such as collagen, elastin, and gelatin; (2) superoxide dismutase (SOD), with a relevant antioxidant activity; and (3) alkaline phosphatase, which is involved in adenosinemonophosphate (AMP) metabolism, which in turn plays a role in suppressing the inflammatory process.

Another protein with structural function relevant to skin is metallothionein (MT), which, besides its weak antioxidant properties, plays a primary role in the storage and possibly in the transport of zinc. Metallothioneins (MTs) complex up to 20% of intracellular zinc [12]. These ubiquitous, cysteine-rich low-molecular-weight proteins regulate the intracellular supply of zinc to enzymes, gene-regulatory molecules, and zinc depots, and protect cells from deleterious effects of exposure to elevated levels of zinc. One MT molecule can bind seven zinc ions.

Another zinc-binding protein quite recently discovered is psoriasine, which is known to strongly bind to epidermal fatty acid binding protein. Psoriasine is the principal and preferentially *E.coli*–killing antimicrobial component of healthy skin. Psoriasine is focally expressed and released from keratinocytes, particularly in areas that have a high bacterial colonization. In addition to keratinocytes, sebocytes also show expression of psoriasine. It was found that psoriasine is induced by interleukin 1-beta or tumor necrosis factor in primary keratinocytes. Structural investigations have clearly shown that psoriasine binds to zinc; the absence of antibacterial activity in zinc-loaded psoriasine suggests that the protein probably kills *E.coli* by zinc sequestration [13].

7.2.1 Cellullar Zinc Homeostasis

Advances in molecular genetics and zinc-specific fluorescent probes have unraveled many of the mechanisms responsible for zinc uptake, intracellular distribution, and elimination [14]. Within cells, 30–40% of the zinc is bound to proteins in the nucleus, 50% is located in the cytoplasm, and the remainder is in plasma membranes [15].

- *Zinc absorption and efflux.* The ZIP family of membranous transporter proteins is mainly involved in cellular zinc uptake, whereas the ZnT family mediates zinc efflux [16]. Energy sources for the zinc channels are elusive, and symport as well as antiport mechanisms have been proposed [17]. Members of the ZIP family consist of 220–650 amino-acyl residues with eight putative membrane-spanning domains. ZIP-1 is the major uptake system in human tissues [18] and is expressed in the small intestine, epidermis, and keratinocytes [19]. The ZnT proteins comprise six putative transmembrane domains with a histidine-rich loop [20]. ZnT-1 is found in plasma membranes and catalyzes efflux from the cytoplasm into the extracellular medium, preventing excessive intracellular zinc concentrations. ZnT-2 translocates zinc ions to an acidic endosomal compartment, another mechanism for cytoprotection. ZnT-4 is constitutively expressed in the mammary gland epithelium, where it controls secretion of zinc into breast milk [21]. Expression of MTs and zinc transporters is transcriptionally regulated by metal-responsive transcription factor-1 that senses zinc levels [22].

7.2.2 Zinc and Calcium in Skin Biochemistry and Physiology

The zinc concentration in the epidermis (50–70 μg/g dry weight) is higher than in the dermis (10–5 μg/g dry weight) in human skin, perhaps reflecting the activity of zinc-dependent RNA and DNA polymerases in mitotically active basal cells [23]. Immunohistochemical and in situ hybridization localization studies on normal skin indicate high levels of MTs in the basal epidermis with reduced concentrations in postmitotic keratinocytes, reticuloendothelial cells, and fibroblasts [24].

There is a critical balance between zinc and calcium in basal cell mitosis and postmitotic functional maturation involving keratohyalin synthesis and keratinization in normal skin [25]. An inverse relationship is seen in the epidermis between the zinc concentration and the state of maturation and keratinization of postmitotic cells. Declining zinc gradients across the epidermis are the reverse of calcium gradients that increase from the basal layer to maximal concentrations in the granular cells [25]. Calcium-binding proteins like calmodulin hold key roles. Heng et al. [26] demonstrated reciprocity between tissue calmodulin and cAMP levels in epidermal cells, and showed that calmodulin levels decline significantly in the presence of excess zinc. Zinc also modulates the activity of calcium/calmodulin-dependent protein kinase II dose dependently [27].

The relative importance of zinc and calcium in cell proliferation and maturation is further illustrated in comparative studies of keratinizing epithelia [28]. In thin hairy skin, where mitosis is inversely proportional to the hair cover, zinc and calcium levels are appreciably lower than in pressure keratinization on the sole of the foot, where the robust epidermis with one to three basal layers is associated with a protracted keratinization and thick compacted stratum corneum. Higher levels of zinc in the sensory epithelia of the nasal mucosa and tongue are not only consistent with high mitotic activity, protracted zones of keratinization, and high levels of protein-bound phospholipids, but reflect the importance of zinc in taste and smell perception [29].

7.2.3 Physiological Role and Mechanism of Action of Zinc

- *Zinc and immune functions.* Zinc is required for a number of immune functions, including T-lymphocyte activity. Zinc deficiency results in thymic involution, depressed delayed hypersensitivity, decreased peripheral T-lymphocyte count, proliferative T-lymphocyte response to phytohemagglutinin, decreased cytotoxic T-lymphocyte activity, T helper lymphocyte function, natural killer cell activity, macrophage and neutrophil functions, and antibody production. Zinc supplementation can restore impaired immune function in patients with zinc deficiency [30], as found in malabsorption syndromes and acrodermatitis enteropathica; but there is little evidence that zinc supplementation will enhance immune responses in subjects who are not zinc-deficient. High doses of zinc may even be immunosuppressive [31]. The mechanism underlying the immune effects of zinc is not fully understood. Some of these effects may be accounted for the membrane-stabilization effect of the ion. This could affect signaling processes involved in cell-mediated immunity. Zinc may also influence gene expression by structural stabilization of different immunological transcription factors. In PBMCs, zinc induces cytokine production, including interleukin (IL)-1 (IL-6 and tumor necrosis factor [TNF]-alpha) [32]. In conclusion, T-lymphocyte activation appears to be delicately regulated by zinc concentrations.
- *Zinc and redox system.* Zinc may have secondary antioxidant activity but does not have redox activity under physiological conditions. It may influence membrane structure by its ability

to stabilize thiol groups and phospholipids. It may also occupy sites that might otherwise contain redox active metals such as iron. These effects may protect membranes against oxidative damage. As mentioned above, zinc plays a structural role in Cu/Zn-SOD. A possible mechanism to explain the antioxidant activity of zinc is its association with the metallothionein in competition with copper [33].

- *Zinc in wound repair and healing.* The requirement of skin for local high concentrations of zinc during wound healing has been demonstrated, or at least implied, in various ways. First, oral zinc supplementation has repeatedly been shown to enhance the rate of wound healing [34]. Second, in a rat wound model, local zinc levels are seen to increase after wounding [35]. Moreover, the expression of MT, which acts as a reservoir of zinc, is higher in healing wounds than in normal skin [36]. MT up-regulation can be induced in vivo by exposure to zinc [37]. Collectively, all of these observations argue that the requirement for zinc is higher in skin during the process of healing.

  Improvement in wound healing on exposure to zinc is not limited to oral supplementation: topically applied zinc compounds are also effective. For example, zinc oxide improves leg ulcers and has beneficial effects in occlusive dressings; it also increases the rate of re-epithelialization in a pig model. Other studies have shown that topically absorbed zinc from wounds promotes the early wound-healing phase and topical zinc oxide promotes cleansing of wounds [34].

7.2.3.1 Overview of the Wound Repair Process and Physiology

The repair of cutaneous injuries is a biologic process designed to regenerate a functional stratum corneum and restore the barrier properties of the skin. The process consists of a sequence of coordinated events that involve cell-cell and cell-matrix interactions that trigger numerous signaling pathways [38].

- *Zinc participation in wound repair.* Zinc and zinc-containing proteins are involved in nearly all stage of cutaneous wound repair; the ion has significant roles in the modification of ECM, cell migration, and protein synthesis and in the reduction of inflammation.

The MMP family of zinc-dependent proteins is actively involved in wound healing. Both the re-epithelialization and granulation tissue formation phases involve cell division and movement of new cells to close the wound. In addition, the initial, fibrous ECM formed for wound closure needs to be remodeled. Degradation of the matrix is carried out by MMPs that specifically digest matrix proteins, such as collagen, elastin, laminin, vitronectin, and fibronectin. Specific MMPs are secreted by keratinocytes, fibroblasts, and other cells in response to local growth factor signals. This degradation is a driving force behind tissue remodeling and facilitates the movement of new cells throughout the healing wound.

Additional zinc proteins critical for the repair process are the zinc finger transcription factors mentioned above, which influence gene expression and thereby regulate cellular behavior. Zinc finger transcription factors cooperate with DNA and RNA polymerases to initiate transcription of key genes involved in wound healing, for example, genes involved in cellular replication and genes encoding ECM proteins. Several zinc finger transcription factors have been identified in the skin, two of which are basonuclin and c-Krox. The former is present mainly in the nuclei of the basal cell layer in a variety of stratified squamous epithelia and it is also found in cultured human keratinocytes [39]. The expression of basonuclin is closely associated with cells that are actively dividing or that have the potential to divide, implying a role in the regulation of cell division. Unlike basonuclin, c-Krox is not involved in cell division but rather in controlling expression of genes encoding ECM proteins. C-Krox is known to coordinate the expression of the two collagen type I genes in skin and may also control expression of fibronectin and other matrix genes [40]. The synthesis of both collagen and fibronectin is central to the wound repair process.

In addition to the role that zinc-containing proteins or enzymes play in wound repair, zinc itself facilitates the process by other mechanisms of action (e.g., the expression of integrins, cell-surface proteins that mediate interactions between the cell and the various extracellular matrix proteins surrounding the cell). The effect of zinc was found to be particularly notable on those integrins affecting cellular mobility in the proliferative phase of wound healing. For a recent review on these topics, see Lansdown et al. [35].

- *Inflammation and zinc.* Inflammation is an important step in wound healing. There is a growing body of evidence that zinc has anti-inflammatory activity. Oral zinc therapy has been shown to be effective in treating the inflammatory conditions of acne, alopecia, rheumatoid arthritis, colitis and Crohn's disease, and psoriatic arthritis [41].

  If the injury is deep enough (dermis) to induce injury to blood vessels, blood clot formation is the first step in healing. Platelets become embedded in a network of cross-linked fibrin fibers designed to stop the loss of blood. The platelets also release a series of growth factors that send two types of messages: recruitment of inflammatory cells and initiation of cellular division.

  The two types of inflammatory cells recruited are macrophages and neutrophils, both of which are involved in cleansing the site of debris and bacteria via phagocytosis. These inflammatory cells bind to specific ECM proteins via integrins. This process of adhesion triggers the release of another group of growth factors that initiate the transition from an inflammatory state to a repair state (re-epithelialization) and that stimulate release or synthesis of materials with anti-inflammatory activity, for example, adenosine [42]. An example of a zinc-dependent enzyme involved in regulation of inflammation is alkaline phosphatase. The enzyme is released from the surface of epithelial cells and, as already previously indicated, dephosphorylates AMP to generate adenosine [43]. Adenosine has potent anti-inflammatory activity and participates in the curtailment of the inflammatory phase of wound healing.
- *Effect of zinc on immunologic cells and cytokines.* The effect of zinc on the different cells involved in inflammation—such as mast cells, platelets, macrophages, neutrophils, natural killer cells, and lymphocytes—is cell type specific; zinc can also influence the cytokine messengers that facilitate communication with these cells [44–47].

One example highlights the importance of zinc in dermatology from this perspective; zinc inhibits mast cell degranulation, thereby reducing the secretion of histamine [48], which is an important mediator of inflammatory response and an inducer of itch. This could represent the pharmacologic basis for the common preparation of calamine lotion (based on zinc oxide or zinc carbonate) to relieve itch.

7.2.3.2 Other Mechanisms behind the Anti-inflammatory Activity of Zinc

Besides its impact on immunologic cells and the messengers secreted by them, zinc has other effects that likely contribute to its anti-inflammatory activity. One such effect is the ability of zinc to reduce keratinocyte activation markers frequently observed in vivo [49]. In particular, zinc is able to reduce the expression of intercellular adhesion molecule 1 on the keratinocyte surface and reduce the secretion of tumor necrosis factor α from keratinocytes in response to various stimuli.

Another effect of zinc is its ability to inhibit the production of nitric oxide (NO) [50], which can interact with superoxide anion (O_2) and form peroxynitrite (ONOO); the last one is a cytotoxic agent causing tissue damage and resulting in skin inflammation [51]. Zinc may inhibit NO production by decreasing the expression of iNOS and/or by inhibiting iNOS activity via effects on calmodulin, which is a subunit-like component of iNOS.

In addition to reducing levels of peroxynitrite via its effects on iNOS, zinc also protects cells from other oxygen radicals by virtue of its antioxidant activity [52]. Zinc has been shown to reduce the cellular and genetic damage caused by exposure to ultraviolet light [53] and enhance resistance of skin fibroblasts to oxidative stress [54]. Although this effect could be due to the influence on zinc-containing enzymes and proteins in skin that have a direct role in scavenging reactive oxygen species, such as SOD and MT, zinc likely has a broader role. It has been proposed that zinc can displace other, more harmful, redox-active metal ions, such as iron and copper, from sites that could cause formation of oxygen-based free radicals [52].

In conclusion, the use of zinc in cosmetics is linked to its role in immune function, in redox system, in wound repairing and healing, and in inflammation.

7.3 Selenium and Skin Health

Selenium is a component of the unusual amino acids selenocysteine and selenomethionine. In humans, selenium is a trace element nutrient that functions as a cofactor for reduction of antioxidant enzymes such as glutathione peroxidases and certain forms of thioredoxin reductase.

Glutathione peroxidase (GSHPx) catalyzes reactions that remove reactive oxygen species such as peroxide:

$$2GSH + H_2O_2 \text{---------GSHPx} \rightarrow GSSG + 2H_2O$$

Selenium also plays a role in the functioning of the thyroid gland by participating as a cofactor for the three known thyroid hormone deiodinases.

7.3.1 Biochemical Functions

For a recent review of selenium metabolism, see Burk and Levandar [55] Twenty-five selenoprotein genes have been identified in the human genome by bioinformatics methods [56]. The selenoproteins that result from expression of these genes are responsible for the biochemical functions of selenium. Some of thc better known selenoproteins are discussed briefly below:

- *Glutathione peroxidases.* Five selenium containing GSHPxs, all separate gene products, have been identified in the human genome [56]. The cellular GSHPx, GSHPx-1, is the most abundant member of this family and is present in all cells. GSHPx-2 is found predominantly in tissues of the gastrointestinal track [57]. GSHPx-3 is present in plasma and milk [58] whereas GSHPx-4 (also known as phospholipid hydroperoxide, GSHPx) is present inside cells and catalyzes reduction of fatty acid hydroxide in the phospholipid cell membrane [59]. This enzyme has two special functions in the spermatozoan; it is a constituent of the mitochondrial capsule [60] and it is involved in chromatin packing in the sperm head [56]. Finally, GSHPx-6 is present in the olfactory apparatus but to date its function is not known [56].
- *Iodothyronine deiodinases.* These enzymes catalyze the deiodination of thyroxine and triiodothyronine and thereby regulate the concentration of the active thyroid hormone [61].

- *Thyoredoxin reductases.* These enzymes are flavin-containing selenoprotein, dependent on reduced NADP that in turn reduce the internal disulfide of thyoredoxin. Three isoforms have been identified; one is present in the cytosol, another in the mitochondrion, the third in the testis [62]. These reductases provide reducing equivalents to a variety of enzymes. Many of them are oxidant defense enzymes, but others operate in DNA synthesis and cell signaling.
- *Selenoprotein P.* This is an extracellular glycoprotein found in plasma and also associated with endothelial cells [63]. One of the functions of selenoprotein P, which contains a large fraction of the plasma selenium (about 45%), is to supply selenium to the brain in order to maintain normal neurological function and to the testis for spermatogenesis [64,65]. Selenoprotein P has been associated with the oxidant defense properties of selenium, and in particular it protects endothelial cells from reactive oxygen species generated by inflammation or by xenobiotic metabolism.

7.3.2 Selenium Supplementation and Skin Cell Well-Being

- *Selenium and acne.* Acne vulgaris is a distressing skin condition that can carry with it significant psychological disability. Patients with acne are more likely to experience anger and are at increased risk of depression, anxiety, and even suicidal ideation. Certain nutrients that have been implicated as influencing the pathophysiology of acne have also been identified as important mediators of human cognition, behavior, and emotions. Zinc, folic acid, selenium, chromium, and omega-3 fatty acids are all examples of nutrients that have been shown to influence depression, anger, and/or anxiety. In a large 2002 review, D. Benton identified at least five studies that show that low selenium intake is associated with lowered mood states [66]. In addition, selenium levels and selenium-dependent enzyme activity have also been shown to be significantly lower in those patients with acne [67].

  For this reason, the deficiency of the above-mentioned nutrients, along with systemic oxidative stress and an altered intestinal microflora, has been implicated in the pathogenesis of acne vulgaris. Probably certain nutritional factors, causing a weakened antioxidant defense system and altered intestinal microflora, may interplay to increase the risk of psychological sequelae in acne vulgaris [68]. Michaelssons et al. [69] have demonstrated that male acne patients have significantly lower GSH-Px levels than the controls. A good clinical result was obtained after supplementation of selenium and tocopheryl succinate, especially in patients with pustular acne and low GSH-Px; the beneficial effect was usually paralleled by a slow rise of the enzyme activity. However 6–8 weeks after withdrawal of the treatment, the GSH-Px value had returned to the pre-treatment level.

- *Selenium and seborrheic dermatitis.* Seborrheic dermatitis is a chronic inflammatory disorder affecting areas where sebaceous glands are most prominent. Yeasts of the *Malassentia* genus, as well as genetic, environmental, and general health factors, contribute to this disorder. Treatment options include application of antifungal preparations containing, for example, selenium sulphide [70].
- *Selenium protection from UVB radiation–induced cell death.* There is much evidence to suggest that selenium (Se) has an important role in protecting skin from the harmful effects of UVB. In humans, subnormal Se status is associated with up to a four-fold increased risk of developing skin cancer [71,72], and topical Se application, as selenomethionine, has been shown to protect subjects from acute skin damage following UVB exposure [73]. The results obtained by Rafferty et al. [74] show that preincubation of primary cultures of keratinocytes, melanocytes, or HaCaT cell lines with Se for 24 hours can provide significant protection from UVB-induced cell death. Selenite was more potent than seleniomethionine at conferring protection. It is widely accepted that Se exerts many of its biochemical actions through the expression of specific selenoproteins in which Se is inserted as specific selenocysteine residues encoded by a TGA triplet. This insertion requires Se to be presented in a chemically active form similar to selenide, and current evidence indicates that selenite is a more potent precursor of selenide than seleniomethionine is. As a consequence, Se seems to be acting through incorporation into selenoproteins rather than by a direct

antioxidant chemical action. In fact, the family of GPXs may mediate the protective effects of Se because these enzymes are capable of detoxifying hydrogen peroxide, lipid hydroxiperoxides, and phospholipid hydroxyperoxides that are produced during UV exposure. The previous findings of Rafferty et al. [74] were more recently confirmed by the same author [75]. Further evidence on the protection offered by selenium against ultraviolet radiation results from the paper of Duthie et al. [76], who ascribed the protective effect of selenium to an increased capacity of the immune system. Moreover, they reported that topical L-selenomethionine application leads to an increased minimal erythema dose. The antioxidant properties of Se-dependent enzymes were demonstrated in the author's laboratory in *Xenopus laevis* embryos [77] and in human blood [78] subjected to oxidative stress by radiation.

- *Selenium and cancer.* Several studies (e.g., the review of Young and Lee [79]) have suggested a link between cancer and Se deficiency. However, these links are still controversial, many for what skin cancers are concerned. After a multicentre double-blind, randomized, placebo-controlled cancer prevention trial, Se treatment did not protect against development of squamous cell carcinomas of the skin. However, results from secondary and point analyses support the hypothesis that supplemental Se may reduce the incidence of, and mortality from, carcinomas in several sites [80]. The Nutritional Prevention of Cancer Trial was a double-blind, randomized, placebo-controlled clinical trial designed to test whether selenium as selenized yeast (200 μg daily) could prevent nonmelanoma skin cancer among 1,312 patients from the eastern United States who had previously had this disease. Although results through the entire blinded period continued to show that selenium supplementation was not statistically significantly associated with the risk of basal cell carcinoma, selenium supplementation was associated with statistically significantly elevated risk of squamous cell carcinoma and of total nonmelanoma skin cancer [81]. The hypothesis of a link between Se and cancer is based on the antioxidant capacity or enhancing immuno-activity of Se. However, not all the studies agree on the cancer fighting effects of Se. One study of naturally occurring levels of Se in over 60,000 participants did not show a significant correlation between those levels and cancer [82]. However, the SU.VI. MAX study [83] concluded that low-dose supplementation with 120 mg of ascorbic acid, 30 mg of vitamin E, 6 mg of beta carotene, 100 μg of selenium, and 20 mg of zinc resulted in a 31% reduction in the incidence of cancer and a 37% reduction in all cause mortality in males, but it did not get a significant result for females.

In conclusion, the possible functional use of Se in cosmetics is linked to its role in preventing oxidative stress due to free radical molecules.

Finally, it must be pointed out that a government report has revealed that most people in the U.K. are not getting sufficient amounts of the trace element selenium in their diet. According to the MAFF report, Food Surveillance Information Sheet Number 51, the average consumption of Se is currently only 34 μg per day, with falls well short of the RNI (References Nutrient Intake, formerly Recommended Daily Allowance) set by the government's COMA report in 1991, which stated that the desired level for men is 75 μg and 60 μg for women.

References

1. Molokhia M.H., Portnoy B. Neutron activation analysis of trace elements in the skin. (III) Zinc in normal skin. *Br. J. Dermatol.* 1969; 81:759–62.
2. Tasaki M., Hanada K., Hashimoto I. Analyses of serum copper and zinc levels and copper/zinc ratios in skin diseases. *J. Dermatol.* 1993; 20:21–4.
3. Kruse-Jarres J. Pathogenesis and symptoms of zinc deficiency. *Am. Clin. Lab.* 2001; 20:17–22.
4. Neldner K.H., Hambridge K.M. Zinc therapy of acrodermatitis enteropathica. *N. Engl. J. Med.* 1975; 292:879–82.
5. Vallee B.L. Zinc in biology and chemistry. In: *Zinc Enzymes*. Spiro T.G. (ed.). Allelbourne, FL: Krieger, 1991; 1–24.
6. Rhodes D., Klug A. Zinc fingers in *Caenorhabditis elegans*: finding families and probing pathways. *Sci. Am.* 1993; 286:56–65.
7. Bao B., Prasad A.S., Beck F.W., Godmere M. Zinc modulates mRNA levels of cytokines. *Am. J. Physiol. Endocrinol. Metab.* 2003; 285:E1095–1102.
8. Sum E.Y., O'Reilly L.A., Jonas N., Lindeman G.J., Visvader J.E. The LIM domain protein Lmo4 is highly expressed in proliferating mouse epithelial tissues. *J. Histochem. Cytochem.* 2005; 53:475–86.
9. Zhu C.H., Ying D.J., Mi J.H., Zhang W., Dong S.W., Sun J.S., Zhang J.P. The zinc finger protein A20 protects endothelial cells from burns serum injury. *Burns* 2004; 30:127–33.

10. Hock H., Orkin S.H. Zinc-finger transcription factor Gfi-1: versatile regulator of lymphocytes, neutrophils and hematopoietic stem cells. *Curr. Opin. Hematol.* 2006; 13:1–6.
11. Slater J.P., Mildvan A.S., Loeb L.A. Zinc in DNA polymerases. *Biochem. Biophys. Res. Commun.* 1971; 44:37–43.
12. Vasak M. Advances in metallothionein structure and functions. *J. Trace Elem. Med. Biol.* 2005; 19:13–17.
13. Glaser R., Harder J., Lange H., Bartels J., Christophers E., Schroeder J.M. Antimicrobial psoriasine (S100A7) protects human skin from *Escherichia coli* infection. *Nature Immunology* 2005; 6:57–64.
14. Beyersmann D., Haase H. Functions of zinc in signaling, proliferation and differentiation of mammalian cells. *Biometals* 2001; 14:331–41.
15. Vallee B.L., Falchuk K.H. The biochemical basis of zinc physiology. *Physiol. Rev.* 1993; 73:79–118.
16. Chimienti F., Aouffen M., Favier A., Seve M. Zinc homeostasis-regulating proteins: new drug targets for triggering cell fate. *Curr. Drug Targets* 2003; 4:323–38.
17. Chao Y., Fu D. Kinetic study of the antiport mechanism of an *Escherichia coli* zinc transporter, ZitB. *J. Biol. Chem.* 2004; 279:12043–50.
18. Gaither L.A., Eide D.J. The human ZIP1 transporter mediates zinc uptake in human K562 erythroleukemia cells. *J. Biol. Chem.* 2001; 276:22258–64.
19. Lioumi M., Ferguson C.A., Sharpe P.T., Freeman T., Marenholz I., Mischke D., Heizmann C., Ragoussis J. Isolation and characterization of human and mouse ZIRTL, a member of the IRT1 family of transporters, mapping within the epidermal differentiation complex. *Genomics* 1999; 62:272–80.
20. Zalewski P.D., Truong-Tran A.Q., Grosser D., Jayaram L., Murgia C., Ruffin R.E. Zinc metabolism in airway epithelium and airway inflammation: basic mechanisms and clinical targets. A review. *Pharmacol. Ther.* 2005; 105:127–49.
21. Huang L., Gitschier J. A novel gene involved in zinc transport is deficient in the lethal milk mouse. *Nat. Genet.* 1997; 17:292–7.
22. Andrews G.K. Cellular zinc sensors: MTF-1 regulation of gene expression. *Biometals* 2001; 14:223–37.
23. Ågren M.S. Percutaneous absorption of zinc from zinc oxide applied topically to intact skin in man. *Dermatologica* 1990; 180:36–9.
24. Lansdown A.B.G. Metallothioneins: potential therapeutic aids for wound healing in the skin. *Wound Repair Regen.* 2002; 10:130–2.
25. Lansdown A.B.G. Calcium: a potential central regulator in wound healing in the skin. *Wound Repair Regen.* 2002; 10:271–85.
26. Heng M.K., Song M.K., Heng M.C. Reciprocity between tissue calmodulin and cAMP levels: modulation by excess zinc. *Br. J. Dermatol.* 1993; 129:280–5.
27. Lengyel I., Fieuw-Makaroff S., Hall A.L., Sim A.T., Rostas J.A., Dunkley P.R. Modulation of the phosphorylation and activity of calcium/calmodulin-dependent protein kinase II by zinc. *J. Neurochem.* 2000; 75:594–605.
28. Lansdown A.B.G. Morphological variations in keratinising epithelia in the beagle. *Vet. Rec.* 1985; 116:127–30.

29. Henkin R.I., Schecter P.J., Friedewald W.T., Demets D.L., Raff M. A double blind study of the effects of zinc sulfate on taste and smell dysfunction. *Am. J. Med. Sci.* 1976; 272:285–99.
30. Duchateau J., Delepesse G., Vrijens R., Collet H. Beneficial effects of oral zinc supplementation on the immune response of old people. *Am. J. Med.* 1981; 70:1001–4.
31. Chandra R.K. Excessive intake of zinc impairs immune response. *JAMA* 1984; 252:1443–6.
32. Wintergerst E.S., Maggini S., Hornig D.H. Contribution of selected vitamins and trace elements to immune function. *Ann. Nutr. Metab.* 2007; 51:301–23.
33. Valko M., Morris H., Cronin M.T. Metals, toxicity and oxidative stress. *Curr Med Chem.* 2005; 12(10):1161–208.
34. Lansdown AB, Mirastschijski U, Stubbs N, Scanlon E, Agren MS. Zinc in wound healing: theoretical, experimental, and clinical aspects. *Wound Repair Regen.* 2007; 15:2–16.
35. Lansdown A.B.G., Sampson B., Rowe A. Sequential changes in trace metal, metallothionein and calmodulin concentrations in healing skin wounds. *J. Anat.* 1999; 195:375–86.
36. Iwata M., Iwata Takebayashi T., Ohta H., Alcade R.E., Itano Y., Matsumura T. Zinc accumulation and metallothionein gene expression in the proliferating epidermis during wound healing in mouse skin. *Histochem. Cell Biol.* 1999; 112:283–90.
37. Andrews G.K. Regulation of metallothionein gene expression by oxidative stress and metal ions. *Biochem Pharmacol.* 2000; 59:95–104.
38. Turchi L., Chassot A.A., Rezzonica R., Yeow K., Loubat A., Ferrua B., Lenegrate G., Ortonne J.P., Ponzio G. Dynamic characterization of the molecular events during in vitro epidermal wound healing. *J. Invest. Dermatol.* 2002; 119:56–63.
39. Vanhoutteghem A., Djian P. Basonuclin 2: an extremely conserved homolog of the zinc finger protein basonuclin. *Proc. Natl. Acad. Sci. USA.* 2004; 101:3468–73.
40. Kypriotou M., Beauchef G., Chadjichristos C., Widom R., Renard E., Jimenez S.A., Korn J., Maquart F.X., Oddos T., Von Stetten O., Pujol J.P., Galéra P. Human collagen Krox up-regulates type I collagen expression in normal and scleroderma fibroblasts through interaction with Sp1 and Sp3 transcription factors. *J. Biol. Chem.* 2007; 282:32000–14.
41. Schwartz J.R., Marsh R.G., Draelos Z.D. Zinc and skin health: overview of physiology and pharmacology. *Dermatol. Surg.* 2005; 31:837–47.
42. Cronstein B.M. Adenosine receptors and wound healing, revised. *ScientificWorldJournal.* 2006; 6:984–91.
43. Gallo R.L., Dorschner R.A., Takashima S., Klagsbrun M., Eriksson E., Bernfield M. Endothelial cell surface alkaline phosphatase activity is induced by IL-6 released during wound repair. *J. Invest. Dermatol.* 1997; 109:597–603.
44. Salgueiro M., Zubillaga M., Lysionek A., Cremaschi G., Goldman C.G., Caro R., DePaoli T., Hager A., Weill R., Boccio J. Zinc status and immune system relationship: a review. *Biol. Trace Elem. Res.* 2000; 76:193–205.

45. Rink L., Haase H. Zinc homeostasis and immunity. *Trends Immunol.* 2007; 28:1–4.
46. Wellinghausen N., Kirchner H., Rink L. The immunobiology of zinc. *Immunol. Today* 1997; 18:519–21.
47. Overbeck S, Rink L, Haase H. Modulating the immune response by oral zinc supplementation: a single approach for multiple diseases. *Arch. Immunol. Ther. Exp. (Warsz).* 2008; 56:15–30.
48. Fivenson D.P. The mechanisms of action of nicotinamide and zinc in inflammatory skin disease. *Cutis.* 2006; 77:5–10.
49. Guéniche A., Viac J., Lizard G., Charveron M., Schmitt D. Protective effect of zinc on keratinocyte activation markers induced by interferon or nickel. *Acta Derm. Venereol.* 1995; 75:19–23.
50. Yamaoka J., Kume J.T., Akaike A., Miyachi Y. Suppressive effect of zinc ion on iNOS expression induced by interferon-gamma or tumor necrosis factor-alpha in murine keratinocytes. *J. Dermatol. Sci.* 2000; 23:27–35.
51. Cui L., Okada A. Nitric oxide and manifestations of lesions of skin and gastrointestinal tract in zinc deficiency. *Curr. Opin. Clin. Nutr. Metab. Care* 2000; 3:247–52.
52. Rostan E., DeBuus H.V., Madey D.L., Pinnell S.R. Evidence supporting zinc as an important antioxidant for skin. *Int. J. Dermatol.* 2002; 41:606–11.
53. Record I.R., Jannes M., Dreosti I.E. Protection by zinc against UVA- and UVB-induced cellular and genomic damage in vivo and in vitro. *Biol. Trace Elem. Res.* 1996; 53:19–25.
54. Richard M.J., Guiraud P., Leccia M.T., Beani J.C., Favier A. Effect of zinc supplementation on resistance of cultured human skin fibroblasts toward oxidant stress. *Biol. Trace Elem. Res.* 1993; 37:187–99.
55. Burk R.F., Levander O.A. Selenium. In: *Modern Nutrition in Health and Disease*, 10th Edition. Shils M.E., et al. (eds.). Lipincott, Williams and Wilkins, 2006; 312–25.
56. Kryukov G.V., Castellano S., Novoselov S.V., Lobanov A.V., Zehtab O., Guigó R., Gladyshev V.N. Characterization of mammalian selenoproteomes. *Science* 2003; 300:1439–43.
57. Chu F.F., Doroshow J.H., Esworthy R.S. Expression, characterization, and tissue distribution of a new cellular selenium-dependent glutathione peroxidase, GSHPx-GI. *J. Biol. Chem.* 1993; 268:2571–6.
58. Avissar N., Slemmon J.R., Palmer I.S., Cohen H.J. Partial sequence of human plasma glutathione peroxidase and immunologic identification of milk glutathione peroxidase as the plasma enzyme. *J. Nutr.* 1991; 121(8): 1243–9.
59. Ursini F., Maiorino M., Brigelius-Flohé R., Aumann K.D., Roveri A., Schomburg D., Flohé L. Diversity of glutathione peroxidases. *Methods Enzymol.* 1995; 252:38–53.
60. Ursini F, Heim S, Kiess M, Maiorino M, Roveri A, Wissing J, Flohé L. Dual function of the selenoprotein PHGPx during sperm maturation. *Science* 1999; 285:1393–6.
61. Pfeifer H., Conrad M., Roethlein D., Kyriakopoulos A., Brielmeier M., Bornkamm G.W., Behne D. Identification of a specific sperm nuclei selenoenzyme necessary for protamine thiol cross-linking during sperm maturation. *FASEB J.* 2001; 15:1236–8.

62. Sun Q.A., Zappacosta F., Factor V.M., Wirth P.J., Hatfield D.L., Gladyshev V.N. Heterogeneity within animal thioredoxin reductases. Evidence for alternative first exon splicing. *J. Biol. Chem.* 2001; 276:3106–14.
63. Burk R.F., Hill K.E., Boeglin M.E., Ebner F.F., Chittum H.S. Selenoprotein P associates with endothelial cells in rat tissues. *Histochem. Cell Biol.* 1997; 108:11–15.
64. Hill K.E., Zhou J., McMahan W.J., Motley A.K., Atkins J.F., Gesteland R.F., Burk R.F. Deletion of selenoprotein P alters distribution of selenium in the mouse. *J. Biol. Chem.* 2003; 278:13640–6.
65. Schomburg L., Schweizer U., Holtmann B., Flohé L., Sendtner M., Köhrle J. Gene disruption discloses role of selenoprotein P in selenium delivery to target tissues. *Biochem. J.* 2003; 370:397–402.
66. Benton D. Selenium intake, mood and other aspects of psychological functioning. *Nutr. Neurosci.* 2002; 5:363–74.
67. Michaelsson G. Decreased concentration on selenium in whole blood and plasma acne vulgaris. *Acta Derm. Venereol.* 1990; 70:92.
68. Katzman M., Logan A.D. Acne vulgaris: nutritional factors may be influencing psychological sequelae. *Medical Hypotheses* 2007; 69:1080–4.
69. Michaelssons G., Edqvist L.E. Erythrocyte glutathione peroxidase activity in acne vulgaris and the effect of selenium and vitamin E treatment. *Acta Derm. Venereol.* 1984; 64:9–14.
70. Johnson B.A., Nunley J.R. Treatment of seborrheic dermatitis. *Am. Fam. Physichan* 2000; 61:2703-10, 2713–14.
71. Clark L.C., Graham G.F., Crounse R.G., Grimson R., Hulka B., Shy C.M. Plasma selenium and skin neoplasms: a case-control study. *Nutr. Cancer* 1984; 6:13–21.
72. Reid M.E., Duffield-Lillico A.J., Slate E., Natarajan N., Turnbull B., Jacobs E., Combs G.F. Jr., Alberts D.S., Clark L.C., Marshall J.R. The nutritional prevention of cancer: 400 mcg per day selenium treatment. *Nutr. Cancer* 2008; 60:155–63.
73. Burke K.E., Burford R.G., Combs G.F. Jr., French I.W., Skeffington D.R. The effect of topical L-selenomethionine on minimal erythema dose of ultraviolet irradiation in humans. *Photodermatol. Photoimmunol. Photomed.* 1992; 9:52–7.
74. Rafferty T.S., McKenzie R.C., Hunter J.A., Howie A.F., Arthur J.R., Nicol F., Beckett G.J. Differential expression of selenoproteins by human skin cells and protection by selenium from UVB-radiation-induced cell death. *Biochem. J.* 1998; 332:231–6.
75. Rafferty T.S., Beckett G.J., Walker C., Bisset Y.C., McKenzie R.C. Selenium protects primary human keratinocytes from apoptosis induced by exposure to ultraviolet radiation. *Clin. Exp. Dermatol.* 2003; 28:294–300.
76. Duthie M.S., Kimber I., Norval M. The effects of ultraviolet radiation on the human immune system. *Br. J. Dermatol.* 1999; 140:995–1009.
77. Rizzo A.M., Rossi F., Zava S., Montorfano G., Adorni L., Cotronei V., Zanini A., Berra B. Antioxidant metabolism in *Xenopous laevis* embryos is affected by stratospheric balloon flight. *Cell. Biol. Inter.* 2007; 31:716–23.
78. Rizzo A.M., Adorni L., Montorfano G., Negroni M., Zava S., Berra B. Blood and oxidative stress (BOS): Soyuz mission "Eneide". *Microgravity Sci. Technol.* 2007; 19:210–14.

79. Young K.J., Lee P.N. Intervention studies on cancer. *Eur. J. Cancer. Prev.* 1999; 8:91–103.
80. Clark LC., Combs G.F. Jr., Turnbull B.W., Slate E.H., Chalker D.K., Chow J., Davis L.S., Glover R.A., Graham G.F., Gross E.G., Krongrad A., Lesher J.L. Jr., Park H.K., Sanders B.B. Jr, Smith C.L., Taylor J.R. Effects of selenium supplementation for cancer prevention in patients with carcinoma of the skin. A randomized controlled trial. Nutritional Prevention of Cancer Study Group. *JAMA* 1996; 276:1957–63.
81. Duffield-Lillico A.J., Slate E.H., Reid M.E., Turnbull B.W., Wilkins P.A., Combs G.F. Jr., Park H.K., Gross E.G., Graham G.F., Stratton M.S.,. Marshall J.R., Clark L.C. Selenium supplementation and secondary prevention of nonmelanoma skin cancer in a randomized trial. *JNCI Journal of the National Cancer Institute* 2003; 95:1477–81.
82. Garland M., Morris J.S., Stampfer M.J., Colditz G.A., Spate V.L., Baskett C.K., Rosner B., Speizer F.E., Willett W.C., Hunter D.J. Prospective study of toenail selenium levels and cancer among women. *J. Natl. Cancer. Inst.* 1995; 87:497–505.
83. Hercberg S., Galan P., Preziosi P., Roussel A.M., Arnaud J., Richard M.J., Malvy D., Paul-Dauphin A., Briançon S., Favier A. Background and rationale behind the SU.VI.MAX Study, a prevention trial using nutritional doses of a combination of antioxidant vitamins and minerals to reduce cardiovascular diseases and cancers. SUpplementation en VItamines et Minéraux AntioXydants Study. *Int. J. Vitam. Nutr. Res.* 1998; 68:3–20.

PART 4
PROTECT YOUR SKIN WITH NATURAL ANTIOXIDANTS

8

Botanical Antioxidants for Protection Against Damage from Sunlight

Mohammad Abu Zaid, PhD, Farrukh Afaq, PhD, Deeba N. Syed, MBBS, and Hasan Mukhtar, PhD*

Department of Dermatology, University of Wisconsin, Madison, WI, USA

Aaron Tabor and Robert M. Blair (eds.), *Nutritional Cosmetics: Beauty from Within*, 159–183, © 2009 William Andrew Inc.

Abstract

Solar ultraviolet (UV) radiation is considered a complete carcinogen, because it initiates a photo-oxidative reaction that impairs the antioxidant status and increases the cellular level of reactive oxygen species (ROS) in the skin. To counteract this, the skin is endowed with efficient antioxidant defense mechanisms, but when the generation of ROS overwhelms its defense capacity, the ability of the skin to protect itself from the damaging effects of ROS is impaired. Excessive exposure to solar UV radiation, particularly its UVB component, leads to the development of various skin disorders including erythema, edema, inflammation, hyperpigmentation, hyperplasia, immunosuppression, skin cancers, and photoaging. Therefore, additional efforts are needed to protect the skin from the deleterious effects of UV exposure. One such approach is to make antioxidants available to the skin at the time of exposure, through topical or oral administration of botanicals through a process that we define as photochemoprevention. Here, we focus on the effects of selected botanical antioxidants for protection against damage to the skin caused by sunlight.

8.1 Introduction

Solar ultraviolet (UV), particularly its UVB component, is responsible for more than 1.2 million cutaneous malignancies diagnosed each year in the U.S. alone [1]. UV radiation causes multifaceted damage to the skin and its adjacent tissues, and it is one of the leading causes of premature skin aging, immunosuppression, and carcinogenesis [2–4]. According to the International Commission on Illumination, UV radiation is divided into three categories: UVA or long wavelength UV (320–400 nm), UVB or medium wavelength UV (280–320 nm), and UVC or short wavelength UV (100–280 nm). UVA constitutes about 90–95% of solar radiation reaching the earth. UVA, due to its longer wavelength, has high penetrating power, and reaches deep into the epidermis and dermis of the skin. Intense exposure of the skin to UVA burns sensitive skin and damages the underlying structures, causing premature aging [5]. UVA exposure also leads to the generation of singlet oxygen, hydrogen peroxide, and hydroxyl free radicals, causing damage to cellular proteins, lipids, and DNA [6,7]. UVB, in contrast, constitutes only about 4–5% of UV radiation but is thought to be the most active constituent of solar radiation reaching the earth. However, even though UVB is more genotoxic and capable of causing much more cell damage than UVA, it has less penetrating power than UVA and acts mainly on the epidermal basal layer of the skin [7]. UVB induces direct adverse biological effects including DNA damage, oxidative stress, free radical production, photoaging, and skin cancer [7]. At the other end of the spectrum, UVC is extremely damaging to the skin, even with a very short exposure. Fortunately, UVC is prevented from reaching the earth, because it is almost completely absorbed by the molecular oxygen and ozone present in the earth's troposphere. Nevertheless, during the past few decades, depletion of the stratospheric ozone layer has led to an increase in the amount of solar UV radiation reaching the earth. This has led to a significant increase in the amount of UV radiation that people receive. In addition, compounded by changes in lifestyle related to excessive outdoor activities, UV radiation has consequently led to an increase in the incidence of skin-related disorders.

Skin, the largest organ of the body in terms of surface area, serves as a competent epithelial barrier that interfaces the environment. The major role of the skin is to provide a protective covering at this crucial interface through a variety of passive and active features. Exposure of the skin to UV radiation initiates a photo-oxidative reaction, which impairs the antioxidant status and increases the cellular level of reactive oxygen species (ROS). This overwhelms the defense capacity of the skin, thereby impairing the ability of skin to protect itself from the damaging effects of UV, resulting in increased oxidative stress. The induction of oxidative stress and subsequent imbalance of the antioxidant defense system results in damage to the cutaneous tissues and has been associated with the onset of several disease states. Therefore, additional efforts are needed to protect the skin against the deleterious effects of UV radiation. One such approach to prevent the occurrence of skin damage is to enhance endogenous photoprotection through topical or oral administration of botanical antioxidants that possess photoprotective properties.

8.2 Photochemoprevention by Botanical Antioxidant

Antioxidants from natural sources may provide new possibilities for the treatment and/or prevention of oxidative stress-mediated diseases [1]. Botanical supplements, specifically dietary botanicals, possessing anti-inflammatory, immunomodulatory, and antioxidant properties are among the promising group of compounds that can be exploited as ideal chemopreventive agents for a variety of skin disorders. Studies have demonstrated that the compounds present in the botanicals can attenuate the damages induced by UV light [8,9]. Botanical compounds typically have a broad spectrum of activities and work against the harmful effects of UV radiation via a number of ways: (i) decreasing UV-induced sunburn and inflammation, (ii) scavenging free radicals and ROS that are harmful to the skin, and (iii) modulating signaling pathways altered as a consequence of UV exposure.

For a variety of reasons, botanical antioxidants are gaining popularity as more and more skin care products containing botanical ingredients are introduced in the market. Individuals can modify their dietary habits and lifestyles in combination with a careful use of skin care products, because exposure to UV radiation is difficult to avoid. It should be emphasized that the approach of using botanical antioxidants could be an add-on to the existing strategies of preventing damage from excessive exposure to sun. Topically, botanical antioxidants can be used as sunscreens to augment the level of photoprotection without causing any skin sensitization. Endogenous photoprotection as nutritional/dietary supplement can add to the body's antioxidant reservoir and protect against oxidative stress and inflammation. In addition, botanical antioxidants are also involved in repair of structural cutaneous photodamage by stimulating collagen synthesis and elastin formation, along with a decrease in the breakdown of structural components [10]. Here, we focus on the effects of selected botanical antioxidants gaining attention these days for their photochemopreventive effect against UV-induced damages (Table 8.1).

8.2.1 Green Tea/EGCG

Tea is a popular beverage consumed by over two-thirds of the world population across all continents. The major commercial types of tea are green, black, and oolong tea, depending on how the young leaves of plant *Camellia sinesis* of theaceae family are dried, fermented, processed, and produced. Green tea is not fermented but steamed and dried, whereas oolong tea is partially fermented before drying and black tea is fully fermented. Of the total tea consumed in the world 78% is black tea, 20% is green tea, and 2% is oolong tea. Black tea is primarily consumed in Western and a few Asian countries; green tea is consumed in Japan, China, and some North African and Middle East countries, whereas oolong tea consumption is limited to southeastern China and Taiwan [11].

Tea contains polyphenolic compounds of which catechins account for 16–30% of its dry weight. During the fermentation process enzymes from

Table 8.1 Phytochemicals for Protection Against UVB-Mediated Damages*

Phytochemicals	Source	Photochemopreventive Action	Reference
Green tea	Leaves of *Camellia sinesis*	Inhibits UV-induced, pro-antioxidant enzymes and depletion of antioxidant	[15]
		Induces phase II antioxidant enzymes; scavenges ROS, H_2O_2, and nitric oxide; and decreases lipid peroxidation	[14,21]
		Inhibits UV-induced sunburns, leukocyte infiltration, dermo-epidermal activity, collagen changes and elastic fiber pathologies	[20]
		Prevents immunosuppression, infiltrates leukocytes, induces IL-12, downregulates IL-10	[23,24]
		Decreases erythema, bi-fold thickness, edema	[28]
		Modulates NF-κB, MAPKs, AP-1 and PI3K	[24–26]
		Prevents tumorigenesis and photocarcinogenesis	[9,31]
		Prevents activation of MMPs, degradation of ECM, and photoaging	[28,38]
		Inhibits UV-induced pigmentation	[39]
Pomegranate	Peel, seed, juice of fruit of *Punica granatum*	Possesses strong antioxidant activity	[37]
		Shows antiproliferative, antiatherogenic, anti-inflamatory, and antitumorigenic properties	[38]
		Inhibits UVB-induced depletion of GSH and lipid peroxidation	[39]
		Modulates MAPK, STAT-3, AKT, AP-1 and NF-κB pathways	[7,38,40]
		Prevents sunburns, DNA damage, protein oxidation, PCNA and various MMPs	[42]
		Delays photoaging and prevents tumorigenesis	[39,41]
		Inhibits UV-induced pigmentation	[43]

(Continued)

Table 8.1 Phytochemicals for Protection Against UVB-Mediated Damages* (*Continued*)

Phytochemicals	Source	Photochemopreventive Action	Reference
Curcumin	Turmeric (rhizome of *Curcuma longa)*	Shows antioxidant, anti-inflammatory, anticancer, antimicrobial, and wound-healing properties	[44]
		Inhibits MMP-2 activity and prevents breakdown of collagen	[50]
		Protects from radiation-induced DNA damage and lipid peroxidation	[51]
		Inhibits activation of NF-κB and AP-1 and downregulates IL-6, IL-1β and TNFα	[52,53,56]
		Enhances GSH and glutathione-S-peroxidase activity, inhibits lipid peroxidation, and metabolizes arachidonic acid	[54]
Resveratrol	Grapes, fruits, nuts, mulberries, red wine	Induces phase II drug–metabolizing enzymes, inhibits COX and hydroperoxidase	[58]
		Delays tumor initiation, promotion, and progression of cancer	[60]
		Blocks UVB-induced NF-κB pathway and prevents photocarcinogenesis	[61]
		Inhibits UVB-mediated bi-fold skin thickness, edema, hyperplasia, COX-2, ODC, lipid peroxidation, cell proliferation, and tumorigenesis	[68–70]
		Downregulates PCNA; cdk-2, -4 and -6; and cyclin-D1 and -D2	[65]
Genistein	Soybean, red clover	Possesses strong antioxidant and anticarcinogenic effects	[66]
		Augments induction of c-jun, jun B, STAT1-binding activity	[68,69]
		Inhibits UV-induced COX-2 and PGE2 synthesis	[70,71]
		Decreases UV-induced cyclobutane pyrimidine dimers, 8OHdG, H_2O_2, and MDA formation	[73,74]
		Inhibits PKT, topoisomerase II, MMP-9, and VEGF, and inhibits cell proliferation with cell growth arrest at G2/M phase	[75]

Apigenin	Endive, clove, apple, cherries, grapes, celery, beans, broccoli, tomatoes, onions, leeks, barley, parsley, tea, wine	Acts as free-radical scavenger, exhibits anti-inflammatory and anticarcinogenic properties	[76]
		Inhibits UVB-induced COX-2 protein expression	[78]
		Scavenges UVA-induced generation of ROS and inhibits collagenase activity	[79]
		Scavenges UVA-induced generation of ROS and inhibits collagenase activity	[81,87]
Carotenoids (β-carotene, lycopene)	Carrot, tomatoes, watermelon, grapefruit, apricots	Protects cells against oxidative damage, acts as scavenger of peroxy radicals	[82,83]
		Protects protein, DNA, lipids, and low-density lipoproteins from solar-induced damages	[84]
		Decreases UV-induced ODC activity and development of melanoma	[85]
Quercetin	Apple, grapes, olives, lemons, tomatoes, onions, broccoli, tea, red wine	Scavenges ROS, inhibits collagenase MMP activity, and prevents photoaging	[79]
		Decreases UVA-mediated increase in 8-OHdG and IL-1α	[86]
		Modulates UVB-induced myeloperoxidase activity and GSH depletion	[87]
		Protects the activity of antioxidant enzymes: catalase, SOD, glutathione peroxidase, and glutathione reductase	[89]
		Inhibits UVB-induced immunosuppression	[90]
Carnosic acid	Rosemary, sage	Suppresses UVA-induced MMP activity and prevents photoaging and photocarcinogenesis	[85]
		Reduces UV-induced erythema	[91]
Caffeic acid & ferullic acid	Grains, vegetables, fruits	Prevents propagation of lipid peroxidative chain reaction	[85]
		Inhibits UVB-induced erythema and protects membrane phospholipids from UV damage	[92]

*This is a representative list and by no means is complete.

tea leaves such as polyphenol oxidase react with polyphenolic tannins and catechins and convert polyphenols to phlobaphanes and form aromatic compounds that give aroma to the beverage. In green tea, leaves are not allowed to oxidize by fermentation and are steamed, thereby inactivating the enzymes and preserving about 90% of the polyphenols from being degraded. The major catechins present in green tea are (–)-epigallocatechin (EGC), (–)-epicatechin (EC), and its gallic acid esters (–)-epigallocatechin gallate (EGCG) and (–)-epicatechin gallate (ECG). These combined units are phenolic structures that can be easily oxidized. Compounds that are easier to oxidize are better antioxidants. Hence, the catechol group reacts readily with oxidants such as free radicals and ROS to form a stable semiquinone radical. EGCG, the predominant polyphenol in green tea, constituting up to roughly 10–50% of the total catechin content, is also the most powerful of catechins with an antioxidant activity about 25–100 times more than the commonly known antioxidants vitamins C and D [12].

Studies investigating the modulatory effects of EGCG on UVA-activated gene expressions in human fibroblasts and keratinocytes indicated that the effect of green tea polyphenols on cellular stress responses is complex and involves direct effects on signal transduction as well as changes that may be associated with its antioxidant activity [13]. Using human epidermal keratinocytes as an *in vitro* model system we have shown that EGCG inhibits UVB-induced generation of hydrogen peroxide (H_2O_2) and phosphorylation of mitogen-activated protein kinase (MAPK) signaling pathways [14]. In addition, EGCG prevented UVB-induced infiltration of leukocytes, depletion of antigen-presenting cells, and reduced oxidative stress in the murine skin [15]. A follow-up study in human subjects further demonstrated that topical application of EGCG to human skin prior to UV exposure markedly decreased UV-induced production of H_2O_2 and nitric oxide in both the dermal and the epidermal cells, along with inhibition of infiltration of inflammatory leukocytes, particularly CD11b(+) cells, the major producers of reactive oxygen species. Furthermore, EGCG restored the UV-induced decrease in the total glutathione level and prevented the decrease in antioxidant glutathione peroxidase enzyme activity [16].

UV irradiation leads to distinct changes in skin connective tissues by degradation of collagen, a major structural component of the extracellular matrix, through activation of matrix metalloproteinases (MMPs). These changes in collagenous skin tissues have been suggested to be the cause of skin wrinkling and, consequently, premature aging of the skin. EGCG was shown to hinder UVB-induced collagenolytic MMP production via interfering with the MAPK-responsive pathways [17]. In addition, UV-induced Nuclear Factor-kappa B (NF-κB) activation and expression of Interleukin (IL)-6 has been shown to be attenuated by EGCG treatment in cultured human keratinocytes [18]. Kim et al. [19] examined the protective effects of EGCG on UV-induced photoaging in various animal models and human dermal fibroblast cultures. They found that lipid peroxidation was significantly reduced in the EGCG-treated group. In addition, the erythema relative index of the treated group was also significantly reduced. Also, EGCG reduced UVA-induced skin damage and lessened the roughness and sagginess of the murine skin through inhibition of decrease of dermal collagen [19]. This was further corroborated by another study that showed that UV-induced oxidative damage and induction of MMPs might be prevented in mouse skin by oral administration of green tea. Treatment with green tea polyphenol (GTP) resulted in the inhibition of UVB-induced protein oxidation in mouse skin, a hallmark of photoaging. GTP treatment also inhibited UVB-induced protein oxidation in human skin fibroblasts, which supported the *in vivo* observations [20]. A double-blinded, placebo-controlled trial was conducted to evaluate the effects of green tea supplementation on the clinical and histologic characteristics of photoaging. Although no significant differences in clinical grading were found between the green tea–treated and placebo groups, other than higher subjective scores of irritation in the green tea–treated group, histologic grading of skin biopsies did show significant improvement in the elastic tissue content of treated specimens [21].

Data indicate that concentrations of EGCG as low as 10 microM can significantly decrease the level of DNA single-strand breaks and alkali-labile sites in UV-irradiated cells. Studies have shown that EGCG has the ability to ablate the mutagenic effects of UVA, reducing the induced hypoxanthine-guanine phosphoribosyl transferase mutant frequency to spontaneous levels [22]. Katiyar et al. [23] showed that topical application of GTP to human individuals protected against UV-induced DNA damage by preventing the formation of cyclobutane pyrimidine dimers. A pilot study was conducted in human volunteers to determine whether ingestion of green tea could afford protection against UV-induced DNA damage. Samples of peripheral blood cells taken after green tea consumption showed lower levels of DNA damage than those taken prior to ingestion, when subjects were exposed to 12 min of UVA radiation [24]. Studies further suggest that the protective effects of GTP appear to be mediated via immunosuppressive cytokine interleukin (IL)-12, most likely through induction of DNA repair [4]. Exposure of skin to UV radiation causes diverse biological effects, including induction of inflammation, alteration in cutaneous immune cells, and impairment of contact hypersensitivity responses. Nonetheless, in spite of the beneficial effect of green tea polyphenols against UV-induced damages, Sevin et al. [25] showed that the protective effects of EGCG were evident only when EGCG was applied prior to UVA exposure to Wistar albino rats.

There is considerable data indicating that green tea has the potential to protect against UV-mediated carcinogenesis. GTP and its constituent epicatechin derivatives were found to interact with hepatic cytochrome P450 and inhibit the P450-dependent mixed-function oxidase enzymes [26]. EGCG is considered a potent antioxidant and major anticarcinogenic component in tea. When low concentrations of EGCG were added to the extracts from green and black tea, the capacities to scavenge H_2O_2 and inhibit UV-induced 8-OHdG were substantially enhanced, suggesting that EGCG may be the predominant component responsible for the antioxidant activities [27]. Induction of skin tumors by UV radiation was significantly reduced by topical, but not by oral, administration of purified EGCG [18]. In SKH-1 hairless mice GTP resulted in inhibition of UVB-induced activation of NF-κB and phosphorylation of MAPKs: ERK1/2, JNK1/2, and p38 [28]. Feeding mice with green tea, black tea, or EGCG resulted in decreased growth of well-established tumors, and a significant regression of tumors was also observed [29]. Oral consumption of tea at concentrations similar to human consumption not only inhibited UVB-induced skin tumorigenesis in mice but also reduced the size of paramaterial fat pad and fatty tissue layer of the dermis both distal and directly under tumors [30]. In-depth studies, however, are required to further delineate the preventive effects of EGCG against UV exposures to human skin.

8.2.2 Pomegranate

Pomegranate (*Punica granatum* L of family Punicaceae) fruit, widely consumed fresh and in beverage as juice or wine, has been extensively used for medicinal purposes in ancient cultures. Pomegranate fruit is a rich source of two types of polyphenolic compounds: hydrolyzable tannins (such as punicallin, pedunculagin, punicalagin, gallagic, and ellagic esters of glucose) and flavanoids (such as anthocyanins, catechins, and other complex flavanoids). Pomegranate contains six types of anthocyanins (delphinidin-3-glycoside, delphinidin-3,5-diglycoside, pelargonidin-3-glycoside, pelargonidin-3,5-diglycoside, cyanidin-3-gylcoside, cyanidin-3,5-digylcoside). The other flavanoids present in pomegranate include quercetin, kaempherol, and luteolin glycosides [31]. The effects of pomegranate are thought to be due to its free radical scavenging and antioxidant properties, attributable to its high polyphenolic content [32]. Extracts from different parts of this plant such as juice, seed, and peel have been reported to exhibit strong antioxidant activity [33].

Recent studies from our laboratory showed that treatment of normal human epidermal keratinocytes (NHEK) with pomegranate fruit extract (PFE) prior to UV exposure inhibited UVB-induced translocation and phosphorylation of NF-κB [5] and MAPK proteins ERK1/2, JNK1/2, and p38 [7,34]. PFE also protected NHEK from UVA-mediated phosphorylation of signal transducers and activators of transcription (STAT)-3, AKT, and ERK1/2 [34]. More recently we have shown that treatment of immortalized human keratinocyte HaCaT cells with PFE prior to UVB exposure protected cells from UVB-mediated depletion of endogenous glutathione, decreased UV-induced lipid peroxidation, and significantly inhibited UVB-induced expression of MMP-2 and MMP-9 [35]. We further showed that delphinidin, one of the major anthocyanins present in pomegranate, protected NHEK from UVB-mediated increased lipid peroxidation, decrease in cell viability, induction of apoptosis, increase in poly(ADP-ribose) polymerase (PARP), and activation of caspases [34]. Treatment of HaCaT cells and mouse skin with delphinidin prior to UVB exposure inhibited UVB-mediated oxidative stress and reduced DNA damage, thereby affording protection from UVB-induced apoptosis [8].

In another study, we showed that oral feeding of PFE to mice, prior to multiple UVB exposure, protected mouse skin against the adverse effects of UVB by modulating NF-κB, MAPK, and AP-1 signal transduction pathways and induction of MMPs [36]. More recently, we reported that PFE also protected against UVB-induced skin tumorigenesis in mouse model by modulating STAT-3 and hypoxia inducible factor-α, leading to a decrease in inflammatory and angiogenic responses [37]. In addition, we found that pomegranate juice, seed oil, and whole fruit extract causes a decrease in UVB-mediated sunburn, DNA damage, proliferating cell nuclear antigen (PCNA), various MMPs, and increase in protein oxidation, in the reconstituted EpiDerm human skin model [38]. There is further evidence that oral intake of pomegranate extract inhibits UV-induced pigmentation in the human skin [39].

8.2.3 Curcumin

Curcumin is a yellow pigment (diferuloymethane) extracted from the rhizome of *Curcuma longa* of family Zingiberaceae, which is commonly known as turmeric. Studies have shown that curcumin possesses anti-inflammatory and antioxidant properties. The rhizome turmeric, used as a spice, especially in the Indian subcontinent, is widely used as an indigenous medicine for the treatment of many inflammatory diseases. Local application of turmeric is a common household remedy for a number of skin-related problems. Cosmetics supplemented with turmeric are marketed in many parts of the world. Due to its antioxidant properties, both curcumin and turmeric have been shown to scavenge ROS, including superoxide anion radicals, hydroxyl radicals, and reactive nitrogen oxide radicals. Curcumin has been shown to possess a wide spectrum of pharmacological activities that include antioxidant, anti-inflammatory, anticancer, antimicrobial, and wound-healing properties, and due to these it has a wide range of clinical applications [40]. The molecular basis for the chemopreventive action of curcumin is due to its effect on altered targets, which include cell signaling molecules, transcription factors, apoptotic genes, angiogenesis regulators, and cell adhesion molecules [41].

Curcumin protects cultured human lymphocytes from radiation-induced DNA damage and lipid peroxidation, due to its strong antioxidant properties [42]. In epidermal keratinocytes, curcumin resulted in decreased production of superoxide radicals, leading to lower levels of cytotoxic hydrogen peroxide [43]. Curcumin treatment was shown to result in decreased cell viability of melanoma cells through a cell membrane–mediated mechanism independent of the p53 pathway [44]. Curcumin also induces apoptosis in human basal cell carcinoma cells, where the p38 mediated pathway is critically involved [45]. Treatment of highly metastatic murine melanoma cells with curcumin inhibited MMP-2 activity, thereby preventing breakdown of collagen in the highly metastatic murine melanoma cells [46]. Curcumin inhibited UVB-induced activations of COX-2, and p38 and JNK MAPKs in HaCaT cells. UVB-induced DNA binding activity of AP-1 was also markedly decreased with curcumin treatment in HaCaT cells [47].

Curcuminoids have been reported to possess multifunctional bioactivities, especially the ability to inhibit pro-inflammatory induction. In the human keratinocyte cell line NCTC, curcumin downregulated UVB-stimulated IL-6, IL-1β, TNF-α, and NF-κB [48,49]. Topical application of curcumin has been shown to inhibit lipid peroxidation and arachidonic acid metabolism and enhance glutathione status and glutathione-S-peroxidase activity in mice skin [50]. Curcumin applied topically inhibited the UVA-mediated increase of ODC activity in the skin of TPA-treated CD-1 mice, thereby preventing UVA-induced aggravation of TPA-mediated dermatitis [51].

Topical treatment of mice skin with curcumin also inhibited activation of NF-κB and AP-1 [52]. Dietary consumption of curcumin caused a significant decrease in expression of *Ras* and *Fas* proto-oncogenes in skin tumor samples [53].

8.2.4 Resveratrol

Resveratrol, a polyphenolic phytoalexin chemically known as 3,4′,5-trihydroxy-*trans*-stilbene, has been identified in about 70 species of plants. It is largely present in the seeds and skin of grapes, various fruits, nuts, mulberries, and red wine and is known to exhibit a wide range of biological and pharmacological properties. Grape skin is a good dietary source of resveratrol, because the fresh skin contains about 50–100 μg of resveratrol per gram, whereas in red wine its concentration is in the range of 1.5–3.0 mg/l. Resveratrol is a potent antioxidant with antimutagenic, anti-inflammatory, and antiproliferative properties. It is also an inducer of phase II drug-metabolizing enzymes, and inhibitor of COX-2 and hydroperoxidase [54]. Recent data suggests that resveratrol has the potential to prevent and slow the progression of various diseases that include ischemic injury, cancer, and cardiovascular disease [55]. In this context, resveratrol has been shown to inhibit diverse cellular events associated with tumor initiation, promotion, and progression of cancer in all organs including skin [56].

In NHEK, resveratrol pretreatment blocked UVB-induced activation of the NF-κB pathway, which plays a critical role in the chemopreventive effects of resveratrol against the adverse effects of UV radiation [57]. Pretreatment of HaCaT cells with resveratrol attenuated UVB-induced ROS production with concomitant decrease of caspase-3 and caspase-8 activation, resulting in an increase in cell survival after UVB irradiation [58]. Topical application of resveratrol to SKH-1 mice prior to UVB exposure resulted in a significant decrease in UVB-mediated bi-fold skin thickness, edema, and hyperplasia. Resveratrol further inhibited UVB-mediated lipid peroxidation, thus reducing oxidative stress [59]. In addition, topical treatment of mouse skin with resveratrol decreased tumor promotion by inhibiting UVB-mediated cellular proliferation, and induction of COX-2 and ODC enzyme activities, well-established markers for tumor promotion [58,59]. Topical application of SKH-1 hairless mouse skin with resveratrol (both pre- and post-UVB exposure) resulted in downregulation of survivin, a critical regulator of survival/death of cells and upregulated proapoptotic, Smac/DIABLO proteins in skin tumors with increased apoptosis, thereby resulting in a decrease in tumor incidence and a delay in the onset of tumorigenesis [60]. Resveratrol treatment resulted in a significant downregulation in UV-induced critical cell cycle regulatory proteins; cyclin-dependent kinase (cdk)-2, -4 and -6; cyclin-D1; and cyclin-D2 in SKH-1 mouse skin [61].

8.2.5 Genistein

Genistein (4′,5,7-trihydroxyisoflavone), an isoflavone occuring as a glycoside (genistin) in the plants of family Leguminosae, has diverse biological activities. Genistein forms a major constituent of soya bean, red clover, ginkgo biloba, Greek oregano, and Greek sage. Genistein has been shown to possess strong antioxidant and anticarcinogenic activity [62]. Accumulating evidence indicates that genistein shows preventive and therapeutic effects against various types of cancers, osteoporosis, and cardiovascular diseases in both humans and animals. Genistein substantially inhibits UV-induced skin carcinogenesis and cutaneous aging in mice, and photodamage in humans. The mechanisms of action involved are through its antioxidant activity, protection of oxidative and photodynamically damaged DNA, and modulation of UVB-altered signal transduction cascades [63].

Cell culture studies show that genistein augmented the induction of c-jun and jun B by UVB irradiation and prevented UVA-induced enhancement of STAT1 binding activity [64,65]. It also suppressed both the basal and stimulated expressions of COX-2 in HaCaT cells [66]. In human epidermal cells, genistein blocked UVB-stimulated prostaglandin-2 synthesis [67]. Topical application of genistein and its endogenous metabolites and derivatives to SKH-1 mice has been shown to reduce inflammatory edema and suppress contact hypersensitivity reaction induced by UV radiation, thereby protecting the immune system from photosuppression [68]. Pretreatment of skin with genistein prior to UVB exposure resulted in a significant decrease in UV-induced generation of 8-OHdG, H_2O_2, and malondialdehyde in the epidermis as well as other internal organs [69]. Studies have demonstrated that genistein minimizes the detrimental effect of UVB radiation by preserving cutaneous proliferation and repair mechanisms and inhibits UV-induced pyrimidine dimer formation in the human reconstituted skin model, EpiDerm [70]. Genistein protected the melanocytes from UVB-induced carcinogenesis by altering the expression of gangliosides and other carbohydrate antigens so as to facilitate their immune recognition. Furthermore, genistein also inhibited the activities of protein tyrosine kinase, topoisomerase II, and MMP-9, and downregulated the expression of a number of genes, including VEGF, with inhibition of cell proliferation and cell cycle growth arrest at the G2/M phase, thereby preventing invasion and angiogenesis [71].

8.2.6 Apigenin

Apigenin is a flavanoid (5,7,4′-trihydroxyflavone), present in the leaves and stems of many vascular plants that include herbs (e.g., endive, clove), fruits (e.g., apples, cherries, grapes), vegetables (celery beans, broccoli, tomatoes, onions, leeks, barley, parsley) and beverages such as tea and wine. This antioxidant-rich compound has proven to be relatively nontoxic, nonmutagenic, free radical scavenger, anti-inflammatory, and anticarcinogenic in nature [72]. Studies showed that apigenin prevents skin tumorigenesis by inducing G2/M phase cell cycle arrest in mouse keratinocytes [73]. Apigenin also inhibited UVB-induced increase in COX-2 protein and mRNA levels in mouse and human keratinocyte cell lines [74]. In human dermal fibroblasts, apigenin scavenged UVA-induced generation of ROS and inhibited collagenase activity [75]. In cultured human dermal fibroblast cells apigenin-8-C-β-D-glucopyranoside protected against UV-induced adverse skin reactions such as free radical production and skin cell damage [76]. Topical application of apigenin to mouse skin prior to UVB exposure significantly inhibited UV-mediated increase in ODC enzyme activity, reduced tumor incidence rate, increased tumor-free survival, and effectively prevented UV-induced skin tumorigenesis [77].

8.2.7 Carotenoids

Carotenoids, the pigments present in all colored plants, are believed to play an important role in protecting plants from photosensitization. The main carotenoids include β-carotene found in carrots and lycopene, an acyclic isomer of β-carotene, found in red fruits and vegetables, like tomatoes, watermelon, papaya, grape fruit, and apricot. The extended system of conjugated double bonds present in carotenoids is crucial for their antioxidant properties [78]. Many of the health benefits of carotenoids are attributed to their ability to protect cells against oxidative damage, their high singlet oxygen–quenching ability, and scavenging of oxygen radicals [79].

Dietary β-carotene protects hairless mice from UV-induced carcinogenesis. Topical application of β-carotene reduced solar radiation–induced erythema in humans. Similar results were obtained with lycopene, one of the most potent naturally occurring antioxidants. which protected critical biomolecules including protein, DNA, lipids, and low density lipoproteins [80]. Topical application of lycopene prior to UVB exposure has been shown to reduce the damaging effects of UV, as evidenced by decreased ODC activity, inflammatory responses, and development of skin tumors. A case-controlled study indicated that persons with high intake of dietary carotenoids showed a significantly lower risk for the development of melanoma on repeated sun exposure [81].

8.2.8 Quercetin

Quercetin (3,5,7,3′,4-pentahydroxyflavone), present in various common fruits (apple, grapes, olives), vegetables (lemons, tomatoes, onions, broccoli), beverages (tea, red wine), herbs, and beehives, is a powerful antioxidant and metal chelator. Studies have shown that quercetin is capable of decreasing the harmful effect of UV radiation and lipid peroxidation. In human dermal fibroblasts quercetin scavenged UVA-induced generation of ROS and inhibited collagenase MMP activity, thereby preventing UV-stressed skin aging [79]. In living skin–equivalent cultures quercetin-3-glucoside inhibited UVA-induced keratinocyte and fibroblast vacuolation and nuclear pyknosis and decreased the UVA-mediated increase in 8-OHdG and IL-1α [82]. Quercetin applied topically to the dorsal skin of SKH-1 mice inhibited UVB-induced increase in the myeloperoxidase activity and depletion of GSH, and it prevented UVB radiation–induced skin damages [83]. Oral administration of quercetin to these mice prevented UVB-induced immunosuppression [84]. It has been shown to protect the endogenous antioxidant enzymes: catalase, superoxide dismutase, glutathione peroxidase, and glutathione reductase activities in rats [85]. Quercetin also inhibited MMP-1, which plays an important role in the unbalanced turnover or rapid breakdown of collagen molecules in human UV-irradiated skin [86].

8.2.9 Carnosic Acid

Carnosic acid, one of the major constituents of rosemary and sage, has powerful antioxidative activity. Carnosic acid prevented photoaging and photocarcinogenesis by suppressing UVA-induced activity of MMP-1 and showed chemopreventive effects against carcinogens in animal models [81]. A randomized double-blind, placebo-controlled study showed that sage extract significantly reduced UV-induced erythema to a similar extent as hydrocortisone, suggesting that sage extract might be useful in the topical treatment of inflammatory skin diseases [87].

8.2.10 Caffeic Acid and Ferullic Acid

Caffeic acid (3,4-dihydrocinnamic acid) and ferullic acid (4-hydroxy-3-methoxycinnamic acid) are present largely in grains, vegetables, and fruits in conjugated form with saccharides. These hydroxycinnamic acids prevent propagation of lipid peroxidative chain reaction and protect membrane phospholipids from UV-induced peroxidation [81]. Studies show that these acids afford significant protection to the skin against UVB-induced erythema and may be successfully employed as topical protective agents against UV-mediated skin damage [88].

8.3 Conclusion

A large number of studies support the notion that dietary botanicals with antioxidant properties show anti-inflammatory, anticarcinogenic, and anti-photoaging effects both *in vitro* (cell/tissue culture) and *in vivo* (animal models and humans). Therefore, the use of antioxidant-rich botanicals as dietary sources, and/or supplementing skin care products with these botanicals for daily use, may be an effective approach for reducing UV-induced photodamage and skin cancer. There is evidence that by increasing the basal antioxidative protection level systematically by a regular intake of antioxidant-rich botanicals UV-dependent skin damage could be prevented.

References

1. Jemal A, Siegel R, Ward E, Hao Y, Xu J, Murray T, Thun MJ. Cancer statistics, 2008. (2008) CA Cancer J Clin 58, 71–96.
2. Afaq F, Mukhtar H. Botanical antioxidants in the prevention of photocarcinogenesis and photoaging. (2006) Exp Dermatol 15, 678–84.
3. Phan TA, Halliday GM, Barnetson RS, Damian DL. Spectral and dose dependence of ultraviolet radiation-induced immunosuppression. (2006) Front Biosci 11, 394–411.
4. Schwarz A, Maeda A, Gan D, Mammone T, Matsui MS, Schwarz T. Green tea phenol extracts reduce UVB-induced DNA damage in human cells via interleukin-12. (2008) Photochem Photobiol 84, 350–5.

5. Grillari J, Katinger H, Voglauer R. Contributions of DNA interstrand cross-links to aging of cells and organisms. (2007) Nucleic Acids Res 35, 7566–76.
6. Sander CS, Chang H, Hamm F, Elsner P, Thiele JJ. Role of oxidative stress and the antioxidant network in cutaneous carcinogenesis. (2004) Int J Dermatol 43, 326–35.
7. de Gruijl FR. Photocarcinogenesis: UVA vs. UVB radiation. (2002) Skin Pharmacol Appl Skin Physiol 15, 316–20.
8. Afaq F, Syed DN, Malik A, Hadi N, Sarfaraz S, Kweon MH, Khan N, Zaid MA, Mukhtar H. Delphinidin, an anthocyanidin in pigmented fruits and vegetables, protects human HaCaT keratinocytes and mouse skin against UVB-mediated oxidative stress and apoptosis. (2007) J Invest Dermatol 127, 222–32.
9. Katiyar SK. UV-induced immune suppression and photocarcinogenesis: chemoprevention by dietary botanical agents. (2007) Cancer Lett 255, 1–11.
10. Baumann L. Skin ageing and its treatment. (2007) J Pathol 211, 241–51.
11. Katiyar SK, Agarwal R, Wang ZY, Bhatia AK, Mukhtar H. (–)-Epigallocatechin-3-gallate in *Camellia sinensis* leaves from Himalayan region of Sikkim: inhibitory effects against biochemical events and tumor initiation in Sencar mouse skin. (1992) Nutr Cancer 18, 73–83.
12. Intra J, Kuo SM. Physiological levels of tea catechins increase cellular lipid antioxidant activity of vitamin C and vitamin E in human intestinal caco-2 cells. (2007) Chem Biol Interact 169, 91–9.
13. Soriani M, Rice-Evans C, Tyrrell RM. Modulation of the UVA activation of haem oxygenase, collagenase and cyclooxygenase gene expression by epigallocatechin in human skin cells. (1998) FEBS Lett 439, 253–7.
14. Katiyar SK, Afaq F, Azizuddin K, Mukhtar H. Inhibition of UVB-induced oxidative stress-mediated phosphorylation of mitogen-activated protein kinase signaling pathways in cultured human epidermal keratinocytes by green tea polyphenol (–)-epigallocatechin-3-gallate. (2002) Toxicol Appl Pharmacol 176, 110–17.
15. Katiyar SK, Mukhtar H. Green tea polyphenol (–)-epigallocatechin-3-gallate treatment to mouse skin prevents UVB-induced infiltration of leukocytes, depletion of antigen-presenting cells, and oxidative stress. (2001) J Leukoc Biol 69, 719–26.
16. Katiyar SK, Afaq F, Perez A, Mukhtar H. Green tea polyphenol (–)-epigallocatechin-3-gallate treatment of human skin inhibits ultraviolet radiation-induced oxidative stress. (2001) Carcinogenesis 22, 287–94.
17. Bae JY, Choi JS, Choi YJ, Shin SY, Kang SW, Han SJ, Kang YH. (–)Epigallocatechin gallate hampers collagen destruction and collagenase activation in ultraviolet-B-irradiated human dermal fibroblasts: involvement of mitogen-activated protein kinase. (2008) Food Chem Toxicol 46, 1298–307.
18. Xia J, Song X, Bi Z, Chu W, Wan Y. UV-induced NF-κB activation and expression of IL-6 is attenuated by (–)-epigallocatechin-3-gallate in cultured human keratinocytes in vitro. (2005) Int J Mol Med 16, 943–50.
19. Kim J, Hwang JS, Cho YK, Han Y, Jeon YJ, Yang KH. Protective effects of (–)-epigallocatechin-3-gallate on UVA- and UVB-induced skin damage. (2001) Skin Pharmacol Appl Skin Physiol 14, 11–19.

20. Vayalil PK, Mittal A, Hara Y, Elmets CA, Katiyar SK. Green tea polyphenols prevent ultraviolet light–induced oxidative damage and matrix metalloproteinase expression in mouse skin. (2004) J Invest Dermatol 122, 1480–7.
21. Chiu AE, Chan JL, Kern DG, Kohler S, Rehmus WE, Kimball AB. Double-blinded, placebo-controlled trial of green tea extracts in the clinical and histologic appearance of photoaging skin. (2005) Dermatol Surg 31, 855–60.
22. Tobi SE, Gilbert M, Paul N, McMillan TJ. The green tea polyphenol, epigallocatechin-3-gallate, protects against the oxidative cellular and genotoxic damage of UVA radiation. (2002) Int J Cancer 102, 439–44.
23. Katiyar SK, Perez A, Mukhtar H. Green tea polyphenol treatment to human skin prevents formation of ultraviolet light B–induced pyrimidine dimers in DNA. (2000) Clin Cancer Res 6, 3864–9.
24. Morley N, Clifford T, Salter L, Campbell S, Gould D, Curnow A. The green tea polyphenol (–)-epigallocatechin gallate and green tea can protect human cellular DNA from ultraviolet and visible radiation-induced damage. (2005) Photodermatol Photoimmunol Photomed 21, 15–22.
25. Sevin A, Oztas¸ P, Senen D, Han U, Karaman C, Tarimci N, Kartal M, Erdoğan B. Effects of polyphenols on skin damage due to ultraviolet A rays: an experimental study on rats. (2007) J Eur Acad Dermatol Venereol 21, 650–6.
26. Wang ZY, Agarwal R, Bickers DR, Mukhtar H. Protection against ultraviolet B radiation–induced photocarcinogenesis in hairless mice by green tea polyphenols. (1991) Carcinogenesis 12, 1527–30.
27. Wei H, Zhang X, Zhao JF, Wang ZY, Bickers D, Lebwohl M. Scavenging of hydrogen peroxide and inhibition of ultraviolet light–induced oxidative DNA damage by aqueous extracts from green and black teas. (1999) Free Radic Biol Med 26, 1427–35.
28. Afaq F, Ahmad N, Mukhtar H. Suppression of UVB-induced phosphorylation of mitogen-activated protein kinases and nuclear factor kappa B by green tea polyphenol in SKH-1 hairless mice. (2003) Oncogene 22, 9254–64.
29. Stoner GD, Mukhtar H. Polyphenols as cancer chemopreventive agents. (1995) J Cell Biochem Suppl 22, 169–80.
30. Conney AH, Lu YP, Lou YR, Huang MT. Inhibitory effects of tea and caffeine on UV-induced carcinogenesis: relationship to enhanced apoptosis and decreased tissue fat. (2002) Eur J Cancer Prev 11, S28–36.
31. Gil MI, Tomás-Barberán FA, Hess-Pierce B, Holcroft DM, Kader AA. Antioxidant activity of pomegranate juice and its relationship with phenolic composition and processing. (2000) J Agric Food Chem 48, 4581–9.
32. Seeram NP, Adams LS, Henning SM, Niu Y, Zhang Y, Nair MG, Heber D. In vitro antiproliferative, apoptotic and antioxidant activities of punicalagin, ellagic acid and a total pomegranate tannin extract are enhanced in combination with other polyphenols as found in pomegranate juice. (2005) J Nutr Biochem 16, 360–7.
33. Aslam MN, Lansky EP, Varani J. Pomegranate as a cosmeceutical source: pomegranate fractions promote proliferation and procollagen synthesis and inhibit matrix metalloproteinase-1 production in human skin cells. (2006) J Ethnopharmacol 103, 311–18.

34. Syed DN, Malik A, Hadi N, Sarfaraz S, Afaq F, Mukhtar H. Photochemopreventive effect of pomegranate fruit extract on UVA-mediated activation of cellular pathways in normal human epidermal keratinocytes. (2006) Photochem Photobiol 82, 398–405.
35. Zaid MA, Afaq F, Syed DN, Dreher M, Mukhtar H. Inhibition of UVB-mediated oxidative stress and markers of photoaging in immortalized HaCaT keratinocytes by pomegranate polyphenol extract POMx. (2007) Photochem Photobiol 83, 882–8.
36. Afaq F, Zaid MA, Khan N, Syed D, Hafeez BB, Yun J, Sarfaraz S, Mukhtar H. Pomegranate fruit extract inhibits UVB-induced activation of NFκB and MAPK leading to decreased expression of matrix metalloprotenaises in SKH-1 mouse skin. American Association for Cancer Research Annual Meeting 2007, Los Angeles, CA, 2571.
37. Afaq F, Zaid MA, Khan N, Syed DN, Yun J, Sarfaraz S, Suh Y, Mukhtar H. Inhibitory effect of oral feeding of pomegranate fruit extract on UVB-induced skin carcinogenesis in SKH-1 hairless mice. American Association for Cancer Research Annual Meeting 2008, San Diego, CA.
38. Zaid MA, Afaq F, Khan N, Mukhtar H. Protective effects of pomegranate juice, oil and byproduct on UVB-induced DNA damage, PCNA expression and MMPs in human reconstituted skin. American Association for Cancer Research Annual Meeting 2007 proceedings 48, 2573.
39. Kasai K, Yoshimura M, Koga T, Arii M, Kawasaki S. Effects of oral administration of ellagic acid-rich pomegranate extract on ultraviolet-induced pigmentation in the human skin. (2006) J Nutr Sci Vitaminol 52, 383–8.
40. Maheshwari RK, Singh AK, Gaddipati J, Srimal RC. Multiple biological activities of curcumin: a short review. (2006) Life Sci 78, 2081–7.
41. Aggarwal BB, Kumar A, Bharti AC. Anticancer potential of curcumin: preclinical and clinical studies. (2003) Anticancer Research 23, 363–98.
42. Srinivasan M, Rajendra Prasad N, Menon VP. Protective effect of curcumin on gamma-radiation induced DNA damage and lipid peroxidation in cultured human lymphocytes. (2006) Mutat Res 611, 96–103.
43. Azuine MA, Bhide SV. Chemopreventive effect of turmeric against stomach and skin tumors induced by chemical carcinogens in Swiss mice. (1992) Nutr Cancer 17, 77–83.
44. Bush JA, Cheung KJ Jr, Li G. Curcumin induces apoptosis in human melanoma cells through a Fas receptor/caspase-8 pathway independent of p53. (2001) Exp Cell Res 271, 305–14.
45. Jee SH, Shen SC, Tseng CR, Chiu HC, Kuo ML. Curcumin induces a p53-dependent apoptosis in human basal cell carcinoma cells. (1998) J Invest Dermatol 111, 656–61.
46. Banerji A, Chakrabarti J, Mitra A, Chatterjee A. Effect of curcumin on gelatinase A (MMP-2) activity in B16F10 melanoma cells. (2004) Cancer Lett 211, 235–42.
47. Cho JW, Park K, Kweon GR, Jang BC, Baek WK, Suh MH, Kim CW, Lee KS, Suh SI. Curcumin inhibits the expression of COX-2 in UVB-irradiated human keratinocytes (HaCaT) by inhibiting activation of AP-1: p38 MAP kinase and JNK as potential upstream targets. (2005) Exp Mol Med 37, 186–92.

48. Liang G, Yang S, Zhou H, Shao L, Huang K, Xiao J, Huang Z, Li X. Synthesis, crystal structure and anti-inflammatory properties of curcumin analogues. (2008) Eur J Med Chem Feb 3 [Epub ahead of print].
49. Grandjean-Laquerriere A, Antonicelli F, Gangloff SC, Guenounou M, Le Naour R. UVB-induced IL-18 production in human keratinocyte cell line NCTC 2544 through NF-κB activation. (2007) Cytokine 37, 76–83.
50. Iersel ML, Ploemen JP, Struik I, van Amersfoort C, Keyzer AE, Schefferlie JG, van Bladeren PJ. Inhibition of glutathione S-transferase activity in human melanoma cells by alpha,beta-unsaturated carbonyl derivatives. Effects of acrolein, cinnamaldehyde, citral, crotonaldehyde, curcumin, ethacrynic acid, and trans-2-hexenal. (1996) Chem Biol Interact 102, 117–32.
51. Ishizaki C, Oguro T, Yoshida T, Wen CQ, Sueki H, Iijima M. Enhancing effect of ultraviolet A on ornithine decarboxylase induction and dermatitis evoked by 12-O-tetradecanoylphorbol-13-acetate and its inhibition by curcumin in mouse skin. (1996) Dermatology 193, 311–17.
52. Surh YJ, Han SS, Keum YS, Seo HJ, Lee SS. Inhibitory effects of curcumin and capsaicin on phorbol ester–induced activation of eukaryotic transcription factors, NF-κB and AP-1. (2000) Biofactors 12, 107–12.
53. Limtrakul P, Anuchapreeda S, Lipigorngoson S, Dunn FW. Inhibition of carcinogen induced c-Ha-ras and c-fos proto-oncogenes expression by dietary curcumin. (2001) BMC Cancer 1, 1.
54. Jang M, Cai L, Udeani GO, Slowing KV, Thomas CF, Beecher CW, Fong HH, Farnsworth NR, Kinghorn AD, Mehta RG, Moon RC, Pezzuto JM. Cancer chemopreventive activity of resveratrol, a natural product derived from grapes. (1997) Science 275, 218–20.
55. Baur JA, Sinclair DA. Therapeutic potential of resveratrol: the in vivo evidence. (2006) Nat Rev Drug Discov 5, 493–506.
56. Cal C, Garban H, Jazirehi A, Yeh C, Mizutani Y, Bonavida B. Resveratrol and cancer: chemoprevention, apoptosis, and chemo-immunosensitizing activities. (2003) Curr Med Chem Anti-Cancer Agents 3, 77–93.
57. Adhami VM, Afaq F, Ahmad N. Suppression of ultraviolet B exposure-mediated activation of NF-κB in normal human keratinocytes by resveratrol. (2003) Neoplasia 5, 74–82.
58. Park K, Lee JH. Protective effects of resveratrol on UVB-irradiated HaCaT cells through attenuation of the caspase pathway. (2008) Oncol Rep 19, 413–17.
59. Afaq F, Adhami VM, Ahmad N. Prevention of short-term ultraviolet B radiation-mediated damages by resveratrol in SKH-1 hairless mice. (2003) Toxicol Appl Pharmacol 186, 28–37.
60. Aziz MH, Afaq F, Ahmad N. Prevention of ultraviolet-B radiation damage by resveratrol in mouse skin is mediated via modulation in survivin. (2005) Photochem Photobiol 81, 25–31.
61. Reagan-Shaw S, Afaq F, Aziz MH, Ahmad N. Modulations of critical cell cycle regulatory events during chemoprevention of ultraviolet B–mediated responses by resveratrol in SKH-1 hairless mouse skin. (2004) Oncogene 23, 5151–60.
62. Wei H, Bowen R, Cai Q, Barnes S, Wang Y. Antioxidant and antipromotional effects of the soybean isoflavone genistein. (1995) Proc Soc Exp Biol Med 208, 124–30.

63. Wei H, Saladi R, Lu Y, Wang Y, Palep SR, Moore J, Phelps R, Shyong E, Lebwohl MG. Isoflavone genistein: photoprotection and clinical implications in dermatology. (2003) J Nutr 133, 3811S–3819S.
64. Isoherranen K, Westermarck J, Kähäri VM, Jansén C, Punnonen K. Differential regulation of the AP-1 family members by UV irradiation in vitro and in vivo. (1998) Cell Signal 10, 91–5.
65. Mazière C, Dantin F, Dubois F, Santus R, Mazière J. Biphasic effect of UVA radiation on STAT1 activity and tyrosine phosphorylation in cultured human keratinocytes. (2000) Free Radic Biol Med 28, 1430–7.
66. Isoherranen K, Punnonen K, Jansen C, Uotila P. Ultraviolet irradiation induces cyclooxygenase-2 expression in keratinocytes. (1999) Br J Dermatol 140, 1017–22.
67. Miller CC, Hale P, Pentland AP. Ultraviolet B injury increases prostaglandin synthesis through a tyrosine kinase–dependent pathway. Evidence for UVB-induced epidermal growth factor receptor activation. (1994) J Biol Chem 269, 3529–33.
68. Widyarini S, Spinks N, Husband AJ, Reeve VE. Isoflavonoid compounds from red clover (*Trifolium pratense*) protect from inflammation and immune suppression induced by UV radiation. (2001) Photochem Photobiol 74, 465–70.
69. Wei H, Zhang X, Wang Y, Lebwohl M. Inhibition of ultraviolet light–induced oxidative events in the skin and internal organs of hairless mice by isoflavone genistein. (2002) Cancer Lett 185, 21–9.
70. Moore JO, Wang Y, Stebbins WG, Gao D, Zhou X, Phelps R, Lebwohl M, Wei H. Photoprotective effect of isoflavone genistein on ultraviolet B–induced pyrimidine dimer formation and PCNA expression in human reconstituted skin and its implications in dermatology and prevention of cutaneous carcinogenesis. (2006) Carcinogenesis 27, 1627–35.
71. Ravindranath MH, Muthugounder S, Presser N, Viswanathan S. Anticancer therapeutic potential of soy isoflavone, genistein. (2004) Adv Exp Med Biol 546, 121–65.
72. Baliga MS, Katiyar SK. Chemoprevention of photocarcinogenesis by selected dietary botanicals. (2006) Photochem Photobiol Sci 5, 243–53.
73. McVean M, Weinberg WC, Pelling JC. A p21(waf1)-independent pathway for inhibitory phosphorylation of cyclin-dependent kinase p34(cdc2) and concomitant G(2)/M arrest by the chemopreventive flavonoid apigenin. (2002) Mol Carcinog 33, 36–43.
74. Van Dross RT, Hong X, Essengue S, Fischer SM, Pelling JC. Modulation of UVB-induced and basal cyclooxygenase-2 (COX-2) expression by apigenin in mouse keratinocytes: role of USF transcription factors. (2007) Mol Carcinog 46, 303–14.
75. Sim GS, Lee BC, Cho HS, Lee JW, Kim JH, Lee DH, Kim JH, Pyo HB, Moon DC, Oh KW, Yun YP, Hong JT. Structure activity relationship of antioxidative property of flavonoids and inhibitory effect on matrix metalloproteinase activity in UVA-irradiated human dermal fibroblast. (2007) Arch Pharm Res 30, 290–8.
76. Kim JH, Lee BC, Kim JH, Sim GS, Lee DH, Lee KE, Yun YP, Pyo HB. The isolation and antioxidative effects of vitexin from *Acer palmatum*. (2005) Arch Pharm Res 28, 195–202.

77. Birt DF, Mitchell D, Gold B. Inhibition of ultraviolet light induced skin carcinogenesis in SKH-1 mice by apigenin, a plant flavonoid. (1997) Anticancer Res 17, 85–91.
78. Stahl W, Heinrich U, Aust O, Tronnier H, Sies H. Lycopene-rich products and dietary photoprotection. (2006) Photochem Photobiol Sci 5, 238–42.
79. Stahl W, Sies H. Carotenoids and flavonoids contribute to nutritional protection against skin damage from sunlight. (2007) Mol Biotechnol 37, 26–30.
80. Southon S. Increased fruit and vegetable consumption: potential health benefits. (2001) Nutr Metab Cardiovasc Dis 11, 78–81.
81. Millen AE, Tucker MA, Hartge P, Halpern A, Elder DE, Guerry D IV, Holly EA, Sagebiel RW, Potischman N. Diet and melanoma in a case-control study. (2004) Cancer Epidemiol Biomarkers Prev 13, 1042–51.
82. Dekker P, Parish WE, Green MR. Protection by food-derived antioxidants from UV-A1-induced photodamage, measured using living skin equivalents. (2005) Photochem Photobiol 81, 837–42.
83. Casagrande R, Georgetti SR, Verri WA Jr, Dorta DJ, dos Santos AC, Fonseca MJ. Protective effect of topical formulations containing quercetin against UVB-induced oxidative stress in hairless mice. (2006) J Photochem Photobiol B 84, 21–7.
84. Steerenberg PA, Garssen J, Dortant PM, van der Vliet H, Geerse E, Verlaan AP, Goettsch WG, Sontag Y, Bueno-de-Mesquita HB, Van Loveren H. The effect of oral quercetin on UVB-induced tumor growth and local immunosuppression in SKH-1. (1997) Cancer Lett 114, 187–9.
85. Erden Inal M, Kahraman A, Köken T. Beneficial effects of quercetin on oxidative stress induced by ultraviolet A. (2001) Clin Exp Dermatol 26, 536–9.
86. Lim H, Kim HP. Inhibition of mammalian collagenase, matrix metalloproteinase-1, by naturally-occurring flavonoids. (2007) Planta Med 73, 1267–74.
87. Reuter J, Jocher A, Hornstein S, Mönting JS, Schempp CM. Sage extract rich in phenolic diterpenes inhibits ultraviolet-induced erythema in vivo. (2007) Planta Med 73, 1190–1.
88. Saija A, Tomaino A, Trombetta D, De Pasquale A, Uccella N, Barbuzzi T, Paolino D, Bonina F. In vitro and in vivo evaluation of caffeic and ferulic acids as topical photoprotective agents. (2000) Int J Pharm 199, 39–47.

9

The Antioxidant Benefits of Oral Carotenoids for Protecting the Skin Against Photoaging

Pierfrancesco Morganti, PhD

Professor of Applied Cosmetic Dermatology, II University of Naples, Visiting Professor of China Medical University Shenyang, R&D Director, Mavi Sud S.r.l., Aprilia (LT), Italy

Summary

Carotenoids, a group of more than 600 natural molecules, are fat-soluble pigments responsible for the yellow color of corn, the orange color of

---

Aaron Tabor and Robert M. Blair (eds.), *Nutritional Cosmetics: Beauty from Within*, 185–198, © 2009 William Andrew Inc.

pumpkins, the red color of tomatoes, and the green color of many vegetables.

There are two commonly accepted classes of carotenoids: (a) carotens, composed of only carbon and hydrogen, and (b) oxycarotenoids, composed of carbon, hydrogen, and oxygen. All of these compounds have an identical backbone structure and their chemical and biochemical activities are related to their unique structure, consisting of an extended system of conjugated double bonds. Although the color of fruits and vegetables is important, the primary role of carotenoids in nature is to protect from ultraviolet rays the chlorophyll found in plant leaves. Carotenoids also help chlorophyll to absorb light energy.

In fact, they act like excellent antioxidants by quenching singlet oxygen, reactive oxygen species, and the free radicals that are by-products of metabolic processes in vegetable and human cells or environmental pollutants. However, the hydrophilic properties of oxycarotenoids, like lutein and zeaxantin, allow them to react with singlet oxygen generated in the water phase more efficiently than nonpolar. Both carotenoids and oxycarotenoids seem capable of preventing UV-induced DNA damage, protecting both the human eyes and skin against photoaging.

In this chapter we try to review the distribution and potential protective activity of carotenoids and oxycarotenoids in the human body and to explore new potential strategies to explain the causal link between oxidative stress and skin aging.

9.1 The Protective Role of Carotenoids

Nutrition plays an important role in the treatment of many diseases, and an appropriate selection of nutrients contributes to the prevention of disorders such as hyperlipidermia, hypertension, or vitamin deficiency [1,2]. Within this context, carotenoids are among the compounds that have attracted a great deal of attention [3].

They are a class of linear all-*trans* (E) form C40 polyenes found in plants, algae, and some bacteria and fungi. The number of naturally occurring carotenoids reported continues to rise and has now reached about 750 (Fig. 9.1). Although animals and humans cannot biosynthesize them

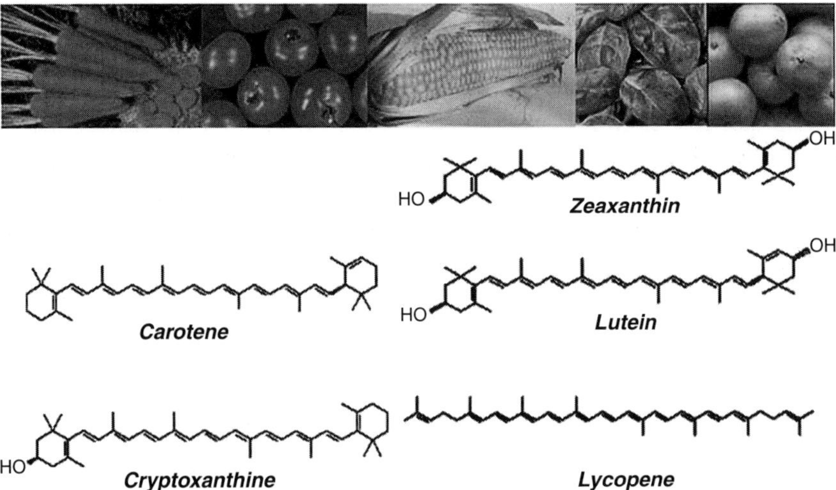

Figure 9.1 Structure of some common carotenoids in the human diet.

de novo, carotenoids are often present, sometimes in high concentrations, in animal tissue. Animals obtain these compounds from diet or perhaps, in some cases, from associated micro-organisms, but they may modify the structure of the ingested carotenoids to produce new metabolites [4]. Chemically, typical carotenoid pigments are tetraterpenoids, consisting of eight 5-carbon isoprenoid units.

Carotenoids, responsible for the yellow color of corn, the orange color of pumpkins, and the red color of tomatoes, have the primary role to protect the chlorophyll found in plant leaves. As is common knowledge, chlorophyll is the primary source of energy generation in plants. However, you might not be aware of the fact that chlorophyll is susceptible to damage caused by exposure to excessive amounts of light, particularly ultraviolet rays. Carotenoids, particularly lutein and zeaxanthin, are present in the chloroplasts to help protect the chlorophyll from such damage. Additionally, carotenoids help chlorophyll to collect light energy.

Moreover, it has been shown that these pigments protect photosynthetic organisms against potentially lethal photosensitization by means of endogenous photosynthetic pigments [5,6]. Therefore, carotenoid pigments

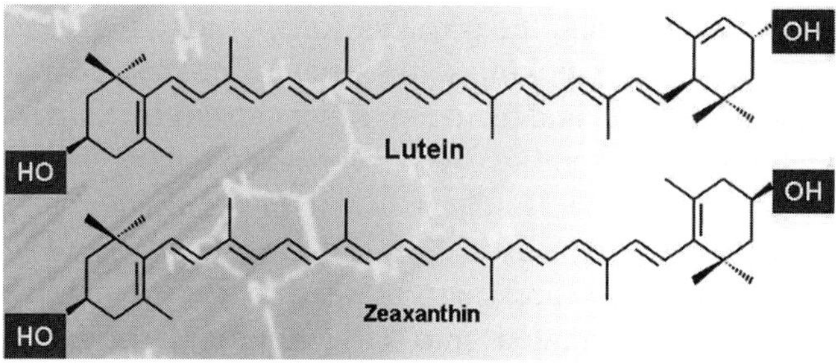

Figure 9.2 Structure of the xanthophylls lutein and zeaxanthin.

may have a protective role not only in plants but in humans as well. There are two known classes of carotenoids:

- carotens, composed of only carbon and hydrogen, including, α-carotene, β-carotene, and lycopene
- xanthophylls, composed of carbon, hydrogen, and oxygen, including lutein, zeaxanthin, and β-cryptoxanthin

Because the hundreds of natural carotenoids contain one of several centers or axes of chirality, they can also occur in various optical isomers. Such configurational changes may have a significant effect on the physical and biochemical properties of the molecules. Hydrocarbon carotenoids, as in carotens, are apolar lipophylic molecules and are not soluble in water but are readily soluble in organic solvents and, to some extent, in fats and oils. The presence of a hydroxy group, as in xanthophylls, gives the molecules some polarity, but such compounds are still predominantly hydrophobic (Fig. 9.2).

9.2 Bioavailability

Because of their hydrophobicity, carotenoids are not soluble in the aqueous environment of the gastrointestinal tract. They need to be dissolved/carried in lipid + bile salt systems to be absorbed at the enterocyte brush border. It is important to remember, in fact, that the uptake of all carotenoids from diet is influenced by many variables such as: (a) the state of the food (raw, cooked, and/or processed); (b) the presence and efficiency of digestive enzymes and other endogenous digestives; and (c) the composition of a meal (presence of fibers, fat, and its physical form).

Moreover, the location and the physical form of all the carotenoids (in addition to age, gender, smoking status, and alcohol composition) influence their bioavailability and their consequent absorption. Thus, absorption of carotenoids and xanthophylls is enhanced by their transfer to the lipid phase during cooking in the presence of oil and by disruption of the cellular matrix during mastication [7,8]. Absorbed by the mucus of the small intestine, they are transported through the enterocyte and hepatocyte, and incorporated into chylomicrons. Finally, they are released into the systemic circulation carried by high- and low-density lipoproteins. It was recently shown that the bioavailability of purified lutein diet supplement is nearly double that of lutein taken by vegetable sources [9–11].

9.3 Biological Activities

The chemical and biochemical activities of carotenoids are related to their unique structure, consisting of an extended system of conjugated double bonds. A number of biological effects therefore have been attributed to carotenoids, including antioxidant activity, influence on the immune system, control of cell growth and differentiation, and stimulatory effects on gap junctional communications. However, recent attempts at dietary manipulation appear to be promising in terms of providing protection against certain solar-induced effects present in photoaged skin. Carotenoids are powerful singlet-oxygen quenchers and exhibit additional antioxidant properties. In fact, their conjugated polyene backbone has the ability to delocalize a charge or an unpaired electron [12]. These physical chemical properties confer the ability to act as an antioxidant and to terminate free radical reactions *in vitro* with the production of resonance-stabilized free radical structures. Termination may be a result of (a) adduct formation, where the free radical joins onto the polyene chain to produce a less reactive free radical; (b) electron transfer from the carotenoid to the free radical to produce a less reactive charged carotenoid radical; or (c) donation of a hydrogen molecule to the free radical to produce a stable carotenoid radical [13]. However, oxygen species that are efficiently scavenged by carotenoids are $^1O_2$ and peroxyl radicals, and physical quenching seems to be the major pathway involved in the deactivation of $^1O_2$.

Moreover, it has been shown that a combination of carotenoids plus vitamins E and C are more effective than β-carotene alone [14–16] and may increase superficial skin lipids (Fig. 9.3), skin hydration (Fig. 9.4),

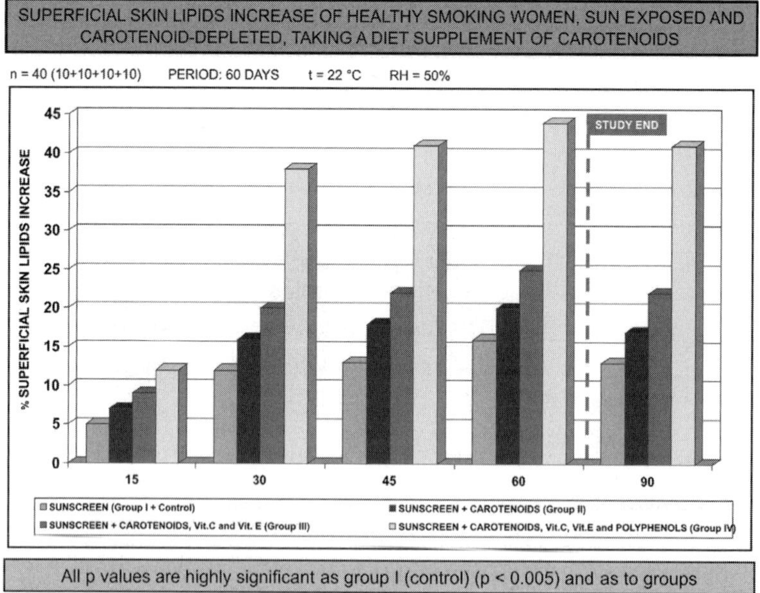

Figure 9.3 Percent increase in superficial skin lipids after dietary supplementation with carotenoids or carotenoids plus other antioxidants.

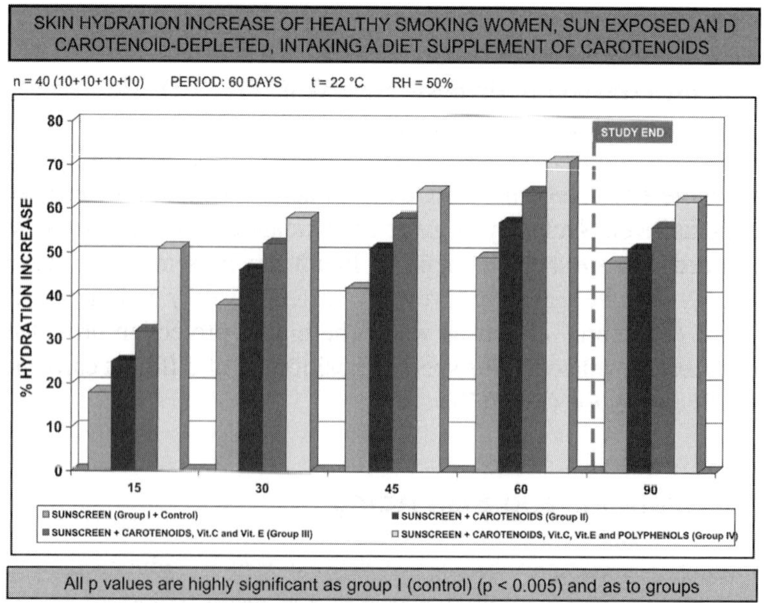

Figure 9.4 Percent increase in skin hydration after dietary supplementation with carotenoids or carotenoids plus other antioxidants.

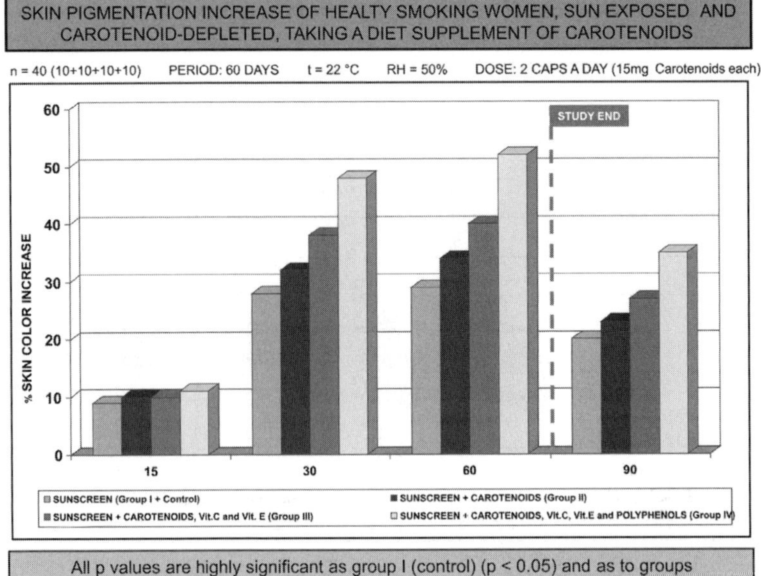

Figure 9.5 Percent increase in skin pigmentation after dietary supplementation with carotenoids or carotenoids plus other antioxidants.

and skin pigmentation of dryness-prone skin (Fig. 9.5), also decreasing oxidative stress at the level of blood serum (Fig. 9.6). These mixtures, in fact, are able to inhibit the formation of thiobarbituric acid reactive compounds more effectively than single components when they are used at the same molar level. Such a synergistic antioxidant effect seems to be more pronounced when either lycopene or lutein are present in the mixture. These data indicate that dose levels of carotenoids may be important and may have differential effects as well. The higher protection provided by mixtures may be related to the specific positioning of different carotenoids in the cell membrane (Fig. 9.7).

9.4 Carotenoids and Skin Aging

Skin aging is a complex biological process that is influenced by both intrinsic and extrinsic factors that lead to a progressive loss of the skin's flexibility and youthful appearance. Natural aging is accelerated by environmental factors and by sun exposure in particular. Macroscopic

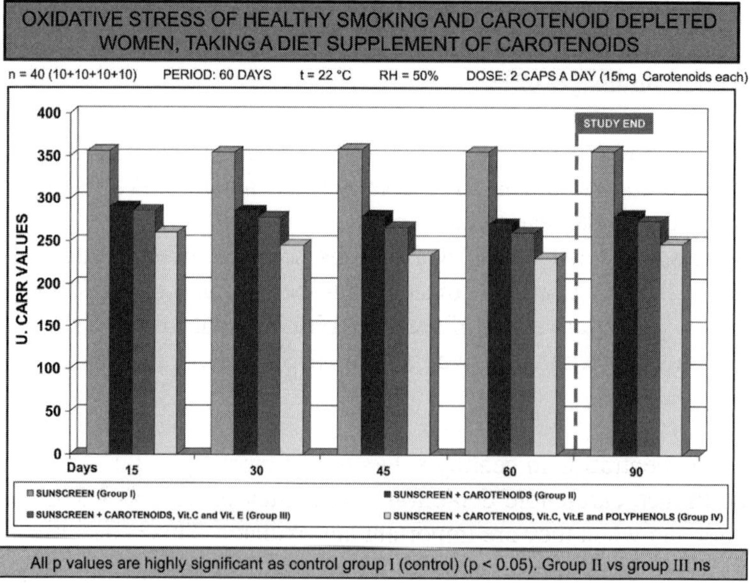

Figure 9.6 Oxidative stress level in blood serum after dietary supplementation with carotenoids or carotenoids plus other antioxidants.

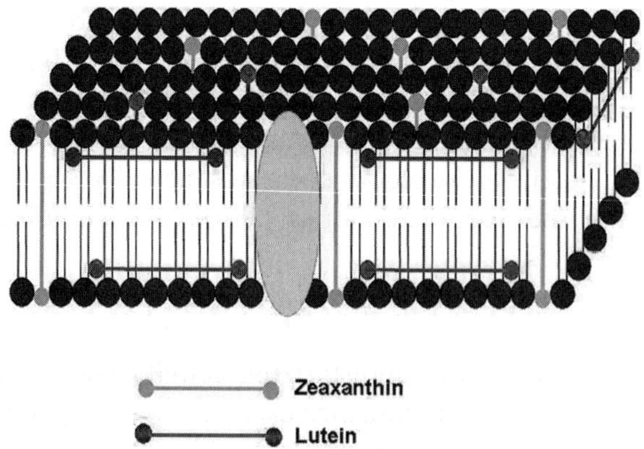

Figure 9.7 Schematic representation of the positioning of carotenoids in the cell membrane.

changes (skin wrinkling, rough skin texture, and irregular pigmentation) and microscopic changes (degradation of extracellular matrix molecules and DNA damage) are the hallmark of photoaging attributed to both UVB and UVA rays [17,18]. All of these processes are involved in the initiation and progression of skin cancers. There is also evidence indicating that prolonged UV exposure depletes the serum and skin of both carotenoids and xanthophylls [19–21].

The protective effects are thought to be related to the antioxidant properties of carotenoids. During ultraviolet (UV) irradiation, skin is exposed to photo-oxidative damage induced by the formation of reactive oxygen and nitrogen species (ROS and RNS). This damage affects cellular lipids, proteins, and DNA, and is considered to be involved in the formation of erythema, premature aging of the skin, photodermatoses, and skin cancer. Carotenoids are efficient scavengers of ROS and RNS [22].

What is important to underline is the necessity to control the baseline carotenoid concentration in healthy subjects participating in a diet supplement study. In fact, carotenoid depletion studies may provide a clear picture of whether and when they are important antioxidants. This is because almost all the conflicting information on the antioxidant activity of carotenoids has been obtained by administering carotenoid supplements to already well-fed individuals. Our double-blind placebo-controlled trial involving the use of lutein/zeaxanthin taken orally and at the same time applied topically has yielded interesting results on different controlled parameters [23] such as skin hydration (Fig. 9.8), superficial skin lipids (Fig. 9.9), skin elasticity (Fig. 9.10), and lipid peroxidation (Fig. 9.11). Test subjects followed a 6-day rotational balanced Mediterranean diet containing no more than 0.5 mg. of β-carotene/day. As a consequence, 15 days before starting, the level of β-carotene in the blood serum was medially 0.35 ± 0.6 μmol/L, whereas during the supplementation period plasma levels increased medially to 2.3 ± 1.7 μmol/L.

What were the unexpected results? Xanthophylls and carotens seem to have not only an interesting moisturizing activity but also a combined metabolic route, and the two influence each other. Thus, playing a specific role as a photoprotective agent thanks to its ability to screen out damaging blue and UV light from the sun, lutein/zeaxanthin has four primary functions:

1. to quench the triplet state of photosensitizer molecules and the singlet state of molecular oxygen
2. to act as an antioxidant against oxygen and nitrogen reactive species

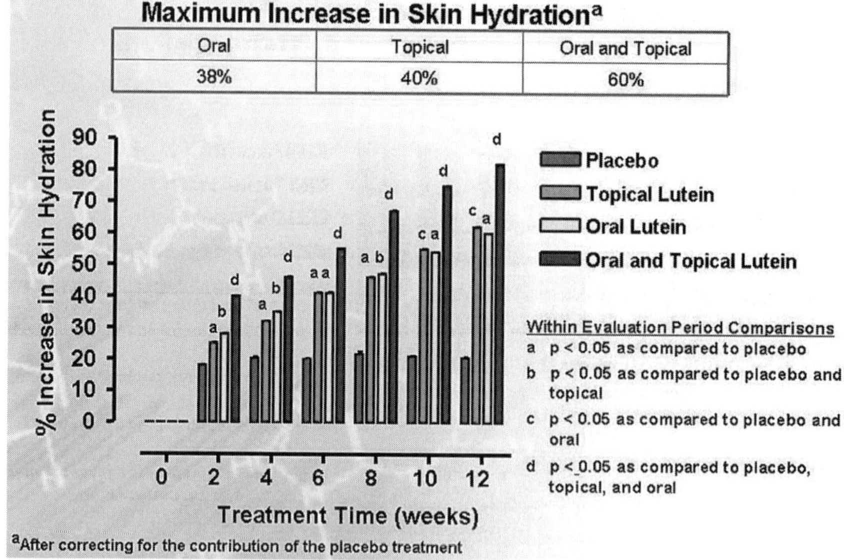

Figure 9.8 Percent increase in skin hydration after treatment with topical lutein, oral lutein, or oral + topical lutein.

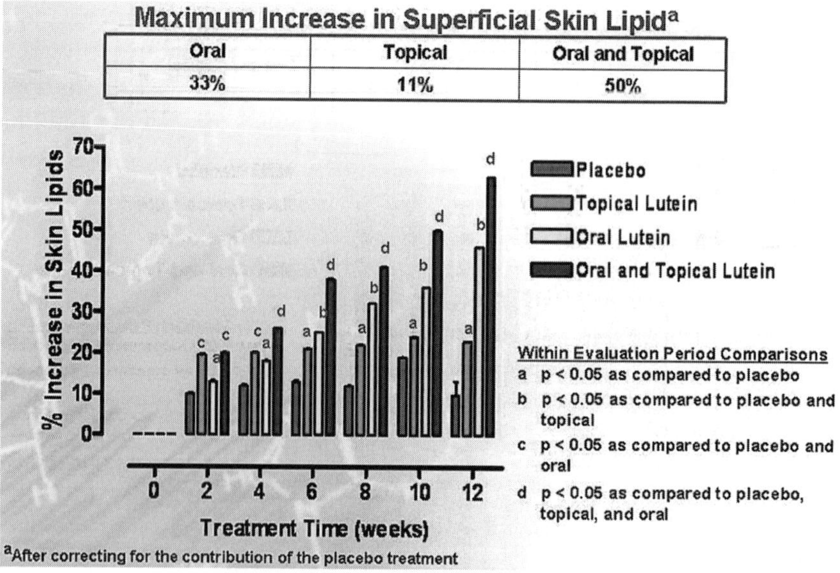

Figure 9.9 Percent increase in superficial skin lipids after treatment with topical lutein, oral lutein, or oral + topical lutein.

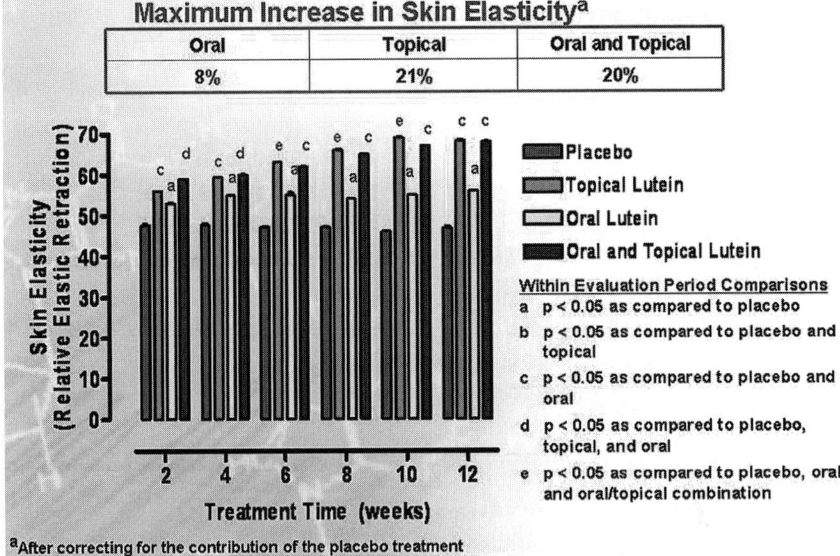

Figure 9.10 Change in skin elasticity (relative elastic retraction) after treatment with topical lutein, oral lutein, or oral + topical lutein.

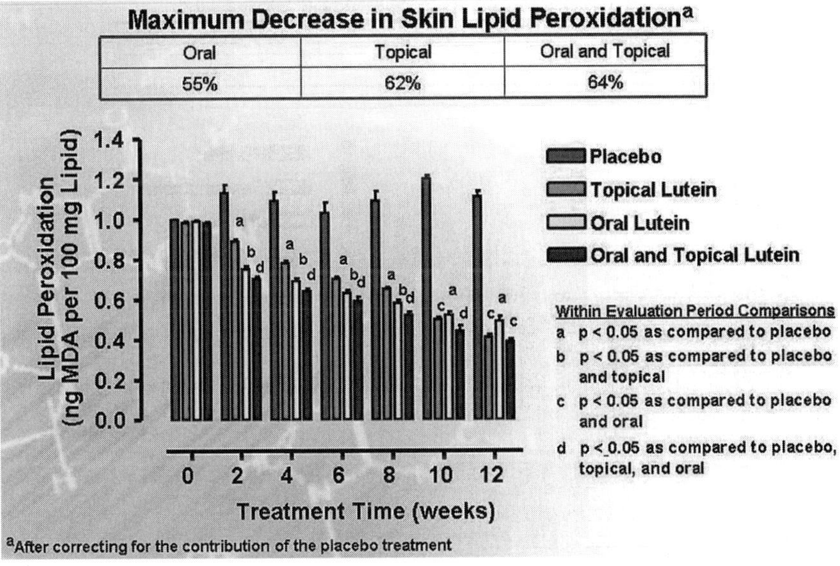

Figure 9.11 Change in skin lipid peroxidation after treatment with topical lutein, oral lutein, or oral + topical lutein.

3. to absorb blue wavelength light, which is currently considered much more detrimental than UV rays
4. to maintain the moisture activity at the level of the human horny layer's lipid lamellae

It therefore appears that these xanthophylls play a specific role as photoprotective agents capable of reducing inflammatory ROS-mediates, modulating skin hydration, decreasing skin aging, and, therefore, improving the quality of our life.

9.5 Concluding Remarks

The large group of plant carotenoids present in the Mediterranean diet attracts major interest because of their potential antiaging and other beneficial properties, presumably based on their function as natural antioxidants. Carotenoids are compounds of particular interest because of their extensive use in dietary supplements. Their regular, long-term consumption may improve antioxidant defence *in vivo* and thus help to lower risks associated with diseases caused by oxidative damage.

However, more information on the photoaging activity of these natural compounds is needed in order to understand how they act on the various target tissues. Systematic pharmacokinetic and dose-response studies are required to determine the different bioavailability of the individual carotenoids and xanthophills, and to estimate the amounts in diet that are likely to induce biological effects. Finally, more long-term carotenoid depletion chemical trials with well-characterized diet supplements are necessary in order to confirm their beneficial antiaging effects in humans.

References

1. Hasler CM. (1998) Functional food: their role in disease prevention and health promotion. *Food Technol* **52**(11): 63–70.
2. Boelsma E, Hendriks HF, Roza L. (2001) Nutritional skin care: health effects of micronutrients and fatty acids. *Am J Clin Nutr* **73**(5): 853–864.
3. Heinrich V, Wielnsch M, Tronnier H. (1998) Photoprotection from ingested carotenoids. *Cosmet Toilet* **113**: 61–70.
4. Britton G, Liacen-Jensen S, Pfander H (eds.) (2004) *Carotenoids Handbook*. Birkhauser Verlag, Basel.

5. Sistrom WR, Griffiths M, Dtanier RY. (1956) A note on the porphyrins excreted by the blue-green mutant of *Rhodopseudomonas spheroides. J Cell Comp Physiol* **48**: 459–515.
6. Mathews-Roth MM, Sistrom WR. (1959) Function of carotenoid pigments in non-photosynthetic bacteria. *Nature* **184**: 1892–1893.
7. Sommerburg O, Keunen JE, Bird AC, van Kuijk FJ. (1998) Fruits and vegetables that are sources for lutein and zeaxanthin: the macular pigment in human eyes. *Br J Ophthalmol* **82**: 907–910.
8. van het Hof KH, Tijburg LB, Pietrzik K, Weststrate JA. (1999) Influence of feeding different vegetables on plasma levels of carotenoids, folate and Vit. C. Effect of disruption of the vegetable matrix. *Br J Nutr* **82**: 203–212.
9. Demming-Adams B, Gilmore AM, Adams III WW. (1996) Carotenoids 3: in vivo function of carotenoids in higher plants. *FASEB J* **10**: 403–412.
10. Baroli I, Niyogi K. (2000) Molecular genetics of xanthophyll-dependent photoprotection in green algae and plants. *Philos Trans R Soc Load B Biol Sci* **355**: 1385–1394.
11. Krinsky NI. (2002) Possible biologic mechanism for a protective role of xanthophylls. *J Nutr* **132**: 540S–542S.
12. Black HS, Rhodes LE. (2001) Systemic photoprotection dietary intervention and therapy. In: Giacomoni P (ed.) *Sun Protection Man.* Elsevier, New York, pp. 573–591.
13. Faulks RM, Southon S. (2001) Carotenoids, metabolism and disease. In: Wildman REC (ed.) *Handbook of Nutraceuticals and Functional Foods*, CRC Press, Boca Raton, FL, pp. 143–156.
14. Postaire E, Jungmann H, Bejot M, Heinrich U, Tronnier H. (1997) Evidence for antioxidant nutrients-induced pigmentation in skin: results of a clinical trial. *Biochem Mol Biol Int* **42**: 1023–1033.
15. Morganti P, Fabrizi G, Morganti G. (2001) Topical and systemic photoprotectants to prevent light-induced reactions. *Eurocosm* **9**(3): 18–21.
16. Morganti P, Fabrizi G, Morganti G. (2000) New data on skin photoprotection. *Int J Cosmet Sci* **22**: 305–312.
17. Kritchevsky SB, Bush AJ, Pahor M, Gross MD. (2000) Serum carotenoids and markers of inflammation in nonsmokers. *Am J Epidemiol* **152**: 1065–1071.
18. Berendschot TT, Goldbohm RA, Klopping WA, van de Kraats J, van Norel J, van Norren D. (2000) Influence of lutein supplementation on macular pigment, assessed with two objective techniques. *Invest Ophthal and Visual Sci* **41**: 3322–3326.
19. White WS, Kim CI, Kalkwarf HJ, Bustos P, Roe DA. (1998) Ultraviolet light-induced reductions in plasma carotenoid levels. *J Clin Nutr* **47**: 879–883.
20. Ribaya-Mercado JD, Garmyn M, Gilchrest BA, Russell RM. (1995) Skin lycopene is destroyed preferentially over β-carotene during ultraviolet irradiation in humans. *J Nutr* **125**: 1854–1859.
21. Sorg O, Tran C, Carraux P, Didierjean L, Falson F, Saurat JH. (2002) Oxidative stress-independent depletion of epidermal vitamin A by UVA. *J Invest Dermato* **118**: 513–518.
22. Heinrich U, Gurtner C, Wiebusch M. (2003) Supplementation with β-carotene or a similar amount of mixed carotenoids protects human from UV-induced erythema. *J Nutr Journ* **133**(1): 98–101.

23. Palombo P, Fabrizi G, Ruocco V, Ruocco E, Flühr J, Roberts R, Morganti P. (2007) Beneficial long-term effects of combined oral/topical antioxidant treatment with carotenoids lutein and zeaxanthin on human skin: a double-blinded, placebo-controlled study in humans. *Skin Pharmacol Physiol* **20**: 199–210.

10

Inhibitory Effects of Coenzyme Q_{10} on Skin Aging

Yutaka Ashida, PhD

Shiseido Co., Ltd., Yokohama, Japan

Aaron Tabor and Robert M. Blair (eds.), *Nutritional Cosmetics: Beauty from Within*, 199–215, © 2009 William Andrew Inc.

10.1 Coenzyme Q_{10}

Coenzyme Q_{10} (CoQ_{10}), which has both energizing and antioxidative effects, was found only a half century ago [1]. It was first isolated in 1957 from beef heart mitochondria by Crane, who named it coenzyme Q based on its function in the electron transfer system of mitochondria [2]. In the same year, Morton independently isolated ubiquinone, which means "ubiquitous quinone," from rat liver [3,4]. In the following year, Isler established the structure of ubiquinone [5], and Folkers determined the structure of CoQ, demonstrating that ubiquinone and coenzyme Q are identical [6]. The "10" in the abbreviation CoQ_{10} indicates that the molecule has a side chain consisting of ten isoprenoid units, although the chain length is different among species. For example, human beings, cattle, horses, pigs, sheep, rabbits, and soybeans contain CoQ_{10}, but rodents, corn, and rice have CoQ_9, *Escherichia coli* has CoQ_8, and various kinds of yeast have CoQ_{10}, CoQ_9, CoQ_7, or CoQ_6. Some kinds of plants, such as sweet pepper, contain CoQ_{11}.

Figure 10.1(a) shows the structure of CoQ_{10} (ubiquinone) in oxidized form. The reduced form, $CoQ_{10}H_2$ (ubiquinol), is shown in Fig. 10.1(b). In general, CoQ_{10} administered orally as a supplement in the oxidized form is reduced to $CoQ_{10}H_2$ by DT-diaphorase [7] or NADPH-dependent CoQ reductase [8], and CoQ_{10} exists as the reduced form, $CoQ_{10}H_2$ or ubiquinol, in the living body. The functions of CoQ_{10} in the mitochondrial respiratory

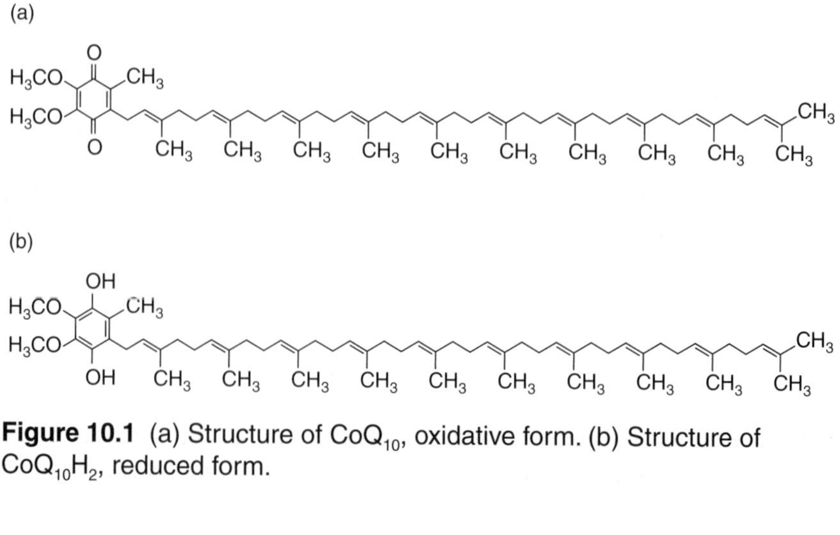

(a)

(b)

Figure 10.1 (a) Structure of CoQ_{10}, oxidative form. (b) Structure of $CoQ_{10}H_2$, reduced form.

system were investigated thoroughly by Mitchell, who won the Nobel Prize for this work in 1978 [9].

CoQ_{10} (30 mg/day) has been used as a medicine for mild heart failure in Japan since 1974. In some other countries, including the U.S., the usage of CoQ_{10} as a health supplement has become popular since the 1990s. Additionally, clinical trials of high-dose CoQ_{10} for various diseases, including Parkinson's [10], Huntington's [11], and Alzheimer's [12] diseases, have been reported. CoQ_{10} is clinically safe [13,14], and few side effects have been reported during long-term trials, even at high doses such as 1,200 mg/day, in humans [10,11]. Chronic oral administration of CoQ_{10} 1,200 mg/kg/day for 52 weeks in rats was also shown to have no adverse effects [15].

Human beings endogenously synthesize CoQ_{10}. Therefore, CoQ_{10} is not a vitamin, since by definition vitamins are not biosynthesized in the human body. Its functions in the body, however, are as important as those of vitamins. For example, it plays a key role in ATP synthesis, that is, biological energy production, in mitochondria. CoQ_{10}, which is hydrophobic, exists not only in the cell membrane and mitochondria, but also in cytosol. It also has an important antioxidative effect, in combination with hydrophobic tocopherols, vitamin E, and hydrophilic ascorbic acid, vitamin C, throughout the body.

The CoQ_{10} content in organs including skin decreases with aging [16], with exercise [17], with various diseases [10–12], and during statin therapy [18]. The CoQ_{10} content in foods, such as vegetables, cereals, meats, and fishes, is not large, and therefore, CoQ_{10} intake as a supplement is recommended. The targets of CoQ_{10} supplementation are: (1) improvement of cardiac function [19,20], (2) improvement of aerobic exercise ability [21–25] and muscle protection [26,27], (3) preventing the decrease of CoQ_{10} owing to statin therapy for hypercholesterolemia [18,28], and (4) antiaging and protective effects in various organs, including skin [29–36], the subject of this book.

Several reports have described antiaging effects of CoQ_{10} in skin [32–35], but the mechanisms have not been elucidated, except the broad concept of "antioxidative and energizing effects." Although topical application of CoQ_{10} may have an antioxidative effect in corneocytes, at the surface of the skin [32], it is not clear whether CoQ_{10} is transferred into deeper regions of the skin after epicutaneous application. Furthermore, the mechanisms of antiaging effects, especially in skin, remain to be established.

10.2 Skin Aging

Skin aging can be subdivided into intrinsic aging and photoaging [37–39]. It is well known that photoaging of areas such as the face or neck skin results in various clinical aging features, including wrinkles. In aging skin, the epidermal-dermal junction becomes flattened [40]. Photoaging also shows features such as cellular atypia, loss of polarity, and a decrease in type I collagen. Furthermore, in sun-exposed skin, disruption and duplication of the lamina densa in the basement membrane were reported [37,41,42]. It is well known that ultraviolet radiation exposure causes premature aging. Tobacco smoking is also an important factor contributing to premature aging of human skin [43–45]. Oxidative stresses inside and outside of the body can cause damage in tissues such as cells, matrices, and basement membranes.

The basement membrane at the dermal-epidermal junction is mainly composed of type IV and VII collagens; several laminins such as laminin 5, 6, and 10; nidogen; and perlecan [40,46–48]. Laminin 5 and type VII collagen are essential for epidermal attachment [40,49–51]. Keratinocytes produce the basement membrane components, except for nidogen [40,52]. Dermal fibroblasts also synthesize the basement membrane components, except for laminin 5 [40,53].

Damage to the basement membranes may result in features of skin aging, such as wrinkle formation. Therefore, inhibition of matrix metalloproteinases (MMP), which degrades the basement membrane components, may reduce wrinkle formation [40]. Laminin 5 appears to be one of the most important factors for wrinkle formation, because Nishiyama et al. showed that laminin 5 initiates hemidesmosome formation, provides stable attachment of the epidermis to the dermis, and also accelerates the assembly of basement membrane and enhances the recovery of damaged skin [54]. Tsunenaga et al. reported that laminin 5 accelerates lamina densa formation along the dermal-epidermal junction [55]. Amano et al. demonstrated that the balance between extracellular matrix synthesis and degradation in the basement membrane is important for its formation [56]. In addition, Takada et al. reported that gelatinase is consistently upregulated by sunlight exposure in daily life and may be an etiological factor in photoaging, including wrinkle formation [57].

10.3 Contribution to Beauty from Within (Benefits for Wrinkles)

We first studied the effects of orally administered CoQ_{10} on wrinkles in humans. To confirm that there is an increase of CoQ_{10} content in skin after oral administration, we developed a simultaneous quantification system for CoQ_9 and CoQ_{10}, and investigated the absorption and distribution to skin of orally administered CoQ_{10}. Then, in order to elucidate the mechanisms through which CoQ_{10} reduces wrinkle formation, we examined the effects of CoQ_{10} on (1) the proliferation of dermal fibroblasts, which produce type I collagen, (2) the antioxidative capacity of epidermal keratinocytes, and (3) the production of laminin 5, and type IV and type VII collagen, which are basement membrane components, in skin cells. These results were presented in the 4th and 5th Conferences of the International CoQ_{10} Association [36,58]. The whitening effects of CoQ_{10} are also under investigation.

Data are presented as mean ± SD. The statistical significance of differences was determined by using ANOVA, Fisher's protected least significant difference (Fisher's PLSD), and the Scheffe, Dunnett, and Wilcoxon tests as appropriate. Analyses were performed with StatView software (version 5). The criterion of statistical significance was set at $P < 0.05$.

Hoppe et al. reported that topical application of 0.3% CoQ_{10} for 6 months reduced the depth of wrinkles in aged human skin [32]. Passi et al. showed that the combined use of oral (50 mg/day) and topical (0.05%) CoQ_{10} for 60 days reduced the wrinkle depth [33]. Furthermore, we have already shown that oral supplementation alone of CoQ_{10} 60 mg/day for at least 2 weeks in humans ($N = 8$, female, 43 ± 3 years old) reduced the wrinkle area rate and wrinkle volume per unit area in the corner of the eyes [34].

Figure 10.2(a) shows the effect of dietary CoQ_{10} on wrinkle area rate. After 2 weeks, the wrinkle area rate was significantly decreased. The improvement rate, in other words, reduction rate of wrinkles, was 33% compared with the pre-administration value, and remained at this level for 3 months. Even after 3 months, the wrinkle area rate was still significantly decreased. Figure 10.2(b) shows the effect of dietary CoQ_{10} on wrinkle volume. After 2 weeks, the wrinkle volume was also significantly decreased. The improvement rate was 38% compared with the pre-administration value. After 3 months, the wrinkle volume was still reduced, although without statistical significance. These results [32–34] suggest that CoQ_{10} is an

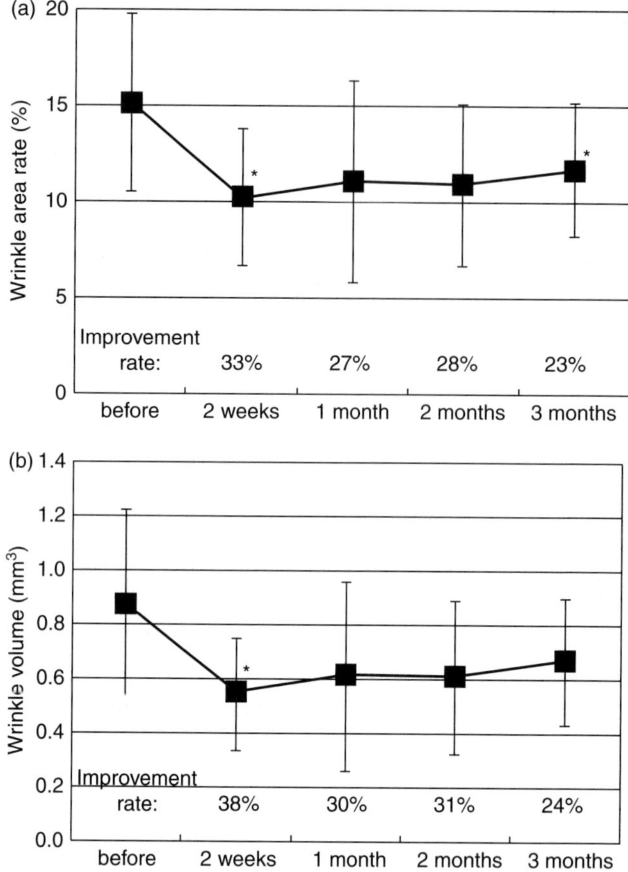

Figure 10.2 Inhibitory effects of nutritional CoQ_{10} (60 mg/day) on wrinkles at the corner of the eye in humans. (a) Wrinkle area rate, (b) wrinkle volume. Mean ± SD, N = 8 (female; 43 ± 3 years old). Wilcoxon-Dunnett test $^{*}P < 0.05$.

effective nutritional supplement for the realization of "beauty from within," especially for wrinkles.

10.4 Mechanisms of Action

10.4.1 Determination of CoQ_{10}

Quantitative determination of CoQ_9 and CoQ_{10} (in the range of 15.6–2,000 ng/ml each) was done with a high-performance liquid

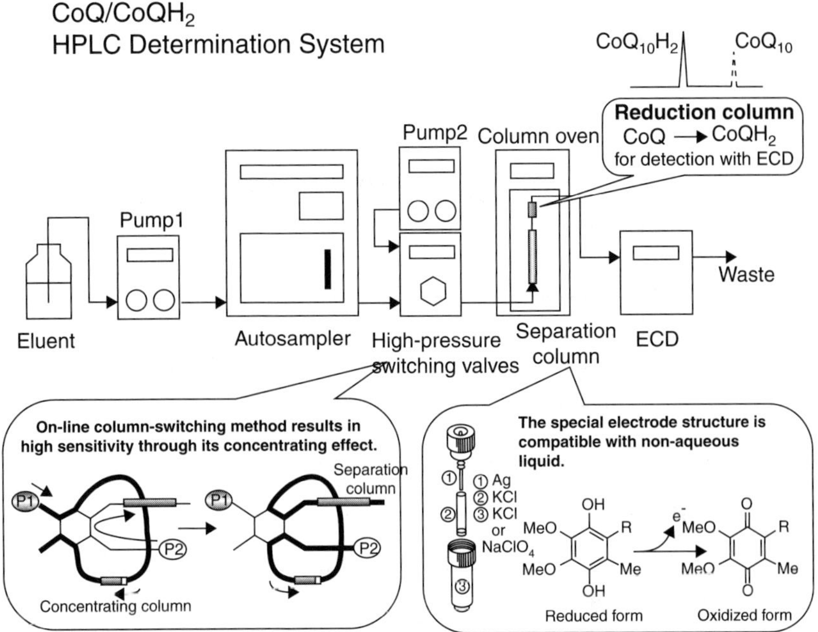

Figure 10.3 High-performance liquid chromatography (HPLC) with electrochemical detector (ECD) for CoQ/$CoQH_2$ quantification [58,59].

chromatography (HPLC)–electrochemical detector (ECD) system, as shown in Fig. 10.3 (NANOSPACE SI-2, Shiseido, Tokyo, Japan), based on the method of Yamashita and Yamamoto [59]. We adopted ECD (instead of a UV detector) and an on-line column-switching method with a concentrating column to increase the sensitivity. Additionally, the special electrode structure in the reduction column makes the nonaqueous reversed phase compatible with the aqueous phase. The peaks of CoQ_4, CoQ_6, CoQ_9, CoQ_{10}, tocopherols, lycopene, and beta-carotene were well separated. We can determine CoQ_9 and CoQ_{10} simultaneously, as well as the quantities of the reduced form and the oxidized form of CoQ_{10}. In this study, however, we measured total CoQ_{10}.

10.4.2 Absorption and Distribution to Skin

Many reports have indicated that oral CoQ_{10} administration increases the CoQ_{10} concentration in serum and organs, such as liver, heart, and brain [60–64]. The concentration in skin, however, had not been examined. In Fig. 10.4, we show that oral administration of CoQ_{10} increases the CoQ_{10}

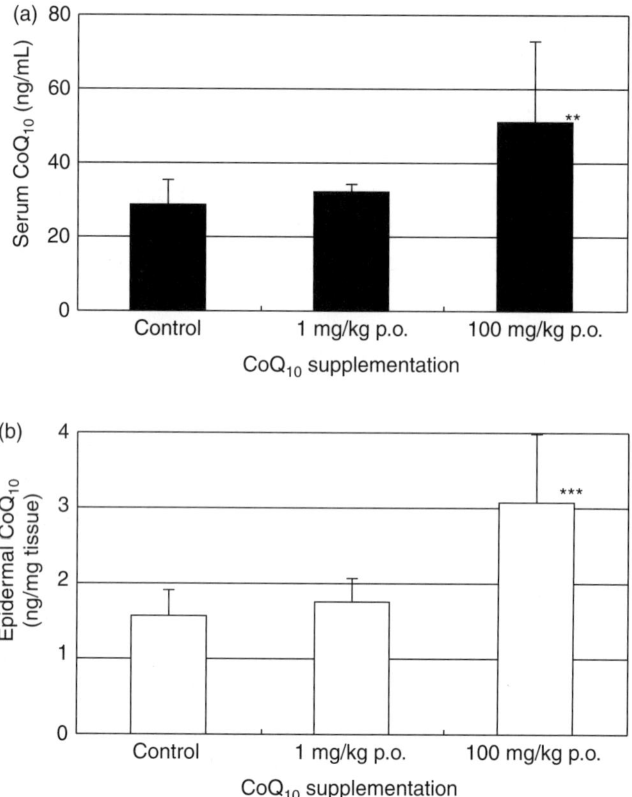

Figure 10.4 Oral administration of CoQ_{10} increases the CoQ_{10} concentration in (a) serum and (b) epidermis in mice. CoQ_{10} (0, 1, and 100 mg/kg) was administered orally for 2 weeks. Mean ± SD, N = 6–7, ANOVA and Fisher's PLSD $^{**}P < 0.01$, $^{***}P < 0.001$.

concentration in serum and epidermis, using the HPLC-ECD system described above [58,65].

The epidermis consists almost totally of keratinocytes, which move progressively from the epidermal basement membrane toward the skin surface, forming four well-defined layers during the transition. These are the stratum basale, stratum spinosum, stratum granulosum, and stratum corneum [66]. CoQ_{10} supplementation may increase the content of CoQ_{10} in the keratinocytes of the stratum basale, and differentiation and migration of these keratinocytes toward the skin surface may account for the increase of epidermal CoQ_{10} following oral CoQ_{10} supplementation. Therefore, oral application of CoQ_{10} may be useful to prevent the decrease of CoQ_{10} in the inner skin with aging.

Passi et al. reported that simultaneous oral and topical application of CoQ_{10} increased the CoQ_{10} levels in both sebum and stratum corneum, but that topical application alone only increased the level in the sebum, not that in the stratum corneum [33]. These findings also indicate that oral CoQ_{10} supplementation would be effective for improving skin condition.

10.4.3 Antioxidant Effects in Skin

It is well known that CoQ_{10} has antioxidative effects, but this has not been demonstrated in skin cells. Therefore, we examined the antioxidant effects of CoQ_{10} in epidermal keratinocytes [36]. As shown in Fig. 10.5, we confirmed that CoQ_{10} significantly prevented 2,2′-azobis-(2-amidinopropane) dihydrochloride (APPH)-induced injury to epidermal keratinocytes. CoQ_{10} also prevented cell damage induced by other oxidants, such as hydroperoxide and *tert*-butyl hydroperoxide, in epidermal keratinocytes (data not shown). The oxidative protein ratio in corneocytes was slightly decreased in humans who took CoQ_{10} 60 mg/day for 3 months (Fig. 10.6). A higher level of supplementation may be needed to produce a significant decrease of protein oxidation in corneocytes, the most external organ in the body.

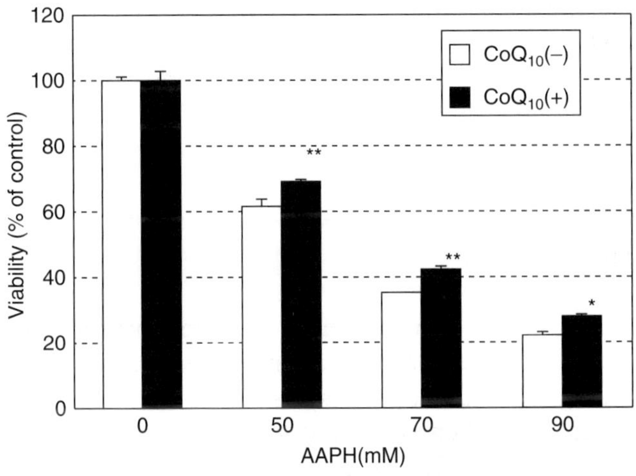

Figure 10.5 CoQ_{10} (25 μM) protects epidermal keratinocytes from AAPH-induced cell damage. AAPH: 2,2′-azobis-(2-amidinopropane) dihydrochloride. Mean ± SD, N = 3, Scheffe test $^*P < 0.05$, $^{**}P < 0.001$.

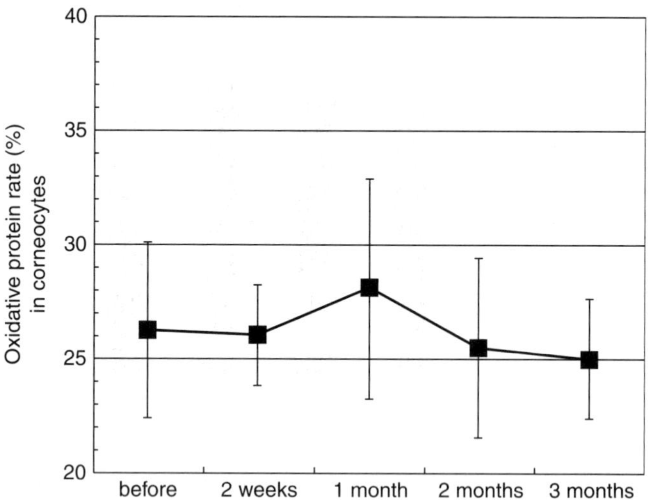

Figure 10.6 Effect of CoQ_{10} supplementation (60 mg/day) on oxidative protein content of human corneocytes. After 3 months, the P value is 0.077 compared with pre-supplementation (before). Student's paired t-test. Mean ± SD, N = 20.

10.4.4 Energizing Effects in Skin

The energizing effects of CoQ_{10} in skin cells were also examined [36]. CoQ_{10} dose-dependently accelerated the proliferation of human dermal fibroblasts *in vitro* (Fig. 10.7). Furthermore, it augmented the production of laminin 5 and type IV and type VII collagen, which are basement membrane constituents anchoring the epidermis and dermis (Figs. 10.8 and 10.9). The contents of laminin 5, type IV, and type VII collagen were determined with ELISA [53,67]. As described in the introduction, damage to the basement membrane is one of the causes of wrinkle formation. These results suggest that CoQ_{10} supplementation is involved in energizing skin cells and improving the synthesis of basement membrane constituents, which may result in the improvement of wrinkles induced by photoaging.

10.5 Current and Future Studies

Hoppe et al. showed that CoQ_{10} inhibits collagenase activity. MMPs are also important factors involved in wrinkle formation. Further investigation of the effects of CoQ_{10} on MMPs is needed.

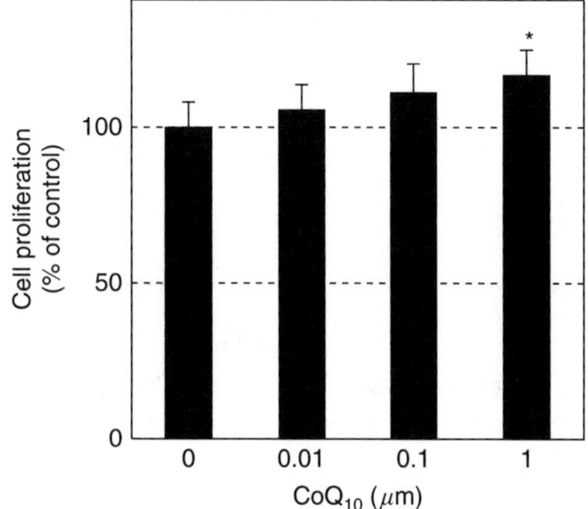

Figure 10.7 CoQ_{10} accelerates cell proliferation significantly in human dermal fibroblasts *in vitro* [32]. Mean ± SD, N = 6, Scheffe test *P < 0.05.

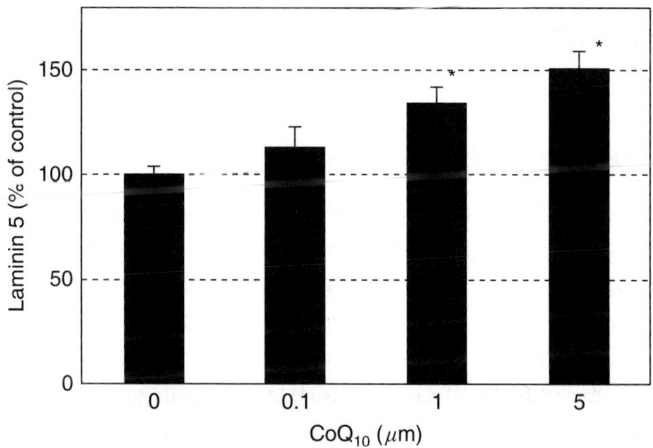

Figure 10.8 CoQ_{10} increases production of laminin 5, one of the major components of basement membrane, in human epidermal keratinocytes *in vitro* [32]. Mean ± SD, N = 6, Scheffe test *P < 0.001.

Whitening effects of CoQ_{10} are also expected as a result of its antioxidative ability. We confirmed that CoQ_{10} does not inhibit tyrosinase activity, but it significantly inhibited the auto-oxidation of 3,4-dihydroxy-L-phenylalanine (L-DOPA), a precursor of melanin, and consequent pigmentation (Fig. 10.10). Oral supplementation of CoQ_{10} over the long term may significantly reduce pigmentation levels *in vivo* (data not shown; submitted for publication).

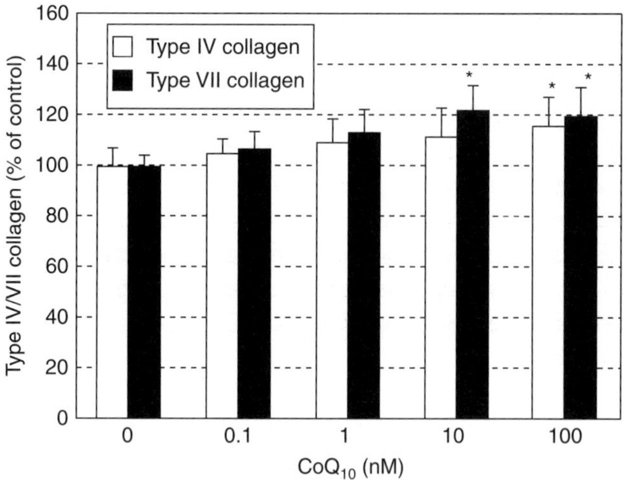

Figure 10.9 CoQ_{10} significantly increases the production of types IV and VII collagen, constituents of basement membrane, in human dermal fibroblasts *in vitro* [32]. Mean ± SD, N = 6, Dunnett test *P < 0.05.

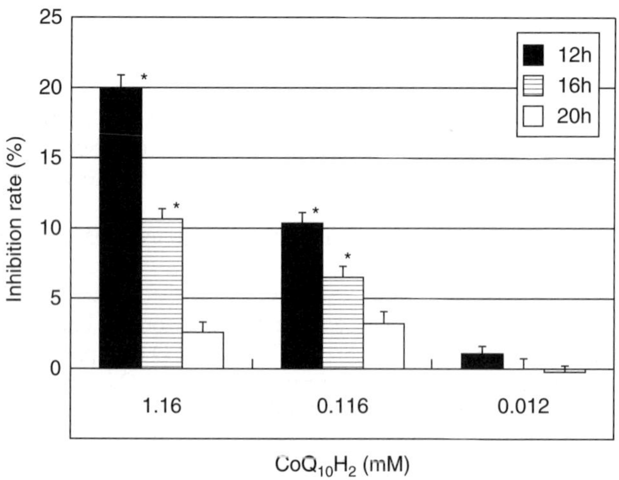

Figure 10.10 $CoQ_{10}H_2$ inhibits L-DOPA auto-oxidation. The absorbance at 475 nm (OD475) of L-DOPA and $CoQ_{10}H_2$ in 0.1 M phosphate buffer was measured after incubation for 12 h, 16 h, and 20 h at room temperature. Mean ± SD, N = 4, Scheffe *P < 0.001.

10.6 Conclusion

CoQ_{10} is considered to be an effective nutritional supplement that ameliorates skin aging, especially wrinkles. The content of endogenously synthesized CoQ_{10} in skin decreases with aging. Oral administration of CoQ_{10} increases the epidermal CoQ_{10} level, and this is considered to have energizing and antioxidative actions in skin cells. Mechanisms of the antiaging effects in skin are (1) promotion of fibroblast proliferation, (2) induction of increased production of basement membrane constituents, and (3) inhibition of cell damage and pigment-generating auto-oxidation through its antioxidative ability.

In conclusion, oral supplementation of CoQ_{10} is expected to be effective for "beauty from within," through its antioxidant and energizing effects in skin.

References

1. Crane FL. Discovery of ubiquinone (coenzyme Q) and an overview of function. Mitchondrion **7**(Suppl 1): S2–S7, 2007.
2. Crane FL, Hatefi Y, Lester RL, Widmer C. Isolation of a quinone from beef heart mitochondria. Biochim Biophys Acta **25**: 220–221, 1957.
3. Morton RA, Wilson GM, Lowe JS, Leat WMF. Ubiquinone. In: Chemical Industry, London, pp. 1649, 1957.
4. Morton RA. Ubiquinone. Nature **182**: 1764–1767, 1958.
5. Morton RA, Gloor U, Schindler O, Wilson GM, Chopard-dit-Jean LH, Hemming FW, Isler O, Leat WMF, Pennock JF, Ruegg R, Schwieter U, Wiss O. Die Structure des Ubichinons aus Schweinherzen. Helv Chim Acta **41**: 2343–2357, 1958.
6. Shunk CH, Linn BO, Wong EL, Wittreich PE, Robinson FM, Folkers K. Coenzyme Q. II. Synthesis of 6-farnesyl- and 6-phytyl-derivatives of 2,3-dimethoxy-5-methylbenzoquinone and related analogs. J Am Chem Soc **80**: 4753, 1958.
7. Beyer RE, Segura-Aguilar J, Di Bernardo S, Cavazzoni M, Fato R, Fiorentini D, Galli MC, Setti M, Landi L, Lenaz G. The role of DT-diaphorase in the maintenance of the reduced antioxidant form of coenzyme Q in membrane systems. Proc Natl Acad Sci USA **93**: 2528–2532, 1996.
8. Takahashi T, Okamoto T, Kishi T. NADPH-dependent coenzyme Q reductase is the most powerful coenzyme Q_{10} reductase in rat liver cytosol [abstract]. In: The 5th Conference of the International CoQ_{10} Association, Kobe, Japan, pp. 21–22, 2007.
9. Mitchell P. Coupling of phosphorylation to electron and hydrogen transfer by a chemi-osmotic type of mechanism. Nature **191**: 144–148, 1961.
10. Shults CW, Oakes D, Kieburtz K, Beal MF, Haas R, Plumb S, Juncos JL, Nutt J, Shoulson I, Carter J, Kompoliti K, Perlmutter JS, Reich S, Stern M,

Watts RL, Kurlan R, Molho E, Harrison M, Lew M. Parkinson Study Group. Effects of coenzyme Q_{10} in early Parkinson disease: evidence of slowing of the functional decline. Arch Neurol **59**: 1541–1550, 2002.

11. Feigin A, Kieburtz K, Como P, Hickey C, Claude K, Abwender D, Zimmerman C, Steinberg K, Shoulson I. Assessment of coenzyme Q_{10} tolerability in Huntington's disease. Mov Disord **11**: 321–323, 1996.
12. Lenaz G, D'Aurelio M, Merlo Pich M, Genova ML, Ventura B, Bovina C, Formiggini G, Parenti Castelli G. Mitochondrial bioenergetics in aging. Biochim Biophys Acta **1459**: 397–404, 2000.
13. Hayashi Y. Methodology for ensuring safety of dietary supplement ingredients with special reference to coenzyme Q_{10} [abstract]. In: The 5th Conference of the International CoQ_{10} Association, Kobe, Japan, p. 72, 2007.
14. Hidaka T, Fujii K, Uchida Y, et al. Overview of safety assessment of coenzyme Q_{10} (CoQ_{10}): animal and human data [abstract]. In: The 5th Conference of the International CoQ_{10} Association, Kobe, Japan, p. 73, 2007.
15. Williams KD, Maneke JD, Abdel Hameed M, Hall RL, Palmer TE, Kitano M, Hidaka T. 52-Week oral gavage chronic toxicity study with ubiquinone in rats with a 4-week recovery. J Agric Food Chem **47**: 3756–3763, 1999.
16. Kalen A, Appelkvist EL, Dallner G. Age-related changes in the lipid compositions of rat and human tissues. Lipids **24**: 579–584, 1989.
17. Okamoto T, Mizuta K, Mizobuchi S, Usui A, Takahashi T, Fujimoto S, Kishi T. Decreased serum ubiquinol-10 levels in healthy subjects during exercise at maximal oxygen uptake. Biofactors **11**: 31–33, 2000.
18. Folkers K, Langsjoen P, Willis R, Richardson P, Xia LJ, Ye CQ, Tamagawa H. Lovastatin decreases coenzyme Q levels in humans. Proc Natl Acad Sci USA **87**: 8931–8934, 1990.
19. Kishi T, Okamoto T, Takahashi T, Goshima K, Yamagami T. Cardiostimulatory action of coenzyme Q homologues on cultured myocardial cells and their biochemical mechanisms. Clin Investig **71**: S71–S75, 1993.
20. Okamoto T, Kobayashi K, Honse Y, Takahashi T, Goshima K, Kishi T. Effect of ubiquinone-10 on fluctuation of beating rhythm of cultured myocardial cell sheets. Int J Vitam Nutr Res **62**: 155–159, 1992.
21. Ylikoski T, Piirainen J, Hanninen O, Penttinen J. The effect of coenzyme Q_{10} on the exercise performance of cross-country skiers. Mol Aspects Med **18**: S283–S290, 1997.
22. Fiorella PL, Bargossi AM, Grossi G, et al. Metabolic effects of coenzyme Q_{10} treatment in high level athletes. In: Folkers K, Littarru GP, Yamagami T (eds.) Biomedical and Clinical Aspects of Coenzyme Q, Volume 6, Elsevier Science Publishers, Amsterdam, pp. 513–520, 1991.
23. Bonetti A, Solito F, Carmosino G, Bargossi AM, Fiorella PL. Effect of ubidecarenone oral treatment on aerobic power in middle-aged trained subjects. J Sports Med Phys Fitness, **40**: 51–57, 2000.
24. Wyss V, Lubich T, Ganzit GP, et al. Remarks of prolonged ubiquinone administration in physical exercise. In: Lenaz G et al. (eds.) Highlights in Ubiquinone Research, Taylor & Francis, London, pp. 303–306, 1990.
25. Linnane AW, Zhang C, Yarovaya N, Kopsidas G, Kovalenko S, Papakostopoulos P, Eastwood H, Graves S, Richardson M. Human aging and global function of coenzyme Q_{10}. Ann N Y Acad Sci **959**: 396–411, 2002.

26. Okamoto T, Takahashi T, Kishi T. Protective effect of coenzyme Q_{10} on cultured skeletal muscle cell injury induced by continuous electric field stimulation. Biochem Biophys Res Commun **216**: 1006–1012, 1995.
27. Kaikkonen J, Kosonen L, Nyyssönen K, Porkkala-Sarataho E, Salonen R, Korpela H, Salonen JT. Effect of combined coenzyme Q_{10} and d-alpha-tocopheryl acetate supplementation on exercise-induced lipid peroxidation and muscular damage: a placebo-controlled double-blind study in marathon runners. Free Radic Res **29**: 85–92, 1998.
28. Silver MA, Langsjoen PH, Szabo S, Patil H, Zelinger A. Effect of atorvastatin on left ventricular diastolic function and ability of coenzyme Q_{10} to reverse that dysfunction. Am J Cardiol **94**: 1306–1310, 2004.
29. Yamamoto Y, Komuro E, Niki E. Antioxidant activity of ubiquinol in solution and phosphatidylcholine liposome. J Nutr Sci Vitaminol **36**: 505–511, 1990.
30. Mukai K, Kikuchi S, Urano S. Stopped-flow kinetic study of the regeneration reaction of tocopheroxyl radical by reduced ubiquinone-10 in solution. Biochim Biophys Acta **1035**: 77–82, 1990.
31. Okada K, Yamada S, Kawashima Y, Kitade F, Okajima K, Fujimoto M. Cell injury by antineoplastic agents and influence of coenzyme Q_{10} on cellular potassium activity and potential difference across the membrane in rat liver cell. Cancer Res **40**: 1663–1667, 1980.
32. Hoppe U, Bergemann J, Diembeck W, Ennen J, Gohla S, Harris I, Jacob J, Kielholz J, Mei W, Pollet D, Schachtschabel D, Sauermann G, Schreiner V, Stäb F, Steckel F. Coenzyme Q_{10}, a cutaneous antioxidant and energizer. BioFactors **9**: 371–378, 1999.
33. Passi S, De Pità O, Grandinetti M, Simotti C, Littarru GP. The combined use of oral and topical lipophilic antioxidants increases their levels both in sebum and stratum corneum. BioFactors **18**: 289–297, 2003.
34. Ashida Y, Nakashima M, Kuwazuru S, Watabe K. Effect of CoQ_{10} as a supplement on wrinkle reduction. Food Style 21 **8**: 52–54, 2004.
35. Ichihashi M, Ooe M, Inui M, Omura K, Fujii K. Efficacy evaluation of coenzyme Q_{10} in aged human skin *in vivo* [abstract]. In: The 5th Conference of the International CoQ_{10} Association, Kobe, Japan, p. 88, 2007.
36. Terada T, Takada K, Ashida Y, et al. Inhibitory effects of coenzyme Q_{10} on skin aging [abstract]. In: The 5th Conference of the International CoQ_{10} Association, Kobe, Japan, pp. 156, 2007.
37. Lavker RM. Structural alterations in exposed and unexposed aged skin. J Invest Dermatol **73**: 59–66, 1979.
38. Gilchrest BA. Skin aging and photoaging: an overview. J Am Acad Dermatol **21**: 610–613, 1989.
39. Griffiths CE. The clinical identification and quantification of photodamage. Br J Dermatol **127**(Suppl 41): 37–42, 1992.
40. Amano S, Ogura Y, Akutsu N, Matsunaga Y, Kadoya K, Adachi E, Nishiyama T. Protective effect of matrix metalloproteinase inhibitors against epidermal basement membrane damage: skin equivalents partially mimic photoageing process. Br J Dermatol **153**(Suppl 2): 37–46, 2005.
41. Constantine VS, Hartley MN. Collagen and elastic fibers in normal dermis and severe actinic (senile) elastosis. A light and electron microscopic study. Ala J Med Sci **3**: 329–342, 1966.

42. Mitchell RE. Chronic solar dermatosis: a light and electron microscopic study of the dermis. J Invest Dermatol **48**: 203–220, 1967.
43. Kadunce DP, Burr R, Gress R, Kanner R, Lyon JL, Zone JJ. Cigarette smoking: risk factor for premature facial wrinkling. Ann Intern Med **114**: 840–844, 1991.
44. Yin L, Morita A, Tsuji T. Skin aging induced by ultraviolet exposure and tobacco smoking: evidence from epidemiological and molecular studies. Photodermatol Photoimmunol Photomed **17**: 178–183, 2001.
45. Helfrich YR, Yu L, Ofori A, Hamilton TA, Lambert J, King A, Voorhees JJ, Kang S. Effect of smoking on aging of photoprotected skin. Arch Dermatol **143**: 397–402, 2007.
46. Marinkovich MP, Keene DR, Rimberg CS, Burgeson RE. Cellular origin of the dermal-epidermal basement membrane. Dev Dyn **197**: 255–267, 1993.
47. Kikkawa Y, Sanzen N, Sekiguchi K. Isolation and characterization of laminin-10/11 secreted by human lung carcinoma cells. Laminin-10/11 mediates cell adhesion through integrin alpha3 beta1. J Biol Chem **273**: 15854–15859, 1998.
48. Kikkawa Y, Sanzen N, Fujiwara H, Sonnenberg A, Sekiguchi K. Integrin binding specificity of laminin-10/11: laminin-10/11 are recognized by alpha 3 beta 1, alpha 6 beta 1 and alpha 6 beta 4 integrins. J Cell Sci **113**: 869–876, 2000.
49. Uitto J, Pulkkinen L, Christiano AM. Molecular basis of the dyatrophic and junctional forms of epidermolysis bullosa: mutations in the type VII collagen and kalinin (laminin 5) genes. J Invest Dermatol **103**: 39S–46S, 1994.
50. Niessen CM, van der Raaij-Helmer MH, Hulsman EH, van der Neut R, Jonkman MF, Sonnenberg A. Deficiency of the integrin beta 4 subunit in junctional epidermolysis bullosa with pyloric atresia: consequences for hemidesmosome formation and adhesion properties. J Cell Sci **109**: 1695–1707, 1996.
51. McGrath JA, Gatalica B, Christiano AM, Li K, Owaribe K, McMillan JR, Eady RA, Uitto J. Mutations in the 180-kD bullous pemphigoid antigen (BPAG2), a hemidesmosomal transmembrane collagen (COL17A1), in generalized atrophic benign epidermolysis bullosa. Nat Genet **11**: 83–86, 1995.
52. Fleischmajer R, Schechter A, Bruns M, Perlish JS, Macdonald ED, Pan TC, Timpl R, Chu ML. Skin fibroblasts are the only source of nidogen during early basal lamina formation *in vitro*. J Invest Dermatol **105**: 597–601, 1995.
53. Amano S, Nishiyama T, Burgeson RE. A specific and sensitive ELISA for laminin 5. J Immunol Methods **224**: 161–169, 1999.
54. Nishiyama T, Amano S, Tsunenaga M, Kadoya K, Takeda A, Adachi E, Burgeson RE. The importance of laminin 5 in the dermal-epidermal basement membrane. J Dermatol Sci **25**(Suppl 1): S51–S59, 2000.
55. Tsunenaga M, Amamo S, Nishiyama T. Laminin 5 can promote assembly of the lamina densa in the skin equivalent model. Matrix Biol **17**: 603–613, 1998.
56. Amano S, Akutsu N, Matsunaga Y, Nishiyama T, Champliaud MF, Burgeson RE, Adachi E. Importance of balance between extracellular matrix synthesis and degradation in basement membrane formation. Exp Cell Res **271**: 249–262, 2001.

57. Takada K, Amano S, Kohno Y, Nishiyama T, Inomata S. Non-invasive study of gelatinases in sun-exposed and unexposed healthy human skin based on measurements in stratum corneum. Arch Dermatol Res **298**: 237–242, 2006.
58. Ashida Y, Terada T, Watabe K, et al. CoQ_{10} intake elevates the epidermal CoQ_{10} level in adult hairless mice [abstract]. In: The 4th Conference of the International Coenzyme Q_{10} Association, Los Angeles, USA, pp. 68–69, 2005.
59. Yamashita S, Yamamoto Y. Simultaneous detection of ubiquinol and ubiquinone in human plasma as a marker of oxidative stress. Anal Biochem **250**: 66–73, 1997.
60. Zhang Y, Aberg F, Appelkvist EL, Dallner G, Ernster L. Uptake of dietary coenzyme Q supplement is limited in rats. J Nutr **125**: 446–453, 1995.
61. Turunen M, Appelkvist EL, Sindelar P, Dallner G. Blood concentration of coenzyme Q_{10} increases in rats when esterified forms are administered. J Nutr **129**: 2113–2118, 1999.
62. Kwong LK, Kamzalov S, Rebrin I, Bayne AC, Jana CK, Morris P, Forster MJ, Sohal RS. Effects of coenzyme Q(10) administration on its tissue concentrations, mitochondrial oxidant generation, and oxidative stress in the rat. Free Radic Biol Med **33**: 627–638, 2002.
63. Bentinger M, Dallner G, Chojnacki T, Swiezewska E. Distribution and breakdown of labelled coenzyme Q_{10} in rat. Free Radic Biol Med **34**: 563–575, 2003.
64. Kamzalov S, Sumien N, Forster MJ, Sohal RS. Coenzyme Q intake elevates the mitochondrial and tissue levels of coenzyme Q and alpha-tocopherol in young mice. J Nutr **10**: 3175–3180, 2003.
65. Ashida Y, Yamanishi H, Terada T, Oota N, Sekine K, Watabe K. CoQ_{10} supplementation elevates the epidermal CoQ_{10} level in adult hairless mice. BioFactors **25**: 175–178, 2005.
66. Eady RAJ, Leigh IM, Pope FM. Anatomy and organization of human skin. In: Champion RH, Burton JL, Burns DA, Breathnach SM, (eds.) Textbook of Dermatology, 6th ed., Blackwell Scientific Publications, UK, pp. 37–111, 1998.
67. Amano S, Ogura Y, Akutsu N, Nishiyama T. Quantitative analysis of the synthesis and secretion of type VII collagen in cultured human dermal fibroblasts with a sensitive sandwich enzyme-linked immunoassay. Exp Dermatol **16**: 151–155, 2007.

11

The Benefits of Antioxidant-Rich Fruits on Skin Health

Francis C. Lau*[1], *PhD, FACN, Manashi Bagchi*[1], *PhD, FACN, Shirley Zafra-Stone*[1], *BS, and Debasis Bagchi*[1,2], *PhD, FACN, CNS

[1]InterHealth Research Center, Benicia, CA, USA
[2]Department of Pharmacological & Pharmaceutical Sciences, University of Houston, College of Pharmacy, Houston, TX, USA

Aaron Tabor and Robert M. Blair (eds.), *Nutritional Cosmetics: Beauty from Within*, 217–232, © 2009 William Andrew Inc.

11.1 Introduction

The desire to look beautiful can be dated back to medieval times when women consumed arsenic and applied bat's blood in an attempt to improve their complexion. Attractive features have been repeatedly selected during evolution in the plant and animal kingdoms. Although Darwin was the first scholar to study human beauty standards from a biological standpoint in the 1800s, it was not until a century later that the cross-cultural beauty standards were validated [1]. Since then the cosmetic business has developed into a $280-billion global industry and has been projected to reach $313 billion by 2011 [2,3]. We now use sophisticated terms such as phytocosmetics, cosmeceuticals, dermaceuticals, and skinceuticals to describe beauty products whose purpose of usage are not at all unlike those utilized in ancient times. In modern society looking good is not only desirable but oftentimes necessary. The pursuit of physical beauty is no longer just for women; the men's grooming business has become the fastest growing segment in the beauty industry. Therefore, physical beauty is more than skin deep because beauty is intimately linked to healthy and radiant skin, which is an external reflection of one's overall inner health status.

Skin is the largest organ that is under constant assault by environmental oxidative stress including ultraviolet radiation (UVR), air pollutants, and chemical oxidants [4]. Thus, skin aging is an inevitable normal process. However, premature skin aging may occur due to factors such as external oxidative stress as well as smoking, imbalanced nutrition, excessive dieting, and mental stress [5]. In vitro and in vivo studies suggest that antioxidants regulate the biomarkers associated with premature aging by reducing oxidative stress including environmental stress such as ozone and cigarette smoking [5]. A double-blinded, placebo-controlled clinical study showed that oral and topical natural antioxidant treatments protect against the development of premature skin aging due to oxidative damage [6]. Therefore, natural fruits and vegetables rich in antioxidants may protect skin against premature aging. This chapter will focus on the beneficial effects of antioxidant-rich berries on skin health.

11.2 Oxidative Stress and Skin Aging

11.2.1 Free Radicals and Oxidative Stress

The free radical theory of aging, proposed by Dr. Harman, states that the reaction of active free radicals with cellular components initiates the changes observed in the aging process [7]. Reactive oxygen species (ROS) is one of the major types of free radicals. ROS is generated during normal aerobic metabolism [8]. Approximately 2–5% of the oxygen consumed by a cell is subsequently reduced to free radicals [9]. ROS production is normally neutralized by cellular antioxidant defense systems so as to maintain an equilibrium between the production and elimination of free radicals [10]. The endogenous antioxidant defense consists of enzymes such as superoxide dismutase, glutathione synthase, and catalase, as well as oxidant scavengers including glutathione (GSH) and uric acid. However, this defense mechanism is not entirely oxidant-proof. In fact, about 1% of the ROS elude the control of endogenous antioxidant defense systems daily, thus tipping the balance in favor of ROS accumulation [11]. This limitation is further exacerbated by the decline of endogenous antioxidant defense during aging. The end result, then, is an imbalance in cellular redox homeostasis, which leads to cellular oxidative damage in macromolecules such as DNA, lipids, and proteins [12]. The gradual accumulation of cellular oxidative damage results in oxidative stress. A myriad of investigations also suggest that oxidative stress may induce the expression of proinflammatory cytokines, which, in turn, may elevate the cellular levels of ROS [13]. This vicious cycle results in the progressive accrual of oxidative stress and inflammation during chronological skin aging [14]. Therefore, oxidative stress and inflammation appear to be a function of age. Although the aging process cannot be reversed as yet, it is possible to delay its onset by combating the accumulation of ROS and inflammation through exogenous sources of antioxidants, which can be obtained from fruits, vegetables, and dietary supplements [15–17].

11.2.2 Skin Aging and Skin Antioxidant Status

Normal skin aging is characterized by thinning of the epidermis, resulting in fine wrinkles, roughness, dry and thin appearance, hyperpigmentation, and seborrhoeic keratoses [5,18]. Although the underlying mechanisms for chronological skin aging remain incompletely understood, several causes including intrinsic factors such as genetic and epigenetic variations and diminished epidermal turnover, as well as extrinsic factors such as UVR, photo-oxidative stress, and ROS-induced collagen degradation and chronic inflammation have been proposed [14,18]. Accordingly, increased exposure to harsh environmental factors as well as reduced overall antioxidant status may accelerate the chronological skin aging process, leading to premature skin aging.

Because the skin is inevitably exposed to environmental oxidative stress, an adaptive enzymatic and nonenzymatic antioxidant defense network has been developed by the skin. The enzymatic defense includes superoxide dismutase, catalase, and glutathione synthase. Superoxide dismutase converts superoxide anion to hydrogen peroxide, which is further detoxified to water by catalase and glutathione synthase [19]. Nonenzymatic endogenous antioxidants in cutaneous cells consist of α-tocopherol (vitamin E), β-carotene (precursor of vitamin A), ascorbic acid (vitamin C), ubiquinon (coenzyme Q10), and glutathione. The lipophilic antioxidants α-tocopherol and β-carotene are found in the cell membranes and serve to protect against lipid peroxidation [20]. This defense system is crucial for normal cellular functions because lipid peroxides and their metabolites such as malonaldehyde and 4-hydroxy-2-nonenal may directly or indirectly induce detrimental consequences by evoking immune and inflammatory response, altering gene expression, and inducing apoptosis [20]. Because lipophilic α-tocopherol and hydrophilic ascorbate antioxidants cannot be synthesized de novo by humans, topical application or systemic administration of these and other antioxidants may be essential to replenish skin with antioxidants that are depleted by environmental assaults or by age-related increase in oxidative stress.

11.3 Cosmeceuticals, Antioxidants, and Skin Health

11.3.1 Cosmeceuticals

The term *cosmeceutical* was first coined by dermatologist Dr. Kligman to describe a product category between cosmetics and pharmaceuticals [21,22]. It is defined as a product that combines common cosmetic preparations with functional nutraceutical ingredients [23]. Although not yet recognized by the U.S. Food and Drug Administration, the term cosmeceutical has been used in the beauty industry for products containing natural ingredients and claiming health or physical improvement benefits beyond aesthetics [23,24]. Cosmeceuticals initially described cosmetic products that enhance beauty from the outside such as topical creams, lotions, and ointments. Today, the term *cosmeceutical* is extended to encompass a variety of drinks or dietary supplements known as skin nutraceuticals or nutricosmetics that provide additional health functions and benefits to enhance beauty from the inside [25].

The pursuit of physical beauty and overall well-being has boosted the demand for cosmeceuticals. In fact, cosmeceuticals have become the fastest growing sector of the natural personal care industry, commanding worldwide annual sales surpassing $14 billion [23]. The U.S. cosmeceutical category is a healthy $4.3 billion business, with a projected annual growth of approximately 8% [24]. Antioxidant ingredients are the largest segment of the cosmeceutical ingredient market. In 2000, antioxidants accounted for about 40% of the total cosmeceutical ingredient sales in the U.S. It is projected that antioxidants will continue to dominate the cosmeceutical market, with an annual growth rate of 7.3% [24].

11.3.2 Cosmeceutical Antioxidants

The use of antioxidants for skin care was well documented during the Renaissance by Paracelsus, who famously noted that "the dose makes the poison" and was often considered to be the father of toxicology [26]. He reportedly showed the beneficial effects of treating skin wrinkles with red wine vinegar, which, as we now know, contains grape polyphenolic antioxidants [21,27,28]. Currently antioxidants are widely incorporated into a variety of antiaging skin care systems [29]. The commonly used cosmeceutical antioxidant ingredients include vitamins A, B, C, and E; CoQ10 and its analogues; and plant polyphenols such as tannins and flavonoids.

11.3.3 Fruit Antioxidants and Skin Health

11.3.3.1 Selected Berry Antioxidants and Health Benefits

Anthocyanins are a subcategory of the flavonoid class that are water-soluble pigments present in all higher plants that give flowers and fruits their bright red, purple, or blue color. Anthocyanins have received a great deal of attention in recent years because of their antioxidant, anti-inflammatory, and antimutagenic effects [15]. There is a significant correlation between anthocyanin content and antioxidant capacity. Berry fruits are generally rich in anthocyanins. Consequently, berries are high in antioxidant capacity, which can be measured by the oxygen radical absorbance capacity (ORAC) assay [30–32].

Wild blueberries, one of the most popular *Vaccinium* species (spp.), have been shown to be among the foods that have the highest anthocyanin concentrations and ORAC values [33,34]. Numerous publications have indicated that blueberry anthocyanins exhibit a wide range of antiaging properties, such as reducing age-enhanced vulnerability to oxidative stress and inflammation, lowering the risk of developing age-related degenerative diseases, and improving overall brain function [15]. Other *Vaccinium* spp., such as bilberries and cranberries, are also enriched with anthocyanins and are capable of decreasing oxidative stress and inflammation [33,35–38].

The antioxidant effect of elderberry (*Sambucus* spp.) anthocyanins on endothelial cells was evaluated by challenging the cells with oxidative stressors such as hydrogen peroxide, azobis(2-amidinopropane) dihydrochloride, and iron sulfate. Endothelial cells preloaded with elderberry extract were significantly protected against insults of the oxidative stressors [39]. Elderberry extract at a low concentration was found to give a considerable amount of antioxidant protection against both copper-induced low-density lipoprotein oxidation and peroxyl radical attack [40]. Therefore, elderberry extract could contribute significantly to the antioxidant capacity of plasma [40].

The Rosaceae family berries, such as strawberries and raspberries, are also high in antioxidant capacity. Ellagic acid is the predominant phenolic antioxidant found in strawberries (*Fragaria* spp.). It has been estimated that ellagic acid accounts for approximately 51% of the total phenolic compounds in strawberries [41]. A comparative study was conducted to determine the total antioxidant capacity of 12 fruits, and strawberries were found to rank the highest in ORAC value among the other nonberry fruits [42]. A clinical study showed that consumption of strawberries increased the serum antioxidant capacity in elderly women [43]. Individual constituents of strawberries appear to afford protective potential against cancer and cardiovascular disease while enhancing mental health and boosting the immune system [44]. Raspberry (*Rubus* spp.) seeds, the byproduct of juice manufacturing, contain a significant amount of phenolic antioxidants and omega-3 unsaturated fatty acids [45]. Omega-3 fatty acids are themselves high in antioxidant capacity [46]. The health benefits of omega-3 fatty acids include antiatherosclerosis, cardioprotection, anticarcinogenesis, antioxidative defense, anti-inflammation, autoimmune disease prevention, and mental health enhancement [47–51].

11.3.3.2 Evidence-Based Approach to Evaluating the Benefits of a Novel Berry Formulation

Because individual berries confer different health benefits, an innovative formulation combining different berry extracts at specific proportions was systematically researched and developed for its health-promoting effects. The comparative antioxidant efficacy was determined for extracts from the fruits of wild blueberries, wild bilberries, cranberries, elderberries, and strawberries, as well as extracts from seeds of raspberries. A total of 20 different combinations comprising varying amounts of each of the extracts from these six berries were evaluated by the ORAC assay [52]. The blend with the highest ORAC value was selected for further analysis. The cytotoxicity of this formulation was assessed by the lactate dehydrogenase assay using human epidermal keratinocyte (HaCaT) cells to ensure the formulation was not toxic to the cells. Keratinocytes are the major cell type of the epidermis, comprising up to 90% of the skin cells. The lactate dehydrogenase assay indicated that the formulation was not cytotoxic to epidermal cell viability as compared to control and each of the six berry extracts [52].

The safety and efficacy properties of this formulation, known as OptiBerry, were further investigated using standard toxicology and whole-body antioxidant status procedures [53]. The results indicated that acute oral LD_{50} of OptiBerry was greater than 5 g/kg body weight whereas acute dermal LD_{50} was greater than 2 g/kg body weight in rats [53]. The antioxidant potential of OptiBerry was evaluated in vitamin E–deficient rats by determining glutathione (GSH) oxidation after exposure to clinically relevant hyperbaric oxygen (HBO) at 2 atmospheric pressure (atm) for 2 hours. Rats fed an OptiBerry-supplemented diet for 8 weeks were significantly protected against HBO-induced GSH oxidation in the lung and liver as compared to placebo-fed rats. The antioxidant property of OptiBerry was further confirmed by a state-of-the-art electron paramagnetic resonance (EPR) technology, which is an imaging system used to examine the whole-body redox status. Vitamin E–deficient mice were subjected to HBO treatment at 2 atm for 2 hours. Mice were then anaesthetized, injected with carbamoyl-PROXYL (nitrosyl radical) solution, and placed into quartz tubes. Antioxidants reduce nitrosyl radicals to hydroxamine, thereby accelerating the EPR signal decay of nitrosyl radicals. To determine the reduction of nitrosyl radicals in vivo, EPR spectra were immediately recorded on the body of the test compound–injected mice at 4-min intervals and a total of 16 projections were taken for each time point. The projections for each interval were deconvoluted to reconstruct images of two-dimension redox status [53]. Mice fed an OptiBerry-supplemented diet for 2 weeks exhibited significant protection against HBO-induced oxidative stress as revealed by the rapid decay of EPR signal intensity [53]. Therefore, the data suggest that OptiBerry is safe and may provide whole-body antioxidant protection.

Angiogenesis is a process involving the generation of new blood vessels from pre-existing vessels in areas of low blood supply [54]. It is a hallmark in tumor growth and cancer metastases. The skin vasculature remains quiescent under normal conditions. However, under external insults such as UVR and heat, skin inflammation, and cancer, as well as during wound healing and hair growth, the skin cells are capable of initiating rapid angiogenesis. The epidermis-derived vascular endothelial growth factor (VEGF) is a potent angiogenic factor. Its expression can be altered by a variety of physiological and pathological conditions of the skin [54]. It has been shown that VEGF expression was stimulated by UVR in epidermal keratinocyte cell lines and in human skin [55,56]. Heat treatment at 43°C for 90 minutes in human volunteers has also been shown to induce an angiogenic switch through upregulation of VEGF protein expression with concomitant increase in vascularization [57]. Therefore, skin angiogenesis is associated with an array of acute external stimuli, which may lead to vascular hyperpermeability, resulting in cutaneous inflammation and progressive loss of skin vessels in aged skin [54].

The inhibitory effect of OptiBerry on angiogenesis was revealed by the level of VEGF expression in HaCaT keratinocytes. In this experiment, HaCaT cells were pretreated with or without Optiberry for 12 hours, followed by treatment with H_2O_2 and tumor necrosis factor-alpha (TNF-α, an inflammatory cytokine). OptiBerry pretreatment was found to significantly inhibit the ROS- and inflammation-induced VEGF protein expression [58,59]. In addition, the formation of endothelial tubes, an indication of angiogenesis, was investigated using Matrigel assay. Human microvascular endothelial cells cultured on Matrigel exhibit complex morphological behavior by reconstructing intricate spider weblike networks resembling the vasculature systems. Treatment with OptiBerry inhibited the construction of this network by reducing the formation of endothelial tubes, corroborating the antiangiogenic property of OptiBerry [58,59]. Unexpectedly, grape seed proanthocyanidin did not elicit the antiangiogenic activity observed for OptiBerry, suggesting that antioxidant capacity alone may not be sufficient to account for the antiangiogenic effect. Therefore, OptiBerry may exert its antiangiogenic property through other means such as transcriptional activation/inhibition in conjunction with its antioxidant potential [58]. This theory was subsequently confirmed in the following experiments.

Hemangiomas are the most common form of infant tumors, affecting 10–12% of normal newborns. Approximately 5% of hemangiomas cause serious tissue damage, and 1–2% of all hemangiomas are life-threatening. Hemangiomas are localized tumors of blood vessels characterized by a rapid proliferation of capillaries during the newborns' first year [60]. The underlying causes for the endothelial cell growth in hemangiomas have not been determined, although proliferating hemangiomas have been shown to be highly angiogenesis-dependent [61,62]. Also, hemangiomas are infiltrated by macrophages, which can initiate angiogenesis. The chemokine monocyte chemoattractant protein-1 (MCP-1) is responsible for recruiting macrophages at the lesion sites, thus facilitating the growth of hemangiomas and vascular malformations [63,64]. Both VEGF and MCP-1 were found to be highly stimulated in proliferative hemangiomas [63,65]. In fact, transgenic rabbits with increased hepatic expression of the human VEGF transgene under the control of the human alpha-antitrypsin promoter were shown to develop liver hemangiomas resembling the Kasabach-Merritt syndrome [66]. In this regard, hemangioma represents a powerful model for the investigation of angiogenesis and for the identification of anti-angiogenic molecules, which arrest endothelial cell growth at a quiescent state under physiological conditions.

Endothelioma cells (EOMA) are derived from the spontaneously arising hemangio-endothelioma in the 129/J murine strain [67]. EOMA cells treated with OptiBerry for 12 hours were significantly protected against TNF-α-induced MCP-1 upregulation [67]. In order to evaluate the molecular mechanisms underlying the observed inhibitory effects of OptiBerry on the expression of MCP-1, EOMA cells were transfected with an MCP-1-controlled luciferase reporter construct. These transfected cells exhibited high basal luciferase expression, indicating an elevated MCP-1 transcription [68]. Pretreatment of the transfected cells with OptiBerry significantly lowered the basal MCP-1-luciferase reporter expression, suggesting that OptiBerry inhibited the basal transcription of MCP-1 in EOMA cells. Because MCP-1 is under the control of transcription factor nuclear factor kappa B (NF-κB), examination of whether OptiBerry affects the activation of NF-κB should provide insight into the upstream events in the cascade leading to the regulation of MCP-1. Consequently, EOMA cells were transfected with NF-κB luciferase construct, pretreated with or without OptiBerry for 24 hours, and then challenged with TNF-α for 6 hours to induce inflammation. Transcriptional activation of NF-κB was assayed by measuring the level of NF-κB–controlled luciferase activity. Transfected cells not pretreated with OptiBerry showed a marked increase in the level of NF-κB–luciferase activity; however, this increase was significantly attenuated by pretreatment of OptiBerry [68]. The findings demonstrated for the first time that berry constituents regulated NF-κB activity in EOMA cells.

Mice injected with EOMA cells develop clinically identifiable hemangiomas resembling the Kasabach-Meritt syndrome within 3–4 weeks, whereas tumor formation only takes days. Therefore, mice injected with EOMA cells provide an in vivo model for the study of angiogenic events. To investigate the in vivo effect of OptiBerry on angiogenesis, mice were injected with EOMA cells with or without OptiBerry pretreatment. Mice (129P3/J) were sacrificed one week post-injection to obtain tissues for histological analyses. Although the OptiBerry treated group tested positive for the presence of hemangiomas, the average mass of such tumor growth was below 50% of those in the untreated control group. Immunohistological analysis revealed a significant reduction in the infiltration of macrophages in hemangioma of treated mice as compared to that of the controls [68]. It appears that the beneficial action of OptiBerry on proliferating hemangiomas may, in part, result from reduced MCP-1 expression and reduced influx of angiogenic macrophages. OptiBerry may therefore represent a potent agent against angiogenesis-induced skin aging.

11.4 Conclusion

The beauty industry has come a long way. The current trend for skin care products is that they not only conceal superficial blemishes but also treat the causes from within. At present the consensus on food is two-fold: it provides essential nutrients to sustain life itself and supplies bioactive agents to promote health and prevent diseases. A large body of evidence has demonstrated the correlation between consumption of fresh fruits and vegetables with delaying and/or preventing chronic degenerative diseases. Fresh fruits and vegetables are enriched with a variety of diverse nutrients, such as vitamins, antioxidants, trace minerals and micronutrients, phytosterols, phytoenzymes, dietary fiber, and potent chemoprotectants [69]. Epidemiological studies have consistently revealed the beneficial effects of nutritional factors on the prevention of chronic diseases and the modulation of human skin conditions [70–72]. In recent years, functional foods and nutraceuticals have gained noticeable popularity. Both oral and topical supplements have become some of the most widely used alternative therapies. Antioxidants make up a majority of the skin health ingredients in the cosmeceutical market because oxidative stress is generally accepted as a major contributing factor to skin aging. Consuming natural antioxidants provides a plethora of health benefits, including lowering age-related oxidative stress and inflammation [15,73].

Anthocyanins are the common components of fruits and vegetables, particularly in berries, which provide the bright red, blue, and purple hues to the plants. Anthocyanins are naturally occurring antioxidants. There is a direct correlation between anthocyanin content and antioxidant capacity in berry fruits [36]. In vitro and in vivo studies have shown that berry anthocyanins possess potent antioxidant activity and many potential health benefits, including cardiovascular protection, anticarcinogenic potential, antidiabetic properties, brain function enhancement, ocular and vision health, urinary tract health, and skin health [74]. Based on these health benefits, a formulation of a synergistic blend (OptiBerry) containing wild blueberry, bilberry, cranberry, elderberry, strawberry, and raspberry seed extracts was developed. Systematic and comprehensive studies have shown that OptiBerry is safer and more potent than individual and all other combinations of the berry extracts tested [58]. Furthermore, OptiBerry has been found to be superior in bioavailability and antioxidative properties, providing whole-body antioxidant protection [53,59].

It has been shown that the skin is capable of exiting the quiescent phase of vasculature to rapidly initiate angiogenesis during external insults such as UVR and thermal stimulus [54]. This surge of acute vessel development incurs profound consequences to the general health of the aging skin because these vessels may be leaky and less mature, which may lead to vascular hyperpermeability and vessel leakage, resulting in skin inflammation and further degradation of the extracellular matrix [54]. Both VEGF and MCP-1 are markedly activated during angiogenesis. Several in vitro and in vivo studies have shown that OptiBerry significantly reduced the expression of VEGF and MCP-1 in highly proliferative hemangioma, an angiogenesis-dependent disease. The inhibitory effect of OptiBerry on angiogenesis was attributable to the modulation of VEGF and MCP-1 through the activation of upstream transcription factor NF-κB [58,59,68]. This finding is of great importance because it demonstrates that, in addition to its antioxidant activities, OptiBerry anthocyanin constituents may serve as signal molecules to modulate gene expression. This observation is consistent with the findings that flavonoids and their metabolites may modulate various protein kinase signaling pathways [75].

Is beauty only skin deep? The answer is apparent. Although visible signs of aging are readily reflected on the surface of the skin as the consequence of time and environmental insults, the fundamental causes may very well be more than skin deep. To achieve overall well-being, antiaging treatment should start from the inside to reduce oxidative stress and inflammation, curtail degenerative diseases, and attenuate age-related decline in cognitive and motor functions, which will then translate to the outside as a vibrant and healthy appearance. In this regard, consuming foods rich in antioxidants and other phytonutrients may help to promote beauty from within.

References

1. Grammer, K., Fink, B., Moller, A. P., and Thornhill, R.: Darwinian aesthetics: sexual selection and the biology of beauty. *Biol Rev Camb Philos Soc*, 2003, *78*: 385–407.
2. http://www.economist.com/science/displaystory.cfm?story_id=10311266 (Accessed May).
3. http://www.gcimagazine.com/marketdata/7862872.html (Accessed May).
4. Zhai, H., Choi, M. J., Arens-Corell, M., Neudecker, B. A., and Maibach, H. I.: A rapid, accurate, and facile method to quantify the antioxidative capacity of topical formulations. *Skin Res Technol*, 2003, *9*: 254–256.
5. Biesalski, H. K., Berneburg, M., Grune, T., Kerscher, M., Krutmann, J., Raab, W., Reimann, J., Reuther, T., Robert, L., and Schwarz, T.: Hohenheimer Consensus Talk. Oxidative and premature skin ageing. *Exp Dermatol*, 2003, *12(Suppl. 3)*: 3–15.
6. Palombo, P., Fabrizi, G., Ruocco, V., Ruocco, E., Fluhr, J., Roberts, R., and Morganti, P.: Beneficial long-term effects of combined oral/topical antioxidant treatment with the carotenoids lutein and zeaxanthin on human skin: a double-blind, placebo-controlled study. *Skin Pharmacol Physiol*, 2007, *20*: 199–210.
7. Harman, D.: Aging: a theory based on free radical and radiation chemistry. *J Gerontol*, 1956, *11*: 298–300.
8. Beckman, K. B., and Ames, B. N.: The free radical theory of aging matures. *Physiol Rev*, 1998, *78*: 547–581.
9. Wickens, A. P.: Ageing and the free radical theory. *Respir Physiol*, 2001, *128*: 379–391.
10. Freeman, B. A., and Crapo, J. D.: Biology of disease: free radicals and tissue injury. *Lab Invest*, 1982, *47*: 412–426.
11. Berger, M. M.: Can oxidative damage be treated nutritionally? *Clin Nutr*, 2005, *24*: 172–183.
12. Junqueira, V. B., Barros, S. B., Chan, S. S., Rodrigues, L., Giavarotti, L., Abud, R. L., and Deucher, G. P.: Aging and oxidative stress. *Mol Aspects Med*, 2004, *25*: 5–16.
13. Lane, N.: A unifying view of ageing and disease: the double-agent theory. *J Theor Biol*, 2003, *225*: 531–540.
14. Thornfeldt, C. R.: Chronic inflammation is etiology of extrinsic aging. *J Cosmet Dermatol*, 2008, *7*: 78–82.

15. Lau, F. C., Shukitt-Hale, B., and Joseph, J. A.: Nutritional intervention in brain aging: reducing the effects of inflammation and oxidative stress. *Subcell Biochem*, 2007, *42*: 299–318.
16. Morganti, P., Bruno, C., Guarneri, F., Cardillo, A., Del Ciotto, P., and Valenzano, F.: Role of topical and nutritional supplement to modify the oxidative stress. *Int J Cosmet Sci*, 2002, *24*: 331–339.
17. Purba, M. B., Kouris-Blazos, A., Wattanapenpaiboon, N., Lukito, W., Rothenberg, E. M., Steen, B. C., and Wahlqvist, M. L.: Skin wrinkling: can food make a difference? *J Am Coll Nutr*, 2001, *20*: 71–80.
18. McCullough, J. L., and Kelly, K. M.: Prevention and treatment of skin aging. *Ann NY Acad Sci*, 2006, *1067*: 323–331.
19. Wlaschek, M., and Scharffetter-Kochanek, K.: Adaptive antioxidant response and photoaging. *Int J Cosmet Sci*, 2004, *26*: 107–108.
20. Briganti, S., and Picardo, M.: Antioxidant activity, lipid peroxidation and skin diseases. What's new. *J Eur Acad Dermatol Venereol*, 2003, *17*: 663–669.
21. Holck, D. E.: What you should know about cosmeceuticals. *Rev Ophthalmol*, 2001, *140*: 140–143.
22. Stanley, J. R.: Albert M. Kligman: 90 years old on March 17, 2006. *J Invest Dermatol*, 2006, *126*: 697–698.
23. Grubow, L.: The market potential of antiaging cosmeceuticals. *Global Cosmetic Industry*, 2008, *176*: 55–56.
24. Morrison, S., and Schmitt, B.: Cosmeceuticals drive healthy growth rates. *Chem Week*, 2001, *163*: 46–48.
25. Matthews, I.: Beauty inside and out. *Global Cosmetic Industry*, 2003, *171*: 76–78.
26. Smith, K. R.: Place makes the poison: Wesolowski Award Lecture—1999. *J Expo Anal Environ Epidemiol*, 2002, *12*: 167–171.
27. Honsho, S., Sugiyama, A., Takahara, A., Satoh, Y., Nakamura, Y., and Hashimoto, K.: A red wine vinegar beverage can inhibit the renin-angiotensin system: experimental evidence in vivo. *Biol Pharm Bull*, 2005, *28*: 1208–1210.
28. Andlauer, W., Stumpf, C., and Furst, P.: Influence of the acetification process on phenolic compounds. *J Agric Food Chem*, 2000, *48*: 3533–3536.
29. Bruce, S.: Cosmeceuticals for the attenuation of extrinsic and intrinsic dermal aging. *J Drugs Dermatol*, 2008, *7*: s17–22.
30. Cao, G., and Prior, R. L.: Comparison of different analytical methods for assessing total antioxidant capacity of human serum. *Clin Chem*, 1998, *44*: 1309–1315.
31. Prior, R. L., Wu, X., and Schaich, K.: Standardized methods for the determination of antioxidant capacity and phenolics in foods and dietary supplements. *J Agric Food Chem*, 2005, *53*: 4290–4302.
32. Wang, S. Y., Chen, C. T., Sciarappa, W., Wang, C. Y., and Camp, M. J.: Fruit quality, antioxidant capacity, and flavonoid content of organically and conventionally grown blueberries. *J Agric Food Chem*, 2008, *56*: 5788–5794.
33. Wu, X., Beecher, G. R., Holden, J. M., Haytowitz, D. B., Gebhardt, S. E., and Prior, R. L.: Lipophilic and hydrophilic antioxidant capacities of common foods in the United States. *J Agric Food Chem*, 2004, *52*: 4026–4037.

34. Wu, X., Beecher, G. R., Holden, J. M., Haytowitz, D. B., Gebhardt, S. E., and Prior, R. L.: Concentrations of anthocyanins in common foods in the United States and estimation of normal consumption. *J Agric Food Chem*, 2006, *54*: 4069–4075.
35. Neto, C. C., Amoroso, J. W., and Liberty, A. M.: Anticancer activities of cranberry phytochemicals: an update. *Mol Nutr Food Res*, 2008, *52(Suppl. 1)*: S18–27.
36. Prior, R. L., Cao, G., Martin, A., Sofic, E., McEwen, J., O'Brien, C., Lischner, N., Ehlenfeldt, M., Kalt, W., Krewer, G., and Mainland, C. M.: Antioxidant capacity as influenced by total phenolic and anthocyanin content, maturity, and variety of Vaccinium species. *J. Agric. Food Chem.*, 1998, *46*: 2686–2693.
37. Neto, C. C.: Cranberry and blueberry: evidence for protective effects against cancer and vascular diseases. *Mol Nutr Food Res*, 2007, *51*: 652–664.
38. Talavéra, S., Felgines, C., Texier, O., Besson, C., Mazur, A., Lamaison, J., and Rémésy, C.: Bioavailability of a bilberry anthocyanin extract and its impact on plasma antioxidant capacity in rats. *J Sci Food Agric*, 2006, *86*: 90–97.
39. Youdim, K. A., Martin, A., and Joseph, J. A.: Incorporation of the elderberry anthocyanins by endothelial cells increases protection against oxidative stress. *Free Radic Biol Med*, 2000, *29*: 51–60.
40. Abuja, P. M., Murkovic, M., and Pfannhauser, W.: Antioxidant and prooxidant activities of elderberry (*Sambucus nigra*) extract in low-density lipoprotein oxidation. *J Agric Food Chem*, 1998, *46*: 4091–4096.
41. Häkkinen, S. H., Kärenlampi, S. O., Mykkänen, H. M., Heinonen, I. M., and Törrönen, A. R.: Ellagic acid content in berries: influence of domestic processing and storage. *Eur Food Res Technol*, 2000, *212*: 175–180.
42. Wang, H., Cao, G., and Prior, R. L.: Total antioxidant capacity of fruits. *J Agric Food Chem*, 1996, *44*: 701–705.
43. Cao, G., Russell, R. M., Lischner, N., and Prior, R. L.: Serum antioxidant capacity is increased by consumption of strawberries, spinach, red wine or vitamin C in elderly women. *J Nutr*, 1998, *128*: 2383–2390.
44. Hannum, S. M.: Potential impact of strawberries on human health: a review of the science. *Crit Rev Food Sci Nutr*, 2004, *44*: 1–17.
45. J. Parry, L. Y.: Fatty acid content and antioxidant properties of cold-pressed black raspberry seed oil and meal. *Journal of Food Science*, 2004, *69*: FCT189–FCT193.
46. Simopoulos, A. P.: Omega-3 fatty acids and antioxidants in edible wild plants. *Biol Res*, 2004, *37*: 263–277.
47. Schachter, H. M., Kourad, K., Merali, Z., Lumb, A., Tran, K., and Miguelez, M.: Effects of omega-3 fatty acids on mental health. *Evid Rep Technol Assess (Summ)*, 2005: 1–11.
48. Ruxton, C.: Health benefits of omega-3 fatty acids. *Nurs Stand*, 2004, *18*: 38–42.
49. Lee, J. H., O'Keefe, J. H., Lavie, C. J., Marchioli, R., and Harris, W. S.: Omega-3 fatty acids for cardioprotection. *Mayo Clin Proc*, 2008, *83*: 324–332.
50. Jude, S., Roger, S., Martel, E., Besson, P., Richard, S., Bougnoux, P., Champeroux, P., and Le Guennec, J. Y.: Dietary long-chain omega-3 fatty

acids of marine origin: a comparison of their protective effects on coronary heart disease and breast cancers. *Prog Biophys Mol Biol*, 2006, *90*: 299–325.

51. Mazza, M., Pomponi, M., Janiri, L., Bria, P., and Mazza, S.: Omega-3 fatty acids and antioxidants in neurological and psychiatric diseases: an overview. *Prog Neuropsychopharmacol Biol Psychiatry*, 2007, *31*: 12–26.
52. Yasmin, T., Sen, C. K., Hazra, S., Bagchi, M., Bagchi, D., and Stohs, S. J.: Antioxidant capacity and safety of various anthocyanin berry extract formulations. *Res Commun Pharmacol Toxicol*, 2003, *2*: 25–33.
53. Bagchi, D., Roy, S., Patel, V., He, G., Khanna, S., Ojha, N., Phillips, C., Ghosh, S., Bagchi, M., and Sen, C. K.: Safety and whole-body antioxidant potential of a novel anthocyanin-rich formulation of edible berries. *Mol Cell Biochem*, 2006, *281*: 197–209.
54. Chung, J. H., and Eun, H. C.: Angiogenesis in skin aging and photoaging. *J Dermatol*, 2007, *34*: 593–600.
55. Blaudschun, R., Brenneisen, P., Wlaschek, M., Meewes, C., and Scharffetter-Kochanek, K.: The first peak of the UVB irradiation–dependent biphasic induction of vascular endothelial growth factor (VEGF) is due to phosphorylation of the epidermal growth factor receptor and independent of autocrine transforming growth factor alpha. *FEBS Lett*, 2000, *474*: 195–200.
56. Kim, M. S., Kim, Y. K., Eun, H. C., Cho, K. H., and Chung, J. H.: All-trans retinoic acid antagonizes UV-induced VEGF production and angiogenesis via the inhibition of ERK activation in human skin keratinocytes. *J Invest Dermatol*, 2006, *126*: 2697–2706.
57. Kim, M. S., Kim, Y. K., Cho, K. H., and Chung, J. H.: Infrared exposure induces an angiogenic switch in human skin that is partially mediated by heat. *Br J Dermatol*, 2006, *155*: 1131–1138.
58. Roy, S., Khanna, S., Alessio, H. M., Vider, J., Bagchi, D., Bagchi, M., and Sen, C. K.: Anti-angiogenic property of edible berries. *Free Radic Res*, 2002, *36*: 1023–1031.
59. Bagchi, D., Sen, C. K., Bagchi, M., and Atalay, M.: Anti-angiogenic, antioxidant, and anti-carcinogenic properties of a novel anthocyanin-rich berry extract formula. *Biochemistry (Mosc)*, 2004, *69*: 75–80, 71 p preceding 75.
60. Takahashi, K., Mulliken, J. B., Kozakewich, H. P., Rogers, R. A., Folkman, J., and Ezekowitz, R. A.: Cellular markers that distinguish the phases of hemangioma during infancy and childhood. *J Clin Invest*, 1994, *93*: 2357–2364.
61. Folkman, J.: Tumor angiogenesis. *Adv Cancer Res*, 1974, *19*: 331–358.
62. Folkman, J., and Cotran, R.: Relation of vascular proliferation to tumor growth. *Int Rev Exp Pathol*, 1976, *16*: 207–248.
63. Isik, F. F., Rand, R. P., Gruss, J. S., Benjamin, D., and Alpers, C. E.: Monocyte chemoattractant protein-1 mRNA expression in hemangiomas and vascular malformations. *J Surg Res*, 1996, *61*: 71–76.
64. Gordillo, G. M., Onat, D., Stockinger, M., Roy, S., Atalay, M., Beck, F. M., and Sen, C. K.: A key angiogenic role of monocyte chemoattractant protein-1 in hemangioendothelioma proliferation. *Am J Physiol Cell Physiol*, 2004, *287*: C866–873.
65. Leaute-Labreze, C., Dumas de la Roque, E., Hubiche, T., Boralevi, F., Thambo, J. B., and Taieb, A.: Propranolol for severe hemangiomas of infancy. *N Engl J Med*, 2008, *358*: 2649–2651.

66. Kitajima, S., Liu, E., Morimoto, M., Koike, T., Yu, Y., Watanabe, T., Imagawa, S., and Fan, J.: Transgenic rabbits with increased VEGF expression develop hemangiomas in the liver: a new model for Kasabach-Merritt syndrome. *Lab Invest*, 2005, *85*: 1517–1527.
67. Hoak, J. C., Warner, E. D., Cheng, H. F., Fry, G. L., and Hankenson, R. R.: Hemangioma with thrombocytopenia and microangiopathic anemia (Kasabach-Merritt syndrome): an animal model. *J Lab Clin Med*, 1971, *77*: 941–950.
68. Atalay, M., Gordillo, G., Roy, S., Rovin, B., Bagchi, D., Bagchi, M., and Sen, C. K.: Anti-angiogenic property of edible berry in a model of hemangioma. *FEBS Lett*, 2003, *544*: 252–257.
69. Heber, D.: Vegetables, fruits and phytoestrogens in the prevention of diseases. *J Postgrad Med*, 2004, *50*: 145–149.
70. Goyarts, E., Muizzuddin, N., Maes, D., and Giacomoni, P. U.: Morphological changes associated with aging: age spots and the microinflammatory model of skin aging. *Ann N Y Acad Sci*, 2007, *1119*: 32–39.
71. Liu, R. H.: Health benefits of fruit and vegetables are from additive and synergistic combinations of phytochemicals. *Am J Clin Nutr*, 2003, *78*: 517S–520S.
72. Cosgrove, M. C., Franco, O. H., Granger, S. P., Murray, P. G., and Mayes, A. E.: Dietary nutrient intakes and skin-aging appearance among middle-aged American women. *Am J Clin Nutr*, 2007, *86*: 1225–1231.
73. Lau, F. C., Bielinski, D. F., and Joseph, J. A.: Inhibitory effects of blueberry extract on the production of inflammatory mediators in lipopolysaccharide-activated BV2 microglia. *J Neurosci Res*, 2007, *85*: 1010–1017.
74. Zafra-Stone, S., Yasmin, T., Bagchi, M., Chatterjee, A., Vinson, J. A., and Bagchi, D.: Berry anthocyanins as novel antioxidants in human health and disease prevention. *Mol Nutr Food Res*, 2007, *51*: 675–683.
75. Williams, R. J., Spencer, J. P., and Rice-Evans, C.: Flavonoids: antioxidants or signalling molecules? *Free Radic Biol Med*, 2004, *36*: 838–849.

12

Olive Fruit Extracts for Skin Health

Aldo Cristoni[1], PhD, Andrea Giori[1], PhD, Giada Maramaldi[1], BSc, Christian Artaria[1], BSc, and Takeshi Ikemoto[2], PhD

[1]Indena S.p.A., Milan, Italy
[2]Kanebo Cosmetics, Kanagawa-Ken, Japan

12.1 Introduction: Origin and History of the Olive Tree

The olive tree has always been the symbol of the Mediterranean culture, because people living around the Mediterranean Sea have grown and used

Aaron Tabor and Robert M. Blair (eds.), *Nutritional Cosmetics: Beauty from Within*, 233–244, © 2009 William Andrew Inc.

its fruits and leaves in everyday life for a very long time. Fossil evidence backs up the hypothesis that the olive tree originated in the Anatolian region, from where the Phoenicians spread it throughout the Greek islands and Greece. Here, the tree gained special relevance, as testified by a law issued by Solon in the 6th century B.C. that made it illegal to cut down more than two olive trees at a time. Indeed, the first large-scale cultivation of the olive tree should probably be credited to the Greeks, who mastered this art and elevated it to a science.

The olive tree was present in Greeks' everyday life, to the point of becoming an object of dispute between Greek deities. According to ancient mythology, when the Greeks were searching for a patron deity for a newly built city in the land of Attica that then became Athens, there was a dispute among Athena, goddess of wisdom, and Poseidon, god of the Sea, who were both interested in the patronage. To solve the dispute, it was decided that the new city would be entitled to the God who could offer the most valuable gift to the citizens. Poseidon offered a well, but the water coming out of it was salty and not very useful, whereas Athena struck her spear on the ground, buried a branch of olive, and generated the olive tree. The council of Gods determined that Athena was the winner, because the olive tree could not only live hundreds of years, but could also provide edible fruits and a precious oil, useful to season food and to heal wounds [1]. The olive tree became a symbol of peace, prosperity, wisdom, and triumph. The relevance of the olive tree was such that it was represented on ancient coins, and the Goddess Athena was represented with an olive wreath on her helmet and an amphorae containing olive oil.

The olive tree was also a symbol of peace for the Jews. In the Book of Genesis, a white dove sent out of Noah's ark to search for land, returned with an olive branch, indicating the end of God's anger, and becoming an important symbol of peace. Olive oil holds also a special role in Christianity, from the celebration of baptism to that of Palm Sunday.

12.2 Epidemiology

While the olive tree has been considered as a universal symbol of love and peace, its fruits and the oil obtained from them hold an important role in the Mediterranean diet. Epidemiological studies suggest that the Mediterranean diet is protective against cancer and coronary heart disease mortality [2] and cardiovascular disease in general [3]. Furthermore,

Mediterranean countries, like Greece, Italy, and Spain, have a lower mortality for colorectal and breast cancer compared to countries where the consumption of olives and olive oil is low (England, Scotland, Denmark) [4]. Recent studies also suggest a correlation between the intake of olives and olive oil and skin wrinkling [5], as well as a link between dietary nutrient intake and skin-aging appearance [6]. It is generally accepted that skin, as the largest surface of the body and because of its high lipidic content, is a major target of oxidative stress. A diet rich in fruits and vegetables with olives and olive oil as the major source of fats is considered an important protection against the main diseases associated with oxidative stress. The antioxidant activity of various polyphenols present in plants is seemingly responsible for this protective activity.

Recent findings indicate that olive oil may have a greater role in disease prevention than previously thought, and it is therefore important to identify the components responsible for these health-promoting effects. Unsurprisingly, olive oil has become a very active area of multidisciplinary research [7].

12.3 The Constituents of Olives and Olive Oil

The biological properties of olive oil are related not only to its fatty acid composition, but also to the presence of some minor polar components endowed with strong antioxidant activity. Among them, phenolic compounds are the most important class of olive antioxidants, and their distribution in olive fruits, olive oil, and olive leaves has been thoroughly investigated. The major phenolics of the olive tree are phenolic alcohols like hydroxytyrosol and tyrosol, secoiridoids like oleuropein, and hydroxycinnamic acid derivatives like verbascoside and caffeic acid (Figs. 12.1–12.5).

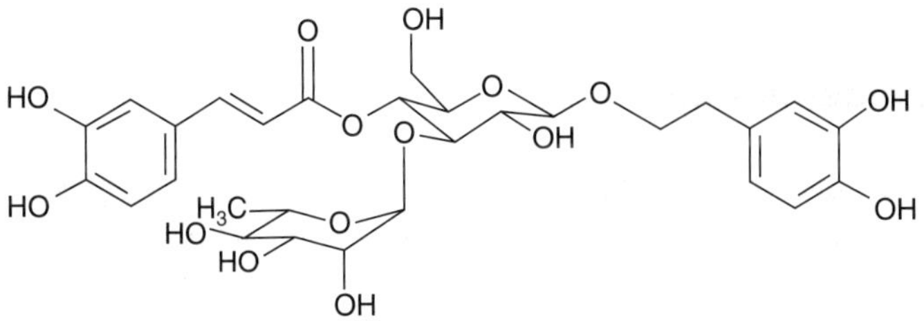

Figure 12.1 Structure of verbascoside.

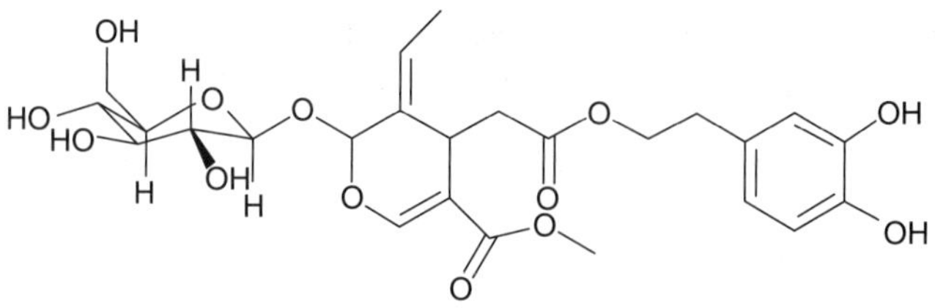

Figure 12.2 Structure of oleuropein.

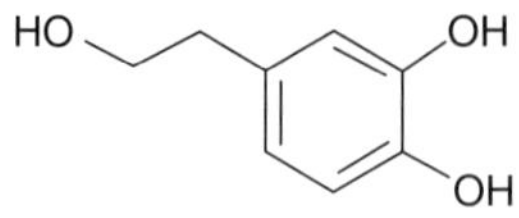

Figure 12.3 Structure of hydroxytyrosol.

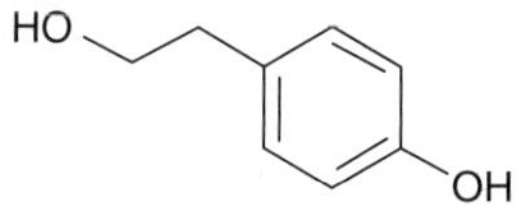

Figure 12.4 Structure of tyrosol.

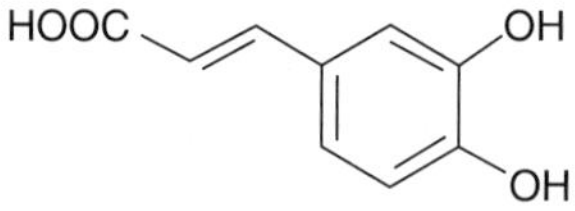

Figure 12.5 Structure of caffeic acid.

In order to better exploit the activity of these antioxidants, botanical extracts have been developed from different parts of the olive tree: fruits, leaves, and by-products of the production of oil, the pomace and the olive mill waste waters. A complete and exhaustive comparison among the phenolic profiles of various *Olea europaea* foodstuff has been conducted at Indena R&D laboratories applying UV analysis for the total phenolic content and HPLC analysis [8].

As an example, the average contents of various polyphenols in fresh fruits from an Italian cultivar is:

- hydroxytyrosol: 0.06–0.41% (w/w by HPLC)
- tyrosol: 0.01–0.12% (w/w by HPLC)

- oleuropein: traces–0.24% (w/w by HPLC)
- verbascoside: 0.02–0.32% (w/w by HPLC) [9]

Olives may be eaten fresh, but they have a very bitter taste, which requires several months of brining and, alternatively, the treatment with strong bases to remove bitterness and make them palatable. These relatively harsh processes reduce the polyphenolic content of the original fresh fruits [10].

The pomace resulting from mechanical defattening of olives shows an average polyphenolic profile almost equivalent to the polyphenolic profile of the starting fruit, because only few polyphenols originally present in the olive are lost in water. Starting from a pomace, having the following profile:

- hydroxytyrosol: 0.12% (w/w by HPLC)
- tyrosol: 0.03% (w/w by HPLC)
- oleuropein: <0.01% (w/w by HPLC)
- verbascoside: 0.35% (w/w by HPLC)
- caffeic acid: 0.01% (w/w by HPLC)

The average content of the extract (Opextan(R)), obtained by optimizing a specific process, is the following:

- hydroxytyrosol: 0.83% (w/w by HPLC)
- tyrosol: 0.28% (w/w by HPLC)
- oleuropein: 0.01% (w/w by HPLC)
- verbascoside: 3.85% (w/w by HPLC)
- caffeic acid: 0.04% (w/w by HPLC)

From a qualitative standpoint, we may observe that the polyphenolic profile is maintained. The total content of polyphenols detected by spectrophotometry is no lower than 10%.

The composition of the olive fruit is very complex and markedly dependent on factors like cultivar, cultivation practices, harvesting method, geographical origin, and degree of maturation. Thus, during fruit development oleuropein is extensively degraded, and is almost undetectable when the fruit is black; conversely, hydroxytyrosol, tyrosol [11], and verbascoside increase [12].

12.4 Daily Consumption

Olive oil consumption is high in the Mediterranean countries, reaching 18 kg/year/person in Greece, 13 in Italy, and 11 in Spain [13]. These amounts correspond to a daily dose of olive oil ranging between 30 g/day and 50 g/day. There is a certain consensus in defining the average polyphenol contents of olive oil as 300 mg/kg. Under this assumption, the estimated consumption of olive oil polyphenols may be evaluated between 9 mg/day and 15 mg/day.

In Italy and in other Mediterranean countries like Spain and Greece, it is also very common to consume olives. In Italy, the average consumption of table olives is about 8 g (two or three fruits) per day [14], corresponding to a polyphenolic content of ca. 16 mg (assuming an average polyphenolic content of 0.2% in edible fruits).

These figures are fully compatible with the intake corresponding to the daily dosage of most olive extracts on the market. Thus, a daily dosage of 100 mg of Opextan® (olive fruit extract) would afford 10 mg polyphenols, perfectly in line with the usual intake of *Olea europaea* products.

Dietary olive polyphenols are absorbed by humans in a dose-dependant way [15], and are excreted in the urine as glucuronide conjugates. Simple phenols like tyrosol and hydroxytyrosol are rather polar, and they are excreted by the kidneys either as such or as their metabolites, with the degree of glucuronidation increasing with the dose.

12.5 Antioxidant Activity (*In Vitro*)

Oxidative stress is one of the main causes for the onset of various degenerative disorders. Lipid peroxidation is a well-known example of oxidative damage in lipid-containing structures [16,17], and epidemiological studies suggest that a high intake of antioxidants is protective in this context.

Olive polyphenols are free radical scavengers with a direct impact on skin health, because they can prevent the oxidative damage involved in the formation of wrinkles and in skin hyperproliferation and dryness. *In vitro* studies have been carried out under the assumption that the antioxidant activity of the olive fruit extract, like its radical scavenging properties, is responsible for the major part of the biological effect attributed to olive polyphenols.

The stable radical DPPH (1,1 diphenyl-2 picrylhydrazyl) method has been used to determine the antioxidant activity of the extract, and of its single polyphenolic constituents [18]. This method measures the hydrogen-donating capacity of a test material. The antioxidant capacity of purified olive polyphenolics (verbascoside, hydroxytyrosol, caffeic acid) has been compared with two references (ascorbic acid and oleuropein), qualifying verbascoside as a very potent antioxidant, five-fold more active than oleuropein (a polyphenolic iridoid typical of olive leaves).

Another very common test for *in vitro* antioxidant activity is based on superoxide anion O^{2-}. This model mimics the *in vivo* situation, employing a physiological oxidant. Thus, the harmful effect of skin exposure to UV rays has been linked to the formation of reactive oxygen species (ROS), including the superoxide radical.

12.6 Reduction of Oxidative Stress

12.6.1 Oral

In a biological system, whenever an imbalance occurs between the natural production of ROS and the capacity to detoxify them or repair the damage they induce, "oxidative stress" results. This condition can be harmful, because it leads to the production of ROS that can cause extensive cellular damage, affecting all cell components, including protein, lipids, and DNA.

Oxidative stress might underlay the onset of chronic degenerative diseases like Parkinson's and Alzheimer's diseases and cancer. It has been linked to cardiovascular disease, and it is certainly important in the aging process. An increased lipid peroxidation is one of the results of oxidative stress and has been correlated to the incidence of a host of diseases that affect the cardiovascular system [7]. There are few suitable biomarkers for lipid peroxidation and oxidative stress, but an increase in both plasma levels and urinary excretion of F2 isoprostanes is believed to be directly related to enhanced oxidative stress. F2 isoprostanes are a family of prostaglandin-like compounds formed *in vivo* by the free radical–catalyzed peroxidation of arachidonic acid in a way independent from the action of cycloxygenase enzymes [19]. Numerous studies have validated isoprostanes as accurate markers of lipid peroxidation *in vivo*.

Epidemiological studies suggest that a high intake of dietary antioxidants can reduce oxidative stress and the effect of an olive fruit extract on isoprostane excretion has been studied in human volunteers [20], quantitating the excretion of 8-isoprostane (8-iso-prostaglandin F2α) (Fig. 12.6).

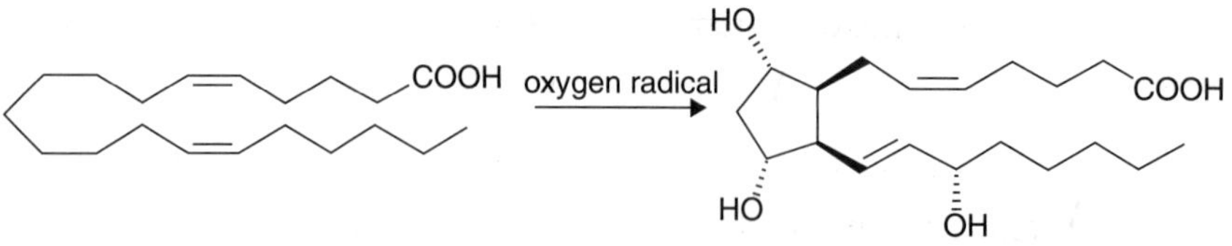

Figure 12.6 Formation of the oxidative stress marker 8-isoprostane.

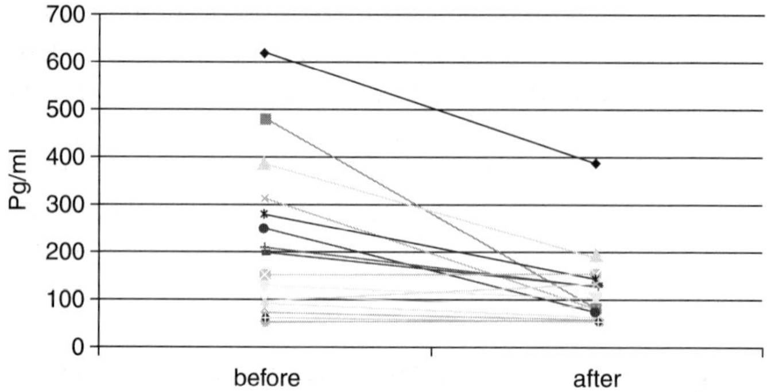

Figure 12.7 Decrease of the oxidative stress marker 8-isoprostane.

Nineteen healthy subjects received an olive fruit extract (Opextan®) at a daily dosage of 400 mg for 4 weeks. The excretion of 8-isoprostane was evaluated in the urine before the beginning of the study and at the end of the 4-week treatment. The oral administration of olive fruit extract significantly decreased the formation of the oxidative stress marker 8-isoprostane (by 47%, mean data, $p < 0.05$, Fig. 12.7). The administration of a polyphenolic olive fruit extract can therefore be recommended in order to control the organism's oxidative stress. Although there is still no evidence that the beneficial impact of antioxidants may have a favorable impact on maximal lifespan, current data indicate that their biological activity on age-related degenerative diseases may produce an improvement in lifespan and enhance quality of life.

12.6.2 Topical

Lipid peroxidation is a well-known example of oxidative damage in lipid-containing structures [17]. Unsaturated phospholipids, glycolipids, and cholesterol present in cell membranes are all targets for oxidation. Lipid peroxidation is a degenerative process that affects both the structure and the function of the target system, and that has been linked to a variety of disorders [17].

The protective effect of the olive fruit extract (Opextan®) against oxidative stress and lipid peroxidation has been assessed experimentally in order to detect its biological activity when applied topically against oxidative damage. Thus, a formulation containing olive fruit extract was applied topically and evaluated in a group of six healthy volunteers, who were asked to wash their face and apply the formulation under investigation and the placebo formulation on each half of their face. Three hours later, they were irradiated with sunlight for about 20 minutes, and sebum was sampled by Sebutape, extracted and measured by dosing luminescence, a direct measurement of lipid peroxidation, in 0.1 μg/ml recovered sebum. The area treated with the olive fruit extract showed a 27.1% (statistically significant, $p < 0.001$ Dunnett's test) decrease of fluorescence compared to the placebo [18].

Clinical data collected so far clearly demonstrates that the olive fruit extract administered either orally or topically can effectively promote healthy skin by reducing oxidative stress.

12.7 UV-Reduced Sensitization

The skin response to sun irradiation can be assessed by measuring the erythemal dose, that is the lowest UV dose that causes enough superficial vasodilatation to be visually perceived as redness. The lowest erythema-inducing UV dose is defined as MED (minimal erythemal dose), and can be visually judged by trained evaluators.

In order to evaluate the effect of orally administered polyphenols from olive (olive fruit extract) on skin sensitivity to UV irradiation, 13 male subjects with high sensitivity to UV light were selected [20]. A dose of 160 mg/day of olive fruit extract (Opextan®) was administered orally for 4 weeks. The subjects were then irradiated with increasing doses of ultraviolet light (UVA + UVB 0.45 mW/cm^2; from 0.054 J/cm^2 to 0.135 J/cm^2) in the dorsal area, and MED was evaluated. The results showed an averaged 16.45% increase of MED compared to before the treatment, confirming the protective effect of olive polyphenols on skin sensitivity to UV irradiation (Fig. 12.8).

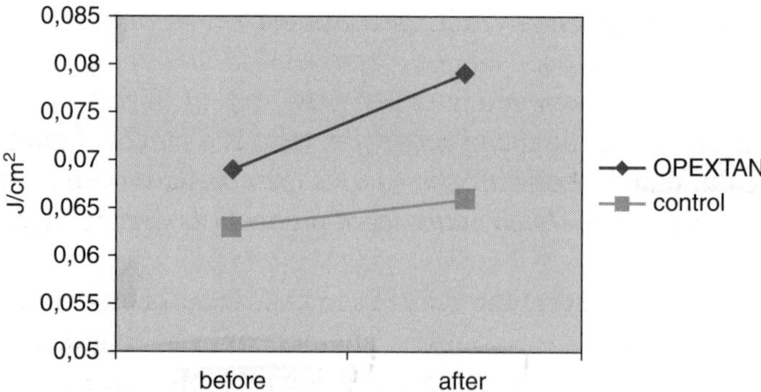

Figure 12.8 MED variation before and after the olive fruit extract treatment.

Other investigations [21] indicate that daily topical use of super virgin olive oil after sun bathing may delay and reduce UV-induced skin cancer development in human skin, possibly by decreasing ROS-induced gene mutations.

12.8 Conclusions

As the world population is growing older, age-related diseases, including skin aging, are becoming major sources of concern. Skin aging may be caused by internal and external causes, defining *internal aging* as the natural physiological process, and *external aging* as a host of conditions (sunlight, pollution, smoke, dietary habits) capable of mimicking the effect of physiological aging. The aging effect of sun on skin can be verified very simply, by comparing the visual appearance of body areas normally exposed to sun (face, hands) with parts normally protected by clothes. Furthermore, as we grow old, we become gradually sensitized to external skin damage by the dwindling healing properties of our organism. For this reason, research of antiaging products is currently hotly pursued at genomic, pharmacological, and cosmetic levels.

Within antiaging strategies, dietary supplementation is getting growing attention, going beyond the topical applications of active principles on the aging skin, and rather focusing on the internal protection against free radicals and oxidative stress, a strategy that can be pursued with specific biomarkers.

Olive derivatives (fruits, olive oil, and olive leaves) are very rich in antioxidant compounds. Polyphenols like oleuropein (in the leaves), verbascoside (in the fruit), and tyrosol and hydroxytyrosol (in all olive foodstuffs) may exert an interesting antioxidant activity in various districts of the organism, including the skin. Unsurprisingly, various extracts have been developed in order to provide olive-based antioxidant products as dietary ingredients.

Over the past few years, the realization that food plants contain small molecules (secondary metabolites, phytochemicals) that can affect our health has fuelled intense research to identify bioactive dietary compounds that have the potential to complement macronutrient-based (protein, lipid, carbohydrates) nutrition. Thus, dietary supplementation has been recognized to play an important role in the prevention of the onset of several pathologies, rationalizing some of the many epidemiological correlations between diet and health. Olive, the hallmark of Mediterranean nutrition, fully qualifies as an interesting source of new dietary supplements to increase health in general and, specifically, to fight skin aging.

References

1. World Olive Encyclopedia—International Olive Council, 1996.
2. Owen R, Giacosa A, Hull WE, Haubner R, Würtele G, Spiegelhader B, Bartsch H, "Olive-oil consumption and health: the possible role of antioxidants," *Lancet Oncol* 1, (2000), 107–112.
3. Covas MI, "Olive oil and the cardiovascular system," *Pharmacol Res* 55, (2007), 175–186.
4. Levi F, Lucchini C, La Vecchia C, "Worldwide patterns of cancer mortality," *Eur J Cancer Prev* 3, (1994), 109–143.
5. Purba MB, Kouris Blazos A, Wattanapenbaiboon N, Lukito W, Rothenberg EM, Steen BC, Wahlqvist ML, "Skin wrinkling: can food make a difference?" *J Am Coll Nutr* 20, (2001), 71–80.
6. Visioli F, Galli C, Plasmati E, Viappiani S, Hernandez A, Colombo C, Sala A, "Olive phenol hydroxytyrosol prevents passive smoking-induced oxidative stress," *Circulation* 102, (2000), 2169–2170.
7. Cosgrove MC, Franco OH, Granger SP, Murray PG, Mayes AE, "Dietary nutrient intakes and skin ageing appearance among middle-aged American women," *J Am Chem Nutr* 86, (2007), 1225–1231.
8. Indena Internal report no. 07/07/LS.
9. Romani A, Mulinacci N, Pinelli P, Vincieri F, Cimato A, "Polyphenolic content in five tuscany cultivars of Olea europaea L.," *J Agruc Food Chem* 47(3), (1999), 964–967.
10. Owen R, Haubner R, Mier W, Giacosa A, Hull WE, Spiegelhalder B, Bartsch H, "Isolation, structure elucidation and antioxidant potential of the major phenolic

and flavonoid compounds in brined olive drupes," *Food Chem Tox* 41, (2003), 703–717.
11. Ryan D, Antolovich M, Prenzler P, Robards K, Lavee S, "Biotransformation of phenolic compounds in *Olea europaea* L.," *Sci Hortic* 92(2), (2002), 147–176.
12. Amiot M, Fleuriet A, Macheix J, "Importance and evolution of phenolic compounds in olive during growth and maturation," *J Agric Food Chem* 34, (1986), 823–826.
13. Tuck KL, Hayball PJ, "Major phenolic compounds in olive oil: metabolism and health effects," *J Nutr Biochem* 13(11), (2002), 636–644.
14. International Olive Council – consumption data. November 2006.
15. Visioli F, Galli C, Bornet F, Mattei A, Patelli R, Galli G, Caruso D, "Olive polyphenols are dose-dependently absorbed in humans," *FEBS Letters* 468, (2000), 159–160.
16. Sakurai H, Yasui H, Yamada Y, Nishimura H, Shigemoto M, "Detection of ROS in the skin of live mice and rats exposed to UV light: a research review on chemiluminescence and trials protection," *Photochem Photobiol Sci* 4(9), (2005), 715–720.
17. Girotti AW, "Lipid peroxide generation, turnover and effector action in biological systems," *J Lipid Res* 39, (1998), 1529–1542.
18. Maramaldi G, Artaria C, Ikemoto T, Haratake A, "Estratto standardizzato di frutti di *Olea europaea*," *L' integratore Nutrizionale* 9(3), (2006), 23–29.
19. Morrow JD, Roberts LJ, II, "The isoprostanes. Current knowledge and directions for future research," *Biochem Pharmacol* 51(1), (1996), 1–9.
20. Yokota T, Haratake A, Komiya A, Suzuki K, Mori M, Ikemoto T, Artaria C, Giori A, Maramaldi G, D'Angiò C, "Antioxidant capacity and skin health studies in human volunteers of a new standardized extract from olive pulp," *Fitoterapia*, article in press.
21. Ichihashi M, Ahmed NU, Budyianto A, Wu A, Bito T, Ueda M, Osawa T, "Preventive effect of antioxidant on ultraviolet-induced skin cancer in mice," *J Dermat Sci* 23(Suppl 1), (2000), S45–S50.

13

Enhancing the Skin's Natural Antioxidant Enzyme System by the Supplementation or Upregulation of Superoxide Dismutase, Catalase, and Glutathione Peroxidase

Nadine Pomarede[1,2], MD, and Meera Chandramouli[2]

[1]Isocell Nutra, Paris, France
[2]PL Thomas & Co., Inc., NJ, USA

Aaron Tabor and Robert M. Blair (eds.), *Nutritional Cosmetics: Beauty from Within*, 245–265, © 2009 William Andrew Inc.

Summary

The skin is the body's largest organ and its first line of defense. It protects against a variety of harmful pathogens and environmental factors such as UV radiation, air pollution, and extreme temperatures. Every square inch of skin contains approximately 20 blood vessels, 60,000 melanocytes (which provide pigmentation and guard against UV radiation), 650 sweat glands, and thousands of nerve endings. In order to preserve its effectiveness, the skin is sustained endogenously (protected from within) through an oxidant-antioxidant balance. Oxidation reactions are crucial factors in many metabolic processes; however, an excess of certain oxidants, classified as reactive oxygen species (ROS), can cause cell death and oxidative stress. Oxidative stress in turn induces cardiovascular disease, inflammation, and cancer, in addition to extensive skin damage. Endogenous and exogenous antioxidants, which inhibit oxidative stress, function to counter these effects in the body.

Examples of endogenous, enzymatic antioxidants include: superoxide dismutase (SOD), catalase (CAT), and glutathione peroxidase (GPx). SOD's role in maintaining the oxidant-antioxidant balance in the body, thus protecting the body against harmful free radicals, was first determined in 1968. Since then, numerous studies into the benefits of SOD's interactions with the free radical superoxide anion (O_2^-) have been conducted. SOD is essential in halting the proliferation of free radicals and in triggering the further antioxidant activity of CAT and GPx. Excess free radical production can be triggered by a variety of external factors (Fig. 13.1), each of which effectively disrupts the oxidant-antioxidant balance in the body.

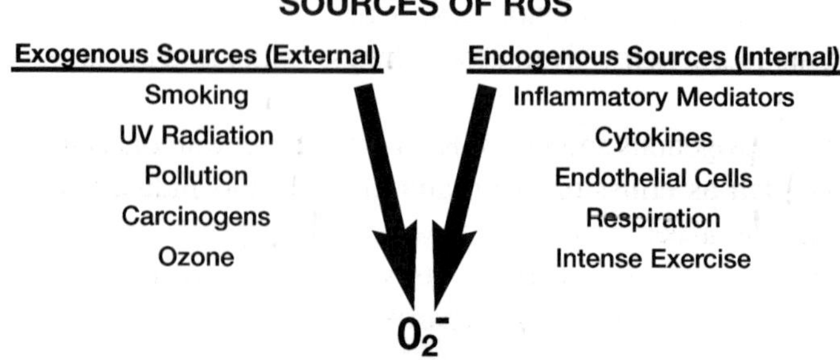

Figure 13.1 Endogenous and exogenous generation of ROS and antioxidant defense system.

These factors include: aging, exposure to UV radiation, smoking, high-intensity exercise, and pollution. Under such conditions, the body will experience oxidative stress.

Nutritional supplements may be effective in increasing the production of endogenous antioxidants. One compound, GliSODin®, has been shown to promote bioactive SOD, CAT, and GPx levels via oral supplementation. Studies have been performed on both humans and animals to demonstrate GliSODin®'s role in the maintenance of antioxidant levels, protection against inflammation, and overall skin health. In helping to preserve the oxidant-antioxidant balance, GliSODin® protects the skin and body from the ill effects of oxidative stress. Other supplements that provide the body extra amounts of the building blocks it requires to make these natural antioxidants, such as manganese, zinc, copper, and selenium, may also be an effective way to increase their presence in the body.

13.1 Introduction: Oxidative Stress

Oxidative stress is damage in a cell, tissue, or organ, caused by ROS. Free radicals and peroxides are generated during metabolism and exist inherently in all aerobic organisms. They are generated through chemical reactions involving both endogenous and exogenous molecules. Endogenous ROS production occurs as cells react with oxygen as part of cellular metabolic processes such as energy generation from mitochondria and detoxification reactions involving the liver cytochrome P-450 enzyme system. The by-products of these processes include hydroxyls, peroxides, and other damaging free radicals. Other endogenous biological indicators of oxidative stress are: lipid peroxidation, protein oxidation (glutamine synthetase activity and protein carbonyl levels), oxidative DNA damage, reduced mitochondrial function, and a decrease in levels of endogenous antioxidants in the heart, liver, blood, lungs, brain, and muscles [1].

Examples of exogenous sources of ROS are: cigarette smoke, environmental pollutants such as emission from automobiles and industries, ultraviolet (UV) radiation, asbestos, excess alcohol consumption, and bacterial, fungal, or viral infections. Oxidative stress can result in lipid peroxidation, which can lead to atherosclerosis (hardening of the arteries) and ultimately cardiovascular disease, damaged or prematurely aged skin, and cancer. Inflammation, which is the key manifestation of rheumatoid arthritis, metabolic syndrome, and diabetes, as well as neurodegenerative diseases like

Alzheimer's, may be accelerated by oxidative stress [2]. Also, oxidative stress may cause tissue toxicity and accelerated aging, which is especially evident in free radical damage to the skin. This process is similar to that of a freshly cut apple turning brown due to exposure to air.

Many of the listed exogenous sources of excess ROS lead to extensive skin damage. Smoking is one such example. In both the gaseous and tar phases of smoke, organic radicals are produced that react with existing molecular oxygen (O_2) to form dangerous free radicals such as hydrogen peroxide (H_2O_2), superoxide anion (O_2^-), and various hydroxyls. Smoking also increases the production of an enzyme that breaks down the protein collagen, which is what keeps skin elastic and supple. After the collagen molecules are broken down, they link back up again in a different way. This process is called cross-linking and causes the normally mobile collagen matrix to become stiff and inflexible. Ten minutes of smoking decreases the skin's O_2 supply for nearly an hour, because nicotine narrows blood vessels and prevents blood from circulating to the capillaries in the dermis. Eventually, the skin starts to exhibit characteristics of accelerated aging, such as premature wrinkles and sagging skin. Smokers are also 3.3% more likely to develop skin cancer [3].

Strong sunlight is another exogenous factor that contributes to skin damage, because it generates free radicals in the skin. The surfaces of the hands, face, neck, and arms are frequently exposed to light and are thus most susceptible to the effects of photoaging. Photoaged skin is characterized by laxity, deep wrinkles, uneven pigmentation, brown spots, and a leathery appearance. In contrast, chronologically aged skin that has been protected from the sun has reduced elasticity but is smooth and unblemished. Fisher et al. (1997) has shown that frequent exposure to UV irradiation leads to an increase in the levels of enzymes that degrade collagen and contribute to photoaging [4]. Sunlight also activates messenger molecules (cytokines), which create inflammatory products in skin cells. Additionally, skin cancer is more likely to develop in people with photoaged skin.

13.2 Role of Endogenous Antioxidants

An antioxidant is a substance that slows or prevents the oxidation of other chemicals in the body. It may be acquired through dietary means or through endogenous production. In order for its role to be completely understood, however, the concept of oxidation must first be discussed. Oxidation

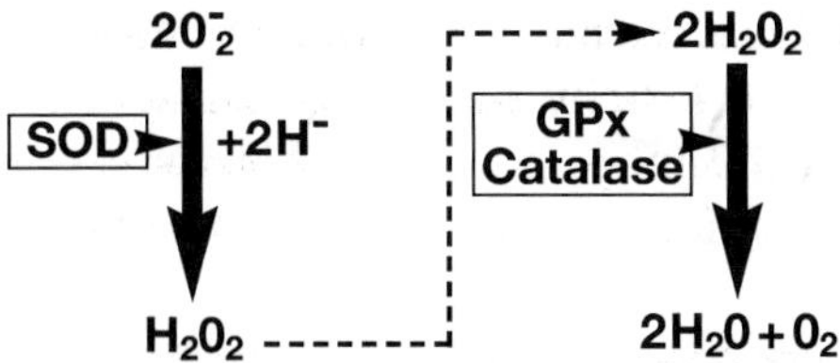

Figure 13.2 Roles of SOD, catalase, and GPx in the inactivation of the superoxide ion.

describes the process of stealing electrons from other atoms for stability. These reactions often involve the production of highly unstable free radicals, which can initiate the formation of multiple destructive ROS. This chain reaction begins with the reduction of O_2. Antioxidants are efficient and effective because they combine with O_2^- at the beginning of the free radical pathway, neutralizing the cascade, and thus preventing damage (Fig. 13.2). Antioxidants can be both nonenzymatic and enzymatic. Nonenzymatic antioxidants include lipid-soluble vitamin E, co-enzyme Q10, phosphatidylserine, and water-soluble vitamin C. Other nonenzymatic antioxidants are beta-carotene, uric acid, and glutathione (GSH). These antioxidants are exogenous and, once ingested, are consumed rapidly or excreted. This occurs because a stoichiometric relationship exists for most vitamins, carotenoids, and thiols.

For example, one vitamin C molecule may halt the progress of only one ROS. Under conditions involving excess production of free radicals, vitamin C must be rapidly consumed to replace lost supplies. Importantly, if the body relies on vitamin C to fight ROS proliferation, vitamin C is no longer available to perform its other crucial duties. These duties include: the production of collagen, increased immunity, fat metabolism, bone health, and the synthesis of certain neurotransmitters.

Endogenous enzymatic antioxidant supplies, on the other hand, are more potent than nonenzymatic antioxidants and are depleted far less quickly. Primary examples of these are SOD, CAT, and GPx (Fig. 13.3). When young skin is exposed to both endogenous and exogenous sources of oxidative stress, there are sufficient levels of ATP (cellular energy) for DNA repair and cell renewal. However, as the individual ages, levels of ATP and endogenous antioxidants are depleted. Additionally, SOD, CAT, and GPx are readily available to scavenge "rogue" ROS. However, supplies of ATP

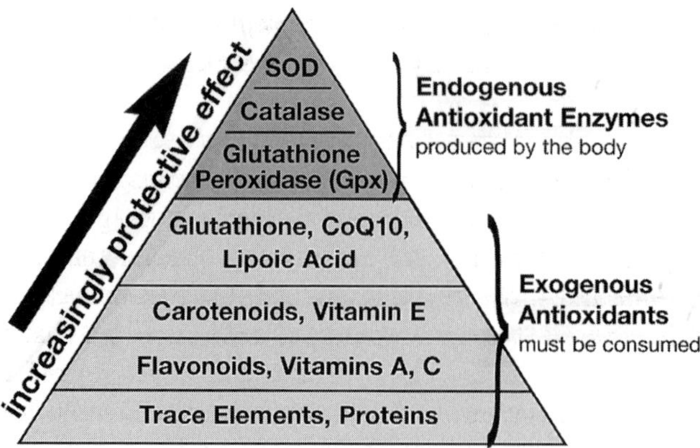

Figure 13.3 Compares the protective effects of the endogenous, enzymatic antioxidants SOD, CAT, and GPx to antioxidants obtained exogenously (i.e., food-sourced).

and endogenous antioxidants decrease with age and therefore supplementation may be desired in order for the body to maintain ROS equilibrium.

13.2.1 Catalase

The enzymatic antioxidant catalase (CAT) is essential in promoting skin health. It is inherent in most aerobic animal cells and is localized in the liver and red blood cells. It uses iron as its trace metal cofactor and is composed of four subunits, each containing a heme group that is responsible for CAT activity. Once SOD catalyzes the reduction of O_2^- to H_2O_2, CAT catalyzes the further reduction of potentially damaging H_2O_2 to O_2 and H_2O. Like SOD, CAT is an extremely efficient enzyme whose high recycling rate demonstrates its ability to detoxify H_2O_2 and to prevent the formation of carbon dioxide (CO_2) bubbles in blood. Good dietary sources of CAT include: clover, lentils, mung beans, radishes, and sunflower seeds.

CAT plays a significant role in skin protection. Rhie et al. (2001) demonstrated the difference between the effects of acute and chronic UV radiation on CAT levels [5]. For acute exposure, human skin samples from a sun-protected area of skin (buttocks) from 13 young Koreans were irradiated with two times the UV radiation necessary to induce sunburn (minimum erythematous dose, MED). Expression of CAT in the epidermis decreased after 24 hours, and was 80% that of the control levels after 48 hours, before levels began to recover. Expression of the enzyme in dermal cells decreased to 72% of control levels after 24 hours, before beginning to recover.

The effect of chronic UV radiation was evaluated using skin samples from sun-exposed (forearm) and sun-protected (upper-inner arm) areas of 13 young (average age 22.8 years) and 8 elderly (average age 70.7 years) Korean subjects. Epidermal CAT activity was greater than dermal CAT activity by 194% and 450% in young and old skin, respectively. CAT activity in the sun-exposed (forearm) epidermal cells exceeded that of the sun-protected (upper-inner arm) cells by 155% and 115% for old and young skin, respectively. However, CAT activity was significantly lower in the dermal cells of the older subjects, being 66% and 51% that of the young skin levels, respectively. These results demonstrated that CAT activity is upregulated on chronic UV exposure during the photoaging period.

Furthermore, a study by Shin et al. (2005) reported that dermal fibroblasts (cells responsible for the synthesis and maintenance of the extracellular matrix) contained lower levels of CAT after photoaging, and subsequently higher H_2O_2 levels than intrinsically aged skin in the same individuals (12 young and 12 elderly Koreans) [6]. Treatment of these photoaged fibroblasts with CAT reduced H_2O_2 levels. A reversal of aging-dependent mitogen-activated protein kinase activities was also reported, coupled with an inhibition of matrix-metalloproteinase (MMP)-1 expression. This study showed that induction and regulation of endogenous antioxidant enzymes could prevent skin aging.

A study performed at the Webb-Waring Lung Institute investigated the effects of CAT on skin burn patients [7]. Prior studies have indicated that during the pathogenesis of lung injury following skin burn, production of ROS such as O_2^- and H_2O_2 increases. Researchers at the Webb-Waring Lung Institute also found that H_2O_2 activity and CAT activity are heightened in adult respiratory distress syndrome patients. Thus, a study was conducted to determine CAT activity in skin burn patients. It was found that rats subjected to skin burn demonstrated increased levels of H_2O_2 and CAT activity in the bloodstream.

In addition to its role as a biomarker in the progression of adult respiratory distress syndrome and skin conditions, CAT is a proven defender of cell health, specifically in its defense against the proliferation of malignant carcinoma cells and nonmelanoma skin cancer. An *in vitro* study was performed on mice to compare CAT levels in papillomas (benign skin tumors) with those in carcinomas (malignant skin tumors) [8]. Researchers applied a carcinogen to normal mouse skin to induce a genetic mutation in epidermal cells. A tumor promoter was then introduced to the cells in which the mutation was activated, causing their proliferation. The proliferation of tumor cells resulted in the formation of benign papillomas. The last step of the protocol involved the application of further genetic changes, stimulating the papilloma cells to become malignant. It was found that the malignant cells contained depleted CAT levels and elevated ROS, specifically H_2O_2, levels. In such instances, H_2O_2 functions as a secondary messenger that activates signal transduction pathways leading to the formation of carcinomas. CAT's role in the prevention of malignant skin tumors is critical, because nonmelanoma skin cancer (over 1 million cases diagnosed every year) is the most common form of malignancy in the United States.

13.2.2 Glutathione Peroxidase

Glutathione peroxidase (GPx) is a tetrameric glycoprotein whose trace metal cofactor is selenium, which promotes antioxidant activity. Selenium plays a huge role in the function of GPx, particularly in the recycling of the protein glutathione (GSH), and can be found in cereals, meat, fish, eggs, and Brazil nuts. Localized in the liver and composed of cysteine, glutamic acid (glutamine), and glycine, GSH is a crucial cofactor in the function of GPx. It plays a role in many biological processes such as enzyme catalysis, protein synthesis, membrane transport, receptor action, cell maturation, and leukotriene synthesis. Using its powerful antioxidant capabilities, GSH defends the cell, specifically the cell membrane and the mitochondria, against harmful free radicals. In addition to its role as an antioxidant, GSH also functions in immune response and in DNA repair and protection. Of the many isozymes of GPx, which vary in cellular location and substrate specificity, GPx 1 is the most abundant and is concentrated in the cytoplasm of most mammalian cells. It uses GSH to reduce H_2O_2 to O_2 and H_2O, and lipid peroxides to their respective alcohols.

GSH is a proven defender against the harmful effects of UVB radiation, which has been linked to melanoma (skin cancer), the breakdown of collagen, and the reddening and burning of the skin. A study was performed on the epidermis of hairless mice to assess the effects of GSH depletion on sunburn cell formation [9]. One group of mice was treated orally with a irreversible inhibitor called buthionine S, R-sulfoximine (BSO) that depleted GSH levels by 10–15%. The other group was not treated with BSO and maintained normal GSH levels. Both groups were then exposed to a moderate amount of UVB radiation. The members of the BSO-treated group were found to contain a much higher sunburn cell count than those of the control group. This study shows that endogenous GSH functions to protect skin cells against moderate levels of UVB radiation.

A 2001 study reported similar findings that the potential of human dermal fibroblast cells to protect against UVA-1 exposure was highly dependent on GSH [10]. Repetitive low-dose UVA irradiation of human dermal fibroblast cells led to a substantial and synchronous upregulation of GPx and SOD activity, and protected cells from a potentially toxic dose of UVA. The study also showed that the increase was dependent on sufficient levels of selenium, the metal cofactor of GPx. Interestingly, the increase in GPx peaks around 12 hours after UVA irradiation to approximately 120% of baseline levels, before decreasing to 80% of baseline.

A 2004 study by Wenk et al. [11] reported that GPx protected against ROS hydroperoxide–induced breakdown of connective tissue in the skin. Human fibroblast cell lines were exposed to UVA radiation. The cells were programmed to overexpress phospholipid-hydroperoxide GPx (PHGPx). In normal human fibroblasts, exposure to UVA led to a NF-κB–mediated increase in interleukin 6 (IL-6) levels, which induced a 4.5-fold increase in matrix metalloproteinase-1 (MMP-1). MMP-1 exhibits substrate specificity for collagens type I and III and is the most important metalloproteinase for the degradation of the extracellular matrix during photoaging. Importantly, no change in MMP-1 levels was recorded after UVA exposure for the cells overexpressing PHGPx.

In further experiments, Wenk et al. treated PHGPx overexpressing cells with phosphatidylcholine hydroperoxides, resulting in a 70% reduction in IL-6 levels, compared to normal cells treated with phosphatidylcholine hydroperoxides. This study was the first to report the effects of GPx on MMP-1 regulation [11].

Selenium, the metal cofactor of GPx, has also proven to be effective in defending the skin against UVB radiation. Past studies have revealed that the proliferation of ROS plays a significant role in UVB radiation–induced cell death. A study was conducted to analyze the effects of selenium in the protection of keratinocytes (main cell type in the epidermis, composing 90% of human epidermal cells) and melanocytes from UVB radiation in human skin cells [12]. Cell cultures that were exposed directly to UVB radiation, without prior selenium supplementation, experienced 80% cell death among keratinocytes. Cells that received selenium supplementation before being exposed to UVB radiation experienced far less cell death among keratinocytes. Similar results were observed for melanocytes in both groups. The body contains an elaborate antioxidant defense system that guards against the free radical damage caused by UVB radiation; however, studies show that UVB exposure depletes SOD and catalase supplies, whereas GPx levels remain constant. As most selenium action is exerted through cytosolic GPx (GPx-1), this demonstrates that GPx plays a major role in protecting the skin against the carcinogenic effects of UVB radiation.

Additionally, GPx-1 is crucial in the prevention of oxidative stress. A 2003 study was conducted among 636 patients with suspected coronary artery disease to assess the cardiovascular risks associated with GPx-1 activity in the red blood cells [13]. It was found that there is an inverse relationship between levels of GPx-1 activity and cardiovascular disease. This study compared the number of cardiovascular injuries in patients to its corresponding level of GPx-1 activity, measured in hemoglobin units per gram. Results showed that the rate of cardiovascular injury in patients in the lowest quartile of GPx-1 activity (20.8%) was almost three times that in patients in the highest quartile (7.0%).

13.2.3 Superoxide Dismutase

SOD is a class of oxido-reductase enzymes, each of which contains copper and zinc, or manganese at its active site. Collectively, they have the fastest recycling rate of any known enzyme. Intracellular SOD-1 uses copper or zinc as its trace metal cofactor and is located in the cytosol of the cell. SOD-2 uses manganese as a cofactor and is located in the mitochondria and bronchial epithelium. Its expression is activated by both endogenous and exogenous oxidants such as inflammatory cytokines, hyperoxia (excess oxygen in the body tissues), and cigarette smoke. Extracellular SOD-3 also uses copper or zinc as a cofactor and is located in pulmonary fluids and the interstitial spaces in the lungs and is crucial in protecting against vascular damage caused by ROS [14].

SOD, CAT, and GPx are the body's lead antioxidant defenses and operate through a feedback mechanism. SOD catalyzes the reduction of the potentially harmful O_2^- to form H_2O_2. Rising H_2O_2 levels result in the gradual inactivation of SOD. As discussed, CAT and GPx catalyze the reduction of H_2O_2 to O_2 and water (H_2O); thus, SOD is conserved. In the same way,

SOD's reduction of O_2^- to form H_2O_2 results in the conservation of CAT and GPx [15]. SOD can also inactivate certain enzymes that control levels of free iron, which can lead to the formation of hydroxyl free radicals in the body.

SOD has numerous health benefits, especially in relation to inflammation and skin protection. SOD-3, for example, protects against oxidative fragmentation of type I collagen in the skin. Immunochemical studies indicate that SOD-3 interacts with type I collagen in the extracellular matrix of blood vessels and airspaces, thus preventing the breakdown of collagen due to oxidative stress [16]. In addition, by neutralizing O_2^-, SOD inhibits the activation of latent collagenases by O_2^-, whose function is to degrade collagen. Once collagen is broken down, the fragments activate chemotactic agents for neutrophils, causing tissue damage and chronic inflammation.

SOD functions by controlling the oxidative reactions that contribute to the pathogenesis of inflammation. Examples of these reactions include: the initiation of lipid peroxidation, the inhibition of mitochondrial respiratory chain enzymes and sodium/potassium ATP-ase activity in the cell membrane, and the inactivation of membrane sodium channels. SOD also halts the proinflammatory effects of O_2^-, which include: damage to the endothelial cells (causing increased microvascular permeability) and the promotion of chemotactic factors such as leukotriene B_4 and the concentration of neutrophils at inflammation sites.

Extensive clinical research into the effects of SOD on inflammation has been conducted. SOD was first used clinically on the degenerative joint condition osteoarthritis. Two early studies were performed confirming SOD's effects on osteoarthritis patients. In both studies, patients were injected with purified bovine SOD (Orgotein), and their progress was charted for varying periods. Significant decreases in pain and subsequent improvements in function, the use of aids, and the use of analgesics were recorded. No adverse side effects were observed [17,18].

SOD also plays a major role in protecting against vascular inflammation, which leads to atherosclerosis and cardiovascular disease. A study was performed on mice lacking apolipoprotein E (responsible for the catabolism of lipoproteins) to examine the effects of SOD-1 and CAT on atherosclerosis [19]. Because the mice were apolipoprotein E–deficient, they developed a build-up of oxidized lipids in the arterial wall, which caused atherosclerosis. Both SOD-1 and CAT were overexpressed in these mice. It was found that the overexpression of SOD-1 together with CAT resulted in reduced levels of F2-isoprostanes (biomarkers of lipid peroxidation and oxidative stress) and the inhibition of atherosclerosis. This study shows that SOD contributes to overall vascular health by protecting against atherosclerosis and oxidative stress.

Several studies with an orally effective delivery of SOD have shown skin health benefits, including the inhibition of UV radiation–induced skin burn, quicker healing, and significant reduction in negative reactions to the sun, particularly for the sun-sensitive and for those with sun allergies [20–22].

13.3 SOD and Dietary Nutrients: GliSODin®

GliSODin® is the first bioactive SOD by the oral route. As the exposure of SOD to the acidic nature of the stomach causes modifications in its quaternary structure and results in a nonfunctioning enzyme, all past oral SOD treatments have been unsuccessful. In order to avoid the denaturation of SOD, GliSODin® combines SOD (derived from SOD-rich melon extract) with gliadin, which is a wheat-based protein that protects SOD during its passage through the gastrointestinal (GI) tract. Gliadin acts in the small intestine by adhering to the sidewall. It then releases the SOD progressively, protecting against intestinal inactivation by digestive enzymes, and enables its recognition by immune-active cells in the GI tract [18]. In was shown *in vitro* that gliadin in combination with SOD improves SOD release, as was demonstrated by the progressive release of SOD in an environment replicating digestive conditions [23]. Animal studies indicate that the oral administration of the SOD-gliadin complex increases the activity of SOD, CAT, GPx, and other antioxidant enzymes in plasma cells, red blood cells, and the liver. Additionally, orally absorbed SOD-gliadin leads to decreased levels of certain biomarkers, such as F2-isoprostanes, which are breakdown markers from the destructive actions of oxidative stress in the body.

A significant *in vivo* study of the effects of GliSODin® with respect to elevated antioxidant levels in the bloodstream was conducted on Balb/c mice [24]. The mice were divided into two groups; each was orally supplemented with standardized SOD melon extract over a period of 28 days. One group was supplemented with SOD alone and the other with GliSODin®. Several biomarkers of oxidative stress, including antioxidant activity, mitochondrial depolarization in hepatic cells, and apoptosis in hepatic cells, were used to analyze the results of both groups. In the group administered with unprotected SOD, there were no significant changes in antioxidant levels. However, the GliSODin® group demonstrated a dramatic increase in the activity of circulating antioxidant enzymes. In addition, hepatic cells isolated from the mice receiving GliSODin® presented a delayed depolarization response and an increase in the resistance to apoptosis induced by oxidative stress. This study demonstrates that GliSODin® supplementation leads to definitive improvements in the antioxidant status of the cells and protects them against the detrimental effects of oxidative stress.

Human studies also demonstrate the role of GliSODin® in promoting the production of SOD and other endogenous antioxidants. Muth et al. (2004) conducted a randomized, double-blind, placebo-controlled trial to determine GliSODin®'s role in the protection against DNA damage induced by hyperbaric oxygen (HBO), or pure oxygen at a pressure of 2.5 atmospheres [25]. Twenty male volunteers were randomly assigned to receive either a daily dose of GliSODin® (1,000 IU) or placebo for 2 weeks, prior to HBO exposure for 60 minutes. A comet assay was then used to measure the resulting DNA damage from HBO exposure (Fig. 13.4). A significantly greater amount of damage was found in the placebo group, as compared to little or no damage in the GliSODin®-supplemented group.

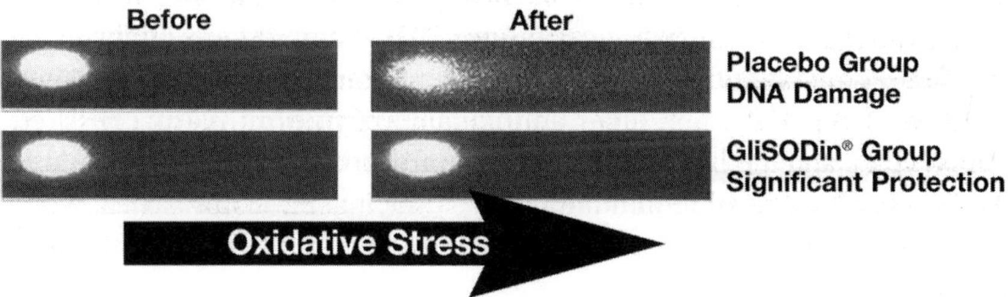

Figure 13.4 Effects of GliSODin® supplementation on DNA damage, measured by the comet assay—a microgel electrophoresis method that allows detection of DNA damage in individual cells. The upper row shows progressive and significant DNA damage in the placebo group, evidenced by the "comet's tail" after the spot. No such effects were observed in the GliSODin® group (1,000 mg/day), showing an intact cell nucleus with no DNA damage.

Additionally, urine levels of F2-isoprostanes, well-established markers of oxidative stress, increased by 33% from baseline in the placebo group but not in the GliSODin®-supplemented group. This study indicates that GliSODin® protects against DNA damage produced by HBO and thus confirms the antioxidant activity of the supplement.

GliSODin® also has proven anti-inflammatory effects. An *in vivo* study by Vouldoukis et al. performed on C57BL/6 mice analyzed these effects [26]. The mice were split into four groups and administered, respectively, with GliSODin®, gliadin alone, SOD from melon extract alone, or a placebo. After 28 days, peritoneal macrophages were activated by the proinflammatory INF-gamma and then collected after 24 hours. The ability of the macrophages to produce free radicals and cytokines such as the proinflammatory tumor necrosis factor (TNF-alpha) and the anti-inflammatory interleukine-10 (IL-10) was then measured. Of the four oral supplementations, only GliSODin® was found to protect cells from the effects of INF-gamma. This result was obtained by observing the significant increase in the production of anti-inflammatory IL-10 and consequent reduction of TNF-alpha in the GliSODin® group.

The impact of GliSODin® on antioxidant status and the vascular inflammatory process was evidenced by a study by Cloarec et al. (2007) [27]. Seventy-six patients at risk for cardiovascular disease were subjected to diet and lifestyle changes for one year, at which point the 34 remaining patients (42 dropped out due to the stringency of study) were randomly split into two groups. One group continued with the prescribed diet and lifestyle, while the other received an additional GliSODin® supplement (500 IU/day) for two more years. After two years, carotid artery intima media thickness (IMT), an indicator of atherosclerosis, was measured using ultrasound-B imaging. A significant reduction in the IMT in the SOD-supplemented group was found, as compared to an increase in the IMT of the nonsupplemented group. Moreover, malondialdehyde levels, a marker of lipid oxidation, were reduced by 34% in the GliSODin® group. This study demonstrates GliSODin®'s important role in cardiovascular health, reducing inflammation and providing improvement in the circulatory system.

13.3.1 GliSODin® and Skin Protection

GliSODin® supplementation is a valuable adjunct to sunscreen for skin protection, especially against harmful UV radiation. Unprotected skin can be severely damaged by constant exposure to the sun's UVB and UVA rays. In addition to burning (erythema), UV radiation can damage collagen fibers and cellular DNA; they corrupt DNA molecules by causing malformations that can lead to mutations and unhealthy cell production. Three human studies demonstrate the benefits of GliSODin® with respect to protecting cells against UV radiation and photo-oxidative stress. The first study revealing the benefits of the antioxidant properties of GliSODin® for skin health was conducted as an open clinical trial conducted by 40 dermatologists in France. The trial involved 150 participants who were chosen based on their susceptibility to sunburn and other photo-oxidative stress reactions. 500 mg of GliSODin® were administered daily over 3–8 weeks of normal sun exposure (including use of sunscreen, SPF 20–100).

Study participants were split into three groups: 75 patients with significant flushing or reddening almost immediately during exposure; 60 patients who experienced negative sun reactions, including sun allergy; and 15 patients who experienced other irritating skin reactions such as pruritus and solar eczema.

Results after 4–8 weeks were as follows: of the 75 patients in the first group, 64 patients reported excellent tolerance, 6 patients had diminished burning, whereas 5 patients had no improvement; in the second group of 60 patients, 44 did not experience allergic reactions to the sun, whereas 6 had a reduced reaction and 10 had a reaction; and in the third group, all of the 15 participants had complete relief of their usual symptoms (Fig. 13.5). In summary, 86% of participants reported significant relief [20].

A follow-up randomized, placebo-controlled, double-blind trial by Mac-Mary et al. involved 49 healthy individuals: 10 phototype II, 19 phenotype III, and 20 phenotype IV [22]. Researchers used UV radiation to induce photo-oxidative stress on the inner forearms of the participants and measured the susceptibility of the participants to sunburn (minimum erythematous dose, or MED) and the resulting redness (actinic erythema). A UV stress test was then conducted before GliSODin® supplementation was administered. Participants received 500 mg doses each week for 4 weeks, and UV stress tests were conducted each week to measure progress. At the end of 4 weeks, an increase of 8 times in the minimum exposure to UV rays necessary to produce skin burn for phototype II participants was found in the GliSODin® group, compared to the placebo group (Fig. 13.6). Additionally, the induced redness decreased faster in the GliSODin® group compared to the control group (Fig. 13.7) [28]. The effects were more marked in the phototype II subjects than the other skin-type groups.

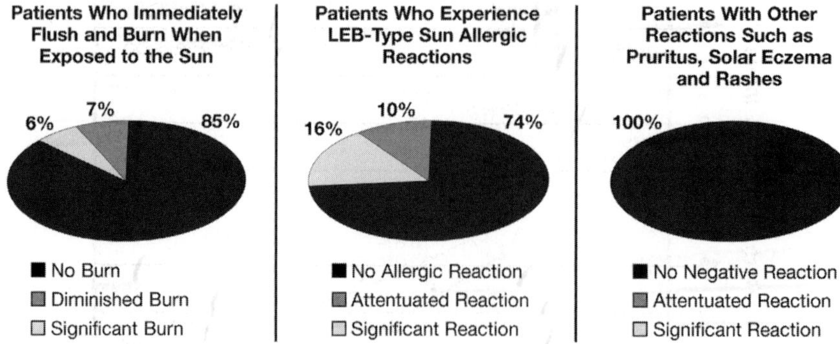

Figure 13.5 Effects of GliSODin® supplementation in 150 sun-sensitive patients divided into three groups; 75 Patients Who Immediately Flush and Burn When Exposed to the Sun; 60 Patients Who Experience LEB-type Sun Allergic Reactions; and 15 Patients With Other Reactions Such as Pruritus, Solar Eczema, and Rashes. This chart shows the scale of relief received in each group by percentage.

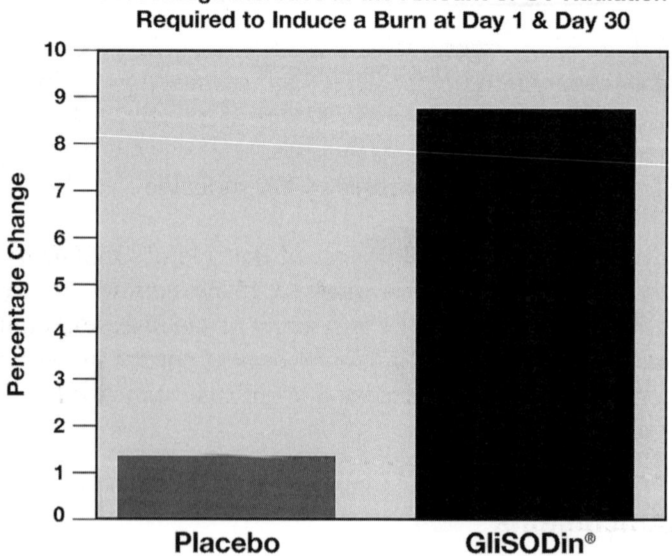

Figure 13.6 Compares the percentage increase of minimum erythematous dose (MED) of type II phototypes in GliSODin® group from day 1 to day 30, to that in placebo group.

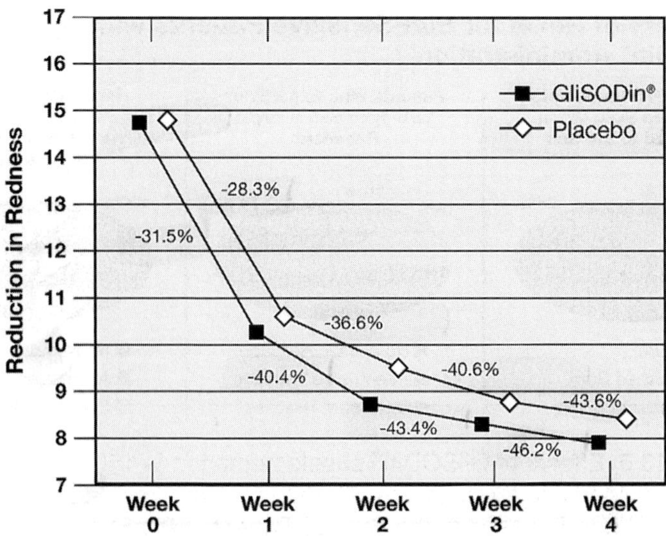

Figure 13.7 The redness produced by UV exposure was reduced in the GliSODin® group in type II phototypes, compared to placebo, suggesting a protective benefit. Figure generously provided by Dr. Humbert, Département de Dermatologie, Centre Hospitalier Universitaire St Jacques 2, Place St Jacques 25030 BESANCON CEDEX.

In addition, capillary regeneration was faster than in the control group. The number of capillaries in the GliSODin® treatment group increased by 37.7% after 4 weeks, compared to an increase of only 18.1% in the placebo group (Fig. 13.8) [28]. It was thus concluded that GliSODin® protects the skin from oxidative stress caused by UV radiation.

One of the first conclusive pilot trials of the role of GliSODin® in skin protection against UV radiation involved 15 patients who were chosen based on their high susceptibility to sunburn or sun disease. Administered daily were 500 mg of GliSODin® for 60 days of normal sun exposure. At the end of 60 days, patients reported significant reductions in sunburn, flushing, and skin rash [21].

13.4 Conclusions

ROS production is an inevitable consequence of metabolic processes in aerobic organisms. Under normal conditions, an oxidant-antioxidant balance—maintained by the defense mechanisms of endogenous, enzymatic

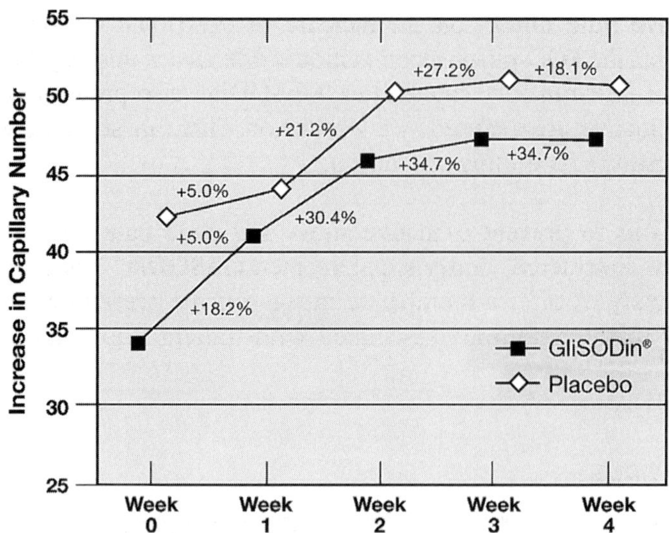

Figure 13.8 Increased evolution of capillary numbers after MED exposure during 4 weeks as a result of GliSODin® supplementation, compared to placebo, suggesting a protective benefit.

antioxidants such as SOD, catalase, and glutathione peroxidase—exists. However, an excess of ROS and the resulting effects of dangerous free radicals such as O_2^- can disrupt this balance and lead ultimately to oxidative stress.

O_2^- is the starting point of a cascade of free radical reactions, resulting in the unchecked proliferation of ROS. First discovered in 1968, SOD is the first antioxidant defense to mobilize against O_2^- and ensuing oxidative stress. For years, attempts at the oral administration of pure SOD to boost natural antioxidant levels failed, because SOD's delicate quaternary structure could not withstand the acidic nature of the GI tract. This issue was solved, however, with the advent of GliSODin®. GliSODin® combines SOD extracted from the cantaloupe melon with the wheat-based biopolymer gliadin, thus making SOD bioactive in the body. The SOD-gliadin combination increases SOD's efficiency, stability, and delivery during its passage into the bloodstream. Additionally, the role of GliSODin® in making SOD bioactive in the body promotes the activity of other crucial endogenous, enzymatic antioxidants such as catalase and GPx.

GliSODin® also produces anti-inflammatory effects. The results of the studies that have been discussed indicate that whereas SOD and gliadin alone have little impact on the increase of IL-10 and the reduction of TNF-alpha, the SOD-gliadin combination is effective in this way. In addition to its anti-inflammatory capabilities, GliSODin® also protects skin from photo-oxidative stress caused by UV radiation. There are several UV studies whose results support this conclusion.

As methods to prevent oxidative stress and contribute toward healthy aging are considered, orally supplemented GliSODin® has been shown scientifically to offer a therapeutic means for the prevention and treatment of many conditions associated with inflammation and oxidative stress.

References

1. Liu, J.; Yeo, H.C.; Övervik-Douki, E.; Hagen, T.; Doniger, S.J.; Chyu, D.W.; Brooks, G.A.; Ames, B.N.; "Chronically and acutely exercised rats: biomarkers of oxidative stress and endogenous antioxidants." *Journal of Applied Physiology*, 2000. Vol. 89, pp. 21–28.
2. Ding, Q.; Dimayuga, E.; Keller, J.N.; "Oxidative damage, protein synthesis, and protein degradation in Alzheimer's disease." *Current Alzheimer Research*, 2007, Vol. 4, pp. 73–79.
3. De Hertog, S.A.E.; Wensveen, C.A.H.; Bastiaens, M.T.; Kielich, C.J.; Berkhout, M.J.P.; Westendorp, R.G.J.; Vermeer, B.J.; Bouwes Bavinck, J.N.; for the members of the Leiden Skin Cancer Study; "Relation between smoking and skin cancer." *Journal of Clinical Oncology*, 2001. Vol. 19, No. 1, pp. 231–238.
4. Fisher, G.J.; Wang, Z.Q.; Datta, S.C.; Varani, J.; Kang, S.; Voorhees, J.J.; "Pathophysiology of premature skin aging induced by ultraviolet light." *New England Journal of Medicine*, 1997. Vol. 337, pp. 1419–1429.
5. Rhie, G.-E.; Seo, J.Y.; Chung, J.H; "Modulation of catalase in human skin in vivo by acute and chronic UV radiation." *Molecules and Cells*, 2001. Vol. 11, pp. 399–404.
6. Shin, M.H.; Rhie, G.-E.; Kim, Y.K.; Park, C.-H.; Cho, K.H.; Kim, K.H.; Eun, H.C.; Chung, J.H.; "H_2O_2 accumulation by catalase reduction changes MAP kinase signaling in aged human skin in vivo." *Journal of Investigative Dermatology*, 2005. Vol. 125, pp. 221–229.
7. Leff, J.A.; Burton, L.K.; Berger, E.M.; Anderson, B.O.; Wilke, C.P.; Repine, J.E.; "Increased serum catalase activity in rats subjected to thermal skin injury." *Inflammation*, 1993. Vol. 17, No. 2, pp. 199–204.
8. Kwei, K.A.; Finch, J.S.; Thompson, E.J.; Bowden, G.T.; "Transcriptional repression of catalase in mouse skin tumor progression." *Neoplasia*, 2004. Vol. 6, pp. 440–448.
9. Hanada, K.; Gange, R.W.; Connor, M.J.; "Effect of glutathione depletion on sunburn cell formation in the hairless mouse." *Journal of Investigative Dermatology*, 1991. Vol. 96, pp. 838–840.

10. Meewes, C.; Brenneisen, P.; Wenk, J.; Kuhr, L.; Ma, W.; Alikoski, J.; Poswig, A.; Krieg, T.; Scharffetter-Kochanek, K.; "Adaptive antioxidant response protects dermal fibroblasts from UVA-induced phototoxicity." *Free Radical Biology and Medicine*, 2001. Vol. 30, No. 3, pp. 238–247. Copyright © 2001 Elsevier Science Inc. Printed in the USA. All rights reserved 0891-5849/01/.
11. Wenk, J.; Schuller, J.; Hinrichs, C.; Syrovets, T.; Azoitei, N.; Podda, M.; Wlaschek, M.; Brenneisen, P.; Schneider, L.A.; Sabiwalsky, A.; Peters, T.; Sulyok, S.; Dissemond, J.; Schauen, M.; Krieg, T.; Wirth, T.; Simmet, T.; Scharffetter-Kochanek, K.; "Overexpression of phospholipid-hydroperoxide glutathione peroxidase in human dermal fibroblasts abrogates UVA irradiation-induced expression of interstitial collagenase/matrix metalloproteinase-1 by suppression of phosphatidylcholine hydroperoxide-mediated NFκB activation and interleukin-6 release." *Journal of Biological Chemistry*, 2004. Vol. 279, No. 44, pp. 45634–45642.
12. Rafferty, T.S.; McKenzie, R.C.; Hunter, J.A.A.; Howie, A.F.; Arthur, J.R.; Nicol, F.; Beckett, G.J.; "Differential expression of selenoproteins by human skin cells and protection by selenium from UVB-radiation-induced cell death." *Biochemical Journal*, 1998. Vol. 332, pp. 231–236.
13. Blankenberg, S.; Rupprecht, H.J.; Bickel, C.; Torzewski, M; Hafner, G.; Tiret, L.; Smieja, M.; Cambien, F.; Meyer, J.; Lackner, K.J.; "Glutathione peroxidase 1 activity and cardiovascular events in patients with coronary artery disease." *New England Journal of Medicine*, 2003. Vol. 349, pp. 1605–1613.
14. Wilcken, D.E.L.; Wang, X.L.; Adachi, T.; Hara, H.; Duarte, N.; Green, K.; Wilcken, B.; "Relationship between homocysteine and superoxide dismutase in homocystinuria." *Arteriosclerosis, Thrombosis, and Vascular Biology*, 2000. Vol. 20, pp. 1199–1202.
15. McCord, J.M.; Fridovich, I.; "Superoxide dismutase: an enzymatic function for erythrocuprein (hemocuprein)." *Journal of Biological Chemistry*, 1969. Vol. 224, pp. 6049–6055.
16. Petersen, S.D.; Oury, T.D.; Ostergaard, L.; Valnickova, Z.; Wegrzyn, J.; Thogersen, I.B.; Jacobsen, C.; Bowler, R.P.; Fattman, C.L.; Crapo, J.D.; Enghild, J.J.; "Extracellular superoxide dismutase (EC-SOD) binds to type I collagen and protects against oxidative fragmentation." *Journal of Biological Chemistry,* 2004. Vol. 279, pp. 13705–13710.
17. Lund-Oleson, K.; Menander, K.B.; "Orgotein: a new inflammatory metalloprotein drug: preliminary evaluation of clinical efficacy and safety in degenerative joint disease." *Current Therapeutic Research, Clinical and Experimental*, 1974. Vol. 16, No.7, pp. 706–717.
18. Menander-Huber, K.B.; Huber, W.; "Orgotein, the drug version of bovine Cu-Zn superoxide dismutase: II. A summary account of clinical trials in man and animals." In: Michelson, A.M.; McCord, J.M.; Fridovich, I. (eds.) *Superoxide and Superoxide dismutases*. New York: Academic Press, 1977, pp. 537–549.
19. Yang, H.; Roberts, L.J.; Shi, M.J.; Zhou, L.C.; Ballard, B.R.; Richardson, A.; Guo, Z.M.; "Retardation of atherosclerosis by overexpression of catalase or both Cu/Zn-superoxide dismutase and catalase in mice lacking apolipoprotein E." *Circulation Research*, 2004. Vol. 95, pp. 1075–1081.

20. Laverdet, C.; Pomarede, N.; Oliveres-Ghouti, C.; "Glisodin and Exposure to the Sun," an open study conducted in France on 150 patients by 40 dermatologists. Sponsored by ISOCELL Nutra, France, March 2005.
21. Laverdet, C.; "Glisodin Sun Pilot Trial," an open study conducted in France on 15 patients presenting fragile skin, hypersensitivity to the sun or even problems of sun disease; Attachée de Consultation des Hopitaux de Paris, July–September 2003.
22. Mac-Mary, S.; Sainthillier, J.M.; Courderotmasuyer, C.; Creidi, P.; Humbert, P.; "Could a photobiological test be a suitable method to assess the antioxidant effect of a nutritional supplement (GliSODin®)?" *European Journal of Dermatology*, 2007. Vol. 17, No. 3, pp. 254–255.
23. Menvielle-Bourg, F.J.; "Superoxide dismutase (SOD), a powerful antioxidant, is now available orally." *Phytothérapie*, 2005. No. 3, pp. 118–121.
24. Vouldoukis, I.; Conti, M.; Krauss, P.; Kamaté, C.; Blazquez, S.; Tefit, M.; Mazier, D.; Calenda, A.; Dugas, B.; "Supplementation with gliadin-combined plant superoxide dismutase extract promotes antioxidant defences and protects against oxidative stress." *Phytotherapy Research*, 2004. Vol. 18, pp. 957–962.
25. Muth, C.M.; Glenz, Y.; Klaus, M.; Radermacher, P.; Speit, G.; Leverve, X.; "Influence of an orally effective SOD on hyperbaric oxygen-related cell damage." *Free Radical Research*, 2004. Vol. 38, No. 9, pp. 927–932.
26. Vouldoukis, I.; Lacan, D.; Kamate, C.; Coste, P.; Calenda, A.; Mazier, D.; Conti, M.; Dugas, B.; "Antioxidant and anti-inflammatory properties of a *Cucumis melo* LC. extract rich in superoxide dismutase activity." *Journal of Ethnopharmacology*, 2004. Vol. 94, pp. 67–75.
27. Cloarec, M.; Caillard, P.; Provost, J.-C.; et al.; "GliSODin®, a vegetal SOD with gliadin, a preventative agent vs. atherosclerosis, as confirmed with carotid ultrasound-B imaging." *European Annals of Allergy & Clinical Immunology*, 2007. Vol. 39, No. 2, pp. 2–7.
28. Mac-Mary, S.; Sainthillier, J.M.; Creidi, P.; Series, J.P.; Vix, F.; Humbert, Ph.; Assessment of oral intake of GliSODin on the intensity of an actinic erythema induced by solar irradiation. Congres Annuel De Recherche Dermatologique (Card), French-Speaking Congress of Dermatological Research, Brest (France), May 27–28, 2005. *Journal of Investigative Dermatology*, 2005. Vol. 125, pp. A13–A24, P19.

PART 5
SUPPORTING A SOLID FOUNDATION FOR FIRMER SKIN

14

Dermal Connective Tissue as the Foundation for Healthy-Looking Skin

James Varani, PhD

Department of Pathology, University of Michigan, Ann Arbor, MI, USA

14.1 Introduction

Chronic exposure to solar radiation, exposure to harsh chemicals, years or decades of cigarette smoking, treatment with corticosteroids or other

Aaron Tabor and Robert M. Blair (eds.), *Nutritional Cosmetics: Beauty from Within*, 267–286, © 2009 William Andrew Inc.

metabolic inhibitors, and underlying metabolic diseases such as diabetes all produce adverse changes in the appearance of the skin. Bed-ridden patients, individuals with underlying conditions that lead to peripheral vascular disease and chronic hypoxic conditions in the skin develop similar changes. All of these effectors of skin damage are superimposed on the deleterious changes that occur simply due to the passage of time (chronological aging). Although the aging process, photodamage, steroid use, etc., each produces a unique constellation of changes, they all result in damage to the connective tissue elements of the skin. This, in turn, speaks to the importance of the connective tissue in maintaining a healthy-appearing skin. There is no question that the dry, wrinkled, mottled appearance of damaged skin is unappealing to many individuals because of how it looks and because of what it implies.

To prevent or repair damaged skin provides important psychological benefits that would justify therapeutic intervention even if there were no other benefits. To suggest, however, that the consequences of skin damage are only cosmetic is incorrect. Skin damage has significant medical consequences [1–5]. The thin, frail skin of elderly individuals is prone to bruising and the bruises often form slow-healing or nonhealing lesions. Minor abrasions occur more frequently and take longer to heal. Chronic ulcers can develop, particularly on the lower extremities, with disastrous consequences in these patients. Anything that can be done to prevent connective tissue damage will benefit not only the appearance of the skin, but its basic function, as well.

Research conducted over the past 25 years has provided definitive evidence that connective tissue damage in the skin—along with its attendant cosmetic and medical consequences—is not simply the inevitable result of living. It is quite the contrary. Damage to the skin can be prevented or delayed, and accumulated damage can be (at least in part) reversed. Repairing skin damage has both cosmetic and medical benefits.

14.2 The Elements of Healthy Skin

Because damage to the connective tissue components of the skin is largely responsible for adverse changes in appearance, we need to understand what constitutes a healthy dermis to begin with. The dermis consists largely of type I collagen. This structural protein accounts for approximately 90% of the dry weight of the skin [6]. Interstitial fibroblasts imbedded in the dermis synthesize and secrete collagen precursors (procollagen peptides), which are processed extracellularly by cleavage of amino acid sequences from both the N- and C-terminal portions of the peptides. Three peptide molecules then bind together noncovalently to form the triple helix that makes up the mature type I collagen molecule. Two of the peptides in the triple helix (the α1 chains) are identical, and the third (the α2 chain) is slightly smaller in size. Once the individual triple helical molecules have formed, they can be cross-linked to form mature collagen fibers, and these can be further cross-linked to form the collagen fiber bundles that are visible at the light- and electron-microscopic levels. The major function of the structural collagen in the skin is strength.

When one thinks about the appearance of the skin, the next important molecule that comes to mind is elastin. Elastic fibers are assembled from monomeric peptides (tropoelastin) synthesized and secreted by interstitial cells during development of the skin. The elastic fibers are much thinner than the type I collagen fibers and, as a percentage of the connective tissue dry weight, constitute a much smaller percentage of the total protein [7]. In contrast to the interstitial collagens, which run in all directions but especially parallel with the plane of the skin, the elastic fibers have a primarily perpendicular orientation. Elastic fibers are uniquely folded or coiled in the resting state. They uncoil when force is applied and then recoil when the force is released. Their capacity to stretch and recoil, coupled with their unique orientation, is primarily responsible for the supple nature of healthy skin. Damage to the elastic fibers during aging and as a consequence of excess sun exposure is responsible for skin sagging. Elastic fibers in the skin not only provide suppleness to the skin as a whole but also associate with the small blood vessels of the skin and allow these vessels to undergo a certain amount of distortion without breaking.

Along with the structural collagens and elastin, there are numerous other extracellular matrix components in the dermis. Fibronectin, laminin, and heparin sulfate proteoglycan are all present in varying amounts and play a variety of roles. Fibronectin is present in the interstitium, closely associated with collagen. Interstitial fibroblasts have specific integrin receptors (primarily α4β1 and α5β1) for fibronectin and bind to this matrix molecule as well as to collagen directly [8]. Laminin consists of a family of large extracellular matrix molecules. Members of this family are found in the basement membrane, separating the epithelium from the adjacent dermis. Epithelial cell attachment to the basement membrane occurs, in part, through interaction with laminin via specific integrin receptors (primarily α6β1 and α6β4) [9]. The function of the proteoglycans is less well understood. These large molecules, consisting of a protein core to which large amounts of carbohydrate (i.e., glycosaminoglycans—repeating disaccharide units) are attached, bind water and are thought to provide volume and "cushion" for the skin.

In addition to these connective tissue elements, there are a large number of other secreted proteins that become incorporated into the dermal matrix in smaller amounts. These molecules play a variety of functions. A detailed description of the extracellular matrix molecules present in the dermis as well as their organization and function is beyond the scope of this chapter. Any number of recent textbooks and reviews are available for those who wish to pursue this subject in more detail. It is sufficient to say the deleterious changes in the appearance of aged skin reflect, in large part, damage to the major connective tissue elements that make up the dermis. Although damage to all of the matrix components occurs during aging and as a consequence of excess sun exposure, etc., because collagen is, by far, the most abundant extracellular matrix component in the skin, skin damage is largely collagen damage. In the remainder of this chapter we will focus on the pathophysiology of collagen destruction (based primarily on what we know from studies on chronological aging and photodamage).

14.3 Histological and Ultrastructural Features of Skin Damage: Relationship to Altered Appearance

Figure 14.1 demonstrates features seen in sun-protected skin of a young adult at the light-microscopic level and the changes that occur as a consequence of the chronological aging process. The collagen fiber bundles are shorter and thinner in the aged skin and there is more space between the fiber bundles. Cells imbedded in the matrix of aged skin have a rounded appearance and many of the cells appear to be separated from the surrounding collagen bundles altogether. In healthy young skin, in contrast, the interstitial cells reside in intimate contact with collagen fibers and have a much more flattened or stretched appearance. In past studies [10,11], we quantified degenerative changes in the skin in groups of individuals of differing age. Using four parameters of damage including fiber length, fiber width, overall connective tissue disorganization, and depth to which the degenerative changes could be seen, we concluded that virtually all individuals below the age of 60 years had a "young skin" phenotype (i.e., longer, thicker collagen fiber bundles and lack of disorganization). Individuals between 60 and

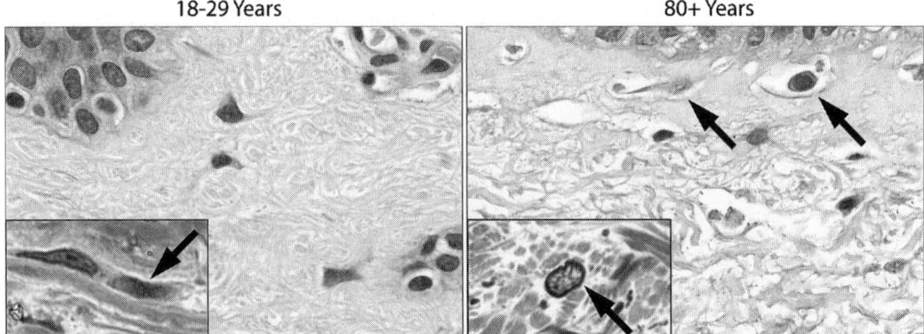

Figure 14.1 Histological features of sun-protected skin from young and old individuals. In young skin (left), thick collagen fiber bundles are present throughout the dermis, extending almost to the epidermis. Interstitial cells imbedded in the collagen bundles have a stretched shape and appear to be in intimate contact with collagen (insert). In the skin sample from the old individual (right), healthy collagen bundles have been replaced with thin, short, disorganized fibers. There is more open space in the dermis. Interstitial cells are round or oblong, and some appear to be surrounded by open space (arrows). Both panels are 5-μm hematoxylin and eosin-stained sections (X490). Insert: Both panels are 1-μm toluidine blue-stained sections (X980). Taken from [11].

79 years of age were mixed in the sense that, in some, phenotypic features were similar to those typically seen in the younger individuals whereas other individuals in this group had an "old skin" phenotype. In the group of individuals who were 80+ years of age, virtually all demonstrated the features associated with the old skin phenotype (Fig. 14.2).

In addition to the semiquantitative approaches used above, there is also direct, biochemical evidence of increased collagen damage in aged/photodamaged skin. The direct biochemical data are based on the fact that intact collagen is resistant to digestion by mammalian enzymes other than the three known collagenases, whereas collagen molecules that have been "clipped" by one of these enzymes becomes sensitive to further degradation by a number of different proteases, including chymotrypsin [12]. We took advantage of this to compare chymotrypsin sensitivity of dermal collagen in healthy young skin versus aged sun-protected skin. The same comparison was made between photodamaged skin and sun-protected skin of the same individuals. With both comparisons, the amount of fragmented collagen in the damaged skin was increased by approximately four-fold over the amount present in healthy skin [13]. In the same studies it was

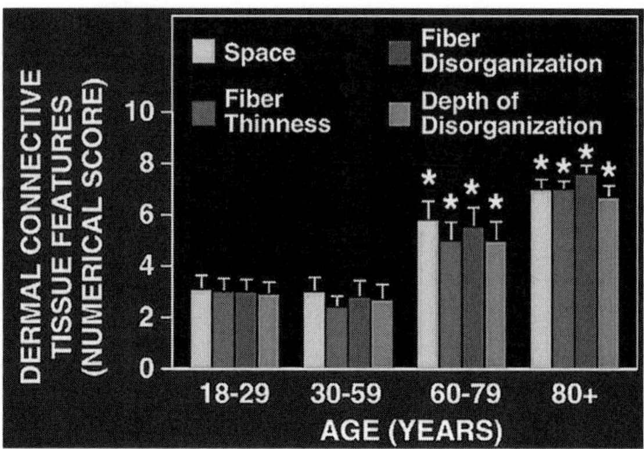

Figure 14.2 Morphometric changes in dermal connective tissue from individuals of increasing age. Sun-protected skin biopsies were obtained from individuals of differing age (9–11 subjects per group). Dermal connective tissue features were evaluated and given a numerical score between 1 and 9 (increasing score indicative of increasing abnormalities). Virtually all of the individuals below the age of 60 years had a young skin phenotype. Individuals between 60 and 79 years of age constituted a mixed population in that some individuals had a young skin phenotype whereas others expressed features associated with aged skin. Virtually all of the individuals 80 years of age and older expressed features associated with aged skin. *Statistical difference from the corresponding value in the 18–29-year-old group. See [10] for details.

shown that total collagen was decreased by 20–30% in the skin of aged (80+ years) individuals as compared to the skin of individuals ages 18–29.

Ultrastructural features of healthy and damaged connective tissue are demonstrated in Fig. 14.3. By scanning electron microscopy (left-hand panels), it can be seen that, in healthy young skin, collagen is organized into a network of fiber bundles that are uniformly distributed throughout the section. In damaged skin, the collagen is fragmented. Some areas are devoid of collagen whereas in other areas, dense clumps of collagen are visible (center of the photograph). At the transmission electron–microscopic level (right-hand panels), it can be seen that in the healthy dermis, interstitial fibroblasts are imbedded in and surrounded by dense bundles of intact collagen. Contact with collagen occurs over much of the cell surface. In the damaged dermis, there are fewer collagen bundles. Interstitial cells are surrounded by and in contact with thin and disorganized collagen fibers, with amorphous material or with empty space.

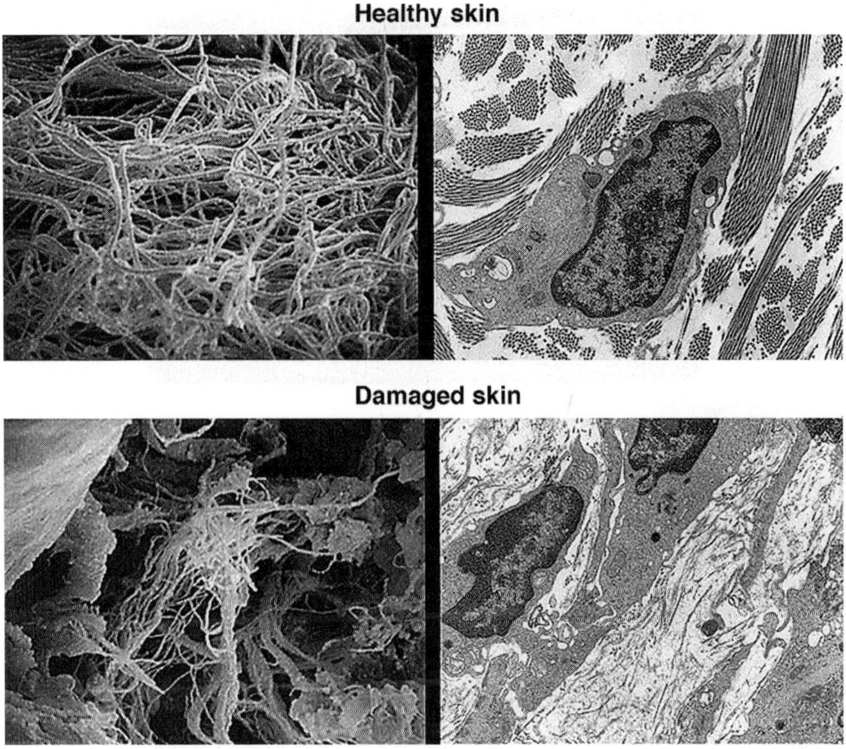

Healthy skin

Damaged skin

Figure 14.3 Scanning electron microscopic– and transmission electron microscopic–features of healthy and damaged skin. Scanning electron microscopy (left) demonstrates uniformly distributed collagen bundles throughout the section of healthy skin. In the damaged skin, fragmented collagen is observed. Instead of a uniform distribution, there is a clump of collagen in the center of the section with a lack of collagen in other areas. At the transmission electron–microscopic level (right), it can be seen that in healthy skin, the cells are in intimate contact with the surrounding collagen. In damaged skin, there is a lack of intact collagen. The cells are surrounded by and in contact with thin, fragmented connective tissue fibers, with amorphous material or with empty space.

The histological and ultrastructural features of damaged skin are readily apparent. The question is this: how do these histological and ultrastructural changes relate to the clinical features of chronological skin aging and/or photodamage? Though acknowledging that correlating histological and ultrastructural features with clinical appearance is never absolute, we are confident that the manifestations of aged and photodamaged skin (especially coarse and fine wrinkles) are a reflection of collagen fragmentation. In healthy skin, the collagen matrix provides a thick, uniform layer of connective tissue. When this material is fragmented and pulled into clumps (leaving some areas devoid of collagen), the clumps and gaps translate directly into the wrinkled contours. Though it may be difficult to prove that this is the case in humans, the formation of a new, thick band of collagen following retinoid treatment in the photodamaged mouse model [14,15] provides direct evidence for the relationship between a layer of intact connective tissue and a smooth surface texture devoid of wrinkles. Compounding the deficits resulting from fragmented collagen are the overall thinning of the skin (primarily in natural aging) and the buildup in elastotic material (photoaging).

In addition to contributing to skin wrinkling, collagen fragmentation and the overall loss of collagen may also contribute to the frequent bruises seen in aged and photodamaged skin insofar as the fragmented matrix provides little support for the microvasculature and results in frequent capillary breakage.

14.4 Pathophysiology of Skin Damage

14.4.1 Cellular and Molecular Events That Bring about Connective Tissue Damage in the Skin

There are fewer fibroblasts in aged skin, with reduced growth capacity, increased production of collagen-degrading matrix metalloproteinases (MMPs) and reduced elaboration of collagen and other components of the extracellular matrix [16–21]. Similar changes are observed in young skin fibroblasts from premature aging conditions such as progeria and Werner's syndrome [22]. Our own studies, which have been published in a series of reports over the past two decades [10,11,23–26], have demonstrated that these age-related changes in fibroblast function that have been so well documented *in vitro* are also present *in vivo*. To reiterate, our *in vivo* studies have demonstrated that the number of interstitial fibroblasts is decreased in skin from 80+-year-old individuals as compared to skin from 18–29-year-old individuals. At the transmission electron–microscopic level, there is less endoplasmic reticulum in the cells from aged skin (indicative of decreased protein synthesis). Synthesis of types I and III procollagen, as well as other components of the extracellular matrix, is lower in the aged skin and levels of MMP-1 (interstitial collagenase) and MMP-9 (gelatinase B) are increased. The fact that cells taken from aged skin show the same deficits *in vitro* (i.e., reduced growth, reduced collagen production, and increased MMP levels) as observed *in vivo* suggests that intrinsic, age-related changes in fibroblast physiology are mechanistically important to the accumulated damage in chronologically aged skin.

Superimposed on the deficits that occur as a consequence of the natural aging process are the effects of photoaging. Acute exposure to ultraviolet (UV) radiation from the sun results in a rapid and strong upregulation of the same MMPs that are gradually upregulated in natural aging [27,28]. There is also a sharp but transient drop in collagen synthesis following acute ultraviolet exposure [29] and a sustained loss of collagen synthesis in late-stage photodamage [30,31]. Thus, though aging and photodamage are unrelated etiologically, they share basic pathophysiological mechanisms (i.e., reduced collagen synthesis and increased collagen breakdown). Metabolic diseases such as diabetes and other conditions in which peripheral vascular function is compromised also lead to decreased collagen production/increased collagen damage in the skin [32–38]. Finally, extended use of potent corticosteroids directly depresses collagen synthesis in the skin [39,40], leading to observable thinning and reduced tissue strength.

In summary, one could end the discussion here and conclude that we know something of the processes that bring about damage to the connective tissue of the skin. Intrinsic changes in fibroblast physiology occurring as a consequence of the passage of time result in a gradual increase in the elaboration of collagen-degrading MMPs, and progressive fragmentation of the major connective tissue elements of the skin. Concomitantly, there is a gradually reduced capacity to repair the damage. Superimposed on these effects of the aging process, other, extrinsic, factors exacerbate the deleterious changes resulting directly from the intrinsic aging process. As a result of these intrinsic and extrinsic factors, the dermis thins measurably. The unsightly appearance that is characteristic of aged skin and the loss of function that leads to easy bruising/wounding and poor wound repair arise. Though correct, this is only a part of the story.

Although intrinsic changes in fibroblast physiology bring about connective tissue damage, recent studies have demonstrated that, at some point, the damaged matrix itself becomes a milieu that no longer supports the optimal functioning by cells that are inherently capable of sustaining a healthy skin phenotype. In contrast, damage to dermal collagen produces an environment that, itself, stimulates further collagen destruction and retards new collagen synthesis by cells that are inherently capable of supporting healthy skin. The result of initial damage, thus, becomes the mediator of further damage.

14.4.2 Damaged Connective Tissue as the Mediator of Impaired Cell Function

In the absence of overt skin injury or active disease, interstitial fibroblasts reside as mostly single cells in the dermal connective tissue matrix. As noted above, the interstitial fibroblasts are physically attached to the connective tissue fibers—primarily type I collagen. This interaction is mediated by specific adhesion receptors on the fibroblast surface (primarily $\alpha1\beta1$ and $\alpha2\beta1$ integrins) binding to specific amino acid sequences in the collagen molecule [41,42]. The interaction between a single integrin receptor molecule and its cognate amino acid sequence in the matrix molecule produces an adhesive force that is insignificantly small compared to the mass of the cell. The force holding the cell to the substratum is magnified, however, as a result of clustering of adhesion receptors at the site of attachment. Cell-substrate attachment and receptor clustering trigger a complex series of molecular events within the cell. Proteins on the inner side of the plasma membrane bind to the cytosolic domain of the integrin molecules as the developing (focal adhesion) site forms. Cellular actin, which is largely monomeric in unattached cells, binds to cytoplasmic proteins in the focal adhesion complex and polymerizes to form the F-actin filaments. Myosin interacts with the actin, and together the actin-myosin complex forms the contractile apparatus of the cell. As the cell's actin-myosin contractile systems shortens, the cell pulls against the rigid substrate to generate mechanical force and produce the stretched appearance [43,44].

A large body of literature demonstrates the importance of mechanical tension to cellular behavior (reviewed in [45–47]). Studies conducted largely *in vitro* on a variety of "deformable" materials have convincingly demonstrated that when there is a sufficient level of mechanical tension on fibroblasts, production of collagen (as well as other components of the extracellular matrix) is high. When tension is reduced, the cells enter a mechanically relaxed state and matrix production declines. Concomitantly, elaboration of matrix-degrading enzymes is stimulated [48–55]. These past studies have shown that reduction in collagen synthesis occurring as a consequence of reduced mechanical tension reflects (i) decreased transcription of genes for interstitial collagens, (ii) effects on enzymes involved in the post-translational processing of procollagen peptides, and (iii) increased elaboration of collagen-degrading MMPs. It has been demonstrated, furthermore, that multiple signaling pathways regulating extracellular matrix gene expression are altered by changes in mechanical tension [56–58].

The three-dimensional collagen matrix that makes up the connective tissue of the skin can be thought of as a rigid but potentially deformable scaffold (much like the steel skeleton of a building), with interstitial fibroblasts imbedded within and attached at many points to the structure. Given the *in vitro* findings described above, we asked what would happen if the collagenous matrix were fragmented by exposure to collagenolytic enzymes. To address this question, we carried out a series of studies in which healthy human dermal fibroblasts were imbedded in a three-dimensional matrix of intact polymerized type I collagen or polymerized collagen that had been exposed to MMP-1 (interstitial collagenase) [11,13,59–61]. As predicted, the cells maintained a well-spread (mechanically stretched) appearance on the intact substrate. Concomitantly, the cells expressed a synthetic phenotype in which collagen and other components of the extracellular matrix were actively synthesized, and collagen-degrading MMPs were suppressed. However, when the collagen matrix was fragmented by exposure to MMP-1 (the major collagen-degrading enzyme in the skin) [62,63], the fragmented matrix collapsed under the tension of the attached cells. As tension on the cells lessened, focal adhesions dissipated and the actin cytoskeleton depolymerized. The attendant fibroblasts lost their mechanically stressed morphology and became spherical in shape. Eventually, these cells detached from the substrate altogether. In contrast to what occurred on the intact collagenous matrix, collagen production by cells on the fragmented matrix decreased (>80% reduction) and MMP levels increased. Thus, the cells "switched" from a synthetic to a degradative phenotype.

Figure 14.4 demonstrates changes in the morphological appearance of dermal fibroblasts as the collagen matrix to which these cells are attached undergoes the transition from intact to fragmented. In the upper part of the cartoon, the cell is depicted as being in contact with intact type I collagen. The cell is attached to collagen bundles at several points. The cell-substrate attachment points represent focal adhesions (shown in green). The red "fibers" running throughout the cell represent filamentous (polymerized) actin. The lower part of the figure depicts a cell imbedded in a matrix containing fragmented collagen. The collagen fiber bundles no longer form an organized superstructure and the cell has a spherical shape. The focal adhesions dissociate and the proteins that make up the focal adhesion in

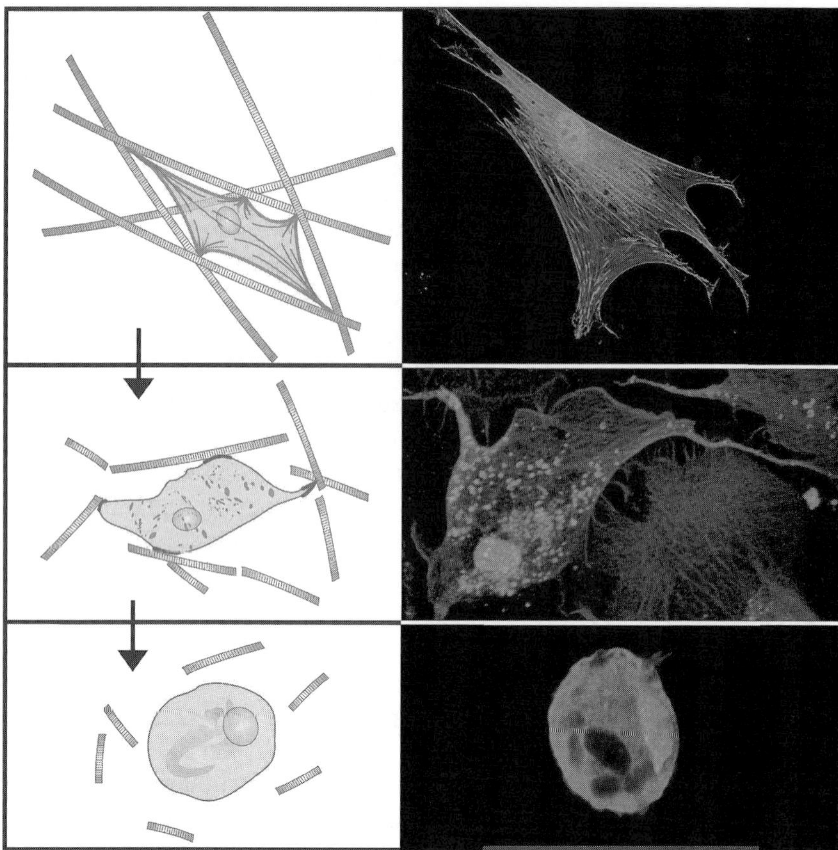

Figure 14.4 Schematic representation of collagen fragmentation and loss of mechanical tension on imbedded cells. The left side of the figure presents a cartoon view of dermal fibroblasts imbedded in a collagen matrix. In the upper panel, the collagen fiber bundles are intact and the cell is attached to the collagen at numerous points. It has a stretched appearance. The middle panel depicts a cell undergoing shape change as a consequence of attachment to collagen that has been fragmented. In the lower panel, the collagen is substantially fragmented and the cell has lost contact with the substratum. It has a spherical appearance. The right-hand part of the figure shows confocal fluorescence microscopic images of cells in contact with intact versus fragmented collagen. Actin stress fibers that begin at focal adhesion proteins are apparent. Nuclei are identified by virtue of being labeled with a fluorescent dye.

the mechanically stressed cell are diffusely expressed throughout the cell. Actin filaments undergo depolymerization. The middle portion of the figure depicts a cell in transition. The right side of the figure presents confocal immunofluorescence images of actual cells (human dermal fibroblasts) in contact with intact or fragmented collagen. The green fluorescence is due to immuno-staining of vinculin, one of the focal adhesion proteins on the intracellular side of the plasma membrane. The red fluorescence is due to fluorescent-tagged phalloidin used as an actin stain. Nuclei are identified by their blue fluorescence (DAPI). Based on these findings, we hypothesize that degradation of collagen *in vivo* (occurring in both natural aging and photodamage) would produce an environment that is unable to support a level of mechanical tension required for efficient collagen-synthetic activity. When one compares the dramatic loss of collagen production seen in aged and photodamaged skin (as compared to controls) and the relatively small decline seen in fibroblasts isolated from aged/photodamaged skin and examined *in vitro* [11,59], the conclusion is inescapable that the damaged connective tissue is, itself, a major contributor to skin damage in aging and photoaging.

14.5 Summary and Conclusions

What is the "take home message?" Simply put, it can be stated that though intrinsic changes in cellular function are (certainly) a part of the skin aging process, decreased cell function in aged or photodamaged skin appears to be due, largely, to interaction of otherwise healthy cells with the damaged connective tissue. In one sense, this is good news. To the extent that resident cells are still capable of expressing a healthy skin phenotype, connective tissue damage in aged/photoaged skin should be reversible. The ability of agents such as 14-all trans retinoic acid (RA) to stimulate collagen production and reduce MMP expression in aged and photodamaged skin (concomitant with their beneficial effects on the appearance of the skin) is supportive of this [27–31]. Likewise, recent studies have suggested that the beneficial effects of skin fillers such as cross-linked hyaluronan are due, at least in part, to providing an artificial (if temporary) support to which resident cells can interact [64]. By filling space in the damaged skin, the injected material produces mechanical stress of the resident fibroblasts and stimulates these cells to begin producing replacement collagen. If a sufficient amount of new collagen is produced, the repair process should continue after the injected material has been cleared from the injection site. Ultimately, then, what was a temporary fix may become long lasting.

References

1. Ashcroft G, Mills SJ, Ashworth JJ. Ageing and wound healing. *Biogerontology* 3:337–345, 2002.
2. Brem H, Tomic-Canic M, Tarnovskaya A, Gill K, Ehrlich HP, Carasa M, Weinberger S, Baskin-Bey E, Entero H. Healing of elderly patients with diabetic foot ulcers, venous stasis ulcers and pressure ulcers. *Surg. Technol. Int.* 11:161–167, 2003.
3. Falanga V. Wound healing and its impairment in the diabetic foot. *Lancet* 366:1736–1743, 2005.
4. Singer AJ, Clark RAF. Cutaneous wound healing. *N. Engl. J. Med.* 341: 738–746, 1999.
5. Strigini L, Ryan T. Wound healing in elderly human skin. *Clin. Dermatol.* 14:197–206, 1996.
6. Fleischmajer R. Type I and type III collagen interaction during fibrillogenesis. *Ann. N. Y. Acad. Sci.* 580:161–164, 1990.
7. Werth VP. Elastic fiber—associated proteins in the skin in development and photoaging. *Photochem. Photobiol.* 63:308–312, 1996.
8. Peltonen J. Localization of integrin receptors for fibronectin, collagen and laminin in human skin: Variable expression in basal cell and squamous cell carcinomas. *J. Clin. Invest.* 84:1916–1923, 1989.
9. Aumailley M, Krieg T. Laminins: A family of diverse multifunctional molecules of the basement membrane. *J. Invest. Dermatol.* 106:209–216, 1996.
10. Varani J, Warner RL, Phan SH, Datta SC, Fisher GJ, Voorhees JJ. Vitamin A antagonizes decreased cell growth, and elevated collagen-degrading matrix metalloproteinases and stimulates collagen accumulation in naturally-aged human skin. *J. Invest. Dermatol.* 114:480–486, 2000.
11. Varani J, Dame MK, Rittie L, Fligiel SEG, Kang S, Fisher GJ, Voorhees JJ. Decreased collagen production in chronologically aged skin: Role of age-dependent alterations in fibroblast function and defective mechanical stimulation. *Am. J. Pathol.* 168:1861–1868, 2006.
12. Bank RA, Krikken M, Beckman B, Stoop R, Maroudas A, Lefeber F, Koppele J. A simplified measurement of degraded collagen in tissues: Application in healthy, fibrotic and osteoarthritic collagen. *Matrix Biol.* 16: 233–243, 1997.
13. Fligiel SEG, Varani J, Datta SH, Kang S, Fisher GJ, Voorhees JJ. Collagen degradation in aged/photoaged skin in vivo and after exposure to MMP-1 in vitro. *J. Invest. Dermatol.* 120:842–848, 2003.
14. Kligman LH. Effects of all-trans retinoic acid on the dermis of hairless mice. *J. Am. Acad. Dermatol.* 15:779–785, 1986.
15. Kligman LH, Duo CH, Kligman AM. Topical retinoic acid enhances the repair of ultraviolet-damaged dermal connective tissue. *Connect. Tissue Res.* 12:139–150, 1984.
16. Plisko A, Gilchrest BA. Growth factor responsiveness of cultured human fibroblasts declines with age. *J. Gerontol.* 38:513–518, 1983.
17. Millis AJ, Hoyle M, McCue HM, Martini H. Differential expression of metalloproteinase and tissue inhibitor of metalloproteinase genes in aged human fibroblasts. *Exp. Cell Res.* 201:373–379, 1992.

18. Millis AJ, Sottile J, Hoyle M, Mann DM, Diemer V. Collagenase production by early and late passage cultures of human fibroblasts. *Exp. Gerontol.* 24: 559–575, 1989.
19. Burke EM, Horton WE, Pearson JD, Crow MT, Martin GR. Altered transcriptional regulation of human interstitial collagenase in cultured skin fibroblasts from older donors. *Exp. Gerontol.* 29:37–53, 1994.
20. Bizot-Foulon V, Bouchard B, Hornebeck W, Dubertret L, Bertaux B. Uncoordinate expressions of type I and III collagens, collagenase and tissue inhibitor of matrix metalloproteinase 1 along in vitro proliferative life span of human skin fibroblasts. Regulation by all-trans retinoic acid. *Cell Biol. Int.* 19: 129–135, 1995.
21. Ricciarelli R, Maroni P, Ozer N, Zingg JM, Azzi A. Age-dependent increase of collagenase expression can be reduced by alpha-tocopherol via protein kinase c inhibition. *Free Radic. Biol. Med.* 27:729–737, 1999.
22. Faragher RG, Kill IR, Hunter JA, Pope FM, Tannock C, Chall S. The gene responsible for Werner syndrome may be a cell division "counting" gene. *Proc. Natl. Acad. Sci. U.S.A.* 90:12030–12034, 1993.
23. Chung JH, Kang S, Varani J, Lin J, Fisher GJ, Voorhees JJ. Decreased extracellular signal-regulated kinase and increased stress-activated MAP kinase activities in aged human skin in vivo. *J. Invest. Dermatol.* 114:177–182, 2000.
24. Varani J, Fisher GJ, Kang S, Voorhees JJ. Molecular mechanisms of intrinsic skin aging and retinoid-induced repair and reversal. *J. Invest. Dermatol.* 3:57–60, 1998.
25. Varani J, Perone P, Griffiths C, Inman D, Fligiel S, Voorhees J. All-trans retinoic acid stimulates events in organ cultured human skin that underlie repair. *J. Clin. Invest.* 94:1747–1756, 1994.
26. Kafi R, Kwak HSR, Schumacher WE, Cho S, Hanft VN, Hamilton TA, King AL, Neal JD, Varani J, Fisher GJ, Voorhees JJ, Kang S. Improvement of naturally aged skin with vitamin A (Retinol). *Arch. Dermatol.* 143:606–612, 2007.
27. Fisher GJ, Datta SC, Talwar HS, Wang ZQ, Varani J, Kang S, Voorhees JJ. The molecular basis of sun-induced premature skin ageing and retinoid antagonism. *Nature (London)* 379:335–338, 1996.
28. Fisher GJ, Wang Z-Q, Datta SC, Varani J, Kang S, Voorhees JJ. Pathophysiology of premature skin aging induced by ultraviolet light. *N. Engl. J. Med.* 337:1419–1428, 1997.
29. Fisher GJ, Datta S, Wang ZQ, Li X-Y, Quan T, Chung JH, Kang S, Voorhees JJ. C-Jun-dependent inhibition of cutaneous procollagen transcription following ultraviolet irradiation is reversed by all-trans retinoic acid. *J. Clin. Invest.* 106: 663–670, 2000.
30. Griffiths CE, Russman AN, Majmudar G, Singer RS, Hamilton TA, Voorhees JJ. Restoration of collagen formation in photodamaged human skin by tretinoin (retinoic acid). *N. Engl. J. Med.* 329:530–535, 1993.
31. Talwar HS, Griffiths CEM, Fisher GJ, Hamilton TA, Voorhees JJ. Reduced type I and type III procollagens in photodamaged adult human skin. *J. Invest. Dermatol.* 105:285–290, 1995.
32. Abdullah KM, Luthra G, Bikski JJ, Grazul-Bilska AT. Cell-to-cell communication and expression of gap junctional proteins in human diabetic and

non-diabetic skin fibroblasts: Effects of basic fibroblast growth factor. *Endocrine* 10:35–41, 1999.

33. Trevisan R, Fioretto P, Barbosa J, Mauer M. Insulin-dependent diabetic sibling pairs are concordant for sodium-hydrogen antiport activity. *Kidney Int.* 55:2383–2389, 1999.
34. Hehenberger K, Hansson A, Heilborn JD, Abdel-Halim SM, Ostensson CG, Brismar K. Impaired proliferation and increased lactate production of dermal fibroblasts in the GK-rat, a spontaneous model of non-insulin dependent diabetes mellitus. *Wound Repair Regen.* 7:65–71, 1999.
35. Loots MA, Lamme EN, Mekkes JR, Bos JD, Middelkopp E. Cultured fibroblasts from chronic diabetic wounds on the lower extremety (non-insulin dependent diabetes mellitus) show disturbed proliferation. *Arch. Dermatol. Res.* 291:93–99, 1999.
36. Stagg AR, Fleming JC, Baker MA, Sakamoto M, Cohen N, Neufeld EJ. Defective high-affinity thiamine transporter leads to cell death in thiamine-responsive megaloblastic anemia syndrome fibroblasts. *J. Clin. Invest.* 103: 723–729, 1999.
37. Teno S, Kanno H, Oga S, Kumakura S, Kanamuro R, Iwamoto Y. Increased activity of membrane glycoprotein PC-1 in the fibroblasts from non-insulin-dependent diabetic mellitus patients with insulin resistance. *Diabetes Res. Clin. Pract.* 45:25–30, 1999.
38. Mendez MV, Stanley A, Phillips T, Murphy M, Menzoian JO, Park H-Y. Fibroblasts cultured from distal lower extremities in patients with venous reflux display cellular characteristics of senescence. *J. Vasc. Surg.* 28:1040–1050, 1998.
39. Hunt TK, Ehrlich HP, Garcia JA, Dunphy JE. Effect of vitamin A on reversing the inhibitory effect of cortisone on healing of open wounds in animals and man. *Ann. Surg.* 170:633–641, 1969.
40. Ulland AE, Shearer JD, Coulter C, Caldwell MD. Altered wound arginine metabolism by corticosterone and retinoic acid. *J. Surg. Res.* 70:84–88, 1997.
41. Knight CC, Morton LF, Peachey AR, Tuckwell DS, Farndale RW, Barnes MJ. The collagen-binding A domains of integrins $\alpha1\beta1$ and $\alpha2\beta1$ recognize the same specific amino acid sequence, GFOGER, in native triple helical collagens. *J. Biol. Chem.* 275:35–40, 1999.
42. Langholz D, Rockel D, Mauch C, Kozlowska E, Bank I, Krieg T, Eckes B. Collagen and collagenous gene expression in three-dimensional collagen lattices are differentially regulated by $\alpha1\beta1$ and $\alpha2\beta1$ integrins. *J. Biol. Chem.* 131:1903–1915, 1995.
43. Grinnell F. Fibroblast-collagen matrix contraction: growth factor signalling and mechanical loading. *Trends Cell Biol.* 10:362–365, 2000.
44. Tomasek JJ, Gabbiani G, Hinz B, Chaponnier C, Brown RA. Myofibroblasts and mechanoregulation of connective tissue remodeling. *Nat. Rev. Mol. Cell Biol.* 3:349–363, 2002.
45. Grinnell F. Fibroblast biology in three-dimensional collagen matrices. *Trends Cell Biol.* 13:264–269, 2003.
46. Silver FH, Siperko LM, Seehra GP. Mechanobiology of force transduction in dermal tissue. *Skin Res. Technol.* 9:3–23, 2003.

47. Geiger B, Bershadsky A, Pankow R, Yamada KM. Transmembrane—extracellular matrix-cytoskeleton crosstalk. *Nat. Rev. Mol. Cell Biol.* 2: 793–805, 2001.
48. Lambert CA, Soudant EP, Nusgens BV, Lapiere CM. Pretranslational regulation of extracellular matrix macromolecules and collagenase expression in fibroblasts by mechanical forces. *Lab. Invest.* 66:444–451, 1992.
49. Geesin J, Brown LJ, Gordon JS, Berg RA. Regulation of collagen synthesis in human dermal fibroblasts in contracted collagen gels by ascorbic acid, growth factors and inhibitors of lipid peroxidation. *Exp. Cell Res.* 206:283–290, 1993.
50. Clark RAF, Nielsen LD, Welch MP, McPherson JM. Collagen matrices attenuate the collagen-synthetic response of cultured fibroblasts to TGF-b. *J. Cell Sci.* 108:1251–1261, 1995.
51. Chiquet M. Regulation of extracellular gene expression by mechanical stress. *Matrix Biol.* 18:417–426, 1999.
52. Tamariz E, Grinnell F. Modulation of fibroblast morphology and adhesion during collagen matrix remodeling. *Mol. Biol. Cell* 13:3915–3929, 2002.
53. Le J, Rattner A, Chepda T, Frey J, Chamson A. Production of matrix metalloproteinase 2 in fibroblast reaction to mechanical stress in a collagen gel. *Arch. Dermatol. Res.* 294:405–410, 2002.
54. Fluck M, Giraud M-N, Tunc V, Chiquet M. Tensile stress–dependent collagen XII and fibronectin production by fibroblasts requires separate pathways. *Biochim. Biophys. Acta* 1593:239–248, 2003.
55. Delvoye P, Wiliquet P, Leveque J-L, Nusgens BV, Lapiere CM. Measurement of mechanical forces generated by skin fibroblasts embedded in a three-dimensional collagen gel. *J. Invest. Dermatol.* 97:898–902, 1991.
56. Lambert CA, Colige AC, Lapiere CM, Nusgens BV. Coordinated regulation of procollagens I and III and their post-translational enzymes by dissipation of mechanical tension in human dermal fibroblasts. *Eur. J. Cell Biol.* 80: 479–485, 2001.
57. Lambert CA, Colige AC, Munaut C, Lapeiere CM, Nusgens BV. Distinct pathways in the over-expression of matrix metalloproteinases in human fibroblasts by relaxation of mechanical tension. *Matrix Biol.* 20:397–408, 2001.
58. Lambert CA, Lapiere CM, Nusgens BV. An interleukin-1 loop I induced in human skin fibroblasts upon stress relaxation in a three-dimensional collagen gel but is not involved in the up-regulation of matrix metalloproteinase 1. *J. Biol. Chem.* 273:23143–23149, 1998.
59. Varani J, Spearman D, Perone P, Fligiel SEG, Datta SC, Wang ZQ, Shao Y, Kang S, Fisher GJ, Voorhees JJ. Inhibition of type I procollagen synthesis by damaged collagen in photoaged skin and by collagenase-degraded collagen in vitro. *Am. J. Pathol.* 158:931–942, 2001.
60. Varani J, Perone P, Fligiel SEG, Fisher GJ, Voorhees JJ. Inhibition of type I procollagen production in photodamage: Correlation between presence of high molecular weight collagen fragments and reduced procollagen synthesis. *J. Invest. Dermatol.* 119:122–129, 2002.
61. Varani J, Fligiel SEG, Schuger L, Kang S, Fisher GJ, Voorhees JJ. Reduced fibroblast interaction with intact collagen as a mechanism for depressed collagen synthesis in photoaged skin. *J. Invest. Dermatol.* 122:1471–1479, 2004.

62. Brennan M, Bhatti H, Nerusu KC, Bhagavathula N, Kang S, Varani J, Voorhees JJ. Matrix metalloproteinase-1 is the major collagenolytic enzyme responsible for collagen damage in UV-irradiated human skin. *Photochem. Photobiol.* 78:43–48, 2003.
63. Yucel T, Mutnal A, Fay K, Fligiel SEG, Wang T, Johnson T, Baker SR, Varani J. Matrix metalloproteinase expression in basal cell carcinoma: Relationship between enzyme profile and collagen fragmentation pattern. *Exp. Mol. Pathol.* 79:151–160, 2005.
64. Wang F, Garza LA, Kang S, Varani J, Orringer JS, Fisher GJ, Voorhees JJ. In vivo stimulation of de novo collagen production caused by cross-linked hyaluronic acid dermal filler injections in photodamaged human skin. *Arch. Dermatol.* 143:155–163, 2007.

15

Amino Acids and Peptides: Building Blocks for Skin Proteins

Ayako Noguchi[1]*, MSc, and David Djerassi*[2]

[1]*Kyowa Hakko USA, Inc., NY, USA*
[2]*Intrachem Technologies, NY, USA*

Aaron Tabor and Robert M. Blair (eds.), *Nutritional Cosmetics: Beauty from Within*, 287–317,

15.1 Amino Acids—General Information

Amino acids are chemical entities that have been recognized as the "building blocks" of life. In humans, nine amino acids are recognized as essential, because the body cannot synthesize them and they are normally required in the diet. The essential amino acids are: isoleucine, leucine, lysine, threonine, tryptophan, methionine, histidine, valine, and phenylalanine. In addition to the nine essential amino acids, there are nonessential amino acids that the human body can synthesize. These amino acids include: alanine, arginine, aspartic acid, cysteine, glutamic acid, glutamine, glycine, ornithine, citrulline, proline, serine, taurine, and tyrosine. The essential amino acids vary from species to species, because different metabolisms can synthesize different substances.

The amino acids form strings of long chains called peptides (1–100 amino acids), polypeptides (100–200 amino acids), and proteins (over 200 amino acids). They are constituents of muscles, tendons, organs, glands, skin, hair, and nails etc. Growth, repair, and maintenance of all cells are dependent on their continuous supply in the body.

15.2 Amino Acids—Function and Benefits (Oral and Topical Administration)

The skin structure consists of various components, with the major part composed of amino acids. It is quite likely that replenishment of constituent amino acids can help retain normal skin structure and thus enhance the beauty of skin. Amino acids have various functions as shown in the following table (Table 15.1). In addition to the simple replenishment of the amino acids that are components of the skin, efficacy in terms of improvement of the skin condition is likely to be manifested in different ways.

Table 15.1 The Key Functions of Amino Acids in Skin

(1) Skin Constituents
NMF (moisture retention): serine, glycine, alanine, citrulline, threonine
Keratin: methionine, cysteine, cystine
Collagen: glycine, proline, alanine, hydroxyproline
(2) Promotion of Skin Tissue Regeneration
Promotion of skin turnover: arginine, citrulline, ornithine, glutamine
Promotion of collagen synthesis: arginine, glycine, ornithine, hydroxyproline
(3) Protection Against UV Damage (Antioxidant Activity)
Suppression of melanogenesis: glutathione, cysteine, methionine, histidine, arginine, citrulline

There are two ways to provide the skin with the necessary components: external application and oral administration. Because the skin is an organ in direct contact with the external environment, the general view is that direct delivery of substances from the outside by external application is the most effective, which has led to the development of numerous topical cosmetic products. In fact, many products are formulated with amino acids. However, because the stratum corneum serves as a barrier against invasion by foreign substances, it is likely that, for substances to be delivered deep into the skin, oral intake—delivery from the interior of the body to the skin through the bloodstream—is more effective. Therefore, two approaches can be suggested for replenishing effective ingredients: external delivery for exerting the effect on the skin surface, such as the stratum corneum, epidermis, etc., and internal delivery for exerting an effect on tissues deep inside, such as the dermis.

Here, the relationship between beautification and amino acids is classified according to their function (Table 15.1), and outlined from each viewpoint: (1) maintenance of constituents of the skin, (2) enhancement of regeneration of skin tissues, and (3) promotion of UV protection (tan reduction and skin whitening).

15.2.1 Skin Constituents

15.2.1.1 Maintenance of the Natural Moisturizing Factor (NMF) (Moisture Retention)

The stratum corneum, the outermost layer of the skin, serves as a barrier that protects the skin. It not only defends the body against invasion by various external stimuli such as bacteria and foreign substances, but also prevents excessive loss of moisture and evaporation of useful materials. This barrier functions by the binding of lipids and proteins to hydrophilic substances, such as the NMF present in the stratum corneum.

The NMF serves as the main complex in retaining water in the stratum corneum [1–3]. It is comprised of free amino acids and amino acid–derived substances. Free amino acids make up to 40% of the NMF, a large proportion of which are serine, glycine, alanine, citrulline, and threonine (Fig. 15.1) [4,5]. The amount of free amino acids in skin keratin correlates with the water content of the epidermis, which can be measured by electrical conductivity (Fig. 15.2) [6]. It has been reported that in skin dehydration due to aging and xeroderma, there is a change in the amino acid composition [7]. In the skin condition known as atopic dermatitis, the amount of free amino acids in the epidermis is significantly smaller and the water content in the stratum corneum is also substantially lower than that seen in healthy individuals (Table 15.2) [8]. Thus, to enhance moisture retention in the stratum corneum, it is important that the correct amounts of amino acids

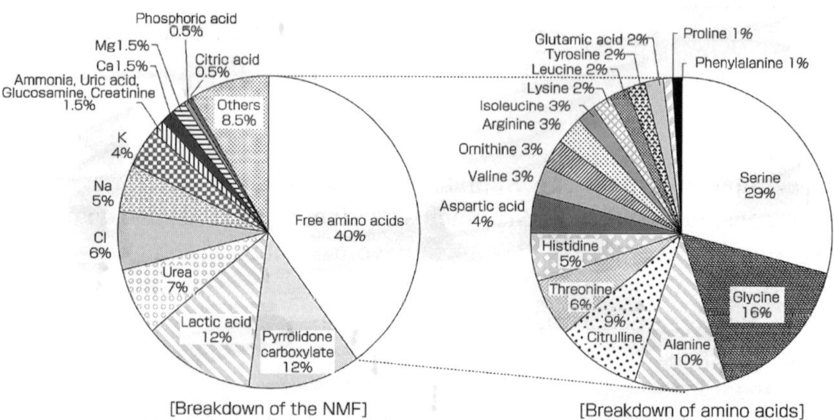

Figure 15.1 Composition of the NMF. Modified from Jacobi 1959 [4] and Koyama et al. 1984 [5].

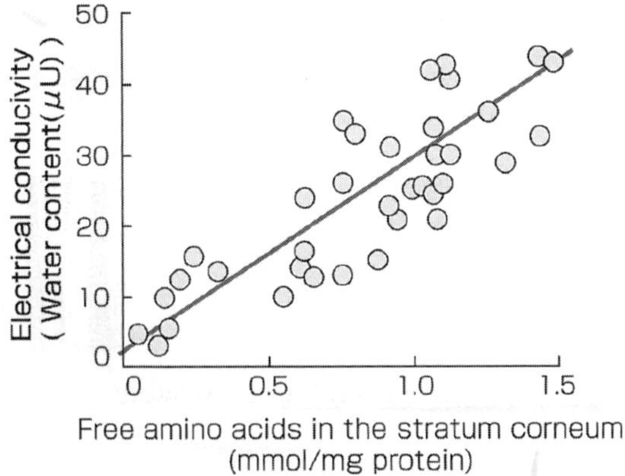

Figure 15.2 Correlation between the levels of free amino acids in the stratum corneum and the water content (electrical conductivity) in the epidermis. Electrical conductivity is used as an indicator of water content. The electrical conductivity of the skin surface of 30 adult patients with various degrees of dermatitis was measured. Using an amino acid analyzer, the amino acid content of the skin was determined by measuring the amino acids contained in samples obtained by consecutive tape striping of the skin. Modified from Horii et al. 1989 [6].

Table 15.2 **Amount of Free Amino Acids in Atopic Dermatitis**

	Atopic Dermatitis	Control
Free amino acids (nmol/mg protein)	186 ± 24* (n = 20)	290 ± 27 (n = 19)

Mean ± SEM, *$P < 0.01$ (vs. control, student's t-test). Modified from Watanabe et al. 1991 [8].

are maintained in the NMF. Replenishment of the amino acids that are the main constituents of the NMF, such as serine, glycine, alanine, citrulline, and threonine, is therefore likely to have the effect of enhancing moisture retention in the stratum corneum of skin. Actually, it was shown that orally administrated citrulline is effective in improving skin moisture retention levels.

In addition, a human clinical study has shown that hydroxyproline is effective for amelioration of dry, rough skin, though hydroxyproline is not a component of the NMF (Kyowa Hakko in-house data, Section 15.3.1).

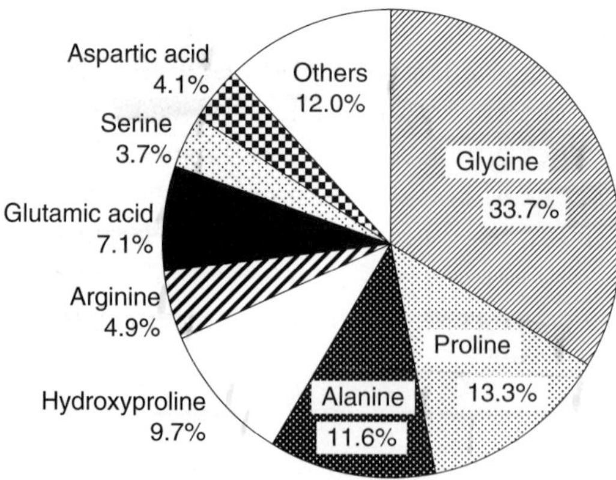

Figure 15.3 Percentages of amino acid constituents of collagen. Modified from Biochemistry Data Book I. 1979 [9].

15.2.1.2 Replenishment of the Constituents of Collagen

While the epidermis is responsible for the barrier function of the skin, the dermis supports the epidermis and serves as a cushion that gives elasticity to skin. UV rays from sunlight, which penetrate deep into the skin, can damage the collagen in the dermis, resulting in wrinkles, loss of firmness, and sagging skin. The metabolic activity of the dermis decreases with age and the amount of collagen decreases due to a loss of balance between the synthesis and degradation of collagen, thereby making the skin thinner and less resilient.

Collagen is an important building block of the skin. Collagen makes up 30% of the protein in the living body and 70% of the protein in the skin. Collagen is composed of 33.7% glycine, 13.3% proline, 11.6% alanine, 9.7% hydroxyproline, and 31.8% of other amino acids (Fig. 15.3) [9]. This composition differs from the amino acid composition of other proteins such as those contained in meat and eggs, which are regarded as nutritionally complete.

15.2.1.3 Replenishment of the Constituents of Keratin

The structure of the stratum corneum comprises layers of tightly adhering cells containing keratin protein. This structure is maintained by the binding of cobblestone-like cell units to water, the NMF, and sebum, greatly enhancing the barrier function. Insoluble and consisting of tightly binding molecules, keratin has a close, firm structure with few crevices, imparting keratin with properties that are ideal for preventing the penetration of foreign substances, or the loss of useful materials. This feature is due to cystine, a sulfur-containing amino acid contained in keratin. Cystine is the cysteine dimer, which acts as the building block of the firm, robust structure of keratin. Because keratin is the chief constituent of nails and hair, in addition to the stratum corneum, it is clear that cystine and cysteine, the building blocks of the keratin structure, and methionine, another sulfur-containing amino acid that is also a cystine precursor, are important for the health of skin, nails, and hair.

15.2.2 Promotion of Skin Tissue Regeneration

15.2.2.1 Skin Turnover

To maintain healthy skin, it is important to maintain normal skin cell turnover. The epidermis, which includes the stratum corneum, has a particularly active metabolism. Elimination of old tissues and production of new tissues are the functions that provide the epidermis with its normal barrier activity. In addition, the epidermis assumes the critical role in skin improvement, including the promotion of melanin excretion. Skin metabolism is called turnover, and the change to horny cells in the epidermis is specifically called keratinization. Keratinization typically occurs over a 28-day period.

Slowed metabolism due to the effects of UV rays, aging, and other factors delays the turnover cycle, allowing a buildup of old tissue. Thickening of the stratum corneum makes the skin less clear, a major contributor to dull-looking skin. It is therefore important to keep active cell division and to promote skin metabolism.

15.2.2.2 Amino Acids Responsible for Regeneration of the Skin (Enhancement of Skin Turnover)

15.2.2.2.1 Promotion of Skin Turnover

The Acceleration of NO Secretion In terms of promotion of skin turnover, one approach to enhancing metabolism is by increasing circulation of blood flow. Nitrogen monoxide (NO) has a vasodilating effect. Arginine and citrulline are involved in the production of NO in the urea cycle and there are numerous reports describing the relationship between arginine/citrulline and the vasodilation effect.

In one report, hypertension induced in salt-sensitive rats was ameliorated by arginine, indicating significant suppression of the elevation of blood pressure and improvement in renal blood flow [10]. In another report, in an arteriosclerosis rabbit model, the vasodilation response that had been reduced by arteriosclerosis was recovered by the administration of arginine into their drinking water [11,12].

Regarding citrulline, it is known that the administration of citrulline significantly decreased blood pressure in a rat model of salt-sensitive hypertension [13] and improved blood flow of rats pretreated with the vasoconstrictor phenylephrine [14].

The Acceleration of the Cell Division The cell division process requires a large amount of energy, and glutamine is frequently used. It appears that glutamine serves to help cell proliferation by participating in the process of donating amino groups, thereby having the effect of promoting wound healing in the skin [15]. In an experiment using acetic acid ulcer rats, glutamine was observed to have the effect of promoting wound healing (collagen synthesis) required for granulation. Glutamine-based therapeutic agents for gastric ulcer and duodenal ulcer are, in fact, available for pharmaceutical use ("Glumin S," Kyowa Hakko Kirin Co., Ltd.).

One group of substances that plays a leading role in cell division is polyamines. Polyamines are believed to be important in stabilization of DNA (chromosomes) during cell division [16]. When cell division occurs, a large amount of energy is required and the amino acids in the body are fully mobilized; therefore, replenishment of various amino acids is important. In terms of polyamine synthesis, ornithine and arginine, polyamine precursors, are important. It has been reported that enteral administration of arginine to burn-model rats increases the urinary polyamine level and enhances protein metabolism [17]. One report indicated that intake of ornithine and arginine promotes polyamine synthesis, thereby enhancing the growth of the intestinal tract [18].

Based on these reports, it is expected that intake of ornithine or arginine will promote skin metabolism via promotion of polyamine synthesis involved in the proliferation and growth of cells, together with the improvement of blood flow.

15.2.2.2.2 Promotion of Collagen Synthesis

It has been reported that some amino acids, including the ones that are not the main constituents of collagen, have a function in enhancing collagen synthesis.

There are some reports on the marked effect that arginine has on collagen synthesis. These functions have been reported in multiple cell lines, such as meniscus cells, fibroblasts, and osteoblasts (Fig. 15.4) [19–21].

Ornithine and arginine are reported to promote wound healing [22,23]. Enteral administration of ornithine to burn patients suppresses the urinary excretion of hydroxyproline, a marker of collagen degradation, throughout the period of treatment with ornithine [24]. It is well known that glutamine also enhances collagen synthesis [25,26].

One report has shown that collagen synthesis in rats is enhanced by combined administration of arginine and glycine, a collagen constituent, in the diet (Fig. 15.5) [27]. These findings indicate that a parallel intake of collagen synthesis–promoting substances such as arginine and other

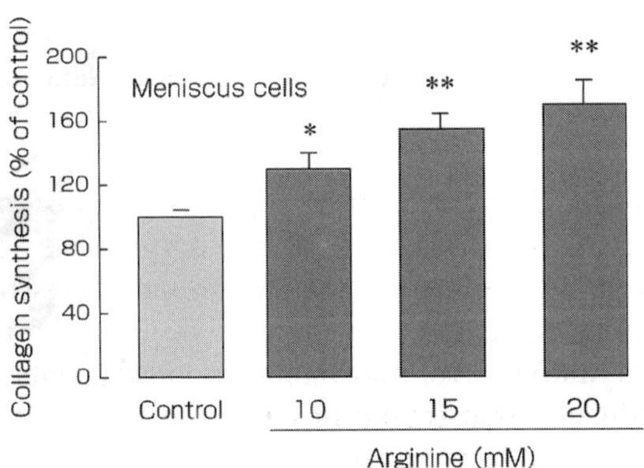

Figure 15.4 Promotion of collagen synthesis by arginine. Arginine (10 mM, 15 mM, and 20 mM) and tritium-labeled proline were added to a confluent meniscus cell culture system. The amount of collagen synthesis was determined by measuring the amount of labeled proline incorporated into the collagen. *$P < 0.05$ (vs. control), **$P < 0.005$ (vs. control). Modified from Cao et al. 1998 [19].

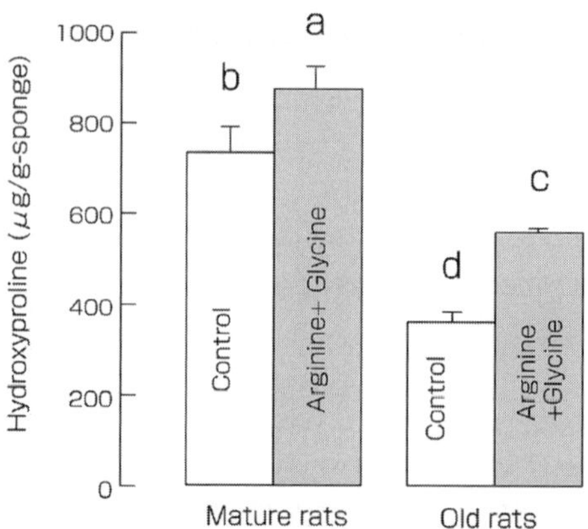

Figure 15.5 Promotion of collagen synthesis by administration of arginine and glycine. Mature rats (4 months old) and old rats (24 months old) received a diet containing 2.4% arginine hydrochloride and 1.0% glycine for 7 days before and after a laparotomy. Polyvinyl alcohol sponges were implanted in the peritoneal cavity during the laparotomy and harvested on the seventh postoperative day. The amount of collagen synthesis was determined by measuring the amount of hydroxyproline in the sponges. *Different letters indicate significant difference. $P < 0.05$ (vs. control, Duncan's multiple-range test). Modified from Chyun and Griminger 1984 [27].

amino acids is required for the optimum enhancement of collagen synthesis in the skin.

Hydroxyproline is a primary constituent of collagen (9–15%) and found almost exclusively in collagen. It is produced by hydroxylation of proline. In several studies, it has been shown that hydroxyproline enhances the collagen synthesis. Hydroxyproline enhances the proliferation activity of dermis-derived cells in vitro (Patent No. US 6692754) and increases the amount of collagen synthesized [28].

The acceleration of the secretion of growth hormone Growth hormone is known for its stimulatory effect on the synthesis of proteins including collagen. It enhances muscle building, accelerates wound healing, and improves the elasticity of skin [29]. It has been reported that oral administration of ornithine, arginine, glutamine, and GABA to healthy human volunteers raises the level of growth hormone in the plasma [30–39].

15.2.3 Protection against UV Damage (Antioxidant Activity)

UV rays from sunlight penetrating deep into the skin can damage the collagen in the dermis, resulting in wrinkles, loss of firmness, and sagging skin. This is referred to as photoaging, with 80% of senescence in sunlight-exposed areas, such as the face, being associated with sun damage. It is recommended that protective measures should be taken, not just during the summer, but also throughout the year.

Skin color is dependent on the type and amount of melanin pigment present. Melanin is a substance created by epidermal melanocytes, which are melanin-producing cells that play an intrinsic role in protecting the skin from skin disorders caused by exposure to UV rays. However, exposure to a large amount of UV radiation triggers overproduction of melanin and simultaneously increases the number of melanocytes, which results in spots, freckles, and a darker complexion. It is further known that when metabolism slows with age, the melanin excretion capability of the epidermis is reduced, causing pigmentation and spots. It is therefore necessary to suppress melanin production. There are two basic mechanisms by which this can be done.

One way to suppress melanin production is through antioxidant activity that removes reactive oxygen species (ROS) generated by UV radiation. ROS are believed to trigger melanin synthesis by melanocytes. Antioxidants such as glutathione, which is present in the cells, prevent melanogenesis in melanocytes by helping to scavenge ROS. Glutathione is a powerful antioxidant. Glutathione is composed of three amino acids: glutamic acid, cysteine, and glycine, bonded together (gamma-glutamyl-cysteinyl-glycine). It is reported that intact glutathione is absorbed Na^+-dependently by small intestinal epithelial cells [40].

Cysteine, methionine, and histidine also have antioxidant effects. Methionine has free radical scavenging ability [41] and thus is regarded as an antioxidant. Histidine has been reported as an antioxidant that suppresses the production of superoxides in a cardiac ischemia–reperfusion model [42]. The antioxidant effects of arginine and citrulline have been also reported [43,44].

Another approach to melanin suppression is the incorporation of substances that inhibit the melanin synthesis reaction. Melanin is synthesized through multiple steps by tyrosinase, using tyrosine as a precursor in melanocytes, suggesting that a tyrosinase inhibitor might be effective as a UV-protection

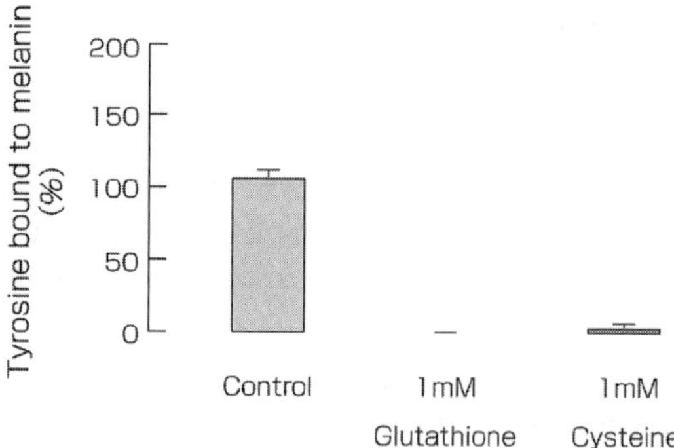

Figure 15.6 Suppression of melanogenesis by glutathione and cysteine. The effect of sulfur-containing compounds during melanogenesis was confirmed using mushroom tyrosinase. Glutathione and cysteine (1 mM each) were added to the medium, with the tyrosinase enzyme at 40 μg/ml, and the amount of tyrosine bound to melanin was measured and compared with that in the control. Modified from Jara et al. 1988 [45].

agent. Glutathione and cysteine have been confirmed not only to be antioxidants but also to be inhibitors of tyrosinase activity in mouse melanoma cells (Fig. 15.6) [45], thus suppressing melanin synthesis in melanocytes. Furthermore, cysteine has been reported to have the effect of suppressing skin pigmentation [46,47].

15.3 The Effects on Skin of Oral and Topical Administration of Amino Acids

Studies on the effects on skin of oral and topical administration of several amino acids have been conducted.

15.3.1 Hydroxyproline

15.3.1.1 Physiological Function

Hydroxyproline is a major component of collagen, present in skin, connective tissues, and bones, accounting for 9–15% of the amino acid composition of collagen. It plays an important role in the formation and stabilization of collagen's triple-helix structure, protecting it against digestion by proteolytic enzymes [48].

15.3.1.2 In Vitro Studies

Kyowa Hakko's own studies on hydroxyproline, over several years, revealed some unique benefits of hydroxyproline as an internal beauty ingredient. Hydroxyproline enhanced the proliferation of cultured human epidermal cells and the effect of hydroxyproline on the proliferation of collagen by cultured human fibroblast cells has been under investigation.

15.3.1.2.1 Hydroxyproline Enhances the Proliferation of Cultured Human Epidermal Cells

The proliferation of human epidermal cells was measured by the MTT method, which involves the colorimetric determination of MTT formazan produced by the incorporation of 3-(4,5-dimethylthiazo2-yl)-2,5-dipheny *2H*-tetrazolium bromide (MTT) into the cells. The measurement was performed at a wavelength of 570 nm (reference = 650 nm) [49].

Epidermal keratinocytes from a human newborn infant were cultivated in KGM medium at 37°C under 5% CO_2. Then, hydroxyproline was added to the medium and the cultivation was continued for an additional 4 days. 100 μl of MTT reagent (5 mg/ml in a phosphate buffer, pH 7.5, calcium-magnesium free) was added to the medium (1 ml), followed by cultivation at 37°C under 5% CO_2 for 4 hours. After removal of the medium, MTT formazan in the cells was extracted with isopropanol containing 0.4 mol/l hydrochloric acid, and the measurement was performed at a wavelength of 570 nm (reference = 650 nm).

An enhanced proliferation of the cultured epidermal cells was observed in the hydroxyproline-treated cells (Fig. 15.7).

15.3.1.3 Oral Administration of Hydroxyproline

The effect of 8 weeks of oral supplementation of hydroxyproline (Kyowa Hakko's Lumistor®, a natural, fermentation-based amino acid) was examined in a double-blind, placebo-controlled study with female volunteers over age 47 who had chronic dry, rough skin. The facial skin moisture content and the perception of their skin condition were evaluated on day 0, week 4, and week 8.

> *Study Objective:* Determine improvement of dry and rough skin conditions after 8 weeks of dietary supplementation of hydroxyproline (Lumistor®).

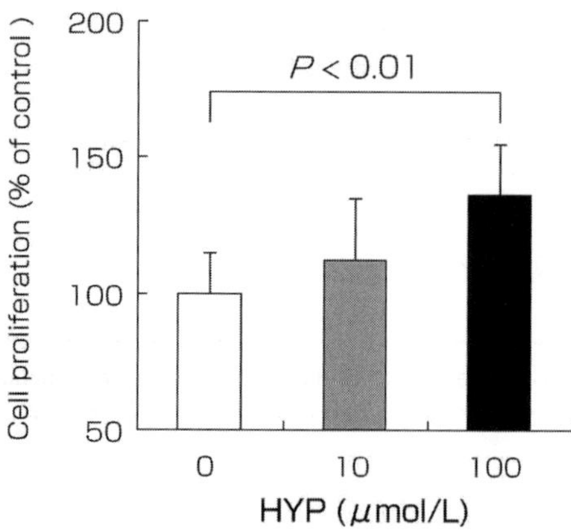

Figure 15.7 Enhancement of proliferation of epidermal cells. The bar indicates the relative value of MTT compared with that (100%) in the absence of hydroxyproline (HYP). Results are given as mean ± SD. Statistical analysis is performed using Dunnett's multiple comparison test after one-way analysis of variance (ANOVA).

Materials: Placebo (corn starch) and Lumistor® capsules, 2 g/day, respectively.

Subjects: Twenty-seven female volunteers (ages between 47 and 66) with chronic dry, rough skin, visually determined by trained technicians, were divided into two groups.

Protocol: Subjects ingested 2 g/day of Lumistor® or placebo, respectively, for 8 weeks. Subjects visited the testing center on day 0, week 4, and week 8, and their facial skin moisture content (cheeks) was examined by trained technicians at baseline and again at weeks 4 and 8 using the Corneometer methodology (Courage and Khazaka Electronic GmbH, Koln, Germany). Participants were also asked to give a subjective assessment of the condition of their skin, using a scale of 1–10.

Results: After 8 weeks of dietary supplementation with Lumistor®, skin moisture content improved significantly as compared to the placebo (Table 15.3). The volunteers' perceptions of facial skin elasticity and other skin parameters, including dryness and wrinkles, were also significantly improved (Table 15.4).

Table 15.3 Comparison of Water Contents of Facial Skin with Dietary Supplementation of Hydroxyproline (Lumistor®) and Placebo

Improvement of Water Content (Arbitrary Unit)		
Test Product	**4 Weeks**	**8 Weeks**
Hydroxyproline (Lumistor®)	10.7 ± 4.8	21.5 ± 4.3*
Placebo	4.1 ± 3.8	4.4 ± 4.9

Values are the change of water content values from baseline, mean ± SEM, $*P < 0.05$, significantly different from the values of the placebo group, unpaired t-test.

Table 15.4 Comparison of Volunteers' Perception of Facial Skin with Dietary Supplementation of Hydroxyproline (Lumistor®) and Placebo for 8 Weeks

Improvement of Perception Score Compared to Baseline		
Skin Parameters	**Lumistor®**	**P lacebo**
Elasticity	2.2 ± 0.7*	0.3 ± 0.5
Wrinkles	1.3 ± 0.5	–0.1 ± 0.5
Firmness	2.1 ± 0.8	1.1 ± 0.4
Dryness	1.9 ± 0.4	0.4 ± 0.7
Dryness around eyes and mouth	2.0 ± 0.7	0.7 ± 0.7
Roughness	0.6 ± 0.5	0.6 ± 0.5
Itchiness due to rough skin	1.5 ± 0.6	1.1 ± 0.3
Red patches, pimples, and other skin imperfections	2.2 ± 0.7	0.6 ± 0.5
Use of foundation creams and make-up make a difference in how you look	–0.2 ± 0.7	1.0 ± 0.8

Volunteers' perception was evaluated by questionnaire on a scale of 1–10, with the larger values showing better feelings on each item. Values are the change of score from baseline, mean ± SEM. $*P < 0.05$ significantly different vs. placebo group, unpaired t-test.

15.3.1.4 Topical Application of Hydroxyproline

In addition to oral application, recent studies conducted with Kyowa Hakko's natural, fermentation-based hydroxyproline (Lumistor®) have shown that, when topically applied, it moisturizes skin, reduces fine lines and wrinkles, and enhances skin elasticity. This effect is enhanced by combining Lumistor®

with a low molecular–weight hyaluronic acid (150,000–250,000 Da) and vitamin C (US Patent 6497889).

15.3.1.4.1 Hydroxyproline and Low Molecular Weight Hyaluronic Acid—Effect on Facial Lines and Wrinkles

Test Objective: Determine the reduction of superficial fine lines and wrinkles after 7 weeks of treatment.

Test Materials: Hydroxyproline (Lumistor®); low molecular–weight hyaluronic acid (150,000–200,000 Da).

Test Products: Placebo cream base; 2% hydroxyproline in a cream base; 2% hydroxyproline; and 1% low molecular–weight hyaluronic acid in a cream base.

Test Subjects: Three panels of 15 female volunteers (ages 45+) with moderate signs of clinical aging (superficial fine lines and wrinkles).

Protocol: Participants applied the treatment products twice a day (am and pm) after thorough cleansing, to designated sites of the face. Subjects visited the testing center at day 0, week 2, week 5, and week 7.

Measurement of Superficial Fine Lines and Wrinkle Reduction: Superficial fine lines and wrinkle reduction were assessed visually according to Packman et al. (1978) [50]. The shallowness, depth, and total number of superficial fine lines and wrinkles of each subject were analyzed and scored via image analysis using replicas.

Results: The addition of 2% hydroxyproline (Lumistor®) to a cream base decreased fine lines and wrinkles by 25.3 and 32.7%, respectively, after 5 weeks and 7 weeks of treatment (Fig. 15.8). Addition of 1% low molecular weight hyaluronic acid (150,000–200,000 Da) to the hydroxyproline- (Lumistor®) containing cream decreased fine lines and wrinkles by 35.8% and 45.0% respectively, after 5 weeks and 7 weeks of treatment, suggesting synergy and enhanced penetration into the skin (Fig. 15.8).

15.3.1.4.2 Hydroxyproline, Low Molecular–Weight Hyaluronic Acid and Vitamin C—Effect on Fine Lines and Wrinkles

Test Objective: Determine antiwrinkling effects after 6 weeks of treatment.

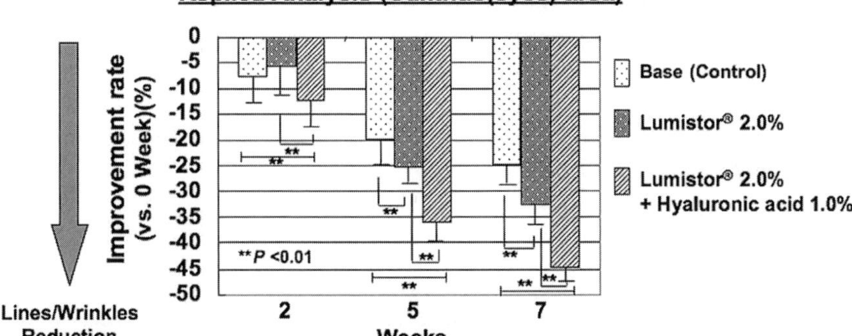

Figure 15.8 The effects of hydroxyproline and a combination of hydroxyproline and low molecular–weight hyaluronic acid on facial line and wrinkles.

Test Materials: Hydroxyproline, low molecular–weight hyaluronic acid (150,000–250,000 Da), and vitamin C.

Test Products: 2% hydroxyproline and 1% low molecular–weight hyaluronic acid in a cream base; 2% hydroxyproline, low molecular–weight hyaluronic acid, and 5% magnesium ascorbyl phosphate in a base cream.

Test Subjects: Two panels of 15 female volunteers (ages 45+), with moderate signs of clinical skin aging (superficial fine lines and wrinkles).

Protocol: Participants applied the treatment products twice a day (am and pm), after thorough cleansing to designated sites of the face. Subjects visited the testing center at day 0, week 2, week 4, and week 6.

Measurement of Superficial Fine Lines and Wrinkle Reduction: Superficial fine lines and wrinkle reduction was assessed visually according to the Packman et al. (1978) [50]. The shallowness, depth, and total number of the superficial fine lines and wrinkles of each subject were analyzed and scored via image analysis using replicas.

Results: The combination of 2% hydroxyproline (Lumistor®) and 1% low molecular–weight hyaluronic acid (150,000–200,000 Da) in a cream base decreased superficial fine lines by 30.4% and 47.6% compared to baseline after 4 weeks and 6 weeks of treatment (Fig. 15.9). The addition of 5% magnesium

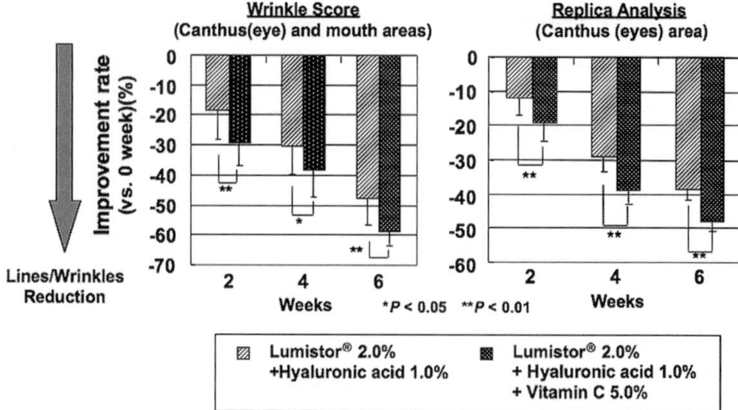

Figure 15.9 The effect of a combination of hydroxyproline and low molecular–weight hyaluronic acid and vitamin C on facial lines and wrinkles.

ascorbyl phosphate to the 2% hydroxyproline and 1% low molecular–weight hyaluronic acid–containing cream, decreased the superficial fine lines by 38.3% and 58.8%, respectively, compared to baseline after 4 weeks and 6 weeks of treatment, suggesting enhanced activity in the skin (Fig. 15.9).

15.3.2 Ornithine

15.3.2.1 Physiological Function

Ornithine is not an amino acid constituent of proteins, but it has many functions. In the body, as a component of the ornithine cycle, it plays a role in ammonia detoxification and it has the enhancing action on the immune system by activating macrophages and NK cells [51].

Ornithine enhances the secretion of growth hormone by stimulating the pituitary [52], thereby stimulating collagen synthesis, accelerating wound healing, aiding in intestinal mucosal repair, and reinforcing muscles. In addition, it is a precursor of poliamines [53]. Poliamine is a cell growth factor that plays a role in tissue growth [53]. It was reported that the administration of ornithine improved intestinal mucosal repair by stimulating polyamine synthesis [54]. It has been known that ornithine promotes proline and collagen synthesis [25,26].

15.3.2.2 Clinical Study

The effect of 3 weeks of oral supplementation of ornithine (Kyowa Hakko's natural fermentation-based ornithine), was examined by a double-blind, placebo-controlled study with healthy men and women ages 45–65. The perception of their skin condition was recorded at the beginning and the end of the test.

Objective: Investigate the volunteers' perception of facial skin conditions after 3 weeks of dietary supplementation of ornithine.

Materials: Placebo (lactose) and ornithine capsules (800 mg of ornithine; 1021 mg of ornithine hydrochloride), manufactured by Kyowa Hakko.

Subjects: Fourteen healthy individuals (ages between 47 and 65) were divided into two groups.

Protocol: Subjects ingested 800 mg/day of ornithine or placebo, respectively, for 3 weeks. A subjective-assessment questionnaire, using a visual analogue scale [55] to evaluate the subjects' perception objectively, was administered at the beginning and the end of the test.

In the questionnaire, the subjects were asked to provide rating on a scale that they felt best described their level of response to the questionnaire items, with descriptions serving as a reference on both ends of the scale. The difference before and after the test was obtained by measuring the distance from the left end of the scale to the points marked by each subject. The proportion of the difference to the baseline value was expressed as a percentage, and the values obtained were expressed in terms of an average improvement rate (%).

Results: The average improvement rate for complexion, facial wrinkles, and elasticity of the facial skin showed significant improvements compared with the placebo group; the average improvements for other questions were also elevated in the ornithine group (Fig. 15.10).

15.3.3 Glutamine, Alanyl Glutamine

15.3.3.1 Physiological Function

Glutamine is an essential nutrient for cells. Glutamine is a major energy source for many cells and a precursor of purines and pyrimidines. As a

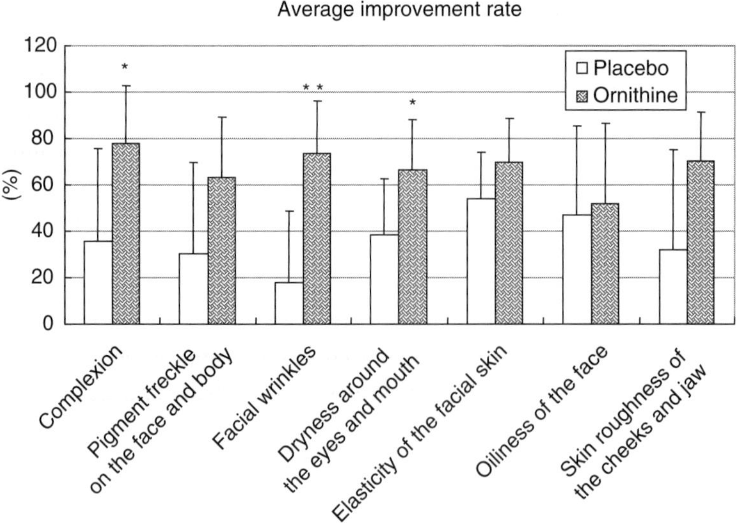

Figure 15.10 Comparison of volunteers' perception of facial skin with dietary supplementation of ornithine and placebo for 3 weeks. Index for skin texture: The difference in the value before and after the test was obtained, and the proportion of the obtained value to the baseline value was expressed as a percentage. The values shown in the graphs are mean value ± SD. An unpaired t-test was conducted to test for a statistically significant difference. $*P < 0.05$, $**P < 0.01$ vs. placebo.

The items contained in the questionnaire are as follows:

- **[Complexion]** Having a good complexion ←→ pale (dark face)
- **[Pigment freckle on the face and body]** Not concerned about ←→ Very concerned about
- **[Facial wrinkles]** Not concerned about ←→ Very concerned about
- **[Dryness around the eyes and mouth]** Moist ←→ Dry
- **[Elasticity of the facial skin]** Elastic ←→ Flabby
- **[Oiliness of the face]** None ←→ Very oily
- **[Skin roughness of the cheeks and jaw]** Not concerned about ←→ Very concerned about

consequence, glutamine sustains proliferation of rapidly dividing cells (e.g., enterocytes, lymphocytes, and fibroblasts). In the body, glutamine is the most common amino acid, found in the blood or intracellular pool, accounting for 50–60% of the amino acids in skeletal muscle and 20% or more in plasma.

Glutamine has a wide variety of important physiological functions in the body such as energy supply, stimulation of protein synthesis (acceleration of growth hormone secretion), wound healing, immuno-stimulation, and anti-inflammatory action. As for glutamine's effects on skin, it accelerates collagen and glycosaminoglycan (e.g., chondroitin sulfate, hyaluronic acid) synthesis. However, problems associated with physical properties of glutamine, such as a poor stability in water and low solubility, limit its use, despite its nutritional effectiveness.

Alanyl glutamine (Ala-Gln) is a stable dipeptide, consisting of the two amino acids alanine and glutamine. Alanyl glutamine is useful as a stable source of glutamine, with high stability to both heat and acids and high solubility (Figs. 15.11 and 15.12, and Table 15.5). In various research studies, it was shown that Ala-Gln has an equal or greater effect than glutamine in the body.

15.3.3.2 In Vitro Study

It has been known that the amount of glutamine required for optimal growth of mammalian cell cultures is 3–10 times greater than the amount of any other amino acids [56] and that the functions of human fibroblasts are highly dependent on glutamine availability [26].

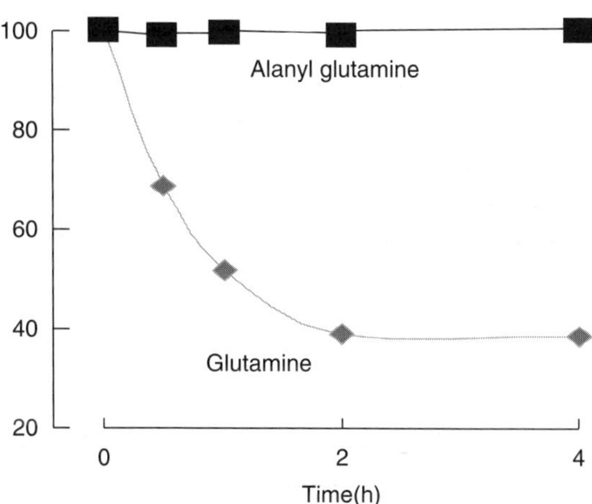

Figure 15.11 Residual ratio at pH 1, 37°C (2g/l).

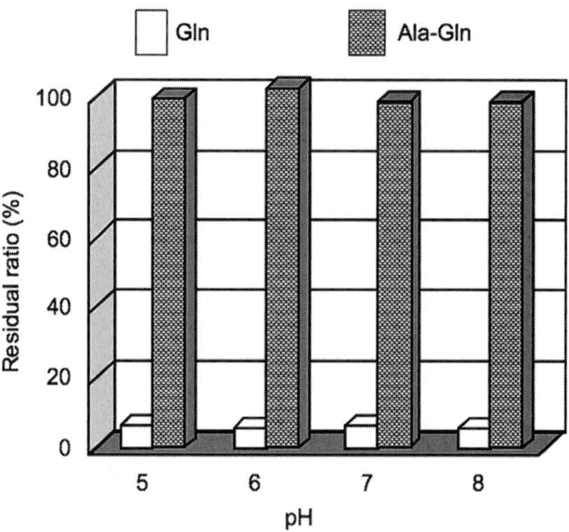

Figure 15.12 Residual ratio after treatment at 122°C for 20 min. The stability of Ala-Gln in aqueous solution was compared with that of glutamine (Gln). Glutamine is unstable, whereas Ala-Gln has an excellent stability in aqueous solution. Ala-Gln can be sterilized by boiling and heating.

Table 15.5 Comparison of the Aqueous Solubility of Ala-Gln with Gln and Gly-Gln

Name of Substance	Solubility (g/l, 25°C)
Gln	53.7
Gly-Gln	154
Ala-Gln	568

The aqueous solubility of Ala-Gln was compared with that of other substitutes for glutamine, including glutamine itself. The solubility of glutamine is 53.7 g/l, whereas that of Ala-Gln is 568.0 g/l. Ala-Gln has a greater solubility than Gly-Gln (154.0 g/l), a substitute for glutamine that is similar to Ala-Gln.

15.3.3.2.1 Acceleration of Collagen Synthesis [25]

The effect of glutamine on collagen biosynthesis in confluent human skin fibroblasts was determined at various times of incubation. Normal human skin fibroblast cells were incubated for 6 hours, 12 hours, and 24 hours with a medium containing 5% serum supplemented with or without 0.25 mM glutamine in the presence of 5-[3H] proline. Incorporation of the

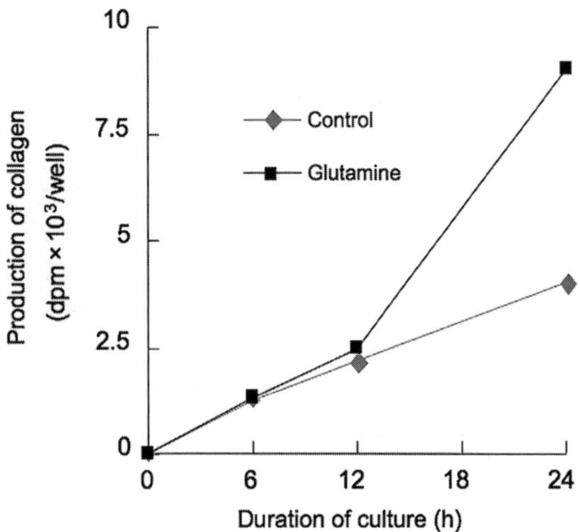

Figure 15.13 Time course experiment for collagen synthesis in fibroblasts treated with glutamine. Modified from Karna et al. 2001 [25].

radioactive material into proteins susceptible to the action of purified collagenase was determined.

It was found that glutamine induced increase in collagen biosynthesis to approximately 112%, 115%, and 230% of control values (Fig. 15.13).

15.3.3.2.2 Acceleration of Glycosaminoglycan Synthesis by Glutamine [57]

The articular cartilage, aorta, and skin tissues derived from chicken (~100 g each) were pre-incubated with a medium (Eagles basal medium containing 0.5 mM glycine and buffered at pH 7.2 with 20 mM HEPES) not containing glutamine in an atmosphere of 100% oxygen. After 1 hour the medium was replaced by fresh medium, either with or without glutamine (1 mM), then 20 μCi of [3H] acetate was added to each medium, which was then incubated for an additional 4 hours. Glycosaminoglycan biosynthesis by the tissues was measured by the uptake of [3H] acetate.

The observations presented in this work suggest that glutamine is required to sustain glycosaminoglycan synthesis in cultured chick chondrocytes (Fig. 15.14).

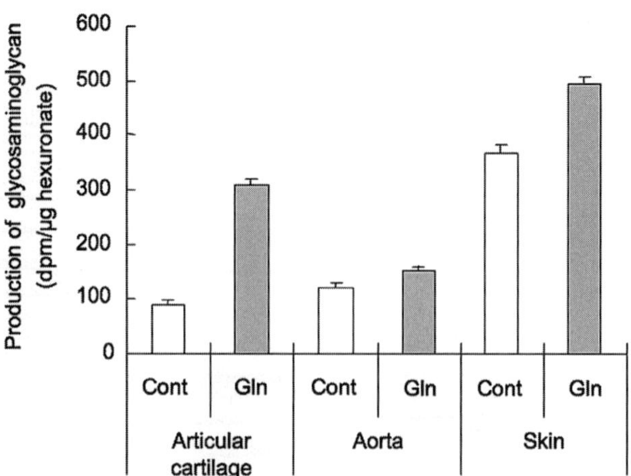

Figure 15.14 Effect of glutamine on glycosaminoglycan synthesis by various avian connective tissues. Modified from Handley et al. 1980 [56].

15.3.3.2.3 Proliferation of Cells from Human Intestinal Mucosal Tissues in Culture Media with Ala-Gln or Gln [58]

Either glutamine (Gln), Ala-Gln, or saline (control) was added at 2 mM to the culture of the normal human ileum, proximal rectum, and sigmoid, and then these samples were incubated for 2 hours. After 2 hours of incubation, 200 mol/L BrdUrd and 20 mol/L fluorodexyuridine were added to the samples for an additional 2 hours. The uptake of BrdUrd into crypt cells was then measured, serving as an indicator of cell proliferation.

As a result, Ala-Gln as well as glutamine stimulated crypt cell proliferation in the ileum, proximal rectum, and sigmoid. No significant difference was observed in cell proliferation between Ala-Gln and glutamine culture media when human intestinal mucosa tissues were incubated (Fig. 15.15).

15.3.3.2.4 Efficacy of Ala-Gln as a Cell Culture Media Component [59]

In this study, four fibroblastic cell lines derived from human amniotic fluid were used. The cells were first cultured in RPMI medium supplemented with sodium bicarbonate, glutamine (2 mM), penicillin (100 U/ml), streptomycin (100 μg/ml), amphotericin B (2 μg/ml), and 10% fetal calf serum until homogeneous fibroblast monolayers devoid of epithelioid

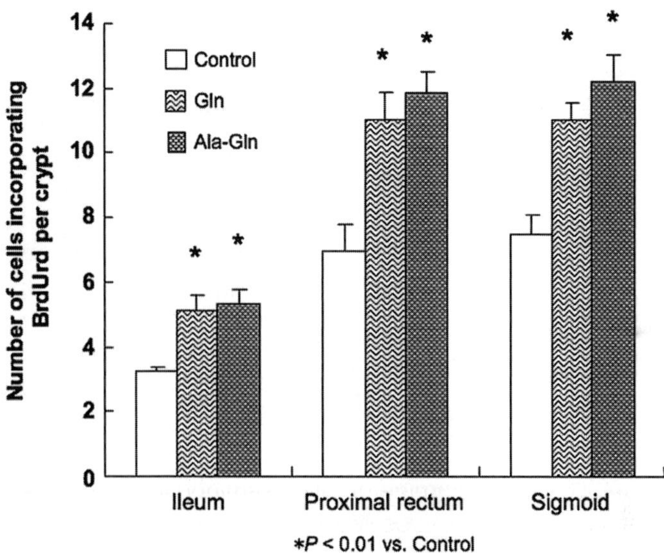

Figure 15.15 Mean number of cells in the human ileum, proximal rectum, and sigmoid incorporating BrdUrd per crypt.

cells were obtained. The cells were dctached by trypsinization and resuspended in glutamine-free supplemented RPMI. Twenty-four hours later, the medium was discarded and replaced by 2% FCS/RPMI with or without 2 mM of alpha-ketoisocaproyglutamine (Kic-Gln), glutamine, Ala-Gln, or acetylglutamine. Protein content was assayed every 24 hours for 72 hours.

As a result, cellular protein content increased progressively throughout the experiment in the glutamine-containing cultures, especially in glutamine and Ala-Gln groups, versus the glutamine-free cultures (Table 15.6).

This result shows that Ala-Gln is as effective as glutamine in promoting growth of cultured fibroblasts.

15.3.3.3 Clinical Study

Alanyl glutamine (Ala-Gln) has been applied to total parenteral nutrition (TPN) because of its effects in the improvement of the metabolic system, maintenance of the function of the gastrointestinal tract, and maintenance of immune functions [60].

Table 15.6 Effect of Various Glutamine-Containing Compounds on Protein Content of Cultured Fibroblasts

	Day 1	Day 2	Day 3	Day 4
Gln-Free	27.7 ± 2.2	31.6 ± 2.5	35.5 ± 1.6	45.2 ± 3.7
Gln		35.2 ± 1.0	48.5 ± 2.9**	93.8 ± 8.1***
Ala-Gln		36.4 ± 1.2	54.7 ± 3.0**	102.2 ± 12.0**
Acetylglutamine		34.2 ± 1.7	40.4 ± 3.4	74.9 ± 8.7*
Kic-Gln		38.1 ± 2.2	48.8 ± 3.8**	79.8 ± 7.7*

Results are expressed as μg protein/dish.
*$P < 0.05$, **$P < 0.001$ vs. glutamine-free cultures.

The effect of the addition of Ala-Gln on intestinal permeability in patients who had undergone excision of extensive dry eschar (scab) in burn wounds was studied. Thirty patients with injuries that covered 30–50% of their total body surface areas, or with third degree burns that covered 15–25% of their total body surface, were divided into an Ala-Gln group and a control group.

The Ala-Gln group was given TPN containing 0.5 g/kg (B.W.)/day of Ala-Gln (equivalent to 0.35 g of glutamine/kg [B.W.]/day) (35 ± 5 kcal/kg [B.W.]/day, 0.25 ± 0.05 gN/kg [B.W.]/day) from post-operative day 1 (POD-1) to day 12 (POD-12), whereas the control group was given an isoenergetic and isonitrogenous standard regimen without Ala-Gln.

The time period required for wound healing was shorter in the Ala-Gln group compared with the control group (32 ± 3 days in the Ala-Gln group vs. 37 ± 6 days in the control group, $P < 0.012$). The infection rate was lower in the Ala-Gln group (13% in the Ala-Gln group vs. 26% in the control group).

15.4 Summary

Amino acids play important roles in "skin beautification." Proteins such as keratin and collagen can act as moisturizers and also serve as protectors and building blocks of the skin.

Skin cell turnover, which is an important process in maintaining and renewing beautiful skin, is enhanced by amino acids, thus promoting skin regeneration. Furthermore, a UV-protection effect is provided by amino acids with antioxidant properties. Amino acids are thus vitally important in the field of beautification.

Cosmetics are applied topically in order to beautify skin, hair, and nails. However, because the stratum corneum is a barrier, penetration of beneficial actives into the skin is limited at times. In view of these studies, for optimal efficacy, a combination of topical and oral supplementation is recommended.

In recent years the concept of "beauty from within," which is based on oral delivery of actives, has become popular in the US, Europe and Asia, because it can be quite effective in beautifying and improving the skin condition.

References

1. Blank IH. Factors which influence the water content of the stratum corneum. Journal of Investigative Dermatology. 18:433–440, 1952.
2. Blank IH. Further observations of factors which influence the water content of the stratum corneum. Journal of Investigative Dermatology. 21:259–271, 1953.
3. Jacobi OK. Nature of cosmetic films on the skin. Journal of the Society of Cosmetic Chemists. 18:149–160, 1967.
4. Jacobi OK. About the mechanism of moisture regulation in the horny layer of the skin. Proceedings of Scientific Section Toilet Goods Association. 31:22–24, 1959.
5. Koyama J. Horii I. Kawasaki K. Nakayama Y. Morikawa Y. Mitsui T. Free amino acids of stratum corneum as a biochemical marker to evaluate dry skin. Journal of Social Cosmetic Chemistry. 35:183–195, 1984.
6. Horii I. Nakayama Y. Obata M. Tagami H. Stratum corneum hydration and amino acid content in xerotic skin. British Journal of Dermatology. 121(5): 587–592, 1989.
7. Jacobson TM. Yuksel KU. Geesin JC. Gordon JS. Lane AT. Gracy RW. Effects of aging and xerosis on the amino acid composition of human skin. Journal of Investigative Dermatology. 95(3):296–300, 1990.
8. Watanabe M. Tagami H. Horii I. Takahashi M. Kligman AM. Functional analyses of the superficial stratum corneum in atopic xerosis. Archives of Dermatology. 127(11):1689–1692, 1991.
9. Japanese Biochemical Society. eds. Biochemical Data Book. Tokyokaga kudojin. Page 1662, 1979.
10. Tomohiro A. Kimura S. He H. Fujisawa Y. Nishiyama A. Kiyomoto K. Aki Y. Tamaki T. Abe Y. Regional blood flow in DahIwai salt-sensitive rats and the effects of dietary arginine supplementation. American Journal of Physiology. 272(4 Pt 2):R1013–1019, 1997.

11. Boger RH. Bode-Boger SM. Phivthong-ngam L. Brandes RP. Schwedhelm E. Mugge A. Bohme M. Tsikas D. Frolich JC. Dietary arginine and alpha-tocopherol reduce vascular oxidative stress and preserve endothelial function in hypercholesterolemic rabbits via different mechanisms. Atherosclerosis. 141(1):31–43, 1998.
12. Cooke JP. Singer AH. Tsao P. Zera P. Rowan RA. Billingham ME. Antiatherogenic effects of arginine in the hypercholesterolemic rabbit. Journal of Clinical Investigation. 90(3):1168–1172, 1992.
13. Chen PY. Sanders PW. Arginine abrogates salt-sensitive hypertension in Dahl/Rapp rats. Journal of Clinical Investigation. 88(5):1559–1567, 1991.
14. Raghavan SA. Dikshit M. Citrulline mediated relaxation in the control and lipopolysaccharide-treated rat aortic rings. European Journal of Pharmacology. 431(1):61–69, 2001.
15. Collins N. Glutamine and wound healing. Advances in Skin & Wound Care. 15(5):233–234, 2002.
16. Rodwell VW. Chapter 33: Conversion of amino acids to specialized products. In: Harper's Biochemistry 25th Edition. Murray RK. Mayes PA. Rodwell EV. Granner DK (eds.) McGraw-Hill/Appleton & Lange. NY. USA. 347–358, 1999.
17. Cui XL. Iwasa M. Iwasa Y. Ohmori Y. Yamamoto A. Maeda H. Kume M. Ogoshi S. Yokoyama A. Sugawara T. Funada T. Effects of dietary arginine supplementation on protein turnover and tissue protein synthesis in scald-burn rats. Nutrition. 15(7–8):563–569, 1999.
18. Wu G. Flynn NE. Knabe DA. Jaeger LA. A cortisol surge mediates the enhanced polyamine synthesis in porcine enterocytes during weaning. American Journal of Physiology-Regulatory Integrative and Comparative Physiology. 279(2):R554–559, 2000.
19. Cao M. Stefanovic-Racic M. Georgescu HI. Miller LA. Evans CH. Generation of nitric oxide by lapine meniscal cells and its effect on matrix metabolism: stimulation of collagen production by arginine. Journal of Orthopaedic Research. 16(1):104–111, 1998.
20. Jalkanen M. Larjava H. Heino J. Vihersaari T. Peltonen J. Penttinen R. Arginine depletion in macrophage medium inhibits collagen synthesis by fibroblasts. Immunology Letters. 4(5):259–261, 1982.
21. Chevalley T. Rizzoli R. Manen D. Caverzasio J. Bonjour JP. Arginine increases insulin-like growth factor-I production and collagen synthesis in osteoblast-like cells. Bone. 23(2):103–109, 1998.
22. Shi HP. Fishel RS. Efron DT. Williams JZ. Fishel MH. Barbul A. Effect of supplemental ornithine on wound healing. Journal of Surgical Research. 106(2):299–302, 2002.
23. Nussbaum MS. Arginine stimulates wound healing and immune function in elderly human beings. JPEN: Journal of Parenteral and Enteral Nutrition. 18(2):194, 1994.
24. De Bandt JP. Cynober LA. Amino acids with anabolic properties. Current Opinion in Clinical Nutrition and Metabolic Care. 1(3):263–272, 1998.
25. Karna E. Miltyk W. Wotczynski S. Palka JA. The potential mechanism for glutamine-induced collagen biosynthesis in cultured human skin fibroblasts. Comparative Biochemistry and Physiology Part B, Biochemistry & Molecular Biology. 130(1):23–32, 2001.

26. Bellon G. Chaqour B. Wegrowski Y. Monboisse JC. Borel JP. Glutamine increases collagen gene transcription in cultured human fibroblasts. Biochimica et Biophysica Acta. 1268:317–323, 1995.
27. Chyun JH. Griminger P. Improvement of nitrogen retention by arginine and glycine supplementation and its relation to collagen synthesis in traumatized mature and aged rats. Journal of Nutrition. 114(9):1697–1704, 1984.
28. Kobayashi A. Shibasaki T. Production and physiological activity of hydroxyproline. Fragrance Journal. 31(3):37–43, 2003.
29. Granner DK. Chapter 45: Pituitary and hypothalamic hormones. In: Harper's Biochemistry 25th Edition. Murray RK. Mayes PA. Rodwell VW. Granner DK (eds.) McGraw-Hill/Appleton & Lange. NY. USA. 550–560, 1999.
30. Welbourne TC. Increased plasma bicarbonate and growth hormone after an oral glutamine load. American Journal of Clinical Nutrition. 61(5):1058–1061, 1995.
31. Besset A. Bonardet A. Rondouin G. Descomps B. Passouant P. Increase in sleep related GH and Prl secretion after chronic arginine aspartate administration in man. Acta Endocrinologica. 99(1):18–23, 1982.
32. Ghigo E. Ceda GP. Valcavi R. Goffi S. Zini M. Mucci M. Valenti G. Cocchi D. Muller EE. Camanni F. Low doses of either intravenously or orally administered arginine are able to enhance growth hormone response to growth hormone releasing hormone in elderly subjects. Journal of Endocrinological Investigation. 17(2):113–117, 1994.
33. Blum A. Cannon RO 3rd. Costello R. Schenke WH. Csako G. Endocrine and lipid effects of oral arginine treatment in healthy postmenopausal women. Journal of Laboratory and Clinical Medicine. 135(3):231–237, 2000.
34. Knopf RF. Conn JW. Fajans SS. Floyd JC. Rull JA. Guntsche EM. Rull JA. Plasma growth hormone response to intravenous administration of amino acids. Journal of Clinical Endocrinology and Metabolism. 25:1140–1144, 1965.
35. Bucci L. Hickson JF. Pivarnik JM. Wolinsky I. McMahon JC. Turner SD. Ornithine ingestion and growth hormone release in bodybuilders. Nutrition Research. 10:239–245, 1990.
36. Evain-Brion D. Donnadieu M. Roger M. Job JC. Simultaneous study of somatotrophic and corticotrophic pituitary secretions during ornithine infusion test. Clinical Endocrinology. 17(2):119–122, 1982.
37. Jeevanandam M. Holaday NJ. Petersen SR. Ornitihe-alpha-ketoglutarate (OKG) supplementation is more effective than its component salts in traumatized rats. Journal of Nutrition. 126(9):2141–2150, 1996.
38. Cavagnini F. Invitti C. Pinto M. Maraschini C. Di Landro A. Dubini A. Marelli A. Effect of acute and repeated administration of gamma aminobutyric acid (GABA) on growth hormone and prolactin secretion in man. Acta Endocrinologica. 93(2):149–154, 1980.
39. Cavagnini F. Benetti G. Invitti C. Ramella G. Pinto M. Lazza M. Dubini A. Marelli A. Muller EE. Effect of gamma-aminobutyric acid on growth hormone and prolactin secretion in man: influence of pimozide and domperidone. Journal of Clinical Endocrinology and Metabolism. 51(4):789–792, 1980.
40. Griffith OW. Bridges RJ. Meister A. Transport of gamma-glutamyl amino acids: role of glutathione and gamma-glutamyl transpeptidase. Proceedings of the National Academy of Sciences of the United States of America. 76(12):6319–6322, 1979.

41. Unnikrishnan MK. Rao MN. Antiinflammatory activity of methionine, methionine sulfoxide and methionine sulfone. Agents and Actions. 31(1–2):110–112, 1990.
42. Cai Q. Takemura G. Ashraf M. Antioxidative properties of histidine and its effect on myocardial injury during ischemia/reperfusion in isolated rat heart. Journal of Cardiovascular Pharmacology. 25(1):147–155, 1995.
43. Lubec B. Hayn M. Kitzmuller E. Vierhapper H. Lubec G. Arginine reduces lipid peroxidation in patients with diabetes mellitus. Free Radical Biology & Medicine. 22(1–2):355–357, 1997.
44. Akashi K. Miyake C. Yokota A. Citrulline, a novel compatible solute in drought-tolerant wild watermelon leaves, is an efficient hydroxyl radical scavenger. FEBS Letters. 508:438–442, 2001.
45. Jara JR. Aroca P. Solano F. Martinez JH. Lozano JA. The role of sulfhydryl compounds in mammalian melanogenesis: the effect of cysteine and glutathione upon tyrosinase and the intermediates of the pathway. Biochimica et Biophysica Acta. 967(2):296–303, 1988.
46. Sasaki M. Clinical experience with Hythiol capsules for facial hyperpig mentation. Skin. 18(3):339–342, 1976.
47. Kon Z. Hori K. Ishikawa K. Iwado T. Goto S. Clinical experience with Hythiol capsules for facial hyperpigmentation. Jpn:Yakuri to Chiryo. 4(8):2132–2141, 1976.
48. Adams E. Frank L. Metabolism of proline and the hydroxyprolines. Annual Review of Biochemistry. 49:1005–1061, 1980.
49. Carmichael J. DeGraff WG. Gazdar AF. Minna JD. Mitchell JB. Evaluation of a tetrazolium-based semiautomated colorimetric assay: Assessment of chemosensitivity testing. Cancer Research. 47(4):936–942, 1987.
50. Packman EW. Gans EH. Topical moisturizers: quantification of their effect on superficial facial lines. Journal of the Society of Cosmetic Chemists. 29:79–90, 1978.
51. Robinson LE. Bussiere FI. Le Boucher J. Farges MC. Cynober LA. Field CJ. Baracos VE. Amino acid nutrition and immune function in tumour-bearing rats: a comparison of glutamine-, arginine- and ornithine 2-oxoglutarate-supplemented diets. Clinical Science. 97(6):657–669, 1999.
52. Evain-Brion D. Donnadieu M. Roger M. Job JC. Simultaneous study of somatotrophic and corticotrophic pituitary secretions during ornithine infusion test. Clinical Endocrinology. 17(2):119–122, 1982.
53. Rodwell VW. Chapter 33: Conversion of amino acids to specialized products. In Harper's Biochemistry 25th Edition. Murray RK. Mayes PA. Rodwell VW. Granner DK. (eds.) McGraw-Hill/Appleton & Lange. NY. USA. 347–358, 1999.
54. Duranton B. Schleiffer R. Gosse F. Raul F. Preventive administration of ornithine alpha-ketoglutarate improves intestinal mucosal repair after transient ischemia in rats. Critical Care Medicine. 26(1):120–125, 1998.
55. Uchiyama Y. Kobayashi T. Shiomi Y. Introduction to Clinical Assessment, Kyodo Isyo Shuppan Co., Ltd. 75–80, 2003.
56. Eagle H. Amino acid metabolism in mammalian cell cultures. Science. 130:432–437, 1959.
57. Handley CJ. Speight G. Leyden KM. Lowther DA. Extracellular matrix metabolism by chondrocytes. 7. Evidence that glutamine is an essential amino

acid for chondrocytes and other connective tissue cells. Biochimica et Biophysica Acta. 627(3):324–331, 1980.

58. Scheppach W. Loges C. Bartram P. Christl SU. Richter F. Dusel G. Stehle P. Fuerst P. Kasper H. Effect of free glutamine and alanyl-glutamine dipeptide on mucosal proliferation of the human ileum and colon. Gastroenterology. 107(2):429–434, 1994.
59. Coudray-Lucas C. Lasnier E. Renaud F. Ziegler F. Settembre P. Cynober LA. Ekindjian OG. Is α-Ketoisocaproyl-glutamine a suitable glutamine precursor to sustain fibroblast growth? Clinical Nutrition. 18(1):29–33, 1999.
60. Zhou Y-P. Jiang Z-M. Sun Y-H. He G-Z. Shu H. The effects of supplemental glutamine dipeptide on gut integrity and clinical outcome after major escharectomy in severe burns: a randomized, double-blind, controlled clinical trial. Clinical Nutrition Supplements. 1(1):55–60, 2004.

16

Natural Products Supporting the Extracellular Matrix: Rice Ceramide and Other Plant Extracts for Skin Health

Hiroshi Shimoda, PhD

Oryza Oil & Fat Chemical Co., Ltd., Ichinomiya, Japan

Summary

Natural cosmeceutical products are attracting increasing attention, especially from the ecological and safety point of view. Collagen, hyaluronic acid,

---

Aaron Tabor and Robert M. Blair (eds.), *Nutritional Cosmetics: Beauty from Within*, 319–334, © 2009 William Andrew Inc.

elastin, and ceramides are components of the skin's extracellular matrix. Ceramides and glucosylceramides are contained in the stratum corneum of the epidermis and play roles in the barrier function and moisturization of the skin. Some of these components can be extracted and purified from plants. Rice ceramide extracted from rice germ and bran consists of at least six types of glucosylsphingolipids. This extract supports skin moisturizing and barrier function, suppresses melanin synthesis, promotes cell proliferation, and acts anti-inflammatorily and antiallergically. Therefore, application of rice ceramide in cosmeceutical products and as dietary supplements should be beneficial for maintaining a healthy skin extracellular matrix. Besides rice ceramide, we found that litchi seed extract, purple rice extract, and grape extract inhibit skin-tissue-degrading enzymes, for example collagenase, hyaluronidase, and elastase. Selective and combined application of these extracts is thus expected to help maintain skin health.

16.1 Introduction

In development of new skin care products, raw materials with novel physiological functions and proven safety play a key role. Choice of synthetic or natural materials for cosmetic composition depends on their cost, safety, and effectiveness. With a safe, novel, and ecological image, more and more consumers are favoring natural cosmetic components recently. In addition, consciousness of environmental aspects has led to the increased demand for organic cultivation, non-solvent extraction, and non-animal- and non-fish-derived materials. We have developed a number of extracts from various plants for application in foods and cosmetics. Ceramide, a skin component with moisturizing activity, can either be synthesized or extracted from plants such as rice, wheat, and corn. The demand for plant-derived ceramides, which can be used for both cosmetics and foods, is increasing. In this chapter, we introduce skin-healthy effects of rice-derived ceramide. We also describe beauty effects of litchi seed extract, purple rice extract, and grape extract.

16.2 Rice Ceramides

16.2.1 Ceramides in Skin

In 1884, ceramides were found in human brain tissues, and later in skin and mucosal tissue. Human skin consists of epidermis, corium, and tela

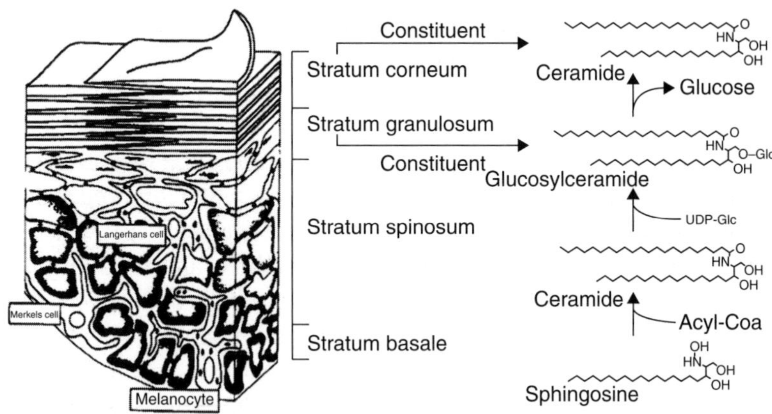

Figure 16.1 Sphingolipids and their biosynthesis scheme in human skin.

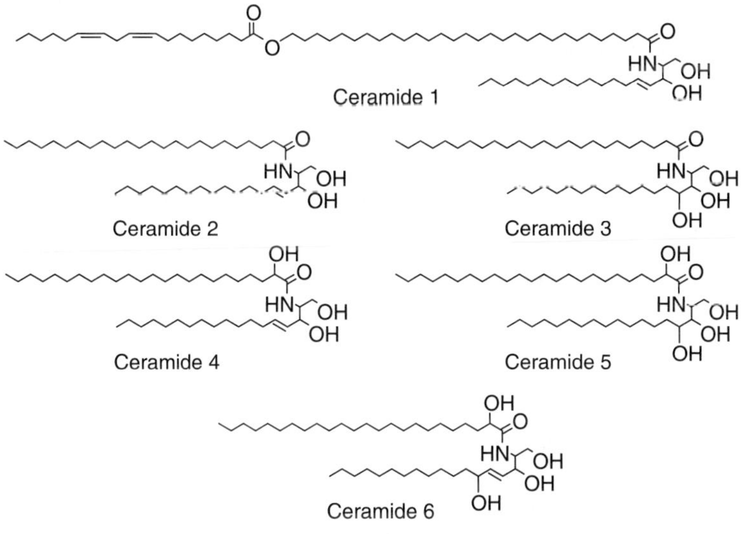

Figure 16.2 Ceramides in stratum corneum.

subcutanea. The epidermis is classified into four layers, namely the stratum corneum, stratum granulosum, stratum spinosum, and stratum basale (Fig. 16.1). More than six types of ceramides have been identified in human skin (Fig. 16.2) [1,2]. These ceramides are produced via several biosynthetic pathways in the epidermis and accumulated in the stratum corneum as the major constituent lipids (40–60% of the total lipids). In the epidermis, ceramides play an important role in forming lamella phases and in maintaining barrier function [3].

Imokawa et al. [4] demonstrated that the content of skin ceramides declines with age (Fig. 16.3). Forearm skin of elderly persons (especially those older than 70 years) is often xerotic, suggesting an association between ceramide decrease and skin drying. Marked reduction of ceramides was found in both lesional and nonlesional forearm stratum corneum of patients with atopic dermatitis (Fig. 16.4). These findings suggest that ceramides are a key factor for moisture maintenance and barrier function of the stratum corneum. A decrease in ceramide content is also thought to be associated with wrinkle formation. Thus, a sufficient amount of ceramides is considered to be essential for maintaining healthy skin.

16.2.2 Rice Ceramide

Rice (*Oryza sativa* L.) has been widely grown in Southeast Asia, not only as a major crop but also as an integral part of traditional culture and lifestyle in many Asian countries. In recent years, attention has been focused on rice bran and rice germ for its unique bioactive compounds and non-GMO profile. We have developed a number of products extracted from rice bran and rice germ as functional ingredients in medicines and cosmetics, as dietary supplements, and as food additives. One such functional compound is rice ceramide, which supports barrier function and moisture of skin. Rice ceramide contains a large amount of glucosphingolipids, which have

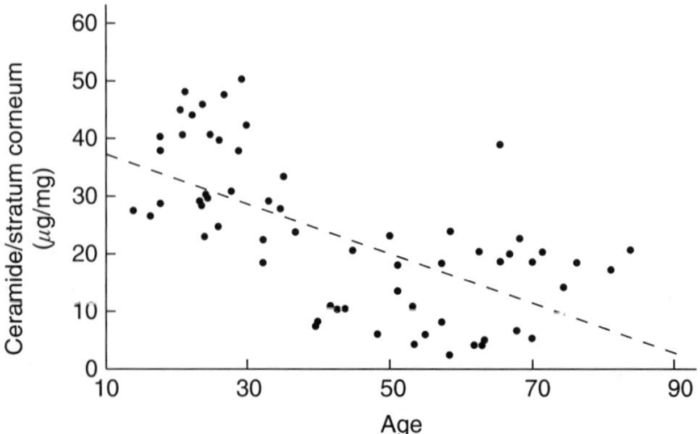

Figure 16.3 Ceramide content of the stratum corneum in healthy subjects of various ages [4].

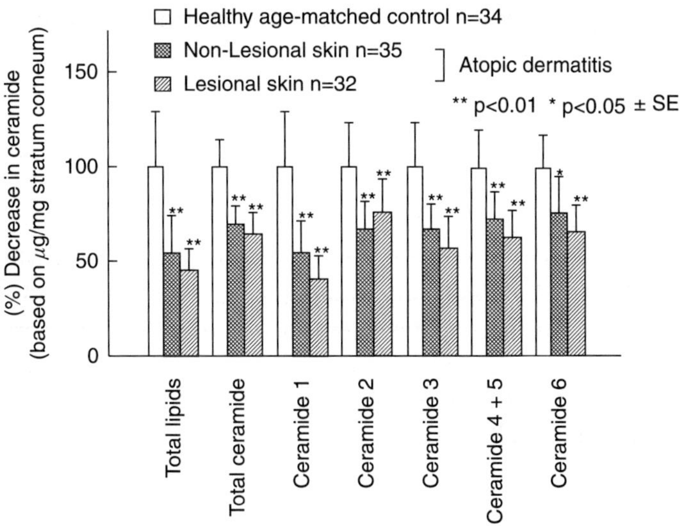

Figure 16.4 Ceramide contents in forearm skin of healthy subjects and atopic dermatitis patients [4].

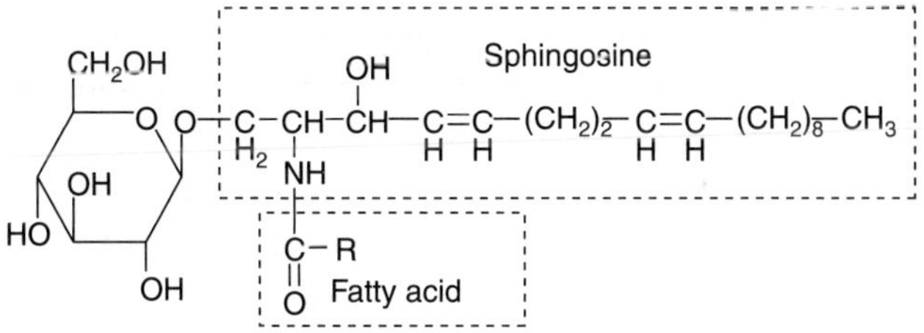

Sphingosine

Fatty acid

R: 1. -$(CH_2)_7CH{:}CHCH_2CH{:}CH(CH_2)_4CH_3$ 2. -$(CH_2)_{14}CH_3$

3. -$(CH_2)_7CH{:}CH(CH_2)_7CH_3$ 4. -$(CH_2)_{16}CH_3$

Figure 16.5 Major glucosphingolipids in rice bran.

similar chemical structures to those from animals. Rice-derived glucosphingolipids consist of sphingoid bases conjugated with fatty acids by amide linkages. The terminal hydroxyl group is substituted to glucose. Glucosphingolipids are classified into different species depending on the structure of their sphingoid bases and fatty acids. Koga et al. identified more than 20 types of sphingolipids in rice bran [5]. We have identified four additional types of ceramides that are the major glucosphingolipids in rice bran (Fig. 16.5).

16.2.3 Digestion, Absorption, and Transport of Ceramides

Digestion, absorption, and metabolism of plant-derived sphingolipids have been studied by Schmelz et al. [6]. They examined the distribution and metabolism of sphingolipids in the intestine by tracing radio-labeled sphingomyelin in mice. Sphingomyelin appeared in all parts of the intestine, and most of it was catabolized to ceramides and their metabolites. Only 1% of undigested sphingomyelin was transferred from intestine to liver 30–60 minutes after administration. Transport of sphingomyelin and its metabolites from the intestinal canal to other tissues is thus not efficient. The absorption and metabolism of sphingolipids vary depending on the types of ceramides. Duan et al. [7] reported that sphingomyelin is digested by alkaline sphingomyelinase mainly in the middle and lower areas of the small intestine and that alkaline sphingomyelinase plays an important role in the first stage of digestion. Other researchers reported the effectiveness of oral application of ceramides for skin diseases. Kimata [8] gave ceramide (1.8 mg/day) to children with moderate atopic dermatitis for 2 weeks and confirmed that the treatment improved skin symptoms and reduced allergic responses. Asai and Miyachi [9] reported that topical application of rice ceramides for 3 weeks enhanced water contents in the stratum corneum. Thus, it is likely that at least a part of digested ceramides can be absorbed and can reach damaged skin, where they improve skin condition by retaining moisture.

16.2.4 Cosmeceutical Function of Rice Ceramide

In the past, mainly synthetic and animal-derived ceramides have been used for cosmetics. In recent years, risk of bovine spongiform encephalopathy has raised a safety concern for using cattle-derived ceramides. Consequently, more attention has focused on plant ceramides such as rice ceramide, which is considered suitable for both food and cosmetics. In this section, we review several cosmeceutical functions of rice ceramide. Most importantly, rice ceramide possesses an excellent moisturizing effect for skin, superior to that of ceramides from other plants, such as elephant foot and wheat, as demonstrated in our *in vitro* experiments (Fig. 16.6). Also, topical application of ceramide has been reported to improve skin moisturization [9,10]. Moreover, ceramide is effective in improving atopic dermatitis in animals and humans [11–13]. These observations suggest that ceramide is effective for supplementation of moisture for dry skin.

In addition, ceramide possesses an antipigmentation activity. We found that rice ceramide (100 μg/ml) inhibited melanin production in B16 melanoma

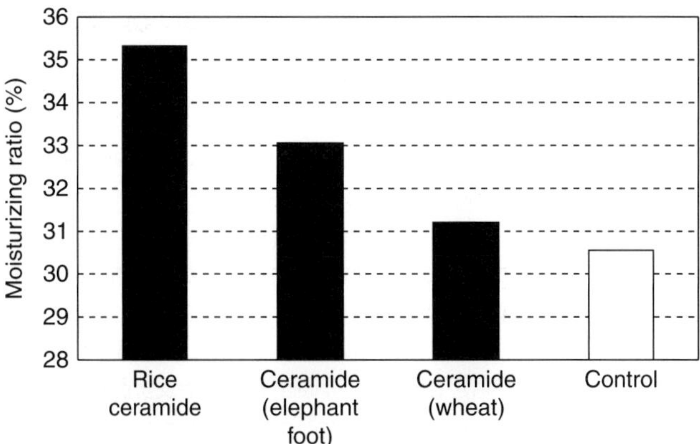

Figure 16.6 Moisturizing effect of plant-derived ceramides. Each ceramide was suspended in water at 3% solution in test tubes, and kept for 8 hours at 35°C and 4% humidity.

cells *in vitro* (Fig. 16.7[a]). At the same concentration, activity of rice ceramide was higher than that of ascorbic acid, arbutin, and ellagic acid. Similar to kojic acid, rice ceramide exhibited a similar concentration-dependent activity in inhibiting melanogenesis (Fig. 16.7[b]).

In collaboration with Professor Igarashi at Hokkaido University, we found that suppression of melanin formation by ceramides involves inhibition of tyrosinase, which is a key enzyme for melanin synthesis. As shown in Fig. 16.8, rice glucosphingolipids exhibited an inhibitory effect on tyrosinase and on melanin production in a concentration-dependent manner. In addition to the tyrosinase inhibitory activity, ceramide has been reported to affect other pathways leading to melanin production. Kim et al. [14] found that C(2)-ceramide suppressed proliferation of mouse melanocytes *in vitro* via inhibition of the Akt/protein kinase B (PKB) activation that produces phosphorylated Akt/PKB. Moreover, using human melanocytes, they found that C(2)-ceramide decreased the protein expression of microphthalmia-associated transcription factor, which is required for tyrosinase expression. They further reported that C(2)-ceramide induced delayed activation of extracellular signal–regulated protein kinase (ERK) and Akt/PKB, which may lead to suppression of cell growth and melanogenesis [15]. These findings suggest that ceramide is effective in preventing skin pigmentation.

Ceramides also enhance proliferation of dermal fibroblasts. As illustrated in Fig. 16.9, except wheat-derived ceramide, all ceramides enhanced cell

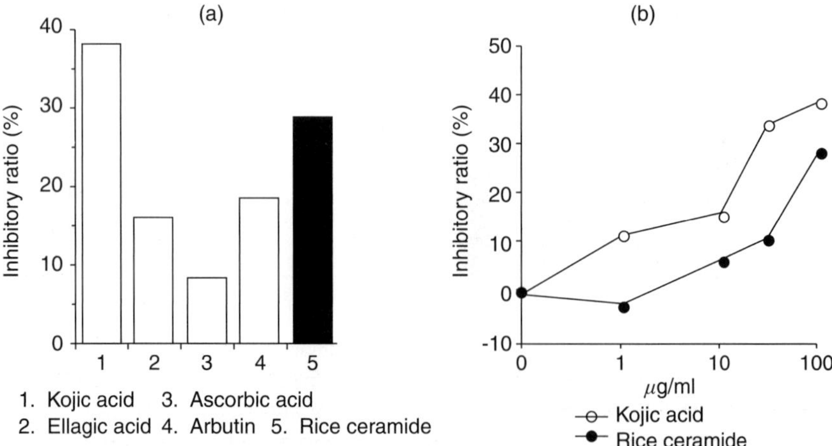

Figure 16.7 Inhibitory effect of rice ceramide on melanin production in melanoma. (a) Comparison of rice ceramide with positive control at 100 µg/ml. (b) Dose response of kojic acid and rice ceramide. B16 melanoma cells (2×10^3 cells/ml) were pre-incubated for 24 hours and the medium was replaced with new media containing 100 µg/ml emulsified glucosphingolipids (>90% of purity) or kojic acid. After 2-day incubation, the medium was replaced with fresh media, followed by another 2-day incubation. Cells were lysed in 2N NaOH, and absorbance was measured at 450 nm. The value was normalized by the cell number.

proliferation at 300 µg/ml whereas rice ceramides exhibited the most potent proliferative effect. The proliferative effect of ceramides seems to be mainly via the mitogenic activity of sphingosine-1-phosphate [16,17]. In contrast, apoptotic activity on murine keratinocytes has been reported for ceramide [18]. Marchell et al. [19] also reported that ceramide inhibits mitosis and induces terminal differentiation and apoptosis in keratinocytes. However, they also found mitogenic activity in glucosylceramide. Rice ceramides contain large amount of glucosylceramides (Fig. 16.5) and are thus expected to have an overall proliferative effect for keratinocytes.

We described in a previous section that topical and oral application of ceramides is effective against atopic dermatitis. Besides their moisturizing effect and barrier function, ceramides exhibit antiallergic and anti-inflammatory activities. We evaluated the effect of rice ceramide against itch in mice induced by compound 48/80 via mast cell degranulation histamine release in skin [20].

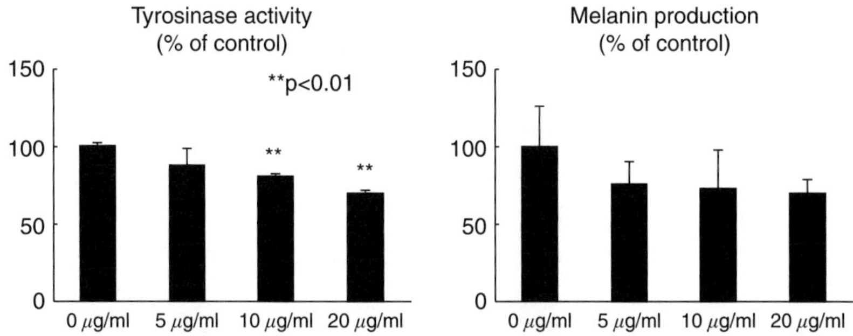

Figure 16.8 Effects of rice glucosphingolipid on tyrosinase activity (left) and melanin production (right) in melanoma. Each column represents mean with the SE. For determination of tyrosinase activity, mouse melan-a cells (1×10^4 cells/well) were incubated for 24 hours in RPMI 1640 medium containing 200 nM phorbor-12-myristate-13-acetate. The medium was replaced with new medium containing ceramide (glucosphingolipids: >95% of purity) and cultured for 2 days. Cells were lysed with PBS and the tyrosinase activity of lysate was determined using L-DOPA as a substrate. For evaluation of melanin production ability, melan-a cells (3×10^5 cells/well) were cultured with ceramide under the same culture condition described above. Cells were lysed in 1 N NaOH, and absorbance was measured at 405 nm.

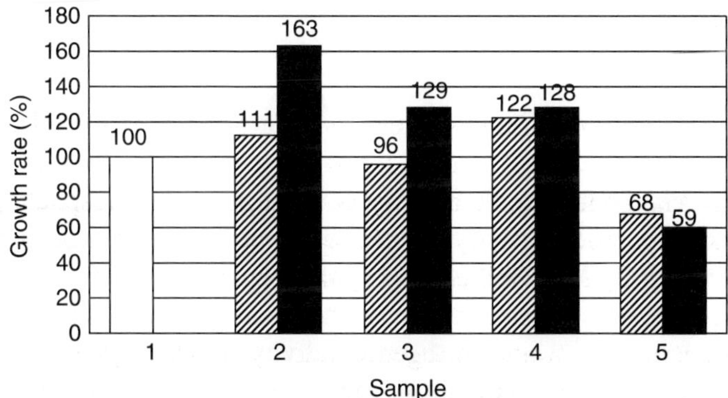

Figure 16.9 Proliferative effect of various glucosphingolipids on human dermal fibroblasts. (1) Control, (2) rice ceramide, (3) elephant foot ceramide, (4) corn ceramide, (5) wheat ceramide. All ceramides contain more than 95% glucosphingolipids. Hatched and solid column represent 100 μg/ml and 300 μg/ml, respectively. Human dermal fibroblasts (HS-K, 1×10^5 cells) were cultured with various ceramides for 72 hours. Cell growth was determined by MTT assay.

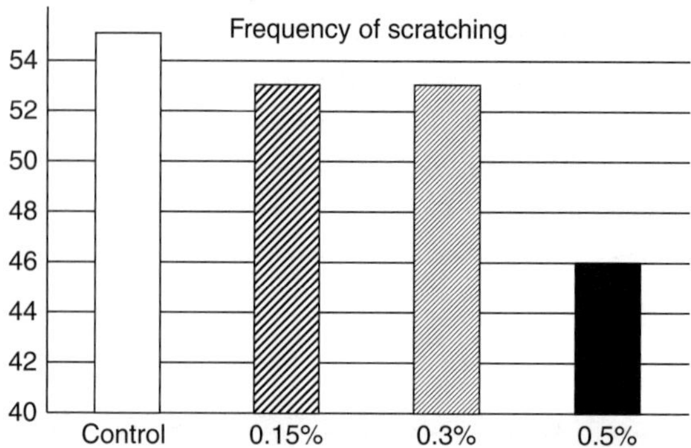

Figure 16.10 Effect of rice ceramide on scratching behavior in mice induced by compound 48/80. Mice (ddY strain, male) were fed with rice ceramide (0.15–0.5%) freely for 3 days. Compound 48/80 solution (3%) was injected into the cervical skin. The scratching action was counted for 30 minutes after they started to scratch.

Table 16.1 Inhibitory Effect of Ceramide on Degranulation from RBL-2H Cells

Origin	Inhibition (%)
Rice	87.3 ± 2.2
Wheat	82.2 ± 5.9
Elephant foot	70.8 ± 5.9
Corn	64.2 ± 5.4

Sample concentration: 1 mg/ml, mean ± S.E., n = 6.

In the model, continuous oral administration of rice ceramide decreased scratching action in mice in a dose-dependent manner (Fig. 16.10), suggesting an antiallergic effect of rice ceramide. We then examined the effect of plant-derived ceramides on degranulation in sensitized mast cells (RBL-2H3). All of the plant ceramides inhibited mast cell degranulation at 1 mg/ml, with rice ceramide being the most potent (Table 16.1). The inhibitory mechanism of ceramides on mast cell degranulation likely involves suppression of phosphorylation of ERK1/2 and p38 mitogen-activation of protein kinase [21]. In contrast, several glycosphingolipids are reported to be mast cell activators. Prieschl et al. [22] reported that galactosylsphingosine enhanced relocation of the tyrosine kinases such as Lyn and Syk, leading to tyrosine phosphorylation followed by mast cell activation. Phosphorylated ceramide (ceramide 1-phosphate) has been reported to play an important role in mast cell degranulation [23]. Masini et al. concluded that ceramide is a proinflammatory agent and that reducing ceramide levels is effective against allergic disease [24]. Principal ceramides contained in plants are glucosylsphingolipids, which are different from ceramides in the above studies. Plant ceramides may thus not act as inducers of allergic responses. However, further investigations are required concerning pro- and antiallergic activities of various ceramides.

16.3 Other Plant Extracts as Inhibitors for Skin Component Degradation

Strength and elasticity of skin are maintained by a balance of collagen, elastin, and hyaluronic acid [25,26]. Distributed in the entire dermis of the skin, collagen constitutes 90% of the dermis [27,28]. Hyaluronic acid is widely distributed in tissues such as skin, synovial fluid, vitreous body, and ligaments [29]. This skin tissue component is involved in cellular adhesion, in cell protection, and in maintenance of moisture and flexibility of the tissue. Skin loses moisture and tension, developing wrinkles and sagging as the level of hyaluronic acid decreases. Elastin is distributed in the dermis and is essential for maintaining appropriate elasticity and strength of the skin [30]. These skin constituents are degraded by collagenase, hyaluronidase, and elastase, respectively.

We have developed several plant extracts with inhibitory activities on these enzymes (Table 16.2). Litchi seed extract is extracted from crushed seed of *Litchi chinensis* with aqueous ethanol and contains saponins [31],

Table 16.2 Inhibitory Effects of Plant Extracts on the Enzyme Activities Related to Skin Degradation

	Specification	IC_{50} (µg/ml)		
		Collagenase	Hyaluronidase	Elastase
Litchi seed extract	Polyphenols: 24%	59	290	45
Purple rice extract	Polyphenols: 30% Anthocyanins: 10%	>1,000	>2,500	180
Grape extract	Polyphenols: 40% Resveratrol: 10%	130	150	5

tannins [31], flavonoid (leucocyanidin [32]), and anthocyanins (cyanidin 3-*O*-glucoside and malvidin 3-*O*-glucoside). The biological effect of litchi seed extract has not yet been well studied. Extract of seeds and pericarps of red grapes contain polyphenols including flavonoids, anthocyanidins, and resveratrol. Purple rice extract contains anthocyanins (cyaniding 3-*O*-glucoside and malvidin 3-*O*-glucoside) as the major constituents. Whereas the former two extracts inhibit all three skin-component-degrading enzymes [33–35], purple rice extract selectively inhibits elastase.

Litchi seed extract exhibited the highest collagenase inhibitory activity (IC_{50} = 59 μg/ml), which is also higher than that of persimmon leaf extract (IC_{50} < 100 μg/ml) [36] and comparable to that of procyanidins isolated from grape (*Vitis vinifera*) seeds (IC_{50} value = 38 μM) [37].

For hyaluronidase, the extract of seeds and pericarps of red grapes was the most potent inhibitor, with an IC_{50} of 150 μg/ml. Litchi seed extract also inhibits this enzyme with an IC_{50} of 290 μg/ml. For comparison, the IC_{50} of myrrha (oleoresin from the *Commiphora mukul* tree), a traditionally natural product used for treatment of arthritis, for inhibiting hyaluronidase is as high as 1,000 μg/ml [38]. Hederagenin and oleanolic acid, sapogenins isolated from horse chestnut (*Aesculus hippocastanum*), have been reported to inhibit hyaluronidase with IC_{50} values of 280.4 μM and 300.2 μM, respectively [39]. Similar types of saponins contained in litchi seeds are likely the principal compounds for the hyaluronidase inhibitory activity. In the case of grape extract, constitutive polyphenols including flavonoids, anthocyanidins, and resveratrol seem to account for the hyaluronic inhibitory activity because this activity has been reported for flavonoid [40].

All three extracts exhibited inhibitory activity for elastase, but litchi seed extract and grape extract are extremely potent (IC_{50} value: 45 μg/ml and 5 μg/ml). For comparison, the IC_{50} of elastase-inhibitory activity in the extract from black currant (*Ribes nigrum* L.) and lady's mantle (*Alchemilla vulgaris* L.) was 560 μg/ml and 160 μg/ml, respectively, as reported by Jonadet et al. after evaluating a number of plant extracts [41–43]. Hence, each of these three extracts is likely effective for elastin stabilization in skin.

16.4 Conclusion

Plant-derived ceramides consist mainly of glucosphingolipids, which are conjugated with glucose. In contrast, skin ceramides distributed in the stratum corneum are mainly sphingolipids. This difference in structures has been the concern for different biological functionalities of plant- and animal-derived and synthetic ceramides. We described the skin-health-promoting effect of rice ceramide in this chapter. In addition to the well-known barrier function and moisturizing effect, we found other novel activities in rice ceramide, such as inhibition of melanin synthesis, promotion of fibroblast proliferation, and anti-inflammatory and antiallergic effects. Supporting evidence for these skin-healthy effects of rice ceramide can also be obtained from a number of studies on various types of sphingolipids. Plant-derived ceramides can improve skin problems by topical and oral application. Further studies on plant ceramides are in progress that will provide more information supporting application of these skin-health-promoting components in development of new cosmetics. We also introduced several other plant extracts that inhibit collagenase, hyaluronidase, and elastase. These plant extracts are expected to be able to improve disturbed skin turnover and suppress excessive degradation of skin components. Finally, as extracts from common crops or cultivated plants, these natural materials are safe, ecological, and environment-friendly. We believe that these plant materials can be widely applied for various skin care products.

References

1. Strömberg N., Karlsson K.A. Characterization of the binding of propionibacterium granulosum to glycosphingolipids adsorbed on surfaces. An apparent recognition of lactose which is dependent on the ceramide structure. *J. Biol. Chem.* **265**, 11244–11250 (1990).
2. Robson K.J., Stewart M.E., Michelsen S., Lazo N.D., Downing D.T. 6-Hydroxy-4-sphingenine in human epidermal ceramides. *J. Lipid Res.* **35**, 2060–2068 (1994).
3. Haugen M., Williams J.B., Wertz P., Tieleman B.I. Lipids of the stratum corneum vary with cutaneous water loss among larks along a temperature-moisture gradient. *Physiol. Biochem. Zool.* **76**, 907–917 (2003).
4. Imokawa G., Abe A., Kumi J., Higaki Y., Kawashima M., Hidano A. Decreased level of ceramides in stratum corneum of atopic dermatitis: An etiologic factor in atopic dry skin? *J. Invest. Dermatol.* **96**, 523–526 (1991).
5. Koga J., Yamauchi T., Shimura M., Ogawa N., Oshima K., Umemura K., Kikuchi M., Ogasawara N. Cerebrosides A and C, sphingolipid elicitors of hypersensitive cell death and phytoalexin accumulation in rice plants. *J. Biol. Chem.* **273**, 31985–31991 (1998).
6. Schmelz E.M., Crall K.J., Larocque R., Dillehay D.L., Merrill A.H. Jr. Uptake and metabolism of sphingolipids in isolated intestinal loops of mice. *J. Nutr.* **124**, 702–712 (1994).

7. Duan R.D., Hertervig E., Nyberg L., Hauge T., Sternby B., Lillienau J., Farooqi A., Nilsson A. Distribution of alkaline sphingomyelinase activity in human beings and animals. Tissue and species differences. *Dig. Dis. Sci.* **41**, 1801–1806 (1996).
8. Kimata H. Improvement of atopic dermatitis and reduction of skin allergic responses by oral intake of konjac ceramide. *Pediatr. Dermatol.* **23**, 386–389 (2006).
9. Asai S., Miyachi H. Evaluation of skin-moisturizing effects of oral or percutaneous use of plant ceramides. *Rinsho Byori.* **55**, 209–215 (2007).
10. Takagi Y., Nakagawa H., Higuchi K., Imokawa G. Characterization of surfactant-induced skin damage through barrier recovery induced by pseudoacylceramides. *Dermatology.* **211**, 128–134 (2005).
11. Kang J.S., Youm J.K., Jeong S.K., Park B.D., Yoon W.K., Han M.H., Lee H., Han S.B., Lee K., Park S.K., Lee S.H., Yang K.H., Moon E.Y., Kim H.M. Topical application of a novel ceramide derivative, K6PC-9, inhibits dust mite extract-induced atopic dermatitis-like skin lesions in NC/Nga mice. *Int. Immunopharmacol.* **7**, 1589–1597 (2007).
12. Asano-Kato N., Fukagawa K., Takano Y., Kawakita T., Tsubota K., Fujishima H., Takahashi S. Treatment of atopic blepharitis by controlling eyelid skin water retention ability with ceramide gel application. *Br. J. Ophthalmol.* **87**, 362–363 (2003).
13. Chamlin S.L., Kao J., Frieden I.J., Sheu M.Y., Fowler A.J., Fluhr J.W., Williams M.L., Elias P.M. Ceramide-dominant barrier repair lipids alleviate childhood atopic dermatitis: changes in barrier function provide a sensitive indicator of disease activity. *J. Am. Acad. Dermatol.* **47**, 198–208 (2002).
14. Kim D.S., Kim S.Y., Moon S.J., Chung J.H., Kim K.H., Cho K.H., Park K.C. Ceramide inhibits cell proliferation through Akt/PKB inactivation and decreases melanin synthesis in Mel-Ab cells. *Pigment Cell Res.* **14**, 110–115 (2001).
15. Kim D.S., Kim S.Y., Chung J.H., Kim K.H., Eun H.C., Park K.C. Delayed ERK activation by ceramide reduces melanin synthesis in human melanocytes. *Cell. Signal.* **14**, 779–785 (2002).
16. van Echten-Deckert G., Schick A., Heinemann T., Schnieders B. Phosphorylated *cis*-4-methylsphingosine mimics the mitogenic effect of sphingosine-1-phosphate in Swiss 3T3 fibroblasts. *J. Biol. Chem.* **273**, 23585–23589 (1998).
17. Hauser J.M., Buehrer B.M., Bell R.M. Role of ceramide in mitogenesis induced by exogenous sphingoid bases. *J. Biol. Chem.* **269**, 6803–6809 (1994).
18. Jung E.M., Griner R.D., Mann-Blakeney R., Bollag W.B. A potential role for ceramide in the regulation of mouse epidermal keratinocyte proliferation and differentiation. *J. Invest. Dermatol.* **110**, 318–323 (1998).
19. Marchell N.L., Uchida Y., Brown B.E., Elias P.M., Holleran W.M. Glucosylceramides stimulate mitogenesis in aged murine epidermis. *J. Invest. Dermatol.* **110**, 383–387 (1998).
20. Kuraishi Y., Nagasawa T., Hayashi K., Satoh M. Scratching behavior induced by pruritogenic but not algesiogenic agents in mice. *Eur. J. Pharmacol.* **275**, 229–233 (1995).

21. Kitatani K., Akiba S., Hayama M., Sato T. Ceramide accelerates dephosphorylation of extracellular signal-regulated kinase 1/2 to decrease prostaglandin D(2) production in RBL-2H3 cells. *Arch. Biochem. Biophys.* **395**, 208–214 (2001).
22. Prieschl E.E., Csonga R., Novotny V., Kikuchi G.E., Baumruker T. Glycosphingolipid-induced relocation of Lyn and Syk into detergent-resistant membranes results in mast cell activation. *J. Immunol.* **164**, 5389–5397 (2000).
23. Kim J.W., Inagaki Y., Mitsutake S., Maezawa N., Katsumura S., Ryu Y.W., Park C.S., Taniguchi M., Igarashi Y. Suppression of mast cell degranulation by a novel ceramide kinase inhibitor, the F-12509A olefin isomer K1. *Biochim. Biophys. Acta.* **1738**, 82–90 (2005).
24. Masini E., Giannini L., Nistri S., Cinci L., Mastroianni R., Xu W., Comhair S.A., Li D., Cuzzocrea S., Matuschak G.M., Salvemini D. Ceramide: a key signaling molecule in a guinea pig model of allergic asthmatic response and airway inflammation. *J. Pharmacol. Exp. Ther.* **324**, 548–557 (2008).
25. Labat-Robert J., Bihari-Varga M., Robert L. Extracellular matrix. *FEBS Lett.* **268**, 386–393 (1990).
26. Baumann L. Skin ageing and its treatment. *J. Pathol.* **211**, 241–251 (2007).
27. Garrone R., Lethias C., Le Guellec D. Distribution of minor collagens during skin development. *Microsc. Res. Tech.* **38**, 407–412 (1997).
28. Smith L.T., Holbrook K.A., Madri J.A. Collagen types I, III, and V in human embryonic and fetal skin. *Am. J. Anat.* **175**, 507–521 (1986).
29. Price R.D., Myers S., Leigh I.M., Navsaria H.A. The role of hyaluronic acid in wound healing: assessment of clinical evidence. *Am. J. Clin. Dermatol.* **6**, 393–402 (2005).
30. Kielty C.M., Sherratt M.J., Shuttleworth C.A. Elastic fibres. *J. Cell Sci.* **115**, 2817–2828 (2002).
31. Edited by Shanghai Scientific and Technical Publishers Li zhi he, Dictionary of Chinese Material Medica, **4**, pp. 2730 (1998). Published by Shogakukan (in Japanese).
32. Turnbull J.J., Nagle M.J., Seibel J.F., Welford R.W.D., Grant G.H., Schofiel C.J. The C-4 stereochemistry of leucocyanidin substrates for anthocyanidin synthase affects product selectivity. *Bioorg. Med. Chem. Lett.* **13**, 3853–3857 (2003).
33. Pilcher B.K., Sudbeck B.D., Dumin J.A., Welgus H.G., Parks W.C. Collagenase-1 and collagen in epidermal repair. *Arch. Dermatol. Res.* **290**(Suppl.1), S37–S46 (1998).
34. Maytin E.V., Chung H.H., Seetharaman V.M. Hyaluronan participates in the epidermal response to disruption of the permeability barrier *in vivo*. *Am. J. Pathol.* **165**, 1331–1341 (2004).
35. Weitzman I., Summerbell R.C. The dermatophytes. *Clin. Microbiol. Rev.* **8**, 240–259 (1995).
36. An B.J., Kwak J.H., Park J.M., Lee J.Y., Park T.S., Lee J.T., Son J.H., Jo C., Byun M.W. Inhibition of enzyme activities and the antiwrinkle effect of polyphenol isolated from the persimmon leaf (*Diospyros kaki folium*) on human skin. *Dermatol. Surg.* **31**, 848–854 (2005).
37. Facino R.M., Carini M., Aldini G., Bombardelli E., Morazzoni P., Morelli R. Free radicals scavenging action and anti-enzyme activities of procyanidines

from *Vitis vinifera*. A mechanism for their capillary protective action. *Arzneimittelforschung*. **44**, 592–601 (1994).

38. Sumantran V.N., Kulkarni A.A., Harsulkar A., Wele A., Koppikar S.J., Chandwaskar R., Gaire V., Dalvi M., Wagh U.V. Hyaluronidase and collagenase inhibitory activities of the herbal formulation *Triphala guggulu*. *J. Biosci.* **32**, 755–761 (2007).
39. Facino R.M., Carini M., Stefani R., Aldini G., Saibene L. Anti-elastase and anti-hyaluronidase activities of saponins and sapogenins from *Hedera helix*, *Aesculus hippocastanum*, and *Ruscus aculeatus*: factors contributing to their efficacy in the treatment of venous insufficiency. *Arch. Pharm.* (*Weinheim*). **328**, 720–724 (1995).
40. Kuppusamy U.R., Das N.P. Inhibitory effects of flavonoids on several venom hyaluronidases. *Experientia*. **47**, 1196–1200 (1991).
41. Jonadet M., Meunier M.T., Bastide J., Bastide P. Anthocyanosides extracted from *Vitis vinifera*, *Vaccinium myrtillus* and *Pinus maritimus*. I. Elastase-inhibiting activities *in vitro*. II. Compared angioprotective activities *in vivo*. *J. Pharm. Belg*. **38**, 41–46 (1983).
42. Jonadet M., Meunier M.T., Villie F., Bastide J.P., Lamaison J.L. Flavonoids extracted from *Ribes nigrum* L. and *Alchemilla vulgaris* L.: 1. *In vitro* inhibitory activities on elastase, trypsin and chymotrypsin. 2. Angioprotective activities compared *in vivo*. *J. Pharmacol.* **17**, 21–27 (1986).
43. Jonadet M., Bastide J., Bastide P., Boyer B., Carnat A.P., Lamaison J.L. *In vitro* enzyme inhibitory and *in vivo* cardioprotective activities of hibiscus (*Hibiscus sabdariffa* L.). *J. Pharm. Belg*. **45**, 120–124 (1990).

17

Asiaticoside Supports Collagen Production for Firmer Skin

Jongsung Lee[1,2], Saebom Kim[1], PhD, Eunsun Jung[1], Juhyeon Lee[2], Su Na Kim[2], Deokhoon Park[1], PhD, and Yeong Shik Kim[2], PhD

[1]*Biospectrum Life Science Institute, Gunpo City, Gyunggi Do, Republic of Korea*
[2]*Natural Products Research Institute, College of Pharmacy, Seoul National University, Seoul, Korea*

Aaron Tabor and Robert M. Blair (eds.), *Nutritional Cosmetics: Beauty from Within*, 335–352,

17.1 Chemical and Pharmacological Properties of Asiaticoside

17.1.1 Introduction

The Araliaceae *Centella asiatica*, which has many common names including gotu kola, hydrocotyle, Indian pennywort, marsh penny, and white rot, is widely cultivated as a vegetable or spice in China, Southeast Asia, India, Sri Lanka, Africa, and Oceanic countries. In Sri Lankan and Indian Ayurvedic traditional medicine, the aerial parts of *C. asiatica* have been used to treat skin diseases, syphilis, rheumatism, mental illness, epilepsy, hysteria, dehydration, and leprosy [1].

17.1.2 Chemistry of Asiaticoside

Analytical studies have shown that *C. asiatica* contains triterpenoids, essential oils, amino acids, and other compounds, such as vellarin. The terpenoids include asiaticoside, centelloside, madecassoside, brahmoside, brahminoside, thankuniside, and centellose, and asiatic, brahmic, centellic, and madecassic acids. The most active chemical constituents in *C. asiatica* are asiatic acid derivatives such as asiatic acid, madecassic acid, asiaticoside, and madecassosides (Fig. 17.1) [1]. Among these terpenoids, asiaticoside is a major ingredient of Madecassol®. Asiaticoside is an oleanane-type triterpene glycoside and classified as an antibiotic. Asiaticoside is a fine white and odorless powder. Its melting point is

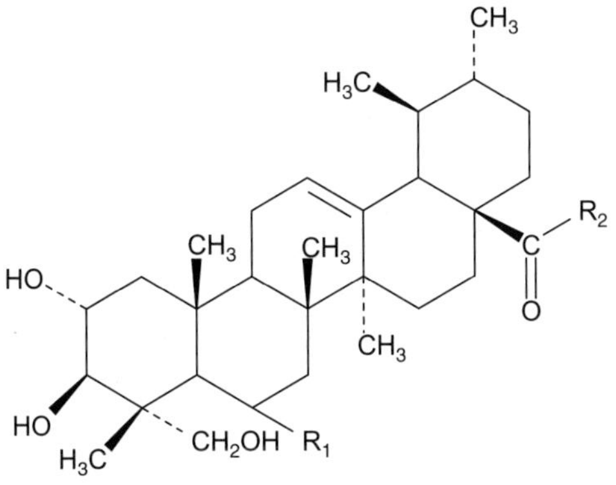

Figure 17.1 Chemical structure of pentacyclic Asiatic acid derivates triterpenes in *C. asiatica* extracts.
R_1 = H, R_2 = OH: asiatic acid.
R_1 = O, R_2 = OH: madecassic acid.
R_1 = H, R_2 = O-glucose-glucose-rhamnose: asiaticoside.
R_1 = OH, R_2 = O-glucose-glucose-rhamnose: madecassoside.

230–232°C. It is easily solubilized in many hydrophilic solvents (propylene glycol, glycerin, ethoxydiglycol) and is commonly used in the manufacture of topical formulations (gels, emulsions). If an emulsion is prepared, asiaticoside is incorporated in the hydrophilic portion before emulsification with the lipophilic phase. According to the safety data, asiaticoside produces no side effects such as localized erythema or any kind of discomfort. In all trials conducted to date, asiaticoside is tolerated extremely well.

17.1.3 Pharmacology of Asiaticoside (Fig.17.2)

17.1.3.1 Antibiotic Character of Asiaticoside

Traditionally, asiaticoside has been used in the treatment of leprosy and tuberculosis in the Far East. Evidence from many studies shows that asiaticoside damages the cell walls of the bacteria that cause leprosy and the weakened bacteria are then easily eliminated by the body's immune system [2].

17.1.3.2 Anti-Herper Simplex Virus Activities of Asiaticoside

A number of Thai medicinal plants that are recommended as remedies for herpes virus infection and used in primary health care were investigated for

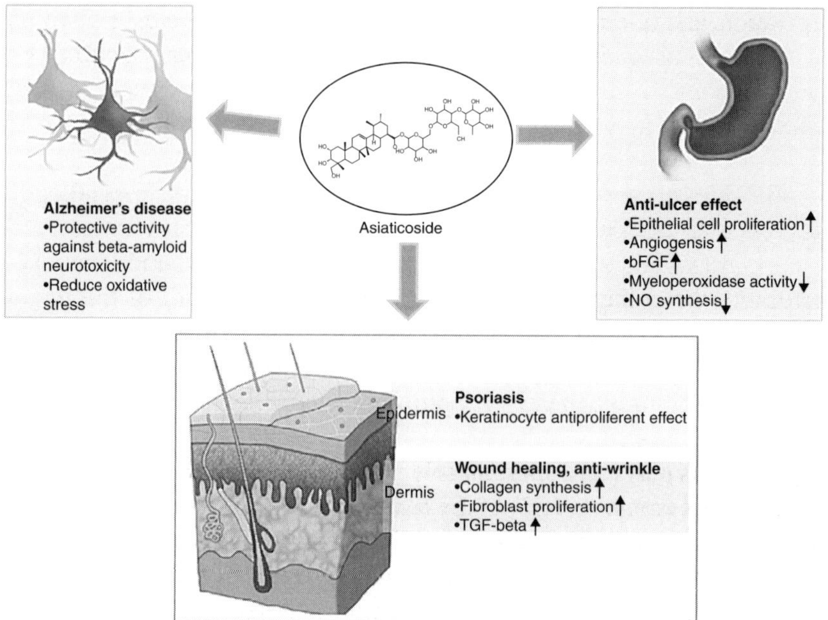

Figure 17.2 Various effects of asiaticoside.

their intracellular activities against herpes simplex viruses (HSV). According to plaque inhibition assays, *C. asiatica* L., *Maclura cochinchinesis Cornor,* and *Mangifera indica* L. have both anti-HSV-1 and -2 activities. Inhibition of the production of infectious HSV-2 virions from infected Vero cells has also been demonstrated.

Combinations of each of these reconstituted extracts with 9-(2-hydroxyethoxymethyl) guanosine (acyclovir; ACV) resulted either in subadditive, additive, or synergistic interactions against HSV-2, depending on the dose of ACV used. Furthermore, the inhibitory effects of these plant extracts were also substantiated by flow cytometric analysis of virus-specific antigens in the infected cells. The active constituent present in *C. asiatica* extract was found to be asiaticoside, whereas that in *M. indica* was mangiferin. Thus, asiaticoside contributes to treatment of HSV infection [3].

17.1.3.3 Wound-Healing Activity of Asiaticoside

Wound healing is a complex biological process that includes inflammation, cell migration, angiogenesis, extracellular matrix synthesis, collagen deposition, and re-epithelialization. However, this process can be disrupted by disease, consequently resulting in chronic wounds or other troublesome complications. *C. asiatica* has been used for hundreds of years in many Asian countries as a traditional medicine to improve wound healing.

Asiaticoside has shown promising wound-healing activity. Topical application of asiaticoside in normal and diabetic animals or oral administration in normal animals significantly enhanced the rate of wound healing by increasing the collagen synthesis and the tensile strength of the wound tissue. Asiaticoside has also been reported to elevate collagen synthesis in fibroblasts [4]. *In vitro* histological findings also revealed enhanced proliferation of fibroblasts, thereby supporting the biochemical results.

Angiogenesis plays an important role in wound healing, and newly formed blood vessels comprise 60% of the repair tissue. Neovascularization helps hypoxic wounds attain normoxic conditions. Asiaticoside promotes angiogenesis in both *in vitro* and *in vivo* models as indicated by histological studies [5]. Asiaticoside also accelerates the healing of gastric kissing ulcers. Gastriculcer healing is a complex orchestrated process involving resolution of inflammation and repair of gastric tissues through granulation tissue formation, re-epithelialization, and extracellular matrix remodeling. Asiaticoside suppresses myeloperoxidase (MPO), which is a marker of neutrophil infiltration during inflammation. The suppressive effects of asiaticoside on MPO activity may result from direct inhibition of the expression of inflammatory cytokines, thereby decreasing neutrophil infiltration in the wound site [6]. Asiaticoside also affects the healing mechanism. Asiaticoside is known to stimulate production of antioxidants in the wound site and provides a favorable environment for tissue healing. Topical application (0.2% solution) twice daily for 7 days to skin wounds in rats resulted in an increase in both enzymatic and nonenzymatic antioxidants, namely superoxide dismutase (SOD) (35%), catalase (67%), glutathione peroxidase (49%), vitamin E (77%), and ascorbic acid (36%) in newly formed tissues. SOD is an intracellular enzyme that catalyzes the conversion of the superoxide anion to molecular oxygen and hydrogen peroxide; glutathione peroxidase converts various peroxides and is mainly present in the cytosol and mitochondria of most tissues. This enhancement of antioxidant levels at the initial stage of healing may contribute to the healing properties of *C. asiatica* [7]. Asiaticoside also affects glycosaminoglycan synthesis in wound tissues. It is well established that glycosaminoglycans, especially hyaluronic acids and small proteoglycans, play a major role in the healing process and contribute to the organization and strength of the fibrillar network of the wounds [8,9]. Therefore, asiaticoside should be developed as a potential antiulcer drug.

Traditionally, *C. asiatica* extracts have been used both topically and internally, mainly for wound healing and treating leprosy [10]. Oral administration of *C. asiatica* or asiaticoside and potassium chloride capsules was reported to be as effective as dapsone therapy in patients with leprosy. In a controlled study of 90 patients with perforated leg lesions owing to leprosy, application of a salve of the plant produced significantly better results than a placebo [11].

17.1.3.4 Anxiolytic Properties of Asiaticoside

C. asiatica has been used for centuries in Ayurvedic medicine and traditional Chinese medicine to alleviate the symptoms of depression and anxiety and to promote a deep state of relaxation and mental calmness during meditation practices. As a dietary supplement, *C. asiatica* is used to treat sleep disorders. The putative anxiolytic activity of asiaticoside was examined in male mice using a number of experimental paradigms of anxiety, with diazepam as a positive anxiolytic control. In the elevated plus-maze test, diazepam (1 and 2 mg/kg) or asiaticoside (5 or 10 mg/kg) increased the percentage of entries into open arms and of time spent on open arms. In the light/dark test, with 1 mg/kg diazepam, asiaticoside (10 and 20 mg/kg) increased the time spent and the movement in the light areas without altering the total locomotor activity of the animals. Asiaticoside interacts with the gamma-aminobutyric acid (GABA) receptor. In the CNS, GABA exerts an inhibitory effect on stress-induced ulcerogenesis. And it is well known that the GABA receptors mediate the anxiolytic effect of benzodiazepines. Thus, recent investigations using animal models of anxiety confirm that *C. asiatica* possesses anxiolytic activity [12].

17.1.3.5 The Antiwrinkle Activity of Asiaticoside

Asiaticoside is used as an ingredient in cosmetic applications. Cosmetic applications include the formulations of lipids and proteins for healthy skin, anticellulitis and antiwrinkle (eyes and facial) properties, skin regeneration, treatment of acne-induced blemishes, rapid renewal of supporting fiber networks, stimulation of collagen synthesis, and enhancement of immune-depressed skin. The assessment of skin wrinkling is essential for evaluating the efficacy of cosmetic products believed to reduce age-related skin changes. After using lipstick containing asiaticoside for 8 weeks, the change in visual grading scores and the replica analyses indicate the wrinkle-improving effect. As the depth and number of wrinkles are reduced, the lipstick appearance also improves significantly by image analysis [13].

17.2 Asiaticoside Induces Human Collagen I Synthesis through TGFβ Receptor I Kinase (TβRI Kinase)-Independent Smad Signaling

17.2.1 Summary

Skin aging appears to be principally related to a decrease in the levels of type I collagen, the primary component of the skin dermis. Asiaticoside, a saponin component that has been isolated from *C. asiatica*, induces type I collagen synthesis in human dermal fibroblast cells. However, the mechanism underlying asiaticoside-induced type I collagen synthesis, especially at a molecular level, is only partially understood. In this study, we have attempted to characterize the mechanism of action of asiaticoside in type I collagen synthesis. We found that asiaticoside induced the phosphorylation of both Smad2 and Smad3. In addition, we detected asiaticoside-induced binding of Smad3 and Smad4. Consistent with these results, nuclear translocation of the Smad3–Smad4 complex was induced by treatment with asiaticoside, pointing to the involvement of asiaticoside in Smad signaling. In addition, SB431542, an inhibitor of TGFβ receptor I (TβRI) kinase, which is an activator of the Smad pathway, did not inhibit either Smad2 phosphorylation or type 1 collagen synthesis induced by asiaticoside. Therefore, our results show that asiaticoside can induce type I collagen synthesis via activation of a TβRI kinase–independent Smad pathway.

17.2.2 Introduction

Aging of the skin is primarily related to reductions in the levels of type I collagen, which is the principal component of skin dermis. Type I collagen is the main structural component of the extracellular matrix, which performs a pivotal function in the maintenance of the structure of the skin dermis. Several molecules have been reported to augment type I collagen synthesis, namely, transforming growth factor-β (TGF-β) and sphingosine 1-phosphate [14,15]. Recently, asiaticoside, which was isolated from

C. asiatica, has been reported to augment the expression of the type I collagen gene [16].

Both the quantity and quality of extracellular collagen are determined by the balance between degradation and synthesis [17]. Degradation appears to be mediated by matrix metalloproteinases and by endogenous tissue inhibitors. Collagen synthesis is both transcriptionally and post-translationally regulated. Several studies have reported that the Smad pathway functions in the activation of type I collagen gene expression. The Smads are a series of proteins that function downstream from the serine/threonine kinase receptors of the TGF-β family, thereby transducing signals to the nucleus. Following binding of TGF-β to its receptors, the receptor-regulated Smads (R-Smads), Smad2 and Smad3, are phosphorylated by the type I receptor and associate with the common partner, Smad4. The resulting heteromultimer then translocates to the nucleus, where it functions as a regulator of type I collagen gene expression [18–20].

Asiaticoside is a saponin component that has been isolated from *C. asiatica*, a plant that has been used for hundreds of years in the traditional medicine of many Asiatic countries, usually to improve wound healing. Recently, several studies have reported that asiaticoside enhances the synthesis of collagen [21,22]. However, the mechanism by which asiaticoside promotes collagen synthesis, particularly at the molecular level, remains somewhat unclear.

In this report, we have demonstrated that treatment with asiaticoside induces the synthesis of type I collagen, and the mechanisms underlying its action may be mediated via a TGFβ receptor I kinase (TβRI kinase)–independent Smad activation pathway in cultured human dermal fibroblast cells.

17.2.3 Asiaticoside Induces Significant Type I Collagen Synthesis

The chemical structure of asiaticoside, one of the primary triterpenic compounds that have been isolated from *C. asiatica*, contains five six-carbon rings. Several reports have shown that asiaticoside induces the synthesis of type I collagen [21,22]. In order to verify this finding, we conducted type I pro-collagen synthesis assays. Our type I pro-collagen test revealed that asiaticoside induces significant type I collagen synthesis (Fig. 17.3). TGF-β (10 ng/ml) was employed as a positive control.

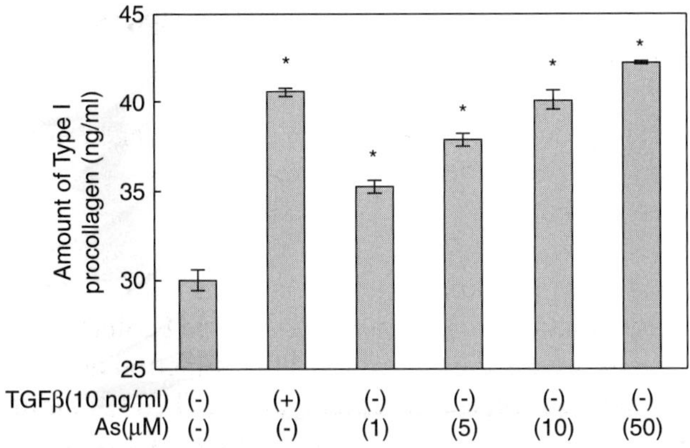

Figure 17.3 Effects of asiaticoside on type I pro-collagen synthesis, as determined with a sandwich immunoassay kit (Takara Bio, Inc., Japan). Data are expressed as the mean ± S.D., $^*P < 0.01$ compared with controls. The results were verified by the repetition of four experiments, each in triplicate. As: asiaticoside.

17.2.4 Asiaticoside Induces Human Collagen I Synthesis through TGFβ Receptor I Kinase (TβRI Kinase)–Independent Smad Signaling

The Smad signaling cascade performs an important function in human collagen production events associated with TGF-β or sphingosine 1-phosphate (S1P) [14,15]. Therefore, as an initial step toward elucidating the mechanism of asiaticoside action in type I collagen synthesis, we examined its effects on phosphorylation of both Smad2 and Smad3. As shown in Fig. 17.4, Smad2 and Smad3 phosphorylation could be induced by treatment with asiaticoside (10 μM). Human dermal fibroblast cells responded to asiaticoside after 10 minutes, and this response persisted for a total of 45 minutes. TGF-β was employed as a positive control.

In order to further substantiate the particular role of asiaticoside, we also assessed the downstream events that occurred after Smad2/3 phosphorylation. The immunoprecipitation of cell lysates with anti-Smad3 antibodies, followed by Western blotting for Smad3 or Smad4, revealed that Smad3 expression levels were not influenced by TGF-β or asiaticoside treatment (data not shown). As expected, no Smad4 co-immunoprecipitated under control conditions. However, Smad3–Smad4 complexes were detected after stimulation with TGF-β and also after exposure to asiaticoside.

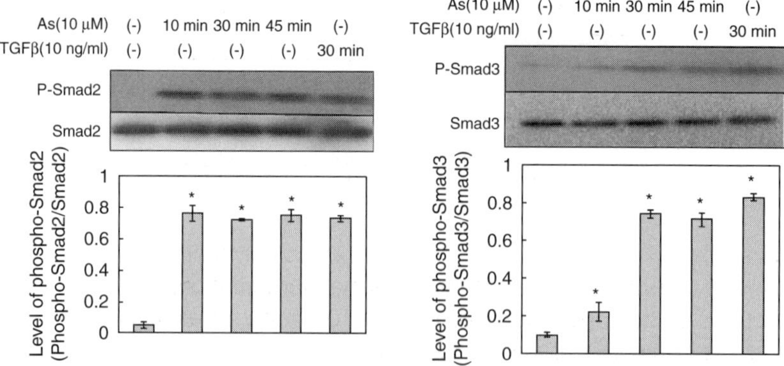

Figure 17.4 Effects of asiaticoside on the phosphorylation of Smad2 (A) and Smad3 (B) in human dermal fibroblast cells. Cells were treated with TGF-β (10 ng/ml) or asiaticoside (10 μM) at different time periods (A and B). The cells were subjected to immunoprecipitation with anti-Smad2 or anti-Smad3 antibodies, followed by Western blot with antiphosphoserine antibodies. As: asiaticoside, $^{*}P < 0.01$ compared with untreated control.

In order to confirm the nuclear translocation of the Smad3–Smad4 complex, as was reported with TGF-β, we performed Western blotting using nuclear extracts. In this experiment, the measurement of marker enzymes localized specifically in the cytosol (lactate dehydrogenase) or the endoplasmic reticulum (α-glucosidase II) allowed us to confirm the purity of these nuclei. Western blotting of Smad3 and Smad4 in the lysates of nuclei revealed the absence of both Smad proteins under control conditions. However, when cells were stimulated for 30 minutes either by TGF-β or asiaticoside, both Smad3 and Smad4 were found in the nuclear fraction (Fig. 17.5), thereby indicating their translocation from the cytosol.

As mentioned previously, although asiaticoside induced Smad2/3 phosphorylation, the detailed mechanism underlying asiaticoside-induced Smad2/3 phosphorylation remains somewhat unclear. Therefore, we characterized the relationship between asiaticoside and TGF-β signaling using SB431542, an inhibitor of the TGF-β type I receptor (TβRI) kinase, which phosphorylates Smad2/3. We examined the effects of SB431542 on the asiaticoside-induced phosphorylation of Smad2 and type I collagen synthesis. As shown in Figs. 17.5 and 17.6, the asiaticoside-induced

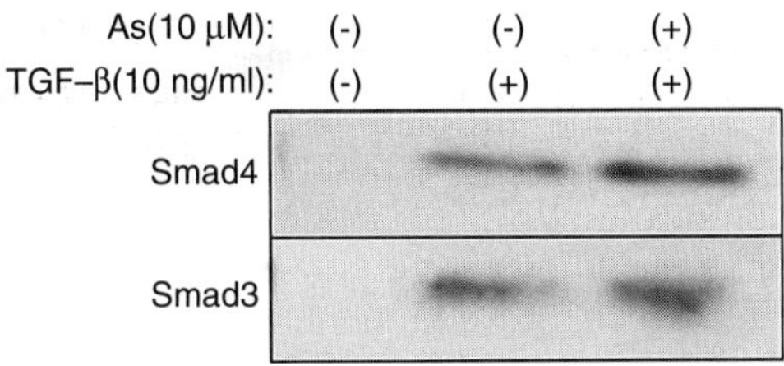

Figure 17.5 Asiaticoside (As) induced the nuclear translocation of Smad3–Smad4 complex. Human dermal fibroblast cells were stimulated using control vehicle, TGF-β (10 ng/ml), or asiaticoside (10 μM) for 45 min. Lysates of the nuclei were immunoprecipitated and immunoblotted.

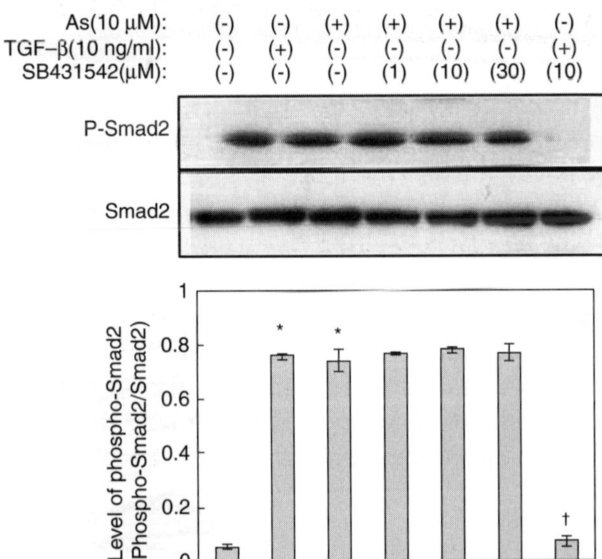

Figure 17.6 Asiaticoside-induced Smad2 phosphorylation was not mediated by TGFβ receptor I (TβRI) kinase. Human dermal fibroblast cells were incubated in either the presence or absence of SB431542, along with asiaticoside (10 μM) or TGF-β (10 ng/ml) for 30 min. The cells were subjected to immunoprecipitation and Western blotting. *$P < 0.01$ compared with untreated control, †$P < 0.01$ versus TGF β treatment. The results were verified by the repetition of four experiments, each in triplicate. As: asiaticoside.

phosphorylaton of Smad2 and type I collagen synthesis were not reduced by administration of SB431542, whereas SB431542 blocked both the phosphorylation of Smad2 and type I collagen synthesis via TGF-β.

17.2.5 Conclusion and Discussion

To the best of our knowledge, this study is the first to attempt to elucidate the detailed mechanisms underlying asiaticoside-induced type I collagen synthesis. We have discovered that asiaticoside induces the synthesis of type I collagen via TGF-β-independent Smad signaling.

Asiaticoside, one of the primary compounds in *C. asiatica*, has recently been reported to increase the formation of extracellular matrix synthesis, including that associated with human 1(I) and 3(III) collagens [22]. We have also demonstrated that asiaticoside induces significant type I collagen synthesis in human dermal fibroblast cells (Fig. 17.3). However, the precise mechanisms underlying asiaticoside-induced collagen synthesis remain somewhat unclear at the molecular level. Therefore, we have attempted to elucidate the molecular changes induced by asiaticoside in human dermal fibroblast cells. We first determined whether asiaticoside can induce both Smad2 and Smad3 phosphorylation, initial molecular events in Smad signaling, as it is well known that Smad signaling is involved in collagen synthesis induced by either TGF-β or sphingosine 1-phosphate. In this study, we determined that Smad2 and Smad3 were both phosphorylated by treatment with asiaticoside. In addition, after asiaticoside treatment, we observed interactions between Smad3 and Smad4, suggesting that asiaticoside is involved in Smad signaling and that it operates upstream of Smad2 and Smad3. This is further strengthened by the fact that the Smad3–Smad4 complex was translocated into the nucleus in response to the administration of asiaticoside.

As previously mentioned, we successfully demonstrated that asiaticoside operates upstream of Smad2 and Smad3. However, the relationship between asiaticoside and the signaling molecules upstream of Smad2/3 must be established in further detail. TGFβ receptor I kinase (TβRI kinase) can phosphorylate Smad2/3 in the process of TGF-β signaling. We thus tested whether asiaticoside-induced Smad signaling is mediated in a TβRI kinase–dependent manner. To that purpose, SB431542, a TβRI kinase inhibitor, was

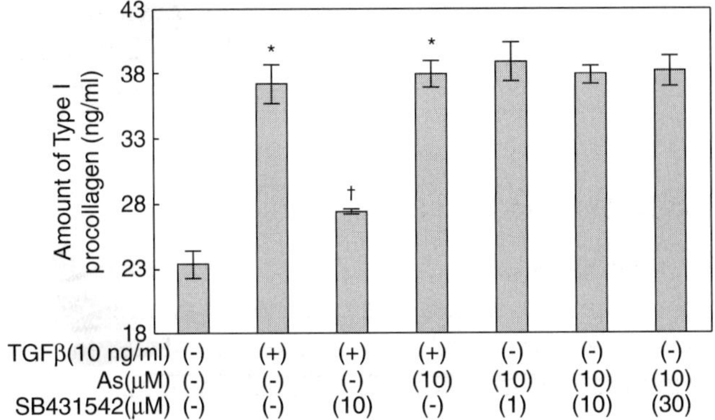

Figure 17.7 Effects of SB431542 on asiaticoside-induced type I pro-collagen synthesis, as determined with a sandwich immunoassay kit (Takara Bio, Inc., Japan). Data are expressed as the mean ± S.D., **P* < 0.01 compared with untreated control, †*P* < 0.01 versus TGF β treatment. The results were verified by the repetition of four experiments, each in triplicate. As: asiaticoside.

introduced during asiaticoside-induced Smad2 phosphorylation and type I collagen synthesis events. As shown in Figs. 17.6 and 17.7, SB431542 did not inhibit asiaticoside-induced Smad2 phosphorylation and type I collagen synthesis. This indicates that asiaticoside does, indeed, induce type I collagen synthesis through the activation of Smad signaling in a TβRI kinase–independent manner. Recently, signals derived from growth factor receptors that exhibited tyrosine kinase activity were also found to modulate Smad-dependent effects. This may occur as the result of the activation of a kinase located downstream of MEK-1 and upstream of MAPK/ERK kinase, resulting in the phosphorylation of Smad2 [23]. In addition, numerous other kinases have been implicated in Smad signaling, including TAK-1 and TAB, although their precise functions have yet to be elucidated [24–26]. When taken together, these data support the conclusion that asiaticoside-induced Smad signaling is mediated by other kinases that play a similar role as TβRI kinase.

In conclusion, the data acquired in this study demonstrate that asiaticoside can induce the synthesis of type I collagen and that the mechanisms underlying its action are mediated via a TβRI kinase–independent Smad activation pathway.

17.3 Effect of a Preparation Containing Asiaticoside on Periocular Wrinkles of Human Volunteers

17.3.1 Summary

Background: Skin aging is accompanied by wrinkle formation. In particular, periorbital wrinkle formation is a relatively early sign of skin aging.

Purpose: We evaluated the effect of a preparation that contained asiaticoside on periorbital wrinkles in a team of volunteers.

Method: The efficiency of a preparation containing asiaticoside as an active compound in a base cream for the treatment of temporal perioribital wrinkles was tested on 27 female volunteers as follows: the women applied the cream twice a day for 12 weeks. Negative replicas were taken of the periorbital skin before and after 4 weeks, 8 weeks, and 12 weeks of cream application. The results were evaluated by semiautomated morphometry of the plastic replicas.

Results: After 12 weeks of treatment, there was significant improvement in the periorbital wrinkles for the majority of the volunteers who tested the cream. Of the 27 periocular wrinkles examined, 65% showed an improvement at the end of the treatment. In two cases, the improvement was 100% on one eye (disappearance of the crow's feet) and 75% and 79% on the other eye. On six eyes, there was no significant change after the end of treatment.

Conclusion: After 12 weeks of treatment with the asiaticoside-containing cream, most volunteers experienced attenuation of their periorbital wrinkles, and some women had a significant improvement of periorbital wrinkles on one of their eyes.

17.3.2 Introduction

Periorbital wrinkle formation is a relatively early sign of skin aging that usually makes women strongly apprehensive. Very few cosmetic preparations have been shown to improve this situation according to objective quantitative methods.

Aging of the skin is primarily related to reductions in the levels of type I collagen, which is the principal component of skin dermis. Type I collagen is the main structural component of the extracellular matrix, which performs a pivotal function in the maintenance of the structure of the skin dermis. In our study, we found that asiaticoside, a saponin that can be isolated from *C. asiatica*, increased the expression of the type I collagen gene through the TGFβ receptor I kinase (TβRI kinase)–independent Smad activation pathway [27].

Based on *in vitro* studies, it has been postulated that asiaticoside could be used topically to prevent and correct skin aging. Considering these effects of asiaticoside, we experimentally evaluated the impact of asiaticoside on the evolution of periorbital wrinkles.

17.3.3 Antiwrinkle Effect of Asiaticoside

Two volunteers dropped out just after the beginning of the trial. The remaining 27 volunteers were evaluated at 4 weeks, 8 weeks, and 12 weeks. Clinical assessment by the dermatologist and self-assessment by the volunteers disclosed an improvement in terms of the "global score." Global score consisted of the sum of six items (hydration, roughness, laxity, suppleness, fine wrinkles, coarse wrinkles). Statistical evaluation of the values from the treated side versus the placebo was performed using a rank test on paired series. This score evolved in a significant manner from 6.7 ± 1.6 to 5.0 ± 1.0, 4.5 ± 1.2, and 4.2 ± 0.7, at baseline, 4 weeks, 8 weeks, and 12 weeks, respectively ($P < 0.05$). According to the assessment by the dermatologist, variance analysis showed a statistically significant improvement in hydration, small wrinkles, wrinkles, glare, and brown spots in each group throughout the trial. Furthermore, roughness, suppleness, and small wrinkle scores improved significantly in the asiaticoside group. Self-assessment by the volunteers disclosed a statistically significant and beneficial evolution of most of these clinical items (hydration, firmness, glare, tonicity, suppleness, brown spots, roughness imperfections, smoothing, and dryness) on the treated area during the 3-month duration of the trial.

The analyses of the skin replicas showed that, compared with placebo, there was a highly significant increase in the density of skin microrelief ($P < 0.05$), as well as a decrease of deep furrows at the region of interest treated with asiaticoside cream over a 3-month period ($P < 0.05$). Replicas of the crow's feet area were measured by image analyzer with no intervening changes in instrumental set-up. The representative data for each group are presented as Ra, Rz, and Rt. Ra represents the arithmetic average roughness, that is, the area surrounded by the profile and the middle line divided by the middle line. Rz is the average roughness, that is, the arithmetic average of the different segment roughnesses calculated from five succeeding measurement segments of the same length. Rt is the distance between the highest and lowest values. In this study, we found that the means of all parameters of the asiaticoside-treated group were lower than those of the placebo group. These findings indicate that asiaticoside exerts an antiwrinkle effect.

17.3.4 Conclusion and Discussion

The results of this clinical trial confirm that topical application of 0.1% asiaticoside over a 3-month period significantly improves the clinical appearance of aged skin compared to application of the vehicle alone.

The dermatologic examination, which was performed in a double-blind manner, allowed us to distinguish differences in the aspect of the skin between the areas treated with asiaticoside cream and placebo. Significant reduction in small and coarse wrinkles, as assessed by both an investigating dermatologist and each volunteer, and improvement in the overall aspect of the skin, especially of firmness, smoothness, and dryness, as assessed by the volunteers, were observed after 3 months of daily treatment.

Considering these results, we conclude that, for the majority of women and most wrinkles (>65%), treatment with cream containing asiaticoside will efficiently decrease or eliminate crow's feet.

References

1. Matsuda, H., Morikawa, T., Ueda, H., & Yoshikawa, M. (2001) Medicinal foodstuffs. XXVII. Saponin constituents of gotu kola (2): structures of new ursane- and oleanane-type triterpene oligoglycosides, centellasaponins B, C, and D, from *Centella asiatica* cultivated in Sri Lanka. *Chem. Pharm. Bull. (Tokyo)*. **49**, 1368–1371.
2. Verhagen, C.E., van der Pouw Kraan, T.C., Buffing, A.A., Chand, M.A., Faber, W.R., Aarden, L.A., & Das, P.K. (1998) Type 1- and type 2-like lesional skin-derived *Mycobacterium leprae*-responsive T cell clones are characterized by coexpression of IFN-gamma/TNF-alpha and IL-4/IL-5/IL-13, respectively. *J. Immunol.* **160**, 2380–2387.
3. Yoosook, C., Bunyapraphatsara, N., Boonyakiat, Y., & Kantasuk, C. (2000) Anti-herpes simplex virus activities of crude water extracts of Thai medicinal plants. *Phytomedicine.* **6**, 411–419.
4. Maquart, F.X., Chastang, F., Simeon, A., Birembaut, P., Gillery, P., & Wegrowski, Y. (1999) Triterpenes from *Centella asiatica* stimulate extracellular

matrix accumulation in rat experimental wounds. *Eur. J. Dermatol.* **9**, 289–296.

5. Shukla, A., Rasik, A.M., Jain, G.K., Shankar, R., Kulshrestha, D.K., & Dhawan, B.N. (1999) In vitro and in vivo wound healing activity of asiaticoside isolated from *Centella asiatica. J. Ethnopharmacol.* **65**, 1–11.
6. Cheng, C.L., Guo, J.S., Luk, J., & Koo, M.W. (2004) The healing effects of *Centella* extract and asiaticoside on acetic acid induced gastric ulcers in rats. *Life Sci.* **74**, 2237–2249.
7. Shukla, A., Rasik, A.M., & Dhawan, B.N. (1999) Asiaticoside-induced elevation of antioxidant levels in healing wounds. *Phytother. Res.* **13**, 50–54.
8. Komarcevic, A. (2000) [The modern approach to wound treatment]. *Med. Pregl.* **53**, 363–368.
9. Maquart, F.X., Bellon, G., Gillery, P., Wegrowski, Y., & Borel, J.P. (1990) Stimulation of collagen synthesis in fibroblast cultures by a triterpene extracted from *Centella asiatica. Connect. Tissue Res.* **24**, 107–120.
10. Chakrabarty, T. & Deshmukh, S. (1976) *Centella asiatica* in the treatment of leprosy. *Sci. Cult.* **42**, 573.
11. Nebout, M. (1974) Resultats d'un essai controle de l'extrait titre de *Centella asiatica* (E.T.C.A) (I) dans une population lepreuse presentant des maux perforants plantaires. *Bulletin de la Société de Pathologie exotique* **67**, 471–478.
12. Wijeweera, P., Arnason, J.T., Koszycki, D., & Merali, Z. (2006) Evaluation of anxiolytic properties of Gotukola—(*Centella asiatica*) extracts and asiaticoside in rat behavioral models. *Phytomedicine.* **13**, 668–676.
13. Ryu, J.S., Park, S.G., Kwak, T.J., Chang, M.Y., Park, M.E., Choi, K.H., Sung, K.H., Shin, H.J., Lee, C.K., Kang, Y.S., Yoon, M.S., Rang, M.J., & Kim, S.J. (2005) Improving lip wrinkles: lipstick-related image analysis. *Skin Res. Technol.* **11**, 157–164.
14. Bitzer, M., von, G.G., Liang, D., Dominguez-Rosales, A., Beg, A.A., Rojkind, M., & Bottinger, E.P. (2000) A mechanism of suppression of TGF-beta/SMAD signaling by NF-κB/RelA. *Genes Dev.* **14**, 187–197.
15. Xin, C., Ren, S., Kleuser, B., Shabahang, S., Eberhardt, W., Radeke, H., Schafer-Korting, M., Pfeilschifter, J., & Huwiler, A. (2004) Sphingosine 1-phosphate cross-activates the Smad signaling cascade and mimics transforming growth factor-beta-induced cell responses. *J. Biol. Chem.* **279**, 35255–35262.
16. Lu, L., Ying, K., Wei, S., Liu, Y., Lin, H., & Mao, Y. (2004) Dermal fibroblast–associated gene induction by asiaticoside shown in vitro by DNA microarray analysis. *Br. J. Dermatol.* **151**, 571–578.
17. Dhalla, A.K., Hill, M.F., & Singal, P.K. (1996) Role of oxidative stress in transition of hypertrophy to heart failure. *J. Am. Coll. Cardiol.* **28**, 506–514.
18. Attisano, L. & Wrana, J.L. (2000) Smads as transcriptional co-modulators. *Curr. Opin. Cell Biol.* **12**, 235–243.
19. Massague, J. & Wotton, D. (2000) Transcriptional control by the TGF-beta/Smad signaling system. *EMBO J.* **19**, 1745–1754.
20. Piek, E., Heldin, C.H., & Ten, D.P. (1999) Specificity, diversity, and regulation in TGF-beta superfamily signaling. *FASEB J.* **13**, 2105–2124.
21. Bonte, F., Dumas, M., Chaudagne, C., & Meybeck, A. (1994) Influence of asiatic acid, madecassic acid, and asiaticoside on human collagen I synthesis. *Planta Med.* **60**, 133–135.

22. Lu, L., Ying, K., Wei, S., Fang, Y., Liu, Y., Lin, H., Ma, L., & Mao, Y. (2004) Asiaticoside induction for cell-cycle progression, proliferation and collagen synthesis in human dermal fibroblasts. *Int. J. Dermatol.* **43**, 801–807.
23. Brown, J.D., DiChiara, M.R., Anderson, K.R., Gimbrone, M.A. Jr., & Topper, J.N. (1999) MEKK-1, a component of the stress (stress-activated protein kinase/c-Jun N-terminal kinase) pathway, can selectively activate Smad2-mediated transcriptional activation in endothelial cells. *J. Biol. Chem.* **274**, 8797–8805.
24. Shibuya, H., Yamaguchi, K., Shirakabe, K., Tonegawa, A., Gotoh, Y., Ueno, N., Irie, K., Nishida, E., & Matsumoto, K. (1996) TAB1: an activator of the TAK1 MAPKKK in TGF-beta signal transduction. *Science.* **272**, 1179–1182.
25. Shirakabe, K., Yamaguchi, K., Shibuya, H., Irie, K., Matsuda, S., Moriguchi, T., Gotoh, Y., Matsumoto, K., & Nishida, E. (1997) TAK1 mediates the ceramide signaling to stress-activated protein kinase/c-Jun N-terminal kinase. *J. Biol. Chem.* **272**, 8141–8144.
26. Yamaguchi, K., Shirakabe, K., Shibuya, H., Irie, K., Oishi, I., Ueno, N., Taniguchi, T., Nishida, E., & Matsumoto, K. (1995) Identification of a member of the MAPKKK family as a potential mediator of TGF-beta signal transduction. *Science.* **270**, 2008–2011.
27. Lee, J., Jung, E., Kim, Y., Park, J., Park, J., Hong, S., Kim, J., Hyun, C., Kim, Y.S., & Park, D. (2006) Asiaticoside induces human collagen I synthesis through TGFbeta receptor I kinase (TbetaRI kinase)–independent Smad signaling. *Planta Med.* **72**, 324–328.

PART 6
NATURAL MOISTURIZERS FOR SMOOTHER SKIN

18

Proper Skin Hydration and Barrier Function

Zoe Diana Draelos, MD

Dermatology Consulting Services, High Point, North Carolina, USA

Water is intrinsic to the function and beauty of the skin. It is the plasticizer of the skin, providing for a flexible body covering of protein mixed with intercellular lipids. Without water, the skin would appear as callus, rough, and lifeless. This chapter examines the interaction between skin hydration and barrier function. It discusses what represents too much, too little, and just right in terms of skin hydration, accompanied by descriptions of

Aaron Tabor and Robert M. Blair (eds.), *Nutritional Cosmetics: Beauty from Within*, 353–363, © 2009 William Andrew Inc.

noninvasive equipment technology used to measure skin hydration. In addition, it explores causes of skin dehydration and methods to restore the proper water balance.

18.1 Hydrated versus Dehydrated Skin

It is important to understand what physiologically constitutes dry skin, medically known as xerosis. Xerosis visually manifests as scaly skin, which consumers associate with dry skin, due to the abnormal desquamation of corneocytes (Fig. 18.1). For the skin to appear and feel normal, the water content must be above 10% and below 30% (Fig. 18.2) [1]. Whereas underhydration leads to premature corneocyte sloughing, overhydration leads to maceration, which destroys the physical strength of the skin and predisposes it to infection in intertriginous areas, such as the groin, buttock, and armpit. Under- and overhydration are both equally damaging to the skin barrier; however, the discussion will now focus only on skin with insufficient water content.

Skin dehydration is common in skin disease and conditions of excessive skin lipid removal. Diseases where skin dehydration is common include psoriasis, atopic dermatitis, xerotic eczema, hand dermatitis, etc. All of these conditions are characterized by a defective skin barrier (Fig. 18.3). Xerosis can also be induced by the use of cleansers and solvents that emulsify the intercellular lipids, destroying barrier architecture. Unfortunately, soaps and detergents cannot distinguish between unwanted lipids on the skin surface, such as those from the application of skin care products and sebum accumulation, and those composing the intercellular lipids. At present, the most common cause of skin dehydration is poor cleanser selection and cleanser overuse.

18.2 The Skin Barrier

The skin barrier regulates skin hydration and maintains the water content. It is composed of protein-rich corneocytes surrounded by carefully organized lamellar lipid bilayers that are covalently bound, forming the intercellular lipids. This organization is much like a brick wall, with the corneocytes representing bricks and the intercellular lipids functioning as mortar. Anything that disorganizes this structure leads to barrier damage

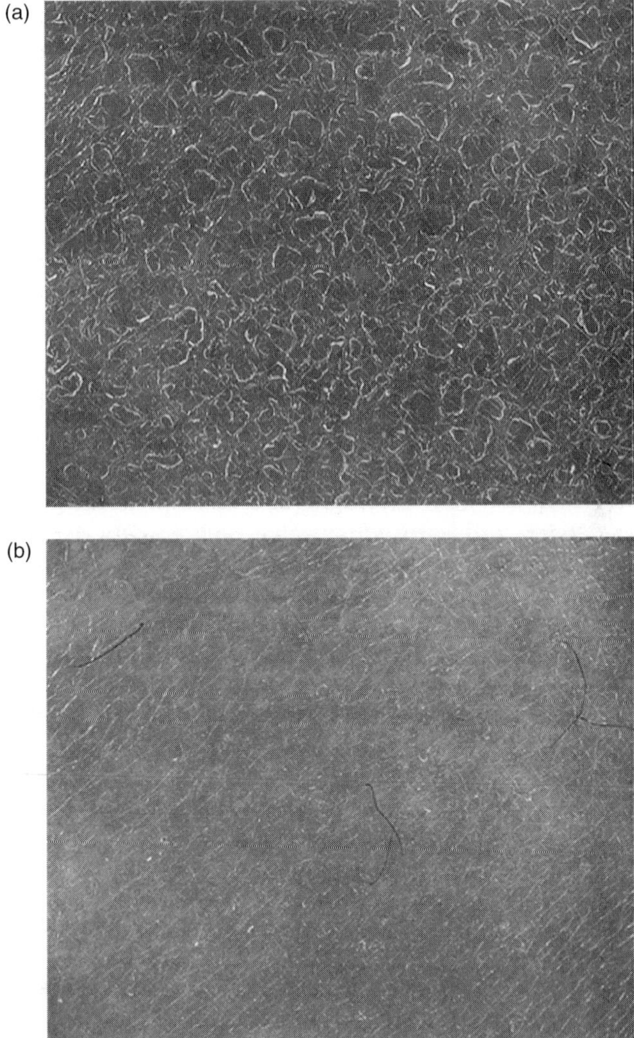

Figure 18.1 Xerotic skin is characterized by scaling created by desquamating corneocytes (a) whereas well-moisturized skin appears smooth (b).

and skin dysfunction. There are three intercellular lipids implicated in epidermal barrier function: sphingolipids, free sterols, and free fatty acids [2]. In addition, it is thought that the lamellar bodies, containing sphingolipids, free sterols, and phospholipids, play a key role in barrier

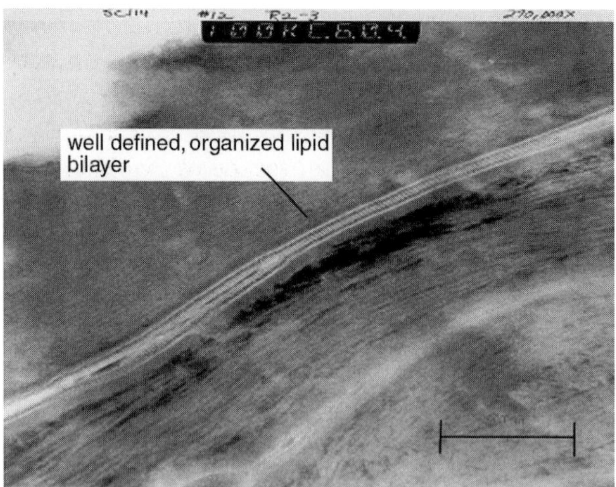

Figure 18.2 Scanning electron micrograph of well-hydrated skin with an intact barrier characterized by an organized lipid bilayer.

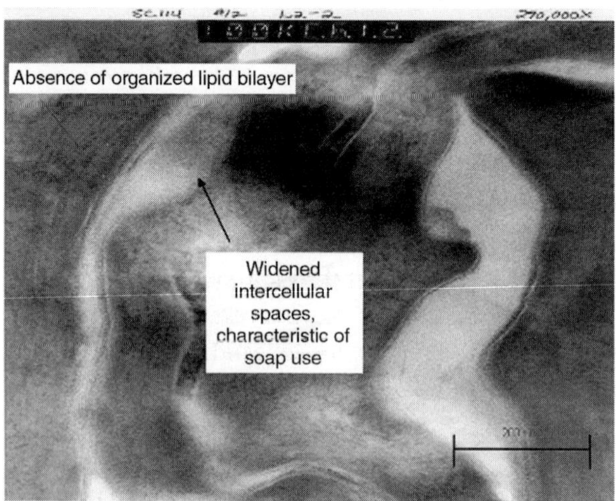

Figure 18.3 Scanning electron micrograph of barrier-damaged skin characterized by loss of the lipid bilayer and the appearance of wide intercellular spaces.

function and are essential to trap water and prevent excessive water loss [3,4]. The lipids are necessary for barrier function because solvent extraction leads to xerosis, directly proportional to the amount of lipid removed [5].

The major lipid by weight found in the stratum corneum is ceramide, which becomes sphingolipid if glycosylated via the primary alcohol of sphingosine [6]. Ceramides possess the majority of the long-chain fatty acids and linoleic acid in the skin. Perturbations within the barrier result in rapid lamellar body secretion and a cascade of cytokine changes associated with adhesion molecule expression and growth factor production [7]. In order to trigger this increased lipid production, a signal must be sent indicating that the skin barrier has been breached.

18.3 Signal for Barrier Repair

Communication must occur between the nonliving skin surface and the viable epidermis and dermis to indicate that the skin barrier has been damaged. The key signal for barrier repair is transepidermal water loss (TEWL). If water loss rises, the skin begins to produce intercellular lipids to replace the barrier to evaporation. Once TEWL returns to normal, the lipid production is halted. It is for this reason that vapor-impermeable wraps, such as plastic food wrap, may 100% halt TEWL, but barrier repair does not occur [8].

An effective moisturizer must allow 1% or more water loss to occur [9,10]. Any rehydration caused by the moisturizer is only temporary. Thus, moisturizers simply create an environment optimal for healing. They do not rehydrate the skin by placing water on the skin surface or attracting water from the environment. Water lost through evaporation to the environment under low humidity conditions from a damaged skin barrier must be replenished by water from the lower epidermal and dermal layers to achieve skin remoisturization [11].

18.4 Skin Remoisturization and the Natural Moisturizing Factor (NMF)

There are several steps in skin remoisturization: initiation of barrier repair, alteration of surface cutaneous moisture partition coefficient, onset of dermal-epidermal moisture diffusion, and synthesis of intercellular lipids [12]. These steps must occur sequentially in order for proper skin barrier repair. Once the barrier has been repaired, there must be some substance that holds and regulates the skin water content. This substance has been termed the natural moisturizing factor (NMF). The constituents of the NMF have been theorized to consist of a mixture of amino acids, derivatives of amino acids, and salts. Artificially synthesized NMF has been constructed from amino acids, pyrrolidone carboxylic acid, lactate, urea, ammonia, uric acid, glucosamine, creatinine, citrate, sodium, potassium, calcium, magnesium, phosphate, chlorine, sugar, organic acids, and peptides [13]. Ten percent of the dry weight of the stratum corneum cells is composed of NMF in well-hydrated skin [14].

18.5 Mechanisms of Moisturization

Rehydration of the skin is the goal of all moisturizer formulations. There are three main methods of retarding TEWL and allowing rehydration through barrier repair. These include placing an oily substance or a large molecular–weight film on the skin surface to retard evaporation or to place substances on the skin surface that attract water to the skin surface [15]. These mechanisms of remoisturization are examined next.

The most common mechanism of moisturization is through the use of occlusive substances. The primary occlusive substance is petrolatum. Petrolatum is a semisolid mixture of hydrocarbons obtained through the dewaxing of heavy mineral oils. Pure cosmetic grade petrolatum is practically odorless and tasteless. Petrolatum first appeared in the U.S. pharmacopoeia in 1880 and has been the most widely used ingredient in skin moisturizers ever since. It is interesting to note that petrolatum has never been duplicated synthetically. Petrolatum is the most effective moisturizing ingredient on the market today, reducing transepidermal water loss by 99%. It functions as an occlusive to create an oily barrier through which water cannot pass. Thus, it maintains cutaneous water content until barrier repair can occur. Petrolatum is able to penetrate into the upper layers of the stratum corneum, creating a reservoir to aid in barrier restoration.

Other commonly used occlusive moisturizers include mineral oil and vegetable oils, such as grapeseed oil, olive oil, jojoba oil, sesame seed oil, etc. These are not as effective at reducing TEWL as petrolatum, because they only afford approximately a 50% reduction, but they do not impart a greasy feel to the skin. Oil-free formulations contain dimethicone or cyclomethicone. These silicone derivatives act as nongreasy occlusive agents that can have an astringent effect on other oily substances. For example, dimethicone can minimize the greasy feel of petrolatum or mineral oil in moisturizers while imparting water-repellent and water-resistant

Table 18.1 Common Occlusive Ingredients to Decrease Transepidermal Water Loss

1. Hydrocarbon oils and waxes: petrolatum, mineral oil, paraffin, squalene
2. Silicone oils
3. Vegetable and animal fats
4. Fatty acids: lanolin acid, stearic acid
5. Fatty alcohol: lanolin alcohol, cetyl alcohol
6. Polyhydric alcohols: propylene glycol
7. Wax esters: lanolin, beeswax, stearyl stearate
8. Vegetable waxes: carnauba, candelilla
9. Phospholipids: lecithin
10. Sterols: cholesterol

properties. Additional occlusive moisturizing ingredients are listed in Table 18.1 [16].

Hydrophilic matrices represent another method of placing a protective barrier over the skin surface. The oldest hydrophilic matrix is colloidal oatmeal, which is raw oatmeal ground to a fine powder. The oatmeal is then placed in bath water, where the small ground oatmeal particles create a "blanket" against water evaporation. Other newer, larger molecular-weight substances, such as proteins and hyaluronic acid, can function in a similar manner. Some of the newer proteins found in skin care products, such as growth factors and hydrolyzed collagen, function in this manner to create an environment for skin rehydration.

Skin hydration can also be enhanced with humectants. Humectants are substances with the ability to attract and hold water much like a sponge. Naturally occurring humectants in the skin include dermal glycosaminoglycans, which function to maintain skin hydration. Glycosaminoglycans are the first substances to see a burst in production following wounding of the skin, because creating a moist environment is key to healing. Humectants are commonly used in skin care products to prevent dehydration of the formulation so that desiccation does not occur on the shelf. This same principle can be used to remoisturize the skin. Commonly used humectants include glycerin, sorbitol, propylene glycol, polyethylene glycol, lactic acid, urea, and gelatin [16,17]. Newer humectant concepts include hyaluronic acid spheres that hydrate on the skin surface to physically fill fine wrinkles on the face, especially around the eyes and on the lips.

A well-designed moisturizer for skin rehydration includes both occlusive and humectant ingredients. It is critical to not only create an artificial barrier to water loss, but to allow rehydration to occur. Humectants in the absence of occlusives will only enhance water loss from the damaged skin surface, further impeding barrier repair [18]. Utilizing a variety of mechanisms of moisturization ultimately results in the best cutaneous result.

18.6 Moisturizers and the Skin Barrier

Healthy skin is well moisturized and maintains a careful organizational structure. Water is intrinsic to the ability of skin to form an attractive, yet pliable, covering for the body. Skin is the initial protection against physical and chemical insults and the first line of defense against infection. It maintains fluid balance, regulates temperature, and allows tactile interaction with the environment. Skin allows human existence in low humidity hot desert conditions and cold rainy winter conditions. It can rapidly adjust to a variety of environments within a matter of days. It is this dynamic nature and the ability to quickly reach homeostasis that is the hallmark of healthy skin. Disease states result when the ability of the skin to provide a barrier for the body is overwhelmed. Thus, the integrity of the skin barrier achieved through proper hydration is key to human health.

References

1. Boisits EK: The evaluation of moisturizing products. Cosmet Toilet 101(May): 31–39, 1986.
2. Elias PM: Lipids and the epidermal permeability barrier. Arch Dermatol Res 270:95–117, 1981.
3. Holleran WM, Man MQ, Wen NG, Gopinathan KM, Elias PM, Feingold KR: Sphingolipids are required for mammalian epidermal barrier function. J Clin Invest 88:1338–1345, 1991.
4. Downing DT: Lipids: their role in epidermal structure and function. Cosmet Toilet 106(December):63–69, 1991.
5. Grubauer G, Elias PM, Feingold KR: Transepidermal water loss: the signal for recovery of barrier structure and function. J Lipid Res 30:323–333, 1989.
6. Petersen RD: Ceramides: key components for skin protection. Cosmet Toilet 107(February):45–49, 1992.
7. Nickoloff BJ, Naidu Y: Perturbation of epidermal barrier function correlates with initiation of cytokine cascade in human skin. J Am Acad Dermatol 30:535–546, 1994.
8. Elias PM: Epidermal lipids, barrier function, and desquamation. J Invest Dermatol 80:44s–49s, 1983.

9. Jass HE, Elias PM: The living stratum corneum: implications for cosmetic formulation. Cosmet Toilet 106(October):47–53, 1991.
10. Holleran W, Feingold K, Man MQ, Gao W, Lee J, Elias PM: Regulation of epidermal sphingolipid synthesis by permeability barrier function. J Lipid Res 32:1151–1158, 1991.
11. Wu MS, Yee DJ, Sullivan ME: Effect of a skin moisturizer on the water distribution in human stratum corneum. J Invest Dermatol 81:446–448, 1983.
12. Jackson EM: Moisturizers: What's in them? How do they work? Am J Contact Dermat 3(4):162–168, 1992.
13. Wehr RF, Krochmal L: Considerations in selecting a moisturizer. Cutis 39:512–515, 1987.
14. Rawlings AV, Scott IR, Harding CR, Bowser PA: Stratum corneum moisturization at the molecular level. Progress in Dermatology 28(1):1–12, 1994.
15. Baker CG: Moisturization: new methods to support time proven ingredients. Cosmet Toilet 102(April):99–102, 1987.
16. De Groot AC, Weyland JW, Nater JP: Unwanted Effects of Cosmetics and Drugs Used in Dermatology. 3rd ed., Elsevier, Amsterdam, pp. 498–500, 1994.
17. Spencer TS: Dry skin and skin moisturizers. Clin Dermatol 6:24–28, 1988.
18. Rieger MM, Deem DE: Skin moisturizers II: the effects of cosmetic ingredients on human stratum corneum. J Soc Cosmet Chem 25:253–262, 1974.

19

Food-Derived Materials Improving Skin Cell Health for Smoother Skin

Hiroshi Shimoda, PhD

Oryza Oil & Fat Chemical Co., Ltd., Ichinomiya, Aichi, Japan

Summary

We have examined a number of plant-derived substances for their effects in protecting skin from damages and in promoting skin health. Oral application

---

Aaron Tabor and Robert M. Blair (eds.), *Nutritional Cosmetics: Beauty from Within*, 365–382, © 2009 William Andrew Inc.

of rice ceramide significantly reduced dryness and exfoliation, enhanced moisture-retention, and improved smoothness and texture of skin in a placebo-controlled double-blind ingestion study in 33 subjects. Oral and topically applied plant-derived tocotrienol was transferred to skin, which acted as a scavenger of reactive oxygen species and was effective in protecting skin from UV damages. In addition, we introduce several other plant-derived extracts with fibroblast-proliferative, skin-turnover-promoting, and pigmentation-suppressive activities.

19.1 Introduction

For protecting skin from damage and for supporting skin functions, various synthetic chemicals and natural substances are added to cosmetics. For example, synthetic and plant-derived ceramides are being used in skin care products, which support barrier function and provide moisture to skin. Human epidermis consists of stratum basale, stratum spinosum, granulosa, and stratum corneum, each with a distinct lipid composition. In stratum basale, phospholipids and cholesterol are the major components. Glucosylceramide gradually increases from stratum basale toward the granulosa layer, where it becomes the major component. In the stratum corneum, glucosylceramide is converted to ceramide, which is the major component of interstitial lipids between corneocytes. Synthetic ceramides have been widely used as a skin-beautifying substance that holds moisture in the skin corneum. In recent years, natural ceramide is attracting more attention, especially because of its safe and healthy image. Plant ceramides are also a functional food for daily ingestion.

Application of antioxidants is regarded as effective for protecting skin from damage induced by reactive oxygen species, and thus preventing sunburn, inflammation, and pigmentation. Hydrophobic compounds are generally known to be effectively absorbed into skin by topical application. So is tocotrienol, a tocopherol-related compound. In this section, we describe cosmeceutical activities of ceramide, tocotrienol, and several other plant-derived substances.

19.2 Rice Ceramide Improves Skin Condition

The stratum corneum protects skin mechanically and maintains moisture. The major lipid components in the stratum corneum are ceramides.

Coderch et al. [1] demonstrated that topical supplementation of ceramide was effective for retaining moisture in skin. They swabbed healthy human skin with wool lipid (IWL) extracts with and without synthetic stratum corneum lipid (SSCL) as liposomes for several months. Transdermal water loss was improved and the water-retaining capacity of the skin was increased by mixed IWL/SSCL, but not by IWL alone. SSCL contains a ceramide, and these results thus suggest that topical application of ceramide is effective for maintaining moisture in the skin [2]. As an intracellular messenger in the sphingomyelin cycle, ceramide provides not only a barrier function to the skin but also regulates lipid biosynthesis [3]. Increase in intracellular ceramide induces skin cell differentiation and/or apoptosis, and suppresses cell proliferation.

Knowing the beneficial effects of topical application of ceramide, we studied the efficacy of oral application of ceramide in a placebo-controlled human study. We gave 40 mg rice ceramide to 33 subjects (6 male, 27 female) aged 25.1 ± 7.8 for 6 weeks and compared skin parameters before, during, and after the supplementation period. As shown in Fig. 19.1, skin moisture was increased and index of skin dryness (SE sc index) was reduced significantly compared to the placebo group. Figure 19.2 shows

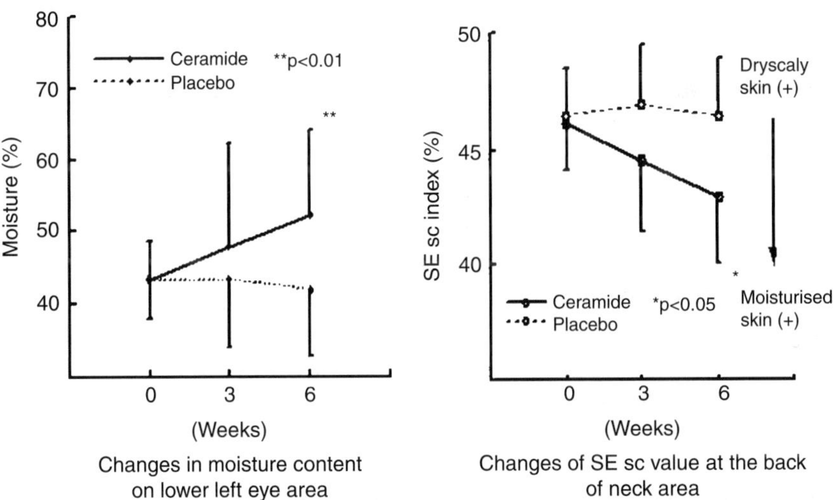

Figure 19.1 Changes in skin parameters by oral treatment with rice ceramide in human subjects. Each point represents mean with standard deviation (SD). Asterisks denote significant differences from placebo.

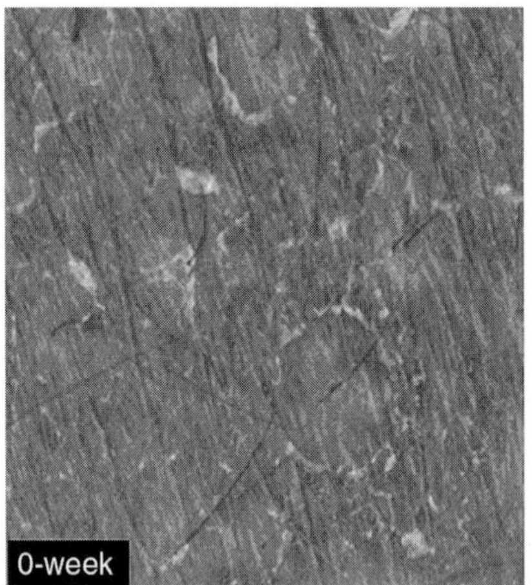

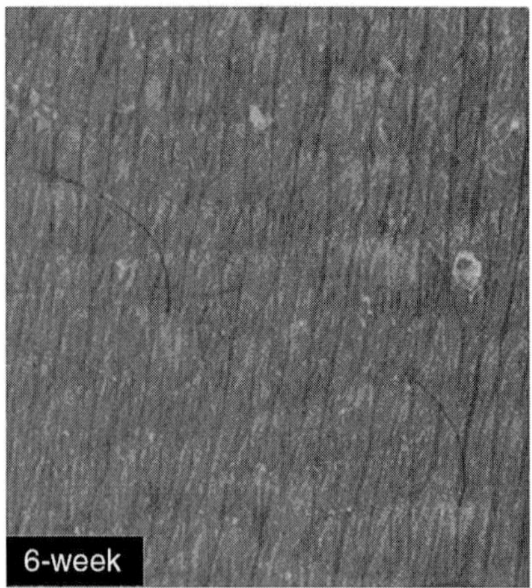

Figure 19.2 Typical images of three-dimensional microscopic illustration of skin surface in the left eye area of a 23-year old female before and after 6 weeks of treatment with rice ceramide.

light microscope images of the skin of a female subject before and after ingestion of rice ceramide. Wrinkles were visibly reduced after 3 weeks of ingestion.

Other dermatological findings before and after ingestion of rice ceramide are summarized in Table 19.1. In facial skin, significant improvement was observed for "dryness," "flush," and "holding of cosmetic" after 3 weeks and 6 weeks of ingestion in the test group. In the placebo group, "dryness" and "flush" were also significantly improved after ingestion, but "holding of cosmetic" did not show significant improvement. Table 19.2 shows the evaluated improvement scores. Although improvement in "facial dryness," "flush," and "holding of cosmetic" was observed in both the ceramide and the placebo groups, the scores were better in the ceramide group. Regarding general skin parameters, significant improvement in "itching," "dryness," and "flush" was observed in both groups. However, the improvement rates were higher in the ceramide group and the overall improvement (64.7%) was only significant in the ceramide group (Table 19.2).

Water content, oil content, and acidity of the skin before and after ingestion in the ceramide and placebo groups are summarized in Table 19.3. Significant increase in water content was found in the face, neck, and forearm of subjects in the ceramide group after 6 weeks of ceramide ingestion. No

Table 19.1 Results of Dermatological Diagnosis Before and After Ingestion of the Test Food

		Ceramide Group				Placebo Group			
		No. of subjects with symptoms	Initial	3 weeks	6 weeks	No. of subjects with symptoms	Initial	3 weeks	6 weeks
Face	Cosmetic rash	4	1.25	1.00	1.00	3	1.00	1.00	1.00
	Dryness	17	2.00	1.29**	1.18**	16	2.13	0.69*	1.63**
	Flush	14	1.86	1.29**	1.21*	15	1.93	1.53*	1.47*
	Holding of cosmetic	8	1.88	1.13*	1.00*	5	1.40	1.20	1.20
General	Itching	17	1.65	1.24*	0.94**	12	1.92	1.50	1.17**
	Dryness	17	2.18	1.47**	1.18**	16	2.00	1.44	1.38*
	Flush	7	1.71	1.29	0.86*	10	1.80	1.30*	1.20*
	Erosion	2	2.00	1.00	0.00	3	2.33	2.00	2.00
	Squamation	4	1.50	1.25	0.75	5	1.80	1.60	1.60
	Papules	3	1.33	1.33	1.33	4	1.75	1.50	1.75
	Blebs	2	1.50	1.50	1.50	2	1.50	1.50	2.00
	Swelling	0				3	2	1.67	1.67
	Overall	17	1.71	1.24*	1.00**	16	1.69	1.25	1.31

The value represents the mean in each group. Asterisks denote significant differences from initial value at $*p < 0.05$, $**p < 0.01$, respectively.
0: no symptoms; –1: mild; –2: moderate; –3: severe.

Table 19.2 Improvement Rate of Each Symptom

Symptoms		No. of subjects with symptoms	Improvement rating					Improvement rate (improved or better)
			Markedly improved	Improved	Unchanged	Aggravated	Markedly aggravated	
Facial symptoms								
Cosmetic rash	Ceramide	4	0	1	3	0	0	25.0%
	placebo	3	0	0	3	0	0	0.0%
Facial dryness	Ceramide	17	3	8	6	0	0	64.7%
	placebo	16	0	8	8	0	0	50.0%
Facial flush	Ceramide	14	2	8	2	2	0	71.4%
	placebo	15	2	3	10	0	0	33.3%
Holding of cosmetic	Ceramide	8	2	3	3	0	0	62.5%
	placebo	11	0	1	4	0	0	9.1%
Somatic symptoms								
Itching	Ceramide	17	5	4	8	0	0	52.9%
	placebo	12	4	1	7	0	0	41.7%

Dryness	Ceramide	17	6	5	6	0	0	64.7%
	placebo	16	1	8	7	0	0	56.3%
Flush	Ceramide	6	2	2	2	0	0	66.7%
	placebo	10	2	3	5	0	0	50.0%
Erosion	Ceramide	2	2	0	0	0	0	100.0%
	placebo	3	0	1	2	0	0	33.3%
Squamation	Ceramide	4	2	1	0	1	0	75.0%
	placebo	5	0	1	4	0	0	20.0%
Papules	Ceramide	3	0	0	3	0	0	0.0%
	placebo	4	0	1	2	1	0	25.0%
Blebs	Ceramide	2	0	1	0	1	0	50.0%
	placebo	2	0	0	1	1	0	0.0%
Overall	Ceramide	17	3	8	5	1	0	64.7%
	placebo	16	0	7	8	1	0	43.8%

Table 19.3 **Measurement Results of Water Content, pH, and Oil Content Before and After Ingestion of the Test Food**

		Ceramide group (n = 17)		
		Initial	**3 weeks**	**6 weeks**
Water content (arbital unit)	Under left eye	43.2 ± 5.5	48.0 ± 14.3	52.2 ± 12.1**
	Left forearm	37.0 ± 5.6	41.1 ± 11.0	43.2 ± 35.7**
	Dorsal neck	43.5 ± 10.8	51.2 ± 11.76**	55.9 ± 11.1**
Acidity (pH)	Under left eye	5.8 ± 0.7	5.6 ± 0.6	5.8 ± 0.5
	Left forearm	5.5 ± 0.5	5.5 ± 0.6	5.8 ± 0.5
	Dorsal neck	5.9 ± 1.1	5.5 ± 0.5	5.4 ± 0.4
Oil content ($\mu g/cm^2$)	Under left eye	42.3 ± 34.8	49.9 ± 35.1	38.1 ± 25.9
		Placebo group (n = 16)		
		Initial	**3 weeks**	**6 weeks**
Water content (arbital unit)	Under left eye	43.4 ± 5.4	43.2 ± 9.2	41.7 ± 9.4
	Left forearm	35.7 ± 5.6	37.7 ± 7.0	35.7 ± 9.0
	Dorsal neck	49.1 ± 8.8	51.0 ± 10.4	56.1 ± 20.5
Acidity (pH)	Under left eye	5.9 ± 0.8	5.8 ± 0.6	5.9 ± 0.7
	Left forearm	5.5 ± 1.0	5.6 ± 0.8	5.9 ± 0.5
	Dorsal neck	5.9 ± 0.8	5.5 ± 0.5	5.8 ± 0.4
Oil content ($\mu g/cm^2$)	Under left eye	58.4 ± 55.8	29.5 ± 22.0	40.8 ± 33.3

Each value represents the mean ± SD. Asterisks denote significant differences from initial value at $^{**}p < 0.01$.

significant change was found in the placebo group. No significant change was found for acidity and oil content in either group (Table 19.3).

Results of imaging analysis of skin using VISIOSCAN are shown in Table 19.4. Significant improvement was obtained for kurtosis (representing smoothness of total skin), SE sm value (an index of skin smoothness calculated from the depth, width, and notch of furrows), and SE r (index of skin roughness) in the ceramide group. No significant improvement was found in the placebo group. SE w (number and width of skin wrinkles) was significantly improved only at the site below the left eye after 3 weeks of ingestion in the ceramide group. These findings show that a dietary supplement of rice-derived ceramide reduces skin dryness and improves general skin health.

19.3 Tocotrienol Prevents UV Damages in the Skin

UV photons induce skin damage mainly through two mechanisms [4]. One is the direct absorption of ultraviolet via cellular chromophores that can lead to photo-induced DNA base damage, with the consequence of increased mutation rate [4]. The other mechanism is photosensitization leading to formation of free radicals including reactive oxygen species (ROS) and reactive nitrogen species (RNS) [5]. ROS include singlet oxygen ($^1O_2$), superoxide anion, H_2O_2, and hydroxyl radical ($OH^{\cdot}$). ROS and RNS are also constantly generated in keratinocytes and fibroblasts, but rapidly neutralized by nonenzymatic (ascorbic acid, tocopherol, ubiquinol, and glutathione) and enzymatic (glutathione peroxidases, superoxide dismutases, catalase, and quinine reductase) antioxidants *in vivo* [5]. However, UV radiation produces excessive free radicals that cannot be neutralized by endogenous antioxidants. These excessive free radicals are the pathogenesis of a number of skin disorders, allergic reactions, and neoplasms. Moreover, ROS induce expression of activator protein-1 (AP-1) and nuclear factor-κB (NF-κB) [5]. Activation of these factors causes inflammation, flare, and pigmentation. Highly reactive peroxynitrite ($ONOO^-$), which is a reaction product of super oxide anion and nitric oxide (NO), damages DNA and thereby causes point mutations, deletions, and chromosomal rearrangements [5].

Plant-derived polyphenols (e.g., caffeic acid, quercetin, genistein, resveratrol, nordihydroguaiaretic acid, carnosic acid, silymarin, catechins, and procyanidin B1) have antioxidative activity [6] and thus can help neutralize

Table 19.4 Parameter Values Measured by VISIOSCAN Before and After Ingestion of the Test Food

		Ceramide group			Placebo group		
		Initial	3 weeks	6 weeks	Initial	3 weeks	6 weeks
Kurtosis	Under left eye	0.38	0.37	0.35	0.39	0.38	0.38
	Left forearm	0.35	0.39	0.40	0.43	0.43	0.40
	Dorsal neck	0.40	0.40	0.30*	0.40	0.40	0.40
SE sm	Under left eye	377.4	364.4	342.1	368.0	354.1	347.7
	Left forearm	339.4	304.8	308.5	326.1	317.3	334.2
	Dorsal neck	386.8	327.8*	333.1*	355.2	349.2	354.5
SE r	Under left eye	0.29	0.26	0.25*	0.30	0.31	0.30
	Left forearm	0.26	0.20	0.16	0.31	0.26	0.25
	Dorsal neck	0.18	0.15	0.14*	0.31	0.30	0.30
SE sc	Under left eye	49.6	47.6*	46.8*	46.6	46.8	46.6
	Left forearm	48.9	47.9*	47.6*	48.3	48.9	48.4
	Dorsal neck	46.1	44.5*	42.9*	46.4	46.9	46.4
SE w	Under left eye	36.1	32.3*	33.9	36.0	33.0	35.5
	Left forearm	26.7	24.7	27.1	27.5	24.5	27.7
	Dorsal neck	28.4	24.7	26.1	28.2	25.9	30.6

The value represents the mean in each group. Asterisks denotes significant difference from initial value at $^*p < 0.05$. The ideal value of kurtosis is 0, and the other SE values are indicated as the ratio of ideal value versus low value.

excessive oxidants and prevent UV damages in the skin. Our laboratory has developed a number of extracts from rice germ and other plants for cosmetics and dietary supplements for supporting skin health. In this section, we introduce functions of tocotrienol, a hydrophobic antioxidant derived from rice bran oil and palm oil.

Rice bran oil is a traditional plant-derived oil widely used as a cooking oil in Southeast Asia. The oil contains large amounts of unsaturated fatty acids, as well as other healthy constituents such as γ-oryzanol, sterols, tocopherols, and tocotrienols. Our product Oryza tocotrienol™ is extracted and refined from rice bran oil. It contains large amounts of tocotrienols and tocopherols. The chemical structure of tocotrienol is similar to that of tocopherols (Fig. 19.3). γ-Tocotrienol is the principal constituent in rice tocotrienols and has been demonstrated to show biological activity in skin. Weber et al. [7] found that topical application of tocopherols and tocotrienols to mice increased content of α-tocopherol and vitamin E derivatives in their skin (Fig. 19.4). Even after UV irradiation, the remaining contents of these components were still significantly higher than that of the control group. The result shows that topically applied tocotrienol and tocopherol can permeate into skin and protect skin from UV damage.

Also, orally applied tocotrienols have been reported to reach skin. Khanna et al. [8] revealed that tocotrienol given orally reached the skin in tocopherol-deficient mice. Ikeda et al. [9,10] reported that continuous supplementation of γ-tocotrienol increased its contents in rat skin. Human epidermis contains 3% tocotrienol and 1% tocopherol [11]. Hence, oral

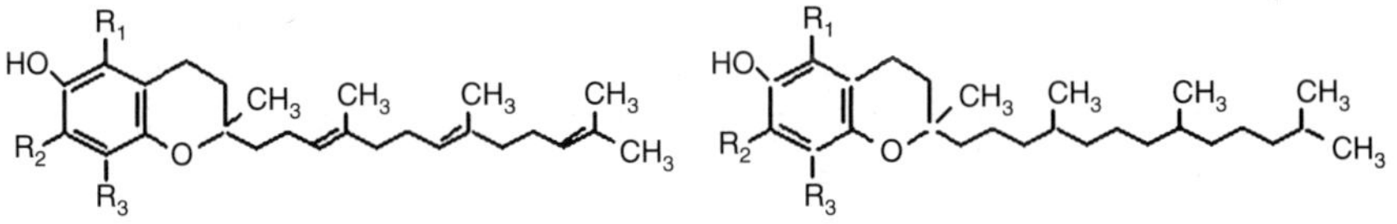

	R_1	R_2	R_3
α-tocotrienol	CH_3	CH_3	CH_3
β-tocotrienol	CH_3	H	CH_3
γ-tocotrienol	H	CH_3	CH_3
δ-tocotrienol	H	H	CH_3

	R_1	R_2	R_3
α-tocopherol	CH_3	CH_3	CH_3
β-tocopherol	CH_3	H	CH_3
γ-tocopherol	H	CH_3	CH_3
δ-tocopherol	H	H	CH_3

Figure 19.3 Structures of tocotrienols and tocopherols.

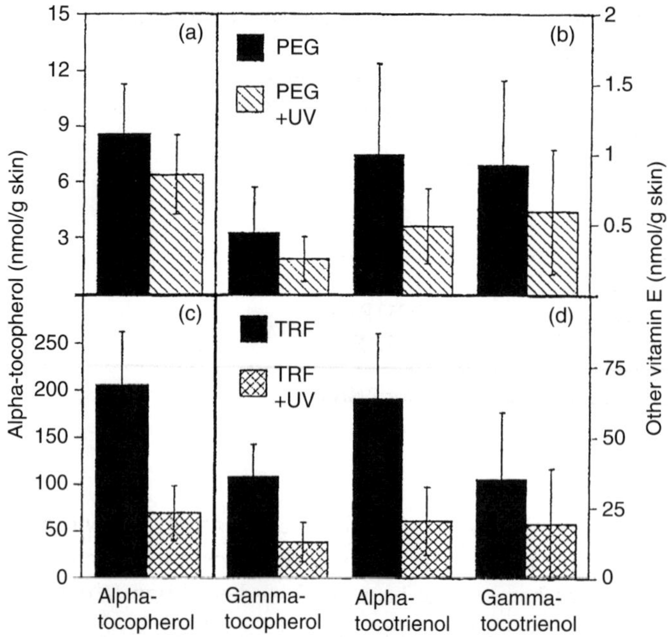

Figure 19.4 The contents of alpha-tocopherol and other vitamin E derivatives in murine skin treated with tocotrienol rich fraction (TRF) containing alpha-tocopherol, gamma-tocopherol, alpha-tocotrienol, and gamma-tocotrienol [7]. Each column represents mean with SD. Polyethylene glycol (PEG) means control group.

application of tocotrienol is expected to be effective for the protection of skin from UV.

19.4 Plant Extracts Enhance Skin Turnover and Inhibit Skin Pigmentation

19.4.1 Enhanced Fibroblast Proliferation

Fibroblasts play significant roles in epithelial-mesenchymal interactions, secretion of various growth factors and cytokines, and differentiation and formation of the extracellular matrix [12]. The interaction of fibroblasts and keratinocytes is essential in the wound-healing process [13]. Keratinocytes stimulate fibroblasts to synthesize growth factors, which in return stimulate keratinocyte proliferation in a double-paracrine manner. We found a number of plant extracts that enhanced fibroblast proliferation

Table 19.5 Proliferative Effect of Food-Derived Materials on Fibroblast Proliferation

	Cell name	Concentration (μg/ml)	Proliferative ratio (%)
Citrus unshiu peel extract	NB1RGB	10	20
Yuzu seed extract	NB1RGB	10	40
Kiwifruit seed extract	NB1RGB	10	70
Rice ceramide	HSK	300	63
α-Lipoic acid	NB1RGB	25	22
Rice tocotrienol	NHDF	25	22

Cells were treated with each sample for 2 days.

in vitro (Table 19.5). Kiwifruit seed extract contains flavonoid glycosides such as quercitrin and kaempferol 3-*O*-rhamnoside. The biological activity of this extract has not yet been well studied. Yuzu (*Citrus junos*) is a citrus species that is mainly cultivated in Japan. Yuzu seed extract contains limonoids (triterpenoids) such as limonin and nomilin. *Citrus unshiu* is also a citrus species fruit mainly cultivated in Japan and China. It contains β-cryptoxanthin and flavonoids (hesperidin). Kiwifruit seed, a food-derived material extract, exhibited the most potent proliferative effect for fibroblasts (70% at 10 μg/ml), followed by yuzu seed extract (40% at 10 μg/ml), *Citrus unshiu* peel extract (20% at 10 μg/ml), and rice tocotrienol (22% at 25 μg/ml). The activity of synthetic α-lipoic acid in enhancing fibroblast proliferation was similar to that of rice tocotrienol. In addition to its fibroblast-proliferative effect, yuzu seed extract also inhibited 5α-reductase and lipase. Application of these plant-derived substances in cosmetics may enhance proliferation of skin fibroblasts.

19.4.2 Enhanced Skin Turnover

The epidermis is constituted and continuously regenerated by terminally differentiating keratinocytes [14]. In the wound-healing process, an increase in cytokine and growth factor production from keratinocytes and fibroblasts is induced to enhance skin cell proliferation [15]. For evaluation of effects of various compounds on skin turnover, a three-dimensional cell culture system has been established by which skin-composing cells are seeded trilaminarily onto membranes that form epidermis, basal lamina, and dermis layers [16–19]. Using this *in vitro* human skin model, we

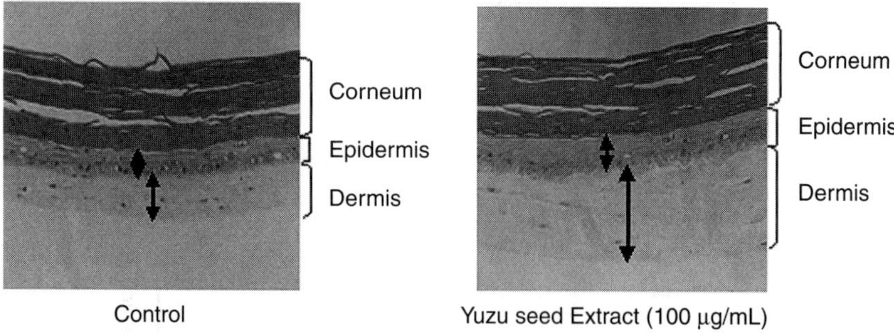

Figure 19.5 Light microscope photograph (×400) of cultured three-dimensional human skin model (TESTSKINTM, Toyobo Co., Ltd., Japan) with and without supplementation of yuzu seed extract.

evaluated the effect of our plant extracts on skin turnover. Yuzu seed extract was found to enhance thickening of the epidermis and dermis (Fig. 19.5), suggesting its proliferative activity for skin cells in these layers. The result suggests that yuzu seed extract promotes skin turnover and may have a softening effect for the skin, and thus can be used as an additive for skin care products.

19.4.3 Inhibited Skin Pigmentation

Melanin is produced from tyrosine by multiple enzymes contained in melanocytes in the basal layer of the epidermis. Because tyrosinase is the rate-limiting enzyme in melanin production, inhibition of this enzyme has been considered as a basic strategy for developing skin care products with whitening effect. We found tyrosinase-inhibiting activity in a number of plant extracts and in α-lipoic acid, as shown in Table 19.6. Litchi seed extract and evening primrose seed extract exhibited the strongest tyrosinase-inhibitory effect, followed by *Citrus unshiu* peel extract, broccoli sprout extract, and kiwifruit seed extract. In addition, melanin formation in melanoma cells was suppressed *in vitro* by most of the extracts we tested to various extents (Table 19.6). Finally, topical or oral application of some of these extracts suppressed UV-induced pigmentation in guinea pigs (Fig. 19.6) [20]. In particular, litchi seed extract, kiwifruit seed extract, and rice ceramide inhibited tyrosinase, suppressed melanin formation *in vitro*, and suppressed UV-induced pigmentation *in vivo*. The former two extracts contain polyphenols: litchi seed extract contains anthocyanins such as cyanidin 3-*O*-glucoside and malvidin 3-*O*-glucoside, and evening primrose

Table 19.6 Suppressive Effects of Food-Derived Extracts on Various Melanin Formation Assays

	Tyrosinase activity inhibition (%)/ concentration (μg/ml)	Melanin formation in B16 melanoma inhibition (%)/ concentration (μg/ml)	Pigmentation in UV-irradiated guinea pig recovery[a] (%)/dose (mg/kg or %), administration route
Litchi seed extract	100/1200	35/100	63/1, topical
Evening primrose seed extract	80/1000	–	–
Citrus unshiu peel extract	50/1000	55/1000	61[b]/800, oral
Yuzu seed extract	–	25/100	–
Broccoli seed extract	49/1000	40/100	–
Kiwifruit seed extract	38/1000	30/100	100/800, oral
Purple rice extract	85/500[c]	49/500	–
Rice ceramide	20/20[d]	30/100	41/1, topical
α-Lipoic acid	–	35/100	51/800, oral
Rice tocotrienols	–	–	39/1[e], topical

[a]Recovery ratio from initial luminosity value; [b]inhibitory ratio; [c]co-treatment with ascorbic acid (5 mg/ml); [d]melan-a cells; [e]as total tocotrienols and tocoopherols.

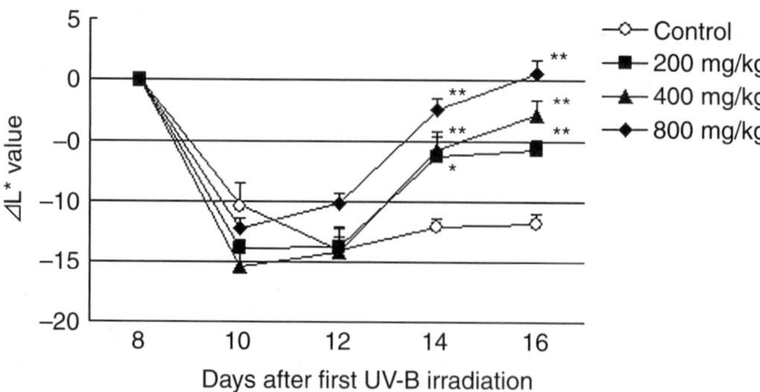

Figure 19.6 Effect of kiwifruit seed extract on skin hyperpigmentation in guinea pigs induced by UV-B. Each point represents mean with the S.E. ($n = 4$). Asterisks denote significant differences from the control, $^*p < 0.05$, $^{**}p < 0.01$, respectively. Decrease in ΔL^* value means change in brightness to darker.

seed extract contains condensed tannins and hydrolyzable tannins. Ceramide is a skin component and its content decreases with age and in dermatitis. These extracts provide effective ingredients for skin-whitening products.

19.5 Conclusion

We described the effects of several of our extracts derived from plants in preventing skin damage induced by UV, in enhancing fibroblast proliferation and skin turnover, and in suppressing skin pigmentation. Oral and topical application of these substances may thus protect skin and promote skin health.

References

1. Coderch L., De Pera M., Fonollosa J., De La Maza A., Parra J. Efficacy of stratum corneum lipid supplementation on human skin. *Contact Dermatitis.* **47**, 139–146 (2002).
2. Coderch L., López O., de la Maza A., Parra J.L. Ceramides and skin function. *Am. J. Clin. Dermatol.* **4**, 107–129 (2003).
3. Geilen C.C., Wieder T., Orfanos C.E. Ceramide signalling: regulatory role in cell proliferation, differentiation and apoptosis in human epidermis. *Arch. Dermatol. Res.* **289**, 559–566 (1997).

4. Svobodova A., Walterova D., Vostalova J. Ultraviolet light induced alteration to the skin. *Biomed. Pap. Med. Fac. Univ. Palacky. Olomouc. Czech. Repub.* **150**, 25–38 (2006).
5. Bickers D.R., Athar M. Oxidative stress in the pathogenesis of skin disease. *J. Invest. Dermatol.* **126**, 2565–2575 (2006).
6. Svobodová A., Psotová J., Walterová D. Natural phenolics in the prevention of UV-induced skin damage. *Biomed. Pap. Med. Fac. Univ. Palacky. Olomouc. Czech. Repub.* **147**, 137–145 (2003).
7. Weber C., Podda M., Rallis M., Thiele J.J., Traber M.G., Packer L. Efficacy of topically applied tocopherols and tocotrienols in protection of murine skin from oxidative damage induced by UV-irradiation. *Free Radic. Biol. Med.* **22**, 761–769 (1997).
8. Khanna S., Patel V., Rink C., Roy S., Sen C.K. Delivery of orally supplemented γ-tocotrienol to vital organs of rats and tocopherol-transport protein deficient mice. *Free Radic. Biol. Med.* **39**, 1310–1319 (2005).
9. Ikeda S., Tohyama T., Yoshimura H., Hamamura K., Abe K., Yamashita K. Dietary α-tocopherol decreases α-tocotrienol but not γ-tocotrienol concentration in rats. *J. Nutr.* **133**, 428–434 (2003).
10. Ikeda S., Toyoshima K., Yamashita K. Dietary sesame seeds elevate α- and γ-tocotrienol concentrations in skin and adipose tissue of rats fed the tocotrienol-rich fraction extracted from palm oil. *J. Nutr*. **131**, 2892–2897 (2001).
11. Fuchs J., Weber S., Podda M., Groth N., Herrling T., Packer L., Kaufmann R. HPLC analysis of vitamin E isoforms in human epidermis: correlation with minimal erythema dose and free radical scavenging activity. *Free Radic. Biol. Med.* **34**, 330–336 (2003).
12. Wong T., McGrath J.A., Navsaria H. The role of fibroblasts in tissue engineering and regeneration. *Br. J. Dermatol*. **156**, 1149–1155 (2007).
13. Werner S., Krieg T., Smola H. Keratinocyte-fibroblast interactions in wound healing. *J. Invest. Dermatol.* **127**, 998–1008 (2007).
14. D'Errico M., Lemma T., Calcagnile A., Proietti De Santis L., Dogliotti E. Cell type and DNA damage specific response of human skin cells to environmental agents. *Mutat. Res*. **614**, 37–47 (2007).
15. Beer H.D., Gassmann M.G., Munz B., Steiling H., Engelhardt F., Bleuel K., Werner S. Expression and function of keratinocyte growth factor and activin in skin morphogenesis and cutaneous wound repair. *J. Investig. Dermatol. Symp. Proc.* **5**, 34–39 (2000).
16. Mohapatra S., Coppola D., Riker A.I., Pledger W.J. Roscovitine inhibits differentiation and invasion in a three-dimensional skin reconstruction model of metastatic melanoma. *Mol. Cancer Res.* **5**, 145–151 (2007).
17. Sugimoto K., Nishimura T., Nomura K., Sugimoto K., Kuriki T. Inhibitory effects of alpha-arbutin on melanin synthesis in cultured human melanoma cells and a three-dimensional human skin model. *Biol. Pharm. Bull.* **27**, 510–514 (2004).
18. Li L.N., Margolis L.B., Hoffman R.M. Skin toxicity determined in vitro by three-dimensional, native-state histoculture. *Proc. Natl. Acad. Sci. U S A*. **88**, 1908–1912 (1991).

19. Fleischmajer R., MacDonald E.D. II, Contard P., Perlish J.S. Immunochemistry of a keratinocyte-fibroblast co-culture model for reconstruction of human skin. *J. Histochem. Cytochem.* **41**, 1359–1366 (1993).
20. Tanaka J., Shan S.J., Kasajima N., Shimoda H. Suppressive effect of defatted kiwi fruit seed extract on acute inflammation and melanin formation: involvement of its flavonol glycosides. *Food Sci. Technol. Res.* **13**, 310–314 (2007).

PART 7
NATURAL SUPPORT FOR A HEALTHIER COMPLEXION

20

A Whey Protein Complex for Skin Beauty from the Inside Out

Petra Caessens, PhD, Wendeline Wouters, PhD, Rick de Waard, PhD, and Angela Walter

DMV International, Delhi, NY, USA

Aaron Tabor and Robert M. Blair (eds.), *Nutritional Cosmetics: Beauty from Within*, 383–398, © 2009 William Andrew Inc.

20.1 Introduction

A smooth, blemish-free, healthy, and youthful skin tone is exactly what consumers want and desperately wish to hold onto. Unfortunately, not everybody is given clear skin at all stages of life, especially teenagers. Adult women can also suffer tremendously from acne, blemishes, and associated poor complexion. The link between eating habits, health, and physical appearance is becoming increasingly evident and has made consumers all over the world open to the concept of ingestible beauty products.

For this highly interesting, fast-growing market segment, DMV International developed a natural ingredient from whey—to be taken orally—that has been shown in several consumer studies to support a healthy skin appearance. Since its introduction on the market, mid-2005, the ingredient has been used in multiple new products launched worldwide. The brand name of this successful ingredient is Praventin™, an ingredient derived from whey predominant in lactoferrin.

20.1.1 Acne Vulgaris

Acne vulgaris is a simply identifiable dermatologic condition. It is one of the most common skin complaints, with prevalence in adolescents reaching nearly 85%. Small noninflammatory acne lesions may not be more than a minor annoyance, but consumers with harsher inflammatory nodular acne can suffer soreness. The resulting social embarrassment, as well as acne-associated physical and psychological scarring, can be life changing [1–3]. Acne vulgaris starts with the formation of a microcomedo. As shown in Fig. 20.1, the next stage comprises noninflammatory lesions, namely closed comedones (whiteheads) and open comedones (blackheads). The non-inflammatory lesions can turn into inflammatory lesions.

Dependent on the degree of severity, the inflammatory lesion range comprises papules, pustules, and nodulocystic lesions. The stimulus for micro-comedo formation is still uncertain. The strongest leads implicate androgen hormones, alterations in follicular linoleic acid levels, and the inflammatory cytokine interleukin-1a. Other factors, such as genetic predisposition, stress, and diet, may also affect the development and severity of acne.

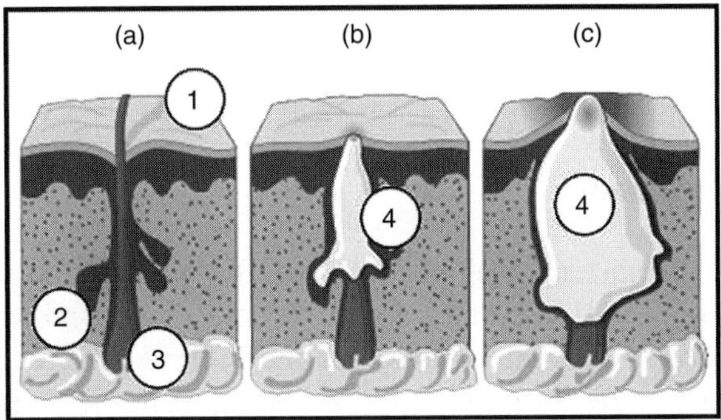

Figure 20.1 Picture of (A) a normal pilosebaceous unit (consisting of hair follicle and sebaceous glands), (B) an open comedone (blackhead), and (C) a closed comedone (whitehead). (1) skin surface and hairs; (2) sebaceous gland; (3) follicle; (4) enlargement of follicle opening.

20.1.2 Current Consumer Strategies for Poor Complexion

Topical and systemic therapies form two of the major therapeutic strategies for the treatment of acne today (Table 20.1). Dependent on the severity of the lesions, a single or combined therapy is applied. In general, the topical and systemic therapies aim at inhibition of hyperkeratosis, follicular plugging, diminishment of sebum production, reduction of bacterial load (e.g., *Propionibacterium acnes*), and inflammatory responses. The main active compounds of therapies (usually in gel, cream, or solution formulations) comprise retinoids and antibiotics. Hormonal treatment is occasionally used for most types of acne in both adult and adolescent females. As depicted in Table 20.1, the therapies are often accompanied by adverse side effects [4–6].

20.2 A Whey Protein Complex Supporting a Healthy Complexion from the Inside

Lactoferrin-predominant whey is created by a proprietary process [7] that derives a bioactive protein and peptide complex. Lactoferrin-predominant whey protein is based on milk's many bioactive components, with specific physiological effects to support a healthy complexion from the inside. Lactoferrin-predominant whey protein is a patented ingredient for the reduction of blemishes and redness associated with a poor complexion [7]. The

Table 20.1 Spectrum of Efficacy of Topical and Systemic Agents in Therapeutic Management of Acne

Therapy	Mode of Action				Reported adverse effects include
	Keratolytic/ comedolytic	Sebum suppression	Anti-microbial	Anti-inflammatory	
Topical antibiotics	+	–	+++	+	Irritation of the skin, antibiotic resistance
Oral antibiotics	+	–	+++	+	Gastrointestinal discomfort, photosensitivity, pseudomembranous colitis, immunological hypersensitivity reactions, vaginal candidiasis, antibiotic resistance
Topical retinoids	+++	+	+	+	Dryness, redness, and irritation of the skin
Oral retinoids	+++	+	+	+	Skin and mucous membrane dryness, tiredness, muscle aches and pains, reduced night vision, headache, teratogenicity, adverse psychiatric event, hepatitis, hyperlipaemia
Oral hormones	–	+++	–	–	Melasma, headache, weight gain, depression, vascular thrombosis

–: none; +: mild; ++: moderate; +++: strong.

mechanisms of action for the internal support of healthy skin are yet to be elucidated; however, several pathways are suspected in supporting healthy skin, including: (a) calming irritated skin by neutralizing the release of lipopolysaccharides to modulate the activity of immune cells; and (b) suppressing the oxidation of a compound found in skin lipids called squalene. The effect of lactoferrin-predominant whey protein in promoting a healthy complexion from the inside has been nicely demonstrated in several consumer test market studies with both American and Asian teenagers.

20.2.1 Human Consumer Studies with Praventin™

The first human consumer study with lactoferrin-predominant whey protein was performed between September and December 2004 in the United States. Forty-four teenagers (23 males and 21 females, with an average age of 15.3 years) who reported having more than 10 acne lesions on their face received 200 mg of lactoferrin-predominant whey protein in a dosage of two chewable tablets twice daily for eight successive weeks. At baseline (week 0 before treatment started), and every 2 weeks thereafter, subjects filled out questionnaires about their skin condition with a final questionnaire at week 12 (4 weeks after the treatment stopped). Every 2 weeks front profile photographs were taken by an independent photographer and analyzed by an independent dermatologist [8]. The number of blackheads (open comedones) and nonblackheads (including whiteheads, closed comedones, papules, pustules, and nodulocystic lesions) were counted on the forehead, left cheek, right cheek, chin, and nose. For each subject blackheads and nonblackheads were summed over the facial regions each week. Data were analyzed by Wilcoxon Signed-Rank test in Stat/SE Version 8.2 (StataCorp, College Station, TX).

Oral supplementation with lactoferrin-predominant whey protein resulted in a considerable decrease in skin blemishes; an example of one of the results with one of the subjects can be seen in Fig. 20.2. The decrease in total blemishes over time for individual subjects is shown in Fig. 20.3 and Table 20.2. The results in Table 20.2 show a median decrease of 71% in complexion blemishes after one month and 95% after two months ($p < 0.001$). Using regression models, it was shown that the changes over time were not affected by gender ($p = 0.165$) or age ($p = 0.667$). Importantly, no adverse side affects were reported after 8 successive weeks of lactoferrin-predominant whey protein supplementation on a daily basis.

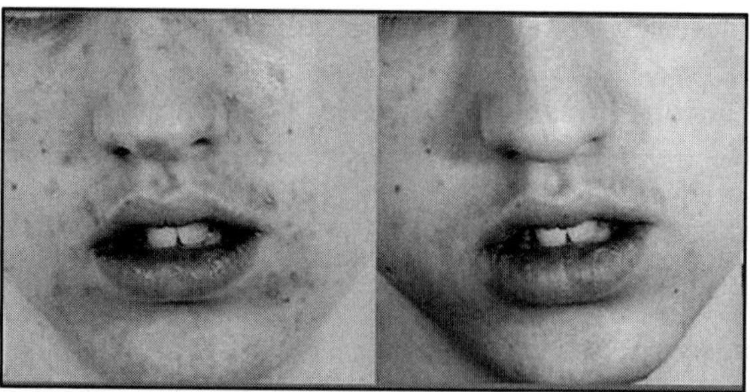

Figure 20.2 Example of one of the subjects of the first American consumer study in Delhi, New York, week 0 (left) compared to week 8 (right).

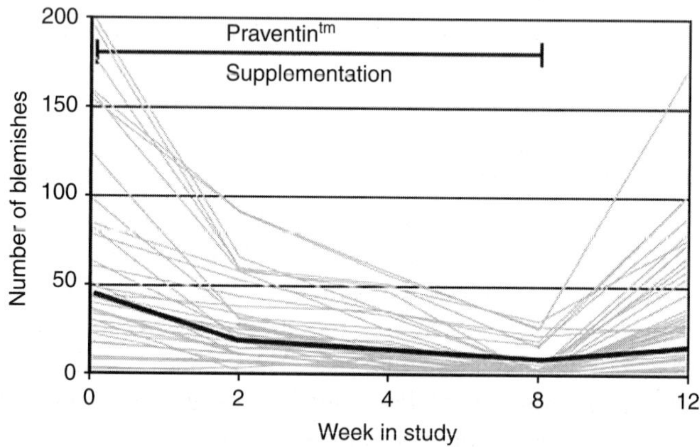

Figure 20.3 Total blemishes (blackheads + whiteheads) across time for individual subjects (grey), with median line superimposed (black) in first American consumer study.

In addition to these objective measurements, the teenagers were asked a set of questions about how they observed that lactoferrin-predominant whey protein had supported the reduction of blemishes and redness. Of all the subjects, 76% saw visible differences in their complexion, such as fewer blemishes, reduction in redness, and less oiliness. The teenagers reported "healthier looking complexion and healthy feeling." The majority (83%) stated they would like to continue taking lactoferrin-predominant whey protein and would advocate its use to their friends. The teenagers

Table 20.2 Overview of the Median Percent Reduction in Total Blemishes (Blackheads + Nonblackheads) of the First American Consumer Study: Overall and Stratified by Level of Severity at Baseline

Level of Severity	Week 2	Week 4	Week 8
All subjects (n = 44)	44.4%*	70.9%*	95.4%*
0 to 10 blemishes at baseline (n = 7)	12.5%	50%	87.5%**
11 to 50 blemishes at baseline (n = 20)	40.0%*	72.8%*	97.3%*
More than 50 blemishes at baseline (n = 17)	66.9%*	71.8%*	95.1%*

*p < 0.001, **p = 0.028.

found the supplementation of lactoferrin-predominant whey protein a safe and natural way that was effective for the support of the reduction in blemishes and redness associated with a poor complexion, which was very appealing to them.

A second follow-up consumer study with lactoferrin-predominant whey protein was performed between January and March 2006 in the United States in the same region as in the above-mentioned study. Forty-two teenagers, who reported having more than 10 acne lesions on their face, were stratified by acne severity and gender and randomized in two groups: group 1 (n = 15, average age 15.1 years) and group 2 (n = 13, average age 15.3 years). They received two chewable tablets twice daily containing a total of 25 mg or 200 mg of Praventin™, respectively, for a period of 8 weeks. Again, pictures were taken by a photographer and analyzed by an independent dermatologist [8]. Data were analyzed by Wilcoxon Signed-Rank test in Stat/SE Version 8.2.

The effects of lactoferrin-predominant whey protein intake on the number of blemishes, as shown in the first consumer study, was repeated in the second consumer study; Fig. 20.4 shows a picture of one of the teenagers as an example of the reduction in blemishes observed. Figure 20.5 shows the results: group 1 (25 mg Praventin™) did not show a decrease of total blemishes, whereas group 2 (200 mg Praventin™) did show a decrease over the 8-week period. Compliance with the chewable intake was again high, and no adverse side effects were reported.

Next to these two consumer studies with American teenagers, a pilot study in India took place between March and May 2006. Fourteen subjects

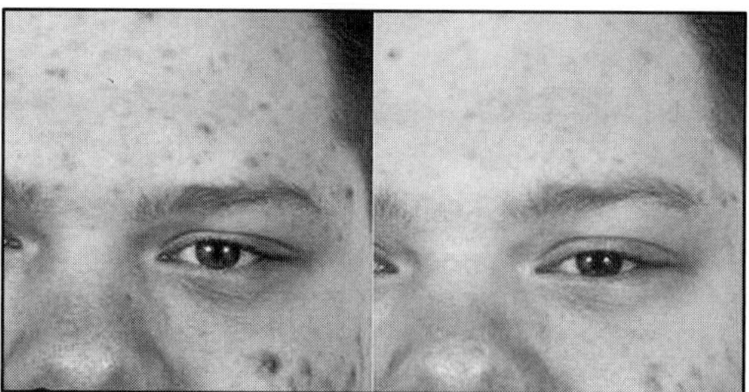

Figure 20.4 Example of one of the subjects of the second American consumer study in Delhi, New York, week 0 (left) compared to week 8 (right).

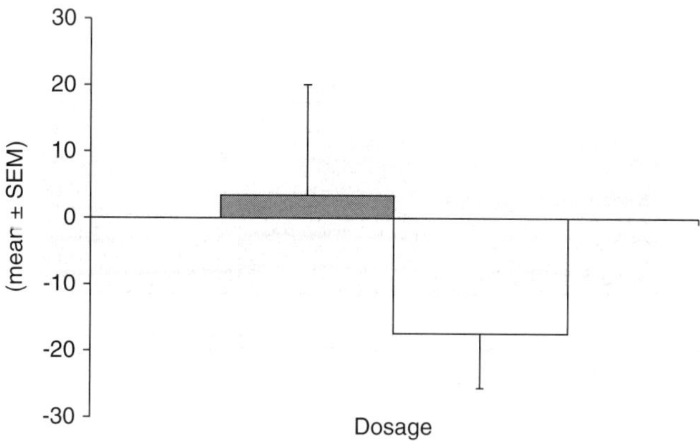

Figure 20.5 Change in total blemishes for individual subjects consuming lactoferrin-predominant whey protein at 25 mg (grey) and 200 mg (grey and white) dosages between baseline and week 8 in the second American consumer study.

having mild to moderate acne were enrolled in the study (4 males and 10 females, with an average age of 20.9 years [range 15–27 years]). The subjects received 200 mg of lactoferrin-predominant whey protein in chewable tablets for 12 successive weeks, to be consumed as two tablets twice daily. Again, front-profile photographs were taken at week 0 (baseline), which occurred before treatment started, as well as at weeks 2, 4, 8, and 12, and they were analyzed by a dermatologist.

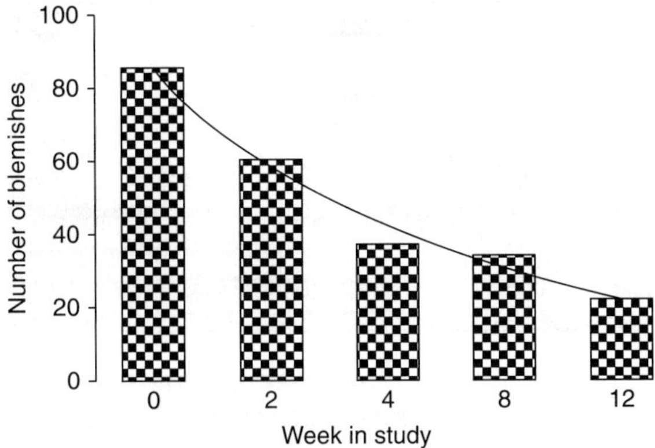

Figure 20.6 Mean total blemishes across time for all subjects in Indian pilot study (third consumer study).

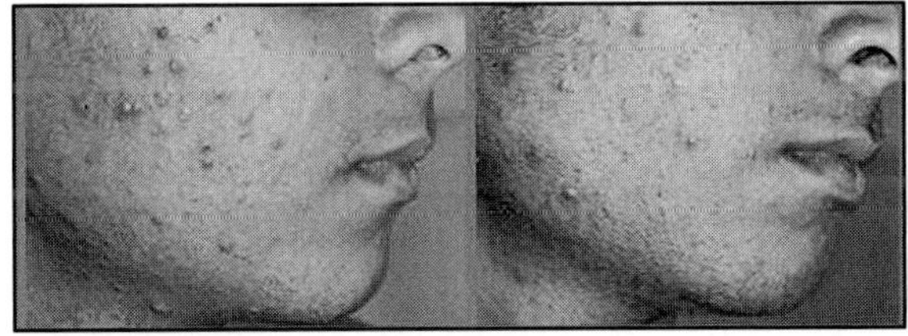

Figure 20.7 Example of one of the subjects of the Indian pilot consumer study in New Delhi, India, week 0 (left) compared to week 4 (right).

In the Indian subjects, the median number of total blemishes decreased by 31% at the end of week 2 and up to 76% at the end of week 12. Figure 20.6 shows the mean of total blemishes plotted against time for all the subjects. There was a significant decline in the number of inflammatory and non-inflammatory blemishes over time. Particularly, inflammatory blemishes (papules, pustules, and nodules) showed a sharper reduction in number and even disappeared in most individuals (results not shown). As an example of the effects of lactoferrin-predominant whey protein supplementation, Fig. 20.7 shows the results obtained for one of the Indian subjects. As per the interview details and subjective observations by the participants, the effect of treatment was fair to very good in 75% of the cases. Though new lesions appeared during the treatment, they were fewer in number, less inflamed, and less erythematous, and they healed in a short duration of 3–4 days.

20.2.2 Conclusion Human Consumer Studies

As demonstrated in three consumer studies, lactoferrin-predominant whey protein can be used to reduce the number and severity of blemishes associated with a bad complexion and can contribute positively to the physical and emotional well-being of teenagers and young adults. The effects of lactoferrin-predominant whey protein are observed in subjects with different skin types (i.e., Caucasian versus Asian) and different local diets. Furthermore, the effects of Praventin appear to be largely independent of gender or seasonal influences.

In contrast to reported side effects for retinoid-, antibiotic-, or hormone-based antiacne therapies (Table 20.2), these studies show that intake of lactoferrin-predominant whey protein on a long-term basis did not result in any adverse side effects.

20.2.3 Proposed Mechanism for Clear Complexion Supporting Effects of Lactoferrin-Predominant Whey Protein

As mentioned, lactoferrin-predominant whey protein is a complex mixture of different bioactive protein components in milk. Lactoferrin is part of this bioactive protein complex, and it is believed that lactoferrin, possibly in synergy with other compounds in the whey protein complex, plays an important role in the effects of lactoferrin-predominant whey protein [9–11]. Bovine lactoferrin has been shown to express several biological properties, some of which are likely relevant to the prevention and treatment of a healthy complexion; these properties include antimicrobial, anti-inflammatory, and antioxidant effects [12–16]. It may be hypothesized that Praventin™ acts via one or a combination of these three different mechanisms in supporting a healthy complexion.

20.2.3.1 Antimicrobial Activity

The antimicrobial action was investigated in a growth inhibition assay. *Propionibacterium acnes* (ATCC 11827) was cultivated in reinforced clostridial medium at 37°C in an anaerobic chamber under an atmosphere of 90% N_2, 5% CO_2, and 5% H_2. Bacteria were suspended at a concentration range of 10^6 CFU/ml in reinforced clostridial medium. Lactoferrin-predominant whey protein solutions (16% w/v) were made in sterile water. In a micro-plate setup, the bacterial suspension (100 μl)

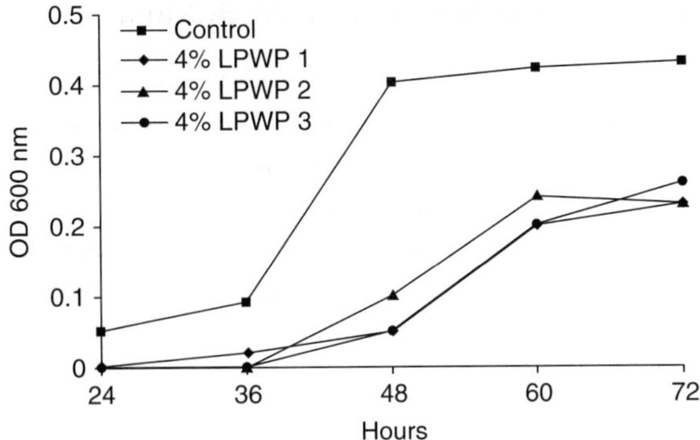

Figure 20.8 Effect of 4% lactoferrin-predominant whey on the growth of *P. acnes in vitro*. Effect of 4% lactoferrin-predominant whey on the growth of *P. acnes in vitro*.

was incubated together with lactoferrin-predominant whey protein (50 μl) and twice-concentrated reinforced clostridial medium (50 μl). The optical density (OD 600 nm) was measured at hours 24, 36, 48, 60, and 72. Figure 20.8 shows that multiple lactoferrin-predominant whey protein batches inhibit the growth of *P. acnes*. Likely, lactoferrin contributes to the antimicrobial properties against propionibacteria or other acne-related bacteria and this may be accomplished by mechanisms such as iron-scavenging, disruption of cell membranes, or inhibition of biofilm formation.

20.2.3.2 Anti-inflammatory Activity

Skin inflammation is associated with redness of the skin. The reduced redness in the human consumer studies is indicative of an anti-inflammatory effect of lactoferrin-predominant whey protein. The reduced erythema in the Indian pilot study also supports the anti-inflammatory effect of lactoferrin-predominant whey protein. Lactoferrin is well known for its anti-inflammatory effects in general. In the skin, lactoferrin may modulate the immune response in the anti-inflammatory direction, through binding of free iron, which catalyzes formation of free radicals. Free radicals play a role in the induction of inflammation. It may also neutralize the effects of the inflammation mediator, lipopolysaccharide (which is released by acne-related bacteria) or modulate the activity of, for instance, skin-specific immune cells. It has been shown in other studies that lactoferrin affects skin immunity [17,18].

20.2.3.3 Antioxidant Activity

Skin surface lipids and lipids from open and closed comedones in acne patients are enriched in polar lipids, as compared with skin surface lipids acquired from acne-free controls [19–21]. In both open and closed comedones, these polar lipids appeared to be derived mainly from the oxidation of squalene (compound of skin lipid), and for that reason it has been suggested that squalene oxidation is the link between comedogenesis and bacterial colonization, and furthermore that it plays an essential role in the pathogenesis of acne. Interestingly, in a recent study with 10 healthy volunteers it was demonstrated that topical application of lactoferrin-predominant whey protein is able to prevent cutaneous lipid peroxidation, as measured by the squalene:squalenehydroperoxide ratio (unpublished results). These results indicate that lactoferrin-predominant whey protein may support a healthy complexion by preventing squalene oxidation, and subsequent comedogenesis. It may therefore be suggested that lactoferrin-predominant whey protein exerts part of its antiacne activity by preventing the oxidation of squalene.

20.3 Conclusions of the Use of Lactoferrin-Predominant Whey Protein for Clear Skin

The statistical data obtained from all three studies with teenagers and young adults demonstrated that orally administered lactoferrin-predominant whey protein is a beneficial way to support the reduction of blemishes and redness associated with a poor complexion. Although the mechanism explaining the positive effects of lactoferrin-predominant whey protein has not yet been fully elucidated, it can be hypothesized that lactoferrin from whey plays an important role. It is likely that lactoferrin, in a synergistic fashion with the other components, plays an important role in its effects. A possible way of action may involve a combination of antimicrobial, anti-inflammatory, and antioxidant activity.

The majority of the study population testified to beneficial effects of lactoferrin-predominant whey protein intake: they noticed reduced blemishes, felt better and healthier, and were determined to continue intake of lactoferrin-predominant whey protein. The outcome of the studies forms a solid base to position lactoferrin-predominant whey protein as a natural ingredient to support a healthy complexion by helping to reduce the blemishes and redness associated with a bad complexion. In contrast to current systemic and topical antiacne therapies (based on antibiotics, retinoids, or other compounds), lactoferrin-predominant whey protein supplementation is not accompanied by side effects and can be used on a continuing basis.

Glossary

Acne vulgaris: A chronic inflammatory disorder of the pilosebaceous unit. The existence of a comedone is critical for the diagnosis.

Microcomedo: First stage in acne vulgaris: follicular plugging of the duct of a pilosebaceous unit (which consists of a hair follicle and a sebaceous gland).

Comedone: A comedone is a sebaceous follicle plugged with sebum, dead cells from inside the sebaceous follicle, tiny hairs, and sometimes bacteria (e.g., *P. acnes*). When a comedone is open, it is often named a blackhead because the exterior of the plug in the follicle has a dark appearance. A closed comedone is generally described as a whitehead because its appearance is usually that of a skin-colored bump (see Fig. 20.1).

Papule: An inflammatory comedone that resembles a small red bump on the skin.

Pustule: An inflammatory comedone that resembles a whitehead with a ring of redness around it.

Nodulocystic lesions: A severe form of acne that is characterized by copious deep, inflamed bumps (nodules) and large, pus-filled lesions that resemble sores (cysts). The severe inflammation can cause the acne to become exceedingly red or purple.

References

1. Pawin, H., C. Beylot, M. Chivot, M. Faure, F. Poli, J. Revuz, and B. Dreno. 2004. Physiopathology of acne vulgaris: recent data, new understanding of the treatments. Eur J Dermatol 14:4–12.
2. Smolinski, K. N., and A. C. Yan. 2004. Acne update: 2004. Curr Opin Pediatr 16:385–391.

3. Zouboulis, C. C., A. Eady, M. Philpott, L. A. Goldsmith, C. Orfanos, W. C. Cunliffe, and R. Rosenfield. 2005. What is the pathogenesis of acne? Exp Dermatol 14:143.
4. Chivot, M. 2005. Retinoid therapy for acne: a comparative review. Am J Clin Dermatol 6:13–19.
5. Simonart, T. 2004. Antibiotic-resistant acne: lessons from good sense. Br J Dermatol 150:369–370.
6. Thiboutot, D. 2004. Acne: hormonal concepts and therapy. Clin Dermatol 22:419–428.
7. De Waard, R., and A. L. Walter. 2006. Dermatological use of milk proteins, WO/2006/098625 A1.
8. Witkowski, J. A., and L. C. Parish. 2004. The assessment of acne: an evaluation of grading and lesion counting in the measurement of acne. Clin Dermatol 22:394–397.
9. German, J. B., C. J. Dillard, and R. E. Ward. 2002. Bioactive components in milk. Curr Opin Clin Nutr Metab Care 5:653–658.
10. Newburg, D. S. 2001. Bioactive components of human milk: evolution, efficiency, and protection. Adv Exp Med Biol 501:3–10.
11. Shah, N. P. 2000. Effects of milk-derived bioactives: an overview. Br J Nutr 84(Suppl 1):S3–S10.
12. Caccavo, D., N. M. Pellegrino, M. Altamura, A. Rigon, L. Amati, A. Amoroso, and E. Jirillo. 2002. Antimicrobial and immunoregulatory functions of lactoferrin and its potential therapeutic application. J Endotoxin Res 8:403–417.
13. Farnaud, S., and R. W. Evans. 2003. Lactoferrin—a multifunctional protein with antimicrobial properties. Mol Immunol 40:395–405.
14. Ward, P. P., and O. M. Conneely. 2004. Lactoferrin: role in iron homeostasis and host defense against microbial infection. Biometals 17:203–208.
15. Brock, J. 1995. Lactoferrin: a multifunctional immunoregulatory protein? Immunol Today 16:417–419.
16. Cumberbatch, M., M. Bhushan, R. J. Dearman, I. Kimber, and C. E. Griffiths. 2003. IL-1beta-induced Langerhans' cell migration and TNF-alpha production in human skin: regulation by lactoferrin. Clin Exp Immunol 132:352–359.
17. Moed, H., T. J. Stoof, D. M. Boorsma, B. M. von Blomberg, S Gibbs, D. P. Bruynzeel, R. J. Scheper and T. Rustemeyer. 2004. Identification of anti-inflammatory drugs according to their capacity to suppress type-1 and type-2 T cell profiles. Clin Exp Allergy 34(12):1868–1875.
18. Crouch, S. P., K. J. Slater, and J. Fletcher. 1992. Regulation of cytokine release from mononuclear cells by the iron-binding protein lactoferrin. Blood 80:235.
19. Chiba, K., K. Yoshizawa, I. Makino, K. Kawakami, and M. Onoue. 2000. Comedogenicity of squalene monohydroperoxide in the skin after topical application. J Toxicol Sci 25:77–83.
20. Downing, D. T., M. E. Stewart, P. W. Wertz, S. W. Colton, W. Abraham, and J. S. Strauss. 1987. Skin lipids: an update. J Invest Dermatol 88:2s–6s.
21. Saint-Leger, D., A. Bague, E. Lefebvre, E. Cohen, and M. Chivot. 1986. A possible role for squalene in the pathogenesis of acne. II. In vivo study of squalene oxides in skin surface and intra-comedonal lipids of acne patients. Br J Dermatol 114:543–552.

21

Nature Knows Best: Where Nature and Beauty Meet

Majda Hadolin Kolar, PhD, Simona Urbančič, BSc, and Dušanka Dimitrijević, MSc

Vitiva d.d., Markovci, Slovenia

Aaron Tabor and Robert M. Blair (eds.), *Nutritional Cosmetics: Beauty from Within*, 399–419,

Preface

Since ancient times, women have been turning to natural sources to help them emphasize their own beauty. Today, cosmetics are one of the most important and fastest growing businesses in the world. We are witnessing a new emerging market of products that are serving the purpose of enhancing a new concept called "beauty from within." Modern cosmetics, even better to say "cosmeceuticals," are not only focusing on topical creams and other preparations that help our skin look younger and slow down the aging process, but also offer solutions for attacking free radicals inside our body by providing simultaneous supplements of natural antioxidants.

There is a growing concern about the potential health hazard of synthetic ingredients (antioxidants, preserving materials, and other functional ingredients) for human health. Therefore, a renewed interest in the use of naturally occurring compounds is showing an increasing trend line. Because they occur in nature and in many cases are derived from plant sources, natural isolated compounds are presumed to be safe. Rosemary (*Rosmarinus officinalis*) extracts are widely used in the food, nutraceutical, and cosmetic areas. Their major bioactive components have shown antioxidant, antimicrobial, anti-inflammatory, anticarcinogenic, and chemopreventive activities. Phenolic diterpenes isolated from rosemary are compounds that have these beneficial properties. Major phenolic diterpenes occurring in fresh rosemary are carnosic and rosmarinic acid and their derivatives.

In this chapter we will present the results of antioxidative activity of rosemary bioactive compounds that have been studied in comparison to synthetic additives and other natural products used in the cosmetic industry and their benefits.

21.1 Introduction

The use of natural ingredients in cosmetic products has a very long history. Since ancient times, women have turned to the goodness of nature to help increase their own beauty. An increasing number of cosmetic products based on plant preparations are available in the market today. Production of bioactive compounds from biological materials and their use in nutrition, cosmetics, and pharmacy is, at the moment, a field in which intense scientific research is being performed. Bioactive ingredients represent compounds that have numerous specific pharmacological and technological values, such as natural antioxidants, natural preservatives, natural coloring agents, antimicrobiological active compounds, and others.

During processing and storing, food, cosmetics, and pharmaceutical products are exposed to environmental factors such as atmospheric composition, light, and temperature. These factors promote their spoilage. The following changes caused by oxidation may occur during cosmetic formulation storage: fragrance profile change, vitamin and active ingredient decomposition, color change, and development of rancidity. A major cause of this quality deterioration is the autoxidation of unsaturated lipids initiated by free radicals.

Antioxidants are crucial additives in cosmetic preparations for increasing their shelf life. Additionally, they can also be useful bioactive cosmetic ingredients to protect the skin against free radical formation induced by UV radiation and chemical environmental stress [1]. Natural extracts with antioxidant properties recently have gained popularity because many studies show that natural ingredients are better and safer than synthetic ones. The interest in natural antioxidants is further heightened by the suggestions that many of these compounds, such as plant phenolics, often possess antioxidant, antimicrobial, anti-inflammatory, chemopreventive, anticarcinogenic, antiatherogenic, and antitumor activity [2].

Rosemary leaves are well known for their essential oils, which are used in oral hygiene products, bath oils and massage fluids, etc. In addition, they are a source of highly active antioxidant compounds belonging to the group of diterpene phenols.

21.2 Review of Bioactive Compounds from Rosemary and Their Activity

The major phenolic diterpene present in fresh rosemary is carnosic acid, which is chiefly responsible for the antioxidant properties of rosemary extracts. Carnosic acid is converted to carnosol upon heating, and carnosol can degrade further to produce other degradation products. Carnosic acid–based rosemary extracts are oil-soluble.

Among various phenolics, rosmarinic acid is an important caffeoyl ester with proven medicinal properties and well-characterized physiological functions. Rosmarinic acid, the major active ingredient in water-soluble extracts, is found in substantial quantities in the family *Labiatae*, with medicinal uses in several cultures [3]. Rosmarinic acid is thought to be a part of the plant defense system against bacterial infections and predators.

Ursolic and oleanolic acids are pentacyclic triterpenoid compounds present in a large number of vegetarian foods, folkloristic botanical herbs, and other plants, and they are an integral part of the human diet. For a long time, they were considered to be pharmacologically inactive. Thus, ursolic acid and its alkali salts were exclusively used as emulsifying agents in pharmaceutical, cosmetic, and food preparations. However, upon closer examination, ursolic acid was found to be medicinally active both topically and internally. For the past 20 years, more than 1,000 articles have been published on the research of ursolic acid and its alkali salts, reflecting tremendous interest and progress in understanding these triterpenoids' effects and mechanisms of action. Ursolic acid's anti-inflammatory, anti-tumor (skin cancer), and antimicrobial properties make it useful in cosmetic applications. Ursolic and oleanolic acids are effective ingredients for internal and external skin care, imparting skin and hair protection. Following the literature data [4], they have the following activities: inhibiting the enzyme leukocyte elastase, which catalyzes rupture of the cell membranes in inflamed tissues while helping to maintain structural integrity of the skin; promoting collagen build-up and elastin synthesis, therefore helping to maintain the integrity of the skin; preventing wrinkling and aging and improving the appearance of photoaged skin; and increasing blood circulation in the skin and scalp. Ursolic and oleanolic acids have been shown to enhance hair growth and prevent scalp irritation, provide alopecia- and dandruff-preventing effects, inhibit the action of pro-inflammatory enzymes (thereby preventing skin inflammation), protect against topical skin cancer, potentially inhibit dental plaque formation and dental caries, and stabilize liposome membranes [4].

Structures of carnosic acid, rosmarinic acid, ursolic acid, and oleanolic acid are presented in Fig. 21.1.

Carnosic acid and its derivative carnosol are powerful inhibitors of lipid peroxidation in microsomal and liposomal systems as reported by Aruoma et al. [5]. This study confirmed effective peroxyl radical scavenging capacity of carnosol and carnosic acid in microsomal and liposomal model systems.

The potent chemoprotective activity of carnosic acid against aflatoxin B1 was reported by Costa et al. [6]. Because oxidative stress plays an important role in the toxicity mechanism of several mycotoxins such as aflatoxin B1, the use of natural or synthetic free radical scavengers could be a potential

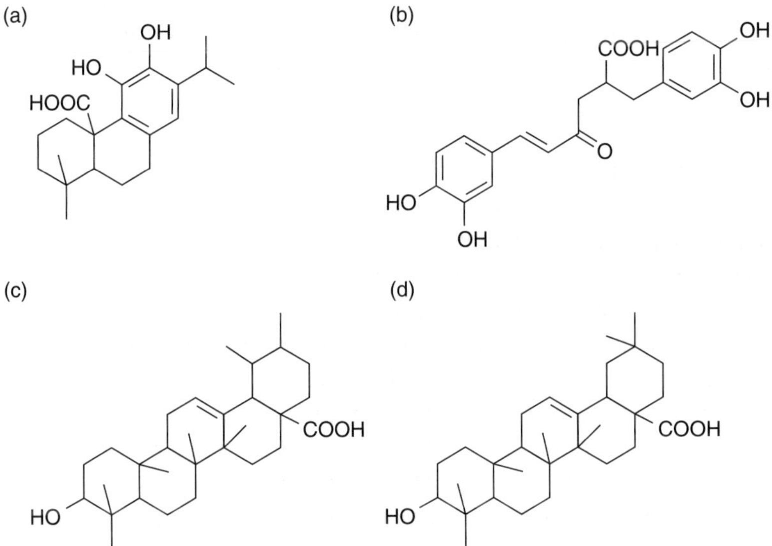

Figure 21.1 Chemical structures of antioxidative components from rosemary: (a) carnosic acid, (b) rosmarinic acid, (c) ursolic acid, (d) oleanolic acid.

chemopreventive strategy. A study made by Costa et al. [6] shows the considerable free radical scavenging capacity of carnosic acid.

Carnosic acid and carnosol were found to potentiate the antimicrobial activity of aminoglycosides [7]. The authors of this study report that a crude extract from *Salvia officinalis* (sage) reduced the minimum inhibitory concentrations (MIC) of aminoglycosides in vancomycin-resistant enterococci (VRE). Carnosol showed a weak antimicrobial activity and greatly reduced the MICs of various aminoglycosides (potentiated the antimicrobial activity of aminoglycosides) and some other types of antimicrobial agents in VRE. Carnosic acid, a related compound, showed similar activity. The effect of carnosol and carnosic acid with gentamicin was synergistic.

It has been reported [8] that carnosic acid and carnosol reduce membrane damage and lower DNA damage induced by oleic acid hydroperoxide (OAHPx). Carnosic acid and carnosol inhibited lipid peroxidation by 88–100% and 38–89%, respectively, under oxidative stress conditions. Both compounds significantly lowered DNA damage induced by OAHPx. Results of this study suggest that antioxidant activities of carnosic acid and carnosol could be partly due to their ability to increase or maintain glutathione peroxidase and superoxide dismutase activities.

Carnosic acid has antiplatelet activity [9]. In a study performed by Lee and co-authors [9], carnosic acid was reported to significantly inhibit collagen-, arachidonic acid-, U46619-, and thrombin-induced washed rabbit platelet aggregation. In agreement with its antiplatelet activity, carnosic acid blocked collagen-, arachidonic acid-, U46619-, and thrombin-mediated cytosolic calcium mobilization. Accordingly, serotonin secretion and arachidonic acid liberation were also inhibited in a similar concentration-dependent manner. However, in contrast to the inhibition of arachidonic acid–induced platelet aggregation, carnosic acid had no effect on the formation of arachidonic acid–mediated thromboxane A_2 and prostaglandin D_2, thus indicating that carnosic acid has no effect on the cyclooxygenase and thromboxane A_2 synthase activity. Overall, these results suggest that the antiplatelet activity of carnosic acid is mediated by the inhibition of cytosolic calcium mobilization and that carnosic acid has the potential of being developed as a novel antiplatelet agent.

Offord et al. [10] reported about photoprotective activity of carnosic acid. The photoprotective potential of the dietary antioxidants vitamin C, vitamin E, lycopene, beta-carotene, and the rosemary polyphenol, carnosic acid, was tested in human dermal fibroblasts exposed to ultraviolet-A

(UVA) light. The carotenoids were prepared in special nanoparticle formulations together with vitamin C and/or vitamin E. The presence of vitamin E in the formulation further increased the stability and cellular uptake of lycopene. UVA irradiation of the human skin fibroblasts led to a 10–15-fold rise in metalloproteinase 1 (MMP-1) mRNA. This rise was suppressed in the presence of low micromolar concentrations of vitamin E, vitamin C, or carnosic acid but not with beta-carotene or lycopene. In conclusion, vitamin C, vitamin E, and carnosic acid showed photoprotective potential.

Numerous laboratory studies indicate that rosemary extract, which is based on carnosic acid and derivatives, can prevent cancer-causing chemicals (carcinogens) from binding to and possibly mutating cellular DNA—two of the earliest steps in cancer development [11,12].

According to Perez-Fons et al. [2], diterpenes and genkwanin from rosemary show membrane-rigidifying effects. This may contribute to their antioxidant capacity through hindering diffusion of free radicals.

Nolkemper et al. [13] reported on the antiviral activity of aqueous rosemary extract. It shows very good results against *Herpes simplex* and offers a chance to use it for topical therapeutic application against recurrent herpes infections.

The application of rosemary extract in cosmetic dermatology was reviewed by Calabrese et al. [14], who concluded that aqueous rosemary extract is endowed with antioxidant activity and that it may therefore constitute an efficient pharmacological tool to control lipoperoxidative changes of the skin, thus highlighting the importance of a natural antioxidant biotechnology in the antiaging treatment of skin. Both *in vitro* and *in vivo* systems were used to confirm that aqueous rosemary extracts are capable of inhibiting oxidative alterations to skin surface lipids.

A rosmarinic acid–based extract may protect human skin against the harmful effects of UV radiation [15]. Because of its antioxidative properties, it could be used against oxidative stress–mediated disorders [16–19]. As an antioxidant, rosmarinic acid prevents cell damage caused by free radical reactions—reactions that are thought to be involved in inflammation, degenerative arthritis, and the aging process in general [16,17,19,20].

Rosmarinic acid could be used in cosmetics for its antioxidant properties and because it supposedly promotes collagen and elastin synthesis [21].

Its anti-inflammatory and antiviral properties make rosmarinic acid also of interest as a nutritional supplement. It is well absorbed from the gastrointestinal tract and from skin [21].

Rosmarinic acid shows promise as a preventive agent for atherosclerosis (the deposition of cholesterol-rich plaques in the artery walls), because rosmarinic acid and other polyphenols prevent the oxidation of LDL (low-density lipoprotein), and oxidized LDL is a primary instigator of plaque formation [21].

Sancheti and Goyal [22] reported about the antitumor-promoting activity of aqueous rosemary extract in Swiss albino mice. Oral administration of rosemary leaf extract, at a dose of 1,000 mg/kg b.wt./day at pre-, peri-, and post-initiational phases, was found to be effective in decreasing the tumor incidence (50%, 41.7%, and 58.3%, respectively) in comparison to the control (100%). Furthermore, the cumulative number of papillomas, tumor yield, and tumor burden were also found to be reduced in rosemary-treated animals. This was associated with significant alteration in liver peroxidation and glutathione levels.

Ursolic acid has been shown to have chemopreventive properties, as reported by Garg et al. [11]. Recent research has suggested that these plant polyphenols might be used to sensitize tumor cells to chemotherapeutic agents and radiation therapy by inhibiting pathways that lead to treatment resistance.

According to Ovesna et al. [23], ursolic and oleanolic acid posses anti-inflammatory, hepatoprotective, gastroprotective, cardiovascular, hypolipidemic, antiviral, antiatherosclerotic, and immunoregulatory effects. Both compounds have been shown to act at various stages of tumor development, including inhibition of tumorigenesis, inhibition of tumor promotion, and induction of tumor cell differentiation.

Liu [24] reported about the anti-inflammatory and antihyperlipidemic properties of ursolic and oleanolic acids in laboratory animals. Both compounds were reported as nontoxic, and their possible use in cosmetics and health products was discussed.

As suggested by Lee et al. [25], triterpenoids such as ursolic and oleanolic acids could be suitable material to develop antiaging reagents for skin. Both compounds induced the differentiation of keratinocytes. Oleanolic acid enhanced the recovery of epidermal permeability barrier function as well as increased ceramides in the epidermis after topical application.

According to Huang et al. [26], rosemary extract possesses skin tumorigenesis–inhibition properties. In this study a methanol extract of rosemary leaves was evaluated for its effects on tumor initiation and promotion in mouse skin. Application of rosemary to mouse skin inhibited the covalent binding of benzo(α)pyrene to epidermal DNA and inhibited tumor initiation by benzo(α)pyrene and 7,12-dimethylbenz(α)anthracene (DMBA). Topical application of 20 nmol benzo(α)pyrene to the backs of mice once weekly for 10 weeks, followed 1 week later by promotion with 15 nmol 12-O-tetradecanoylphorbol-13-acetate (TPA) twice weekly for 21 weeks, resulted in the formation of 7.1 tumors per mouse. In a parallel group of animals that were treated topically with 1.2 mg or 3.6 mg of rosemary 5 minutes prior to each application of benzo(α)pyrene, the number of tumors per mouse was decreased by 54% or 64%, respectively. Topical application of 0.1 μmol, 0.3 μmol, 1 μmol, or 2 μmol ursolic acid together with 5 nmol TPA twice weekly for 20 weeks to DMBA-initiated mice inhibited the number of tumors per mouse by 45–61%.

A study performed by Both et al. [27] demonstrated that ursolic acid incorporated into liposomes (URA liposomes) increases both the ceramide content of cultured normal human epidermal keratinocytes and the collagen content of cultured normal human dermal fibroblasts.

Based on before-mentioned activities of rosemary extract's main active ingredients (carnosic acid, carnosol, rosmarinic acid, ursolic acid, and oleanolic acid), incorporation of rosemary extracts or the individual active ingredients into food products, cosmetic products, or food supplements may contribute significant health benefits to consumers and also to the oxidative stabilization of products.

21.3 ORAC Values of Oil- and Water-Soluble Rosemary Extracts Compared to Vitamin C, Vitamin E, and Dried Fruits and Vegetables

Oxygen radical absorbance capacity (ORAC) is a method of measuring antioxidant capacities of different foods. One benefit of using the ORAC method to evaluate a substance's antioxidant capacity is that it takes into account samples with and without lag phases of their antioxidant capacities. This is especially beneficial when measuring foods and supplements that contain complex ingredients with various slow- and fast-acting antioxidants, as well as ingredients with combined effects that cannot be calculated [28]. The ORAC analysis provides a measure of the scavenging capacity of antioxidants against the peroxyl radical, which is one of the most common reactive oxygen species found in the body. RMCD, a cyclodextrin derivative, is utilized to enhance the solubility of the lipophilic samples in the aqueous solution. Trolox, a water-soluble vitamin E analog, is used as the calibration standard and the ORAC result is expressed as micromole Trolox equivalent (TE) per gram.

Figure 21.2 shows ORAC values of oil-soluble rosemary extract with 20% of carnosic acid as active ingredient (RE-20% CA), water-soluble rosemary extract with 40% rosmarinic acid as active ingredient (RE-40% RA) in comparison to vitamin E (90% of total tocopherols), vitamin C (100% pure), and dried fruits and vegetables. ORAC values of rosemary extracts were determined by Brunswick Laboratories. Data about vitamin C, vitamin E, and fruit and vegetables ORAC values were taken from the literature [29].

Oil-soluble rosemary extract RE-20% CA with 20% carnosic acid has an ORAC value of 3,199 μmol TE/g. As can be seen from Fig. 21.2, it

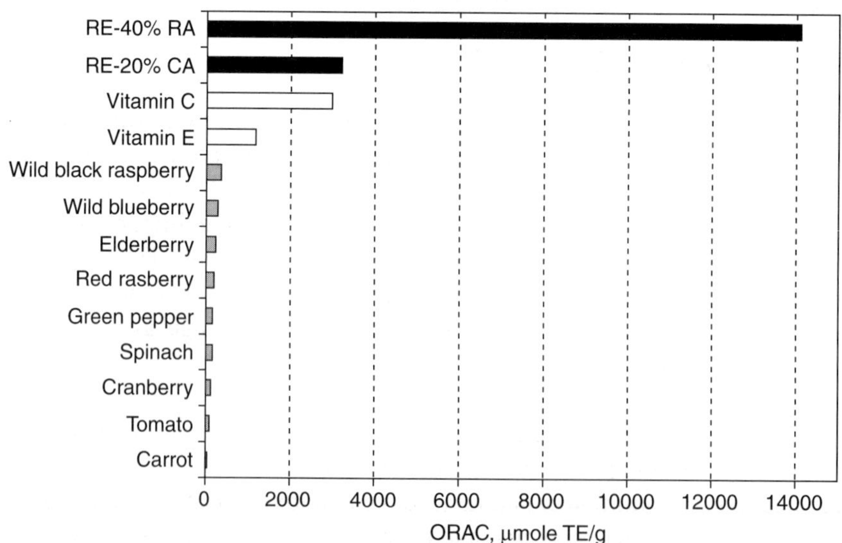

Figure 21.2 ORAC values of water-soluble rosemary extract RE-40% RA with 40% of rosmarinic acid as active ingredient and oil-soluble rosemary extract RE-20% CA with 20% of carnosic acid as active ingredient compared to vitamin C, vitamin E, and dried fruit and vegetable ORAC values.

possesses an ORAC value that is 10–60 times better than those of the dried fruits and vegetables evaluated, 2.7-fold better than that of vitamin E, and equal to vitamin C.

Water-soluble rosemary extract RE-40% RA with 40% rosmarinic acid has an ORAC value of 14,111 μmol TE/g and, following the ORAC value data from the literature, shows much better antioxidant capacity than dried fruits and vegetables. Its ORAC value was 12 times higher than the vitamin E ORAC value and more than 4 times higher than the ORAC values of vitamin C and oil-soluble rosemary extract RE-20% CA with 20% of carnosic acid as the active ingredient.

From these ORAC values we may conclude that oil- and water-soluble rosemary extracts, based on carnosic and rosmarinic acid as active ingredients, are suitable materials for food supplement antiaging preparations, and these properties are widely used in the supplement industry today.

21.4 Cosmetic Oil Rancidity Protection with Addition of Rosemary Extracts

21.4.1 Introduction

Many common natural oils such as echium oil, safflower oil, evening primrose oil, virgin walnut oil, and virgin borage oil are rich in polyunsaturated fatty acids, mainly linoleic (18:2n-6) and α-linolenic (18:3n-3) acids. These natural oils are of significant nutritional importance and are also desirable emollients for skin care applications. However, the weakness of unsaturated oils is their high sensitivity to oxidation and many of them represent limited shelf life during storage as well as in application [30].

In addition, natural oils are good sources of tocopherols and phytosterols, components offering both antioxidant activity and bioactivity for skin care applications. By addition of natural rosemary antioxidants, these compounds could be additionally protected and even deployed to synergistically work with rosemary extract in the final application.

The aim of the present study was to test the antioxidant efficiency of the rosemary extract formulation INOLENS 4, which contains 4% of carnosic acid as the active ingredient, compared to butylated hydroxytoluene (BHT), butylated hydroxyanisole (BHA), and mixed tocopherols (min. 90% of total tocopherols) in virgin borage oil, virgin walnut oil, evening primrose oil, virgin safflower oil, and refined echium oil. The rosemary extract was tested at final carnosic acid concentration of 40 mg/kg of oil. Dosage of BHT and BHA was 0.02% (0.2 g/kg). Mixed tocopherols (min. 90% of total tocopherols) were added at a concentration of 0.03% (0.3 mg/kg). For comparison, control samples without added antioxidants were also prepared and tested. A Rancimat test was done immediately after sample preparation.

21.4.2 Materials and Methods

21.4.2.1 Materials

Refined echium oil, evening primrose oil, virgin borage oil, virgin safflower oil, and virgin walnut oil were supplied by a local pharmacy. BHA and BHT were supplied by Merck Darmstadt, Germany. Mixed tocopherols with a purity of 90% of total polyphenols were purchased at Xinguang Technology Co., Ltd. of Sichuan Province, China. INOLENS 4, a rosemary extract–based formulation with 4% carnosic acid as the main active ingredient, was produced in Vitiva d.d., Slovenia.

21.4.2.2 Methods

Sample preparation. Rosemary extract–based formulation INOLENS 4 was added to the oil samples at concentration, which was equal to 0.004% (0.04 g/kg) of carnosic acid based on final product weight. Mixed tocopherols were added at a concentration of 0.03% (0.3 g/kg). Samples were well homogenized. Synthetic antioxidants BHA and BHT were added in a concentration of 0.02% (0.2 g/kg), and a sample without addition of antioxidant (the control) was tested.

Rancimat test. The oxidative stability was determined by the Rancimat test at a temperature of 100°C and air flow 25 l/h. Method is based on the conductometric determination of volatile products and features automatic plotting of the conductivity against time as well as automatic determination of induction times. The Rancimat test measures conductivity of low molecular–weight fatty acids produced during auto-oxidation of fats. The result is given as induction time. A higher induction time is indicative of better oxidative stability of the product.

Statistical analysis. Fisher's least significant difference (LSD) procedure was used to determine significant differences among the means of treatments at $P < 0.05$. With this method, there is a 5% risk of calling each pair of means significantly different when the actual difference equals zero. Statgraphics Centurion XV (Statgraphics Co., Tulsa, OK, USA) was used for statistical analysis.

21.4.3 Results and Discussion

Following the Rancimat test results, the rosemary extract–based formulation INOLENS 4 significantly increased ($P < 0.05$) the induction times of tested cosmetic oils compared to control samples. Rosemary extract protected refined echium oil 3.8 times better than mixed tocopherols ($P < 0.05$). Induction times of virgin safflower oil, evening primrose oil, virgin walnut oil, and virgin borage oil with added INOLENS 4 were 20%, 30%, 70%, and 45%, respectively, better ($P < 0.05$) than induction times of the same oils with added mixed tocopherols. Results of Rancimat test measurements of refined echium oil, virgin safflower oil, evening primrose oil, virgin walnut oil, and virgin borage oil are presented in Table 21.1 and Fig. 21.3.

Addition of rosemary extract–based formulation to the cosmetic oils extends their shelf life naturally and outperforms golden standards BHA, BHT, and mixed tocopherols. Beside its good antioxidant protection, the use of natural rosemary extract–based formulation INOLENS 4 may have other benefits such as: there are no legal restrictions on its use in the European Union, United States, and Japan; It has GRAS status in the United States; it is GMO- and allergen-free. In general, the application of rosemary extract that possesses antioxidant activity in cosmetic oils may be promising and useful for prolonging their shelf life.

21.5 Cosmetic Cream Rancidity Protection with Addition of Rosemary Extracts

21.5.1 Introduction

Lipid oxidation is the major form of deterioration in personal care preparations, even when the lipid content is very small. The rancidity that develops in cosmetic products causes changes in quality that effect their odor, color, texture, and appearance. The changes due to oxidation occur at

Table 21.1 Changes in Rancimat Test Results (Induction Time, Hours) of Refined Echium Oil, Virgin Safflower Oil, Evening Primrose Oil, Virgin Walnut Oil, and Virgin Borage Oil With and Without Addition of Antioxidants

	Samples			
Control	**0.02% BHA**	**0.02% BHT**	**0.03% mixed tocopherols**	**0.1% INOLENS 4**
Refined echium oil 0.46^{a}	0.57^{b}	0.64^{b}	0.78^{c}	2.96^{d}
Virgin safflower oil 4.74^{a}	6.02^{c}	5.17^{b}	5.29^{b}	6.31^{d}
Evening primrose oil 5.18^{a}	6.02^{b}	6.39^{c}	6.52^{c}	8.51^{d}
Virgin walnut oil 5.02^{a}	6.29^{b}	6.87^{c}	6.74^{c}	11.27^{d}
Virgin borage oil 8.43^{a}	10.80^{d}	9.82^{c}	8.64^{b}	12.57^{e}

Mean values ($n = 3$) with different letters in the same row are significantly different ($P < 0.05$).

different stages of cosmetics production and shelf life period, ranging from the raw materials, through production, packaging, storage, and costumer use period. The inhibition or retardation of oxidation by the addition of antioxidants is of considerable practical importance in preserving personal care products against spoilage. Due to negative effects of synthetic additives like cancerous properties, developmental/reproductive toxicity, allergies, immunotoxicity, organ system toxicity, endocrine disruption, and neurotoxicity [31], application of natural-based products becomes more and more important. Additionally, application of natural-based products may offer other benefits as described in Section 21.2.

The aim of the present study was to test the antioxidant efficiency of the oil-soluble rosemary extract formulation INOLENS 4 and water-soluble rosemary extract formulation AquaROX 6 compared to methyl parabene and BHA/methyl parabene combination in oil in water (O/W) and water in oil (W/O) cosmetic emulsions. The rosemary extract–based formulations INOLENS 4 and AquaROX 6 were tested at a concentration of 0.1% (1 g/kg) based on final product weight. Dosage of BHA was 0.1% (1 g/kg),

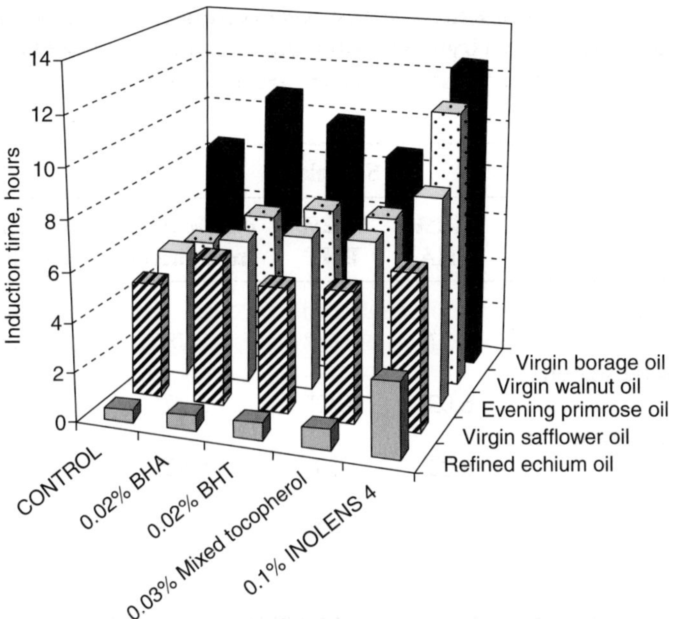

Figure 21.3 Results of Rancimat test (performed at 100°C) of refined echium oil, virgin safflower oil, virgin walnut oil, evening primrose oil, and virgin borage oil, with addition of synthetic antioxidants BHA and BHT, mixed tocopherols, rosemary extract–based formulation INOLENS 4, and without any antioxidants (control). BHA: butylated hydroxytoluene; BHT: butylated hydroxyanisole.

as per GMP. Concentration of added methyl parabene was 0.4%, which is the maximum allowed concentration in cosmetic preparations. For comparison control, samples without added antioxidants were also prepared and tested. The samples were stored at room temperature (22 ± 2°C) for 3 months and peroxide values were determined after storage time.

21.5.2 Materials and Methods

21.5.2.1 Materials

BHA was supplied by Merck Darmstadt, Germany. Methyl parabene was purchased at Sigma-Aldrich. INOLENS 4, a rosemary extract–based formulation with 4% carnosic acid as its main active ingredient, and AquaROX 6, a rosemary extract–based formulation with 6% of rosmarinic acid as its active ingredient, were produced in Vitiva d.d., Slovenia. Cosmetic creams (O/W and W/O emulsions) were prepared in a local pharmacy without addition of any preservative.

21.5.2.2 Methods

Sample preparation. Rosemary extract–based formulations INOLENS 4 and AquaROX 6 were added to the cosmetic cream samples at a concentration of 0.1% (1 g/kg) based on final product weight. Final concentrations of carnosic and rosmarinic acids in the cosmetic preparations were 0.04 g/kg and 0.06 g/kg, respectively. Samples were well homogenized. Comparisons were made against samples with addition of 0.1% (1 g/kg) of synthetic antioxidant BHA in combination with 0.4% (4 g/kg) of methyl parabene, samples with addition of 0.4% (4 g/kg) of methyl parabene, and samples without addition of antioxidant (the control). Methyl parabene and BHA are usually used as preservatives in cosmetic preparations. Samples were stored at room temperature for 3 months.

Peroxide value measurement. Peroxide value is a well-established method for the determination of primary oxidation products in fats and oils. The widely used iodometric titration method [32] for peroxide value determination is based on the measurement of the iodine produced from potassium iodide by peroxides present, and the result is expressed as millimoles of iodine per kg of lipid (mmol/kg).

Statistical analysis. Fisher's LSD procedure was used to determine significant differences among the means of treatments at $P < 0.05$. With this method, there is a 5% risk of calling each pair of means significantly different when the actual difference equals zero. Statgraphics Centurion XV (Statgraphics Co., Tulsa, OK, USA) was used for statistical analysis.

21.5.3 Results and Discussion

Addition of rosemary extract–based formulations INOLENS 4 and AquaROX 6 to the W/O emulsion extended its shelf life in terms of oxidative stability as shown by peroxide value measurement results after 3 months' storage at room temperature. Results are presented in Table 21.2 and Fig. 21.4. Oil-soluble rosemary extract–based formulation INOLENS 4 outperformed ($P < 0.05$) synthetic additives BHA and methyl parabene.

Table 21.2 Changes in Peroxide Values (mmol O_2/kg) of Water in Oil (W/O) and Oil in Water (O/W) Cosmetic Creams With and Without Addition of Antioxidants

Samples				
Control	**0.4% methyl parabene**	**0.1% BHA + 0.4% methyl parabene**	**0.1% INOLENS 4**	**0.1% AquaROX 6**
W/O				
12.50[a]	10.42[b]	10.48[b]	8.08[c]	10.07[b]
O/W				
9.97[a]	8.38[b]	7.02[c]	5.97[d]	5.70[e]

Mean values (n = 3) with different letters in the same row are significantly different ($P < 0.05$).

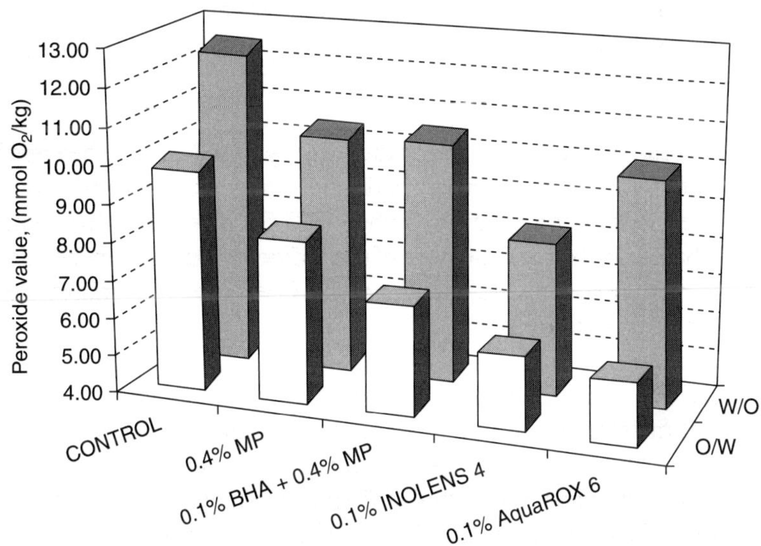

Figure 21.4 Peroxide value measurement results of cosmetic creams (oil in water [O/W] and water in oil [W/O] emulsions) with addition of methyl parabene (MP), BHA/methyl parabene (MP) combination, rosemary extract based–formulations AquaROX and INOLENS 4, and without any additives (control).

The peroxide value of W/O emulsion sample with addition of INOLENS 4 was 30% lower ($P < 0.05$) than peroxide values of samples with addition of methyl parabene and BHA/methyl parabene combination. The W/O emulsion sample with the addition of rosemary extract–based formulation

INOLENS 4 had a 50% lower peroxide value ($P < 0.05$) than the control sample after 3 months' storage at room temperature.

As can be seen from the results in the W/O emulsion, oil-soluble rosemary extract-based formulation INOLENS 4 possesses better antioxidative properties ($P < 0.05$) than the water-soluble rosemary extract-based formulation AquaRox 6. Opposite results were observed at O/W emulsion, where the water-soluble rosemary extract–based formulation showed slightly better results. These results are expected because a W/O emulsion contains more fats, which are very susceptible to oxidation and need to be protected with oil-soluble antioxidants.

Addition of rosemary extract–based formulations INOLENS 4 and AquaROX 6 to the O/W emulsion extended its shelf life 67% and 75% ($P < 0.05$), respectively. Samples with the addition of AquaROX 6 had a 47% lower peroxide value ($P < 0.05$) than samples with the addition of methyl parabene and a 23% lower peroxide value $P < 0.05$) than samples with the addition of BHA/metyl parabene combination. The O/W emulsion sample with the addition of rosemary extract–based formulation AquaROX 6 had a 75% lower peroxide value ($P < 0.05$) than the control sample after 3 months' storage at room temperature. As can be seen from these results, the O/W emulsion is more susceptible to oxidation than the W/O and, consequently, contribution of antioxidants to the final shelf life is more significant in the O/W emulsion.

21.6 Conclusion

Both tested rosemary extract–based formulations INOLENS 4 and AquaROX 6 showed antioxidant activity when added to cosmetic oil and/or cosmetic creams. In retarding lipid oxidation in cosmetic oils and cosmetic creams, rosemary extract–based formulations were, in general, more effective than commercial preservatives BHA, BHT, methyl parabene, and mixed tocopherols. Besides those results, rosemary extract–based formulations possess many beneficial properties, which are separately described in Section 21.2.

This study concluded that oil- and water-soluble rosemary extract formulations exhibit antioxidative properties when used in cosmetic oils and emulsions and may be promising and useful for prolonging their shelf life. In addition, rosemary extract formulations offer a natural (no legal restrictions on use in European Union, United States, and Japan; GRAS status in the United States; GMO- and allergen-free) alternative to synthetic preservatives, which possess negative effects on human health [31].

In general, the application of a natural rosemary extract in cosmetic preparations may improve their quality during time and offer positive influences on the skin itself. Based on many studies performed, some cited in Section 21.2, and relatively high ORAC value compared to vitamin C and vitamin E, rosemary extracts may be also promising active ingredients in "beauty from within" concept supplements.

References

1. Quirin K-W. Herbal CO2 extracts for skincare cosmetics. *Technology & Services, Business Briefing: Global Cosmetic Manufacturing*, 2004, 1–4.
2. Perez-Fons L., Aranda F.J., Guillen J., Villalain J. and Micol V. Rosemary (*Rosmarinus officinalis*) diterpenes affect lipid polymorphism and fluidity in phospholipid membranes. *Arch Biochem Biophys*, 2006, 453(2), 224–236.
3. Shetty K. Biosynthesis and medical application of rosmarinic acid. *Journal of Herbs, Spices & Medicinal Plants*, 2001, 8, 161–181.
4. http://www.ursolicacid.com
5. Aruoma O.I., Halliwell B., Aeschbach R. and Löligers J. Antioxidant and pro-oxidant properties of active rosemary constituents: carnosol and carnosic acid. *Xenobiotica*, 1992, 22(2), 257–268.
6. Costa S., Utan A., Speroni E., Cervellati R., Piva G., Prandini A. and Guerra M.C. Carnosic acid from rosemary extracts: a potential chemoprotective agent against aflatoxin B1. An in vitro study. *J Appl Toxicol*, 2006, 27(2), 152–159.
7. Horiuchi K., Shiota S., Kuroda T., Hatano T., Yoshida T. and Tsuchiya T. Potentiation of antimicrobial activity of aminoglycosides by carnosol from *Salvia officinalis*. *Biol Pharm Bull*, 2007, 30(2), 287–290.
8. Wijeratne S.S.K. and Cuppett S.L. Potential of rosemary (*Rosmarinus officinalis* L.) diterpenes in preventing lipid hydroperoxide-mediated oxidative stress in Caco-2 cells. *J Agric Food Chem*, 2007, 55(4), 1193–1199.
9. Lee J., Jin Y., Lee J., Yu J., Han X., Oh K., Hong J.T., Kim T. and Yun Y. Antiplatelet activity of carnosic acid, a phenolic diterpene from *Rosmarinus officinalis*. *Planta Med*, 2007, 73(2), 121–127.
10. Offord E.A., Gautier J.C., Avanti O., Scaletta C., Runge F., Krämer K. and Applegate L.A. Photoprotective potential of lycopene, beta-carotene, vitamin E, vitamin C and carnosic acid in UVA-irradiated human skin fibroblasts. *Free Radic Biol Med*, 2002, 32(12), 1293–1303.
11. Garg A.K., Buchholz T.A. and Aggarwal B.B. Chemosensitization and radiosensitization of tumors by plant polyphenols. *Antioxid Redox Signal*, 2005, 7(11–12), 1630–1647.
12. http://newgenerationlabs.com/immupro/

13. Nolkemper S., Reichling J., Stintzing F.C., Carle R. and Schnitzler P. Antiviral effect of aqueous extract from species of the Lamiacae family against *Herpes simplex* virus Type 1 and Type 2 in vitro. *Planta Med*, 2006, 72(15), 1378–1382.
14. Calabrese V., Scapagnini G., Catalano C., Dinotta F., Geraci D. and Morganti P. Biochemical studies of a natural antioxidant isolated from rosemary and its application in cosmetic dermatology. *Int J Tissue React*, 2000, 22(1), 5–13.
15. Psotova J., Svobodova A., Kolarova H. and Walterova D. Photoprotective properties of *Prunella vulgaris* and rosmarinic acid on human keratinocytes. *J Photochem Photobiol B*, 2006, 84(3), 167–174.
16. Economou K.D., Oreopoulou V. and Thomopoulos C.D. Antioxidant activity of some plant extracts in the family Labiatae. *J Am Oil Chem Soc*, 1991, 68, 109–113.
17. Hippolyte I., Marin B., Baccou J.C. and Jonard R. Growth and rosmarinic acid production in cell suspension cultures of *Salvia officinalis* L. *Plant Cell Rep*, 1992, 11, 109–112.
18. Morillas-Ruiz J.M., Villegas García J.A., López F.J., Vidal-Guevara M.L. and Zafrilla P. Effects of polyphenolic antioxidants on exercise-induced oxidative stress. *Clin Nutr*, 2006, 25(3), 444–453.
19. Soobrattee M.A., Neergheen V.S., Luximon-Ramma A., Aruoma O.I. and Bahorun T. Phenolics as potential antioxidant therapeutic agents: mechanism and actions. *Mutat Res*, 2005, 579(1–2), 200–213.
20. Kochan E., Wysokinska H., Chmiel A. and Grabias B. Rosmarinic acid and other phenolic acids in hairy roots of *Hyssopus officinalis*. *Z Naturforsch* 1999, 54c, 11–16.
21. http://www.lifelinknet.com/siteResources/Products/Rosmarinic-acid-complex.asp
22. Sancheti G. and Goyal P.K. Modulatory influence of *Rosmarinus officinalis* on DMBA-induced mouse skin tumorigenesis. *Asian Pac J Cancer Prev*, 2006, 6, 331–335.
23. Ovesná Z., Vachálková A., Horváthová K. and Tóthová D. Pentacyclic triterpenoic acids: new chemoprotective compounds. Minireview. *Neoplasma*, 2004, 51(5), 327–333.
24. Liu J. Pharmacology of oleanolic acid and ursolic acid. *J Ethnopharmacol*, 1995, 49(2), 57–68.
25. Lee H.K., Nam G.W., Kim S.H. and Lee S.H. Phytocomponents of triterpenoids, oleanolic acid and ursolic acid, regulated differently the processing of epidermal keratinocytes via PPAR-α pathway. *Exp Dermatol*, 2006, 15, 66–73.
26. Huang M.-T., Ho C.-T., Wang Z.Y., Ferraro T., Lou Y.-R., Stauber K., Ma W., Georgiadis C., Laskin J.D. and Conney A.H. Inhibition of skin tumorigenesis by rosemary and its constituents carnosol and ursolic acid. *Cancer Res*, 1994, 54, 701–708.
27. Both D.M., Goodtzova K., Yarosh D.B. and Brown D.A. Liposome-encapsulated ursolic acid increases ceramides and collagen in human skin cells. *Arch Dermatol Res*, 2002, 293, 569–575.
28. http://en.wikipedia.org/wiki/Oxygen_radical_absorbance_capacity
29. http://www.acaiberryjuice.org/orac.htm

30. Alander J., Andersson A.C. and Lindström C. Cosmetic emollients with high stability against photo-oxidation. *Lipid Technology*, 2006, 18(10), 226–230.
31. http://www.cosmeticdatabase.com
32. AOCS Official Method Cd 8-53, AOCS Press, 1996.

22

Probiotics for Skin Benefits

Audrey Gueniche[1], PharmD, PhD, Jalil Benyacoub[2], PhD, Stephanie Blum[2], PhD, Lionel Breton[1], PhD, Isabelle Castiel[1], PharmD, PhD

[1]*L'Oréal Recherche, Centre C. Zviak, Clichy Cedex, France*
[2]*Nestlé Research Center, Nestec Ltd., Lausanne, Switzerland*

Aaron Tabor and Robert M. Blair (eds.), *Nutritional Cosmetics: Beauty from Within*, 421–439, © 2009 William Andrew Inc.

22.1 Probiotics, Definition, and General Health Benefits

The term probiotic, popularized by R. Fuller in 1989 (Fuller, 1989), was recently defined by an expert committee as "living microorganisms, which, when consumed in adequate amounts, confer a health effect on the host" (FAO/WHO expert consultation, 2001).

Specific strains of probiotic lactic acid bacteria have been shown to beneficially influence the composition and/or metabolic activity of the endogenous microbiota (Langhendries et al., 1995; Macfarlane and McBain, 1999; Isolauri, 2001; Isolauri et al., 2001; Mohan et al., 2006) and some of these have been shown to inhibit the growth of a wide range of enteropathogens (Bernet-Camard et al., 1997; Coconnier et al., 1998). Competition for essential nutrients, aggregation with pathogenic micro-organisms (Rolfe, 2000), competition for receptor sites (Coconnier et al., 1993), and production of anti-microbial metabolites (Bernet-Camard et al., 1997; Coconnier et al., 1998) have all been reported to play a role.

Probiotics can be consumed in various forms of fermented or nonfermented food products. As a common feature, after ingestion, probiotics become transient constituents of the gut microbiota capable of exerting their biologic effects, thus giving a rationale for their use as a component of functional foods. Weaning, stress, dietary changes, use of antibiotics, and intestinal infections are all conditions that affect the natural balance of the intestinal microbiota for which the application of probiotics might be beneficial.

The most often used probiotic genera in humans and animals are enterococci, lactobacilli, and bifidobacteria, which are natural residents of the intestinal tract.

Multiple criteria have been defined for the selection of probiotic strains (reviewed by Ouwehand et al., 2002). Obviously, the most important criterium is that the selected strains has to be safe for use in the host and for the environment. One of the most commonly reported selection criteria is the ability to survive during passage through the gastrointestinal tract (GIT), for which the capacity of a strain to remain unaltered over conditions prevailing in the stomach (acidity) and intestinal tract (bile acids, pancreatic enzymes, and other digestive enzymes) is crucial. Adhesion to intestinal epithelial cells is considered important for immune modulation, pathogen exclusion, and prolonged residence time in the GIT. Viability of the probiotic strain is assumed to be important, and metabolic activity may be crucial for the expression of antipathogenic activity. There is increasing evidence that bacterial compounds such as DNA (some CpG motifs) or cell-wall fragments and/or dead bacteria can elicit certain immune responses (Lammers et al., 2003; Watson and McKay, 2006; Tejada-Simon and Pestka, 1999; Matsuzaki et al., 1998).

Although species-specific origin is thought to be important for host-specific interactions with the probiotic, there are examples of efficiency of non-species-specific probiotic strains. Indeed, health benefits have been shown using the yeast *Saccharomyces boulardii* in humans (Guslandi et al., 2000). Finally, probiotics need to have good technological properties to achieve high cell counts after fermentation at an industrial scale and to survive downstream processing, drying, food manufacturing, and long-term storage under adverse environmental conditions (e.g., humidity, high temperature, presence of oxygen).

The first aim of using probiotics has been to improve the composition of the intestinal microbiota from a potentially harmful composition towards a composition that would be beneficial to the host. Indeed, this approach is particularly relevant because the intestinal microbiota is known to play a major role in the physiological balance, intestinal development, and maturation of the host immune system (Macfarlane and McBain, 1999; Cebra, 1999; Isolauri, 2001; Isolauri et al., 2001). An adequate balance of the microbiota is crucial in maintaining good health conditions. In that sense, a decrease in clostridia and coliforms and an increase in lactobacilli and/or bifidobacteria has often been seen as evidence of healthy gut conditions (Roberfroid et al., 1995). In contrast, changes of the intestinal microbiota composition are associated with certain types of pathologies, in particular gastrointestinal infections, inflammation, and allergies (Roberfroid et al., 1995; Ouwehand et al., 2002).

Different studies reported that specific probiotic strains are able to positively influence the microbiota composition (Benno et al., 1996). In that respect, some probiotic strains have been successfully used to improve the outcome of gastrointestinal diseases, in particular diarrhea or *H. pylori* infections and related gastritis (Szajewska and Mrukowicz, 2001; Sarker et al., 2005; Midolo et al., 1995; Bergonzelli et al., 2005; Pochapin, 2000).

Beyond their capacity to promote a healthy pattern of the intestinal microbiota, several lines of evidence suggest that some probiotic bacteria can modulate the immune system both at the local and systemic levels (Cebra, 1999; Isolauri et al., 2001), thereby improving immune defense mechanisms, and/or downregulate immune disorders such as allergies or intestinal inflammation (Kalliomaki et al., 2001; Isolauri et al., 2000; Isolauri, 2001; Rautava and Isolauri, 2002).

Indeed, several strains of lactic acid bacteria were shown to modulate cytokine and growth factor production *in vitro* and *in vivo* (Miettinen et al., 1996; Borruel et al., 2003; Von der Weid et al., 2001; Christensen et al., 2002). Moreover, results from different preclinical and clinical trials have shown the capacity of various probiotic strains to enhance nonspecific and specific immunity (Perdigon et al., 1988; Schiffrin et al., 1995; Haller et al., 2000; Meydani and Ha, 2000; Kaila et al., 1992; Link-Amster et al., 1994).

22.2 Probiotics and Skin

Different human trials suggest that probiotic supplementation might be useful in the management of atopic dermatitis (Rautava and Isolauri, 2002; Kalliomaki et al., 2001, 2003, 2007). Based on these properties it appears that, beyond the gut, probiotics might exert their benefit at the skin level. Thus, it is assumed that some specific probiotic strains may be useful for the maintenance of cutaneous homeostasis and regulation of the skin immune system.

The skin plays a crucial role in protecting against dehydration and damage or insults from external aggressions, for example, chemical (pollution, tobacco, xenobiotics), mechanical, physical (UV radiation, changes in temperature and hygrometry), or infections. It is composed of a stratified epithelium with various cell types, including keratinocytes, whose differentiation results in building barrier function and, in a lower proportion, dendritic cells, melanocytes, and Langerhans cells. Each of these cell types contributes to skin protection. Moreover, the underlying dermal compartment harbors leukocytes, mastocytes, and macrophages that are key actors in cell defense.

Skin reflects the general health status and age of the host. Although skin aging is genetically programmed, the health and functions of the skin are also influenced by environmental factors, especially in exposed areas such as the face. Indeed, lifestyle, food, climate conditions, the extent and frequency of UV exposure, free radicals, toxins and allergens, xenobiotics, and mechanical damage are all exogenous factors suspected to alter skin health. Furthermore, hormonal status, immunological status, and psychological stress are endogenous factors that can alter skin quality and biological functions.

In this context, the skin can undergo various changes including immune dysfunction, inflammation, photoaging, dryness, wrinkles, dyschromia, and a variety of hyperplasia (Krutmann et al., 1996; Scharffeter-Kochanek et al., 2000).

22.2.1 Probiotics and Oral Route

22.2.1.1 Skin Allergic Reactions

The capacity of certain probiotic strains to modulate immune functions was a rationale for using probiotics to prevent and/or improve clinical outcome of diseases related to immune disorders such as allergies.

Recently, several researchers have focused on the suppressive effect of probiotic agents on allergic response, and have evaluated their prophylactic and/or therapeutic efficacy on atopic dermatitis, asthma, and food allergies (Rautava and Isolauri, 2002; Isolauri, 2001; Isolauri et al., 2001; Sawada et al., 2007; Kukkonen et al., 2007). Especially, perinatal administration of the probiotic *Lactobacillus rhamnosus* GG (LGG) has been shown to reduce the incidence of atopic dermatitis in children at risk during infancy (Kalliomaki et al., 2001, 2003, 2007). It is postulated that a boost in Th1 response, mainly IFN-γ production, may be associated and/or responsible for the beneficial effect of LGG on atopic dermatitis (Pohjavuori et al., 2004; Viljanen et al., 2005).

Moreover, another study showed that a *Lactobacillus casei* strain decreased specific-allergen contact hypersensitivity reaction in mice (Chapat et al., 2004).

22.2.1.2 Skin Aggression: Environmental Stress

Apart from pathological cutaneous disorders such as atopic dermatitis, the skin is continuously challenged by diverse environmental stress, which can later induce important alterations of the cutaneous homeostasis.

Indeed, the skin is known to be an immune-competent tissue and thus it is important for the protection of the host against infections and the control of cell malignancies (Woods et al., 2005; Euvrard et al., 2003). Several epidemiological studies demonstrated that UV radiation induces dramatic change in immune functions. This UV-induced alteration of the immune system is considered as one of the major risk factors for the development of certain skin cancers associated with sun overexposure (Ullrich, 1995, 2002, 2005). Among these changes, a decrease in number and morphological modifications of the Langerhans cells as well as an alteration of their capacity to present antigens have been proved (Cooper et al., 1985; Ullrich, 2005; Cooper et al., 2005; Seité et al., 2003). An increase in immunosuppressive cytokine levels such as IL-10 was also reported (Vink et al., 1996).

These skin disorders associated with dysregulation of immunological and/or neurosensitive mechanisms could be modulated or prevented by nutritional support and, in particular, by the use of certain probiotics (Salminen et al., 2005).

In this context, preclinical studies were performed to evaluate the effect of a diet supplemented with *Lactobacillus johnsonii* on the cutaneous immune system. Supplementation with this probiotic modulated the production of IL-10 in the serum and maintained Langerhans cell density at the site of UV exposure. Overall, the results showed that supplementation with *L. johnsonii* could prevent the deleterious effects of UV radiation on the skin immune system and reinforce host skin defense against antigenic challenges (Gueniche et al., 2006).

The same probiotic was tested in a randomized double-blind placebo-controlled study to evaluate its capacity to maintain skin homeostasis under UV exposure. Fifty-four volunteers were randomized into two groups (n = 27 per group) taking either *L. johnsonii* or a placebo daily for 6 weeks before UV exposure to 2 × 1.5 Minimal Erythemal Dose (MED). Biopsies of skin were analyzed to investigate the effects on the phenotype of skin immune cells and the mixed epidermal cell lymphocyte reaction (MECLR).

Lactobacillus johnsonii supplementation did not prevent UV-induced phenotypic maturation of dendritic Langerhans cells (LCs) or the decrease in MECLR in irradiated skin samples, one day post-irradiation. On day 4 after UV exposure, however, MECLR was still decreased in the placebo group with a parallel reduction in the CD1a LC marker in irradiated epidermis, whereas the allostimulatory capacity of epidermal cells was totally recovered in the La1 group correlating with normalization of

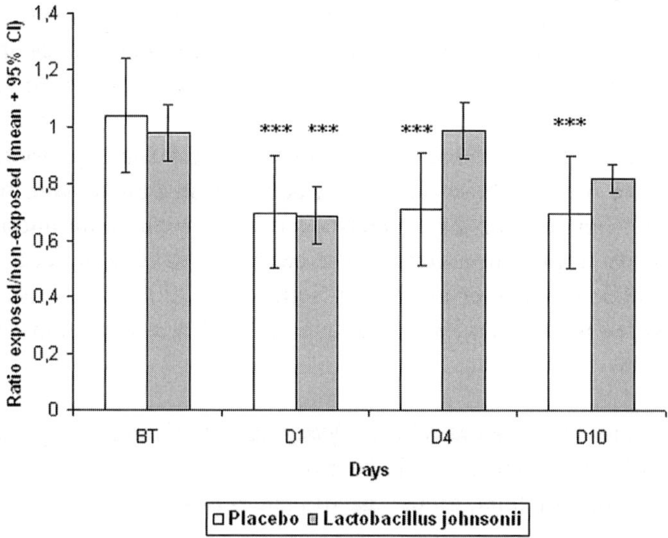

Figure 22.1 Results from MECLR: cpm ratios from exposed versus nonexposed skin samples. The ratios were calculated from 54 volunteers distributed among the *Lactobacillus johnsonii* and placebo groups. ***statistically significant differences at $p < 0.001$ between exposed and nonexposed sides. BT: before treatment.

CD1a expression within the epidermis (Fig. 22.1). Moreover, CD36+ monocytic cells colonized the epidermis one day post-irradiation in all subjects but disappeared faster in the La1 group, which may suggest that they differentiated into CD1a+ dendritic cells (Péguet Navarro et al., 2006).

These results show for the first time that ingested probiotic bacteria can accelerate the recovery of cutaneous immune homeostasis after UV exposure in humans and may therefore play a role in UV-induced skin damage prevention and photoprotection (Cooper et al., 1985; Euvrard et al., 2003; Ullrich, 1995, 2002, 2005; Vink et al., 1996; Woods et al., 2005).

22.2.1.3 Reactive Skin

Another indication for which the probiotics could be beneficial is to improve reactive skin symptoms. Reactive skin is characterized by marked sensitivity of the skin to physical (heat, cold, wind) or chemical (topical product application) stimuli and impaired ability to rebuild skin barrier function. Clinically, reactive skins are generally associated with important skin dryness (de Lacharrière et al., 2001; Pons-Guiraud, 2004).

Several epidemiological studies conducted in Europe and the United States reported that about half of women and a third of men showed sensitive skin (Jourdain et al., 2002; Misery et al., 2005; Pons-Guiraud, 2004; Primavera and Berardesca, 2005; Seidenari et al., 1998; Willis et al., 2001). Subjects with sensitive skin primarily complain of cutaneous discomfort. The main manifestations of this "cutaneous discomfort" are neurosensory signs such as feelings of heat, burning, stinging, or itching (de Lacharrière et al., 2001; Hosipovitch, 1999; Jourdain et al., 2002; Misery et al., 2005; Pons-Guiraud, 2004; Primavera and Berardesca, 2005; Seidenari et al., 1998; Willis et al., 2001). The signs may remain isolated or be associated with fleeting erythema. In most cases, the symptoms are limited to the face. Sometimes, other areas of the body (most often the scalp) could also be affected by a hyper-reactivity.

Several factors are implicated in the onset of reactive skin symptoms. These include environmental factors (temperature changes, heat, cold, wind, sun, air pollution, etc.), contact with certain products such as "hard" water, or internal factors (emotional factors, menstrual cycle, dietary factors, etc.) (de Lacharrière et al., 2001; Hosipovitch, 1999; Jourdain et al., 2002; Misery et al., 2005; Pons-Guiraud, 2004; Primavera and Berardesca, 2005; Seidenari et al., 1998; Willis et al., 2001). In most cases, skin hyper-reactivity is constitutional. In certain situations, a lowering of the cutaneous tolerance threshold may be acquired (sensitive skin induced by application of irritant products) or concomitant with an episode of skin disease.

Even though the skin of a person suffering from eczema is hyper-reactive (episodes of seborrheic dermatitis, rosacea, atopic dermatitis, or contact eczema), many cases of sensitive skin are not entirely related to allergic skin diseases. The etymological relationship between the terms "sensitive" and "sensitization" is undoubtedly responsible for the confusion that still exists between "allergic skin" and "sensitive skin."

Two types of mechanisms underline skin sensitivity: (a) the exacerbated reactivity of sensory nerves and/or (b) impaired barrier function, which contributes to a better accessibility of nervous fiber by exogenous potentially irritant compounds (de Lacharrière, 2002; Marriott et al., 2005).

During acute phases of skin sensitivity, neurogenic inflammation may be triggered (Schmeltz and Petersen, 2001; Steinhoff et al., 2003; Zegarska et al., 2006). This could have long-term consequences, because it may contribute to the maintenance of inflammatory conditions leading to chronicity.

Different studies confirm that there is an association between sensitive skin, the propensity to erythema, and dry skin (de Lacharrière et al., 2001; Misery et al., 2005; Pons-Guiraud, 2004; Primavera and Berardesca, 2005; Willis et al., 2001). Several authors have demonstrated that an impaired barrier function is associated with the onset of sensitive skin (Misery et al., 2005; Ohta et al., 2000; Yokota et al., 2003).

Recently, reactive or sensitive skin was classified into three different types according to physiological characteristics (Primavera and Berardesca, 2005). Type I was defined as the group with low barrier function. Type II was defined as the inflammatory group with a normal barrier function but an increased inflammatory status. Type III was defined as a pseudo-normal group in terms of barrier function and inflammatory status. It is interesting to note that the three types of reactive skin exhibit much slower restoration of barrier function following a skin lesion than that observed in subjects with nonreactive skin (Primavera and Berardesca, 2005).

Epidermal sensory nerves whose endings are in the stratum corneum play a key physiological role in sensitive skin symptom development. Thus, sensitive skin is considered as reflecting cutaneous sensory hyper-reactivity. Subjects with sensitive skin have an elevated neuropeptide level in the stratum corneum compared to the levels found in subjects with normal skin. Moreover, subjects of types I, II, and III reactive skin present a high sensitivity to electrical stimuli (Primavera and Berardesca, 2005).

Homeostatic hydration level of the epidermis is related to the status of the skin barrier and the interrelationships between the cells and their lipid environment. Lipids are involved in the rate of transepidermal water loss (TEWL). Disruption of the skin barrier primarily gives rise to an increase in TEWL (Rogiers and the EEMCO Group, 2001; Franz and Lehman, 2000; Wertz and Michniak, 2000; Idson, 1978; Nilsson, 1977; Pinnagoda et al., 1990).

Impairment of the skin barrier is most frequently associated with "dry" skin, or xerosis. The skin exhibits a dull color and appears fragile, with visible scaling. The smoothness of the skin is impaired. The feelings of tightness and tension may eventually result in pruritus. Skin penetration of

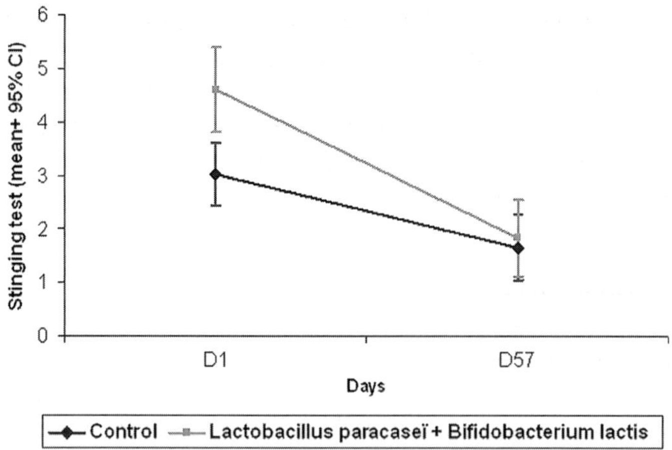

Figure 22.2 Results from the skin sensitivity evaluated by stinging test between volunteers distributed among the combination *Lactobacillus paracaseï* and *Bifidobacterium lactis* and placebo groups.

various compounds is increased compared to normal skin and the topical risks of infection and allergy are enhanced.

A mixed preparation of *Lactobacillus paracaseï* CNCM I-2116 and *Bifidobacterium lactis* CNCM I-3446 was tested in a randomized, double-blind, placebo-controlled trial. Sixty female volunteers (18–35 years old) with reactive skin ingested daily either probiotic (10^{10} cfu of each probiotic strain) (n = 33) or placebo (n = 33) powder suspended in drinking water for 8 weeks. Skin reactivity was assessed by a stinging test performed at the start of supplementation, middle, and end of the study. The results showed a significant decrease in cutaneous neurosensitivity ($p = 0.02$) in volunteers receiving the probiotic mix compared to those taking control powder at the end of the treatment (Fig. 22.2). These results provide the first clinical evidence that specific probiotic strains can have beneficial effects on skin reactivity and afford new opportunity to develop strategies to improve sensitive skin (Gueniche et al., 2007a, b).

22.2.1.4 Mechanisms

22.2.1.4.1 Indirect Action

The maintenance and protection of the gastrointestinal tract contributes to the overall host equilibrium. Although a direct relationship between probiotics and the bioavailability of nutrients has not yet been established, probiotics nonetheless positively influence gastrointestinal homeostasis, which contributes to promote the absorption of dietary nutrients at the intestinal mucosal level. This may help to provide essential nutrients for cell metabolism and the synthesis of the various functional and structural components of the skin.

22.2.1.4.2 Direct Action

The mechanisms whereby probiotics may play a role in skin physiology are not fully elucidated. However, it is proposed that, as shown for other commensal bacteria, probiotics could be directly sampled in the lumen by mucosal dendritic cells, which express tight junction proteins and penetrate the gut epithelial monolayer (reviewed by Uhlig and Powrie, 2003). It is postulated that upon interaction of the probiotic bacteria (or their components) with the intestinal epithelium and/or direct interaction with dendritic cells, other immune cells, such as B and T lymphocytes may be activated (primed) and immune mediators, including cytokines, may subsequently be released. These cytokines, bacterial fractions, and primed immune cells may be transported via the blood to other organs, including the skin, where they could modulate the immune status.

In addition, the improvement of reactive skin after probiotic supplementation could also result from a direct activity of the ingredient on neurosensitve mechanisms. On the one hand, immunoregulating properties of probiotics at the skin level could modulate the inflammatory reactions generated by the release of neuromediators involved in skin neurosensitivity (Nickoloff and Naidu, 1994; Ohta et al., 2000; Thurin and Baumann, 2003). On the other hand, the capacity of certain probiotics to modulate the production of regulating cytokines and growth factors (as Transforming growth factor beta) may play a role in the proliferation and differentiation of skin keratinocytes, which are important for skin barrier repair (Von der Wied et al., 2001). Such possible effects on the process of generating the stratum corneum allow the quality of the cutaneous barrier function and skin dryness to be improved (Thurin and Baumann, 2003).

22.2.2 Probiotics and Topical Application

If live probiotics appear as essential for applications in which their metabolic properties are important, some scientific groups showed that certain semiactive or dead oral probiotic preparations maintained some activities as compared to live forms (Bautista-Garcia et al., 2001; Galdeano and Perdigon, 2004; Sawada et al., 2007). Such semiactive and dead forms seemed to be interesting for topical applications aimed at skin beauty and health.

22.2.2.1 Antibacterial Activity

A number of authors have demonstrated that some bacterial extracts (*Bacillus coagulans, Lactobacillus johnsonii, Lactobacillus casei, Lactobacillus plantarum,* and *Lactobacillus acidophilus*) have antiadhesion and antimicrobial properties when applied to cutaneous and mucosal surfaces (Ouwehand et al., 2003).

The team of Schneedorf underlined that the association of bacterial strains contained in Kefir extracts (mixture of lactic bacteria and yeast) had antimicrobial effects, preferentially against *Streptococcus pyogenes*, that may support skin healing (Rodriguez et al., 2005).

The skin is naturally populated with different types of micro-organisms. Ideally, those with beneficial health effects, known as commensal micro-organisms, dominate over likewise occurring harmful bacteria. If this system gets out of balance, the appearance, health, and well-being of the skin are adversely affected. These disorders can occur, for example, after repeated washing or showering. In such cases *L. stimulans* can promote the rapid regeneration of the skin's protective microbial flora. By secreting growth-promoting substances, it stimulates colonization by health-promoting micro-organisms. Because the rapid alleviation of skin irritations is of high cosmetic relevance, these micro-organisms are interesting candidates for use in lotions or creams and also in medicinal ointments.

Another example of potential use of probiotics for skin health is *Lactobacillus pes-odoris*, which specifically inhibits the odor-producing bacteria of the feet and *L. ala-odoris*, which prevents the formation of odor in armpits. Both *Lactobacilli* cultures can improve the effectiveness of deodorants, foot sprays, or lotions.

22.2.2.2 Skin Integrity: Homeostasis

A recent study showed the benefit of topical application of a novel bacterial extract in aqueous solution on skin reactivity. A topical cream containing

Bifidobacterium longum lysate was tested in a randomized, double-blind, placebo-controlled trial. Sixty-six female volunteers with reactive skin were randomly given either the cream with the bacterial extract at 10% (n = 33) or control cream (n = 33). The volunteers applied the cream twice a day to the face, arms, and legs for 2 months. Skin sensitivity was assessed by stinging test (lactic acid), and skin barrier recovery was evaluated by measuring TEWL at days 1, 29, and 57 following barrier disruption induced by repeated tape-stripping. The results showed that the volunteers who applied the cream with bacterial extract had a significant decrease in skin sensitivity at the end of the treatment (day 57, $p = 0.0024$). Moreover, the treatment led to increased skin resistance against physical and chemical aggression compared to the group of volunteers who applied control cream. Noteworthy, the number of strips to obtain barrier function disruption was significantly increased in volunteers treated with the active ingredient compared to the control-treated group ($p = 0.0044$).

Clinical and self-assessment showed a significant decrease in skin dryness after 29 days in volunteers treated with the cream containing the 10% bacterial extract ($p = 0.03$) (Gueniche et al., 2007c).

Other authors have reported that certain extracts of lactic bacteria, such as those from *Streptococcus salivarium sp thermophilus* (commonly used in milk fermentation for the manufacture of yogurts and cheeses) when applied to the skin after sonication increased the level of ceramides, thus supporting the preservation of skin integrity (barrier function and flexibility of the stratum corneum) (Di Marzio et al., 2003). Altogether, these properties help fight against skin dryness. The same authors reported that when the extracts were introduced into a cream and then applied topically to patients affected by atopic dermatitis, skin ceramide levels were increased. This increase could be related to the hydrolysis of sphingomyelin by sphingomyelinases present in these bacterial extracts. These topical applications supported the disappearance of the signs and characteristic symptoms of the skin in atopic dermatitis patients (erythema, pruritus) (Di Marzio et al., 2003).

The results of these studies demonstrate that topical applications of a specific bacterial lysate or extract have a beneficial effect on reactive skin. These findings suggest that new approaches based on a bacterial lysate could be developed for the treatment and/or prevention of symptoms related to reactive and dry skin.

22.3 Conclusions and Perspectives

In conclusion, the current experimental and clinical data strengthen the assumption that certain probiotic strains or specific bacterial lysates or extracts exert their effects beyond the gut or topically applied directly to the skin and confer benefits at the skin level. There is indeed emerging evidence that such probiotics alive or in extract forms can contribute to the reinforcement of skin barrier function and modulate skin immune system, leading to the preservation of skin homeostasis.

Altogether the data afford the possibility of designing new strategies based on a nutritional approach for the treatment and/or prevention of UV-induced damaging effects, and of symptoms related to reactive skin or atopic skin or changes in skin homeostasis.

References

Bautista-Garcia CR, Ixta-Rodriguez O, Martinez-Gomez F, Lopez MG, Aguilar-Figueroa BR. Effect of viable or dead *Lactobacillus casei* organisms administered orally to mice on resistance against *Trichinella spiralis* infection. Parasite 8: S226–S228, 2001.

Benno Y, He F, Hosoda M, Hashimoto H, Kojima T, Yamazaki K, Lino H, Mykkanen H, Salminen S. Effects of *Lactobacillus GG* yogurt on human intestinal microecology in Japanese subjects. Nutr Today 31: 9S–11S, 1996.

Bergonzelli GE, Blum S, Brussow H, Corthezy-Theulaz IE. Probiotics as a treatment strategy for gastrointestinal diseases? Digestion 72: 57–68, 2005.

Bernet-Camard MF, Lievin V, Brassart D, Neeser JR, Servin AL, Hudault H. The human *Lactobacillus acidophilus* strain LA1 secretes a nonbacteriocin antibacterial substance(s) active in vitro and in vivo. Appl Environ Microbiol 63: 2747–2753, 1997.

Borruel N, Casellas F, Antolin M, Llopis M, Carol M, Espiin E, Naval J, Guarner F, Malagelada JR. Effects of nonpathogenic bacteria on cytokine secretion by human intestinal mucosa. Am J Gastroenterol 98: 865–870, 2003.

Cebra JJ. Influences of microbiota on intestinal immune system development. Am J Clin Nutr 69: 1046S–1051S, 1999.

Chapat L, Chemin K, Dubois B, Bourdet-Sicard R, Kaiserlian D. *Lactobacillus caseï* reduces CD8+ T cell–mediated skin inflammation. Eur J Immunol 34: 2520–2528, 2004.

Christensen HR, Frokiaer H, Pestka JJ. Lactobacilli differentially modulate expression of cytokines and maturation surface markers in murine dendritic cells. J Immunol 168: 171–178, 2002.

Coconnier MH, Bernet MF, Kerneis S, Chauviere G, Fourniat J, Servin AL. Inhibition of adhesion of enteroinvasive pathogens to human intestinal Caco-2 cells by *Lactobacillus acidophilus* strain LB decreases bacterial invasion. FEMS Microbiol Lett 110: 299–305, 1993.

Coconnier MH, Lievin V, Hemery E, Servin AL. Antagonistic activity against *Helicobacter* infection in vitro and in vivo by the human *Lactobacillus acidophilus* strain LB. Appl Environ Microbiol 64: 4573–4580, 1998.

Cooper KD, Neises G, Katz SI. Effects of ultraviolet radiation on human epidermal cell alloantigen presentation: initial depression of Langerhans cell-dependent function is followed by the appearance of T6-DR+ cells that enhance alloantigen presentation. *J Immunol.* 134: 129–137, 1985.

Cooper KD, Fox P, Neises G, Katz SI. Effects of ultraviolet radiation on human epidermal cell alloantigen presentation: initial depression of Langerhans cell-dependent function is followed by the appearance of T6- Dr+ cells that enhance epidermal alloantigen presentation. J Immunol 134: 129–137, 1985.

de LaCharrière O, Jourdain R, Bastien P, Garrigue JL. Sensitive skin is not a sub clinical expression of contact allergy. Contact Dermatitis 44: 131–132, 2001.

de LaCharrière O. Peaux sensibles, peaux réactives. Encycl Med Chir (ed. Scientifiques et Médicales, Elsevier SAS, Paris) Cosmétologie et Dermatologie Esthétique, 50-220-A-10, p. 4, 2002.

Di Marzio L, Centi C, Cinque B, Masci S, Giuliani M, Arcieri A, Zicari L, De Simone C, Cifone MG. Effect of lactic acid bacterium *Streptococcus ther mophilus* on stratum corneum ceramide levels and signs and symptoms of atopic dermatitis patients. Exp Dermatol 12: 615–620, 2003.

Euvrard S, Kanitakis J, Claudy A. Skin cancers after organ transplantation. N Eng J Med 348: 1681–1691, 2003.

FAO/WHO expert consultation. Joint expert consultation on evaluation of health and nutritional properties of probiotics in food including powdered milk with live lactic acid bacteria, 2001.

Franz TJ, Lehman PA. The skin as a barrier: structure and function. In: *Biochemical Modulation of Skin Reactions*. Kydonieus AF, Wille JJ (Eds.), CRC Press, Boca Raton, Florida, US, pp. 15–33, 2000.

Fuller R. Probiotics in man and animals. J Appl Bacteriol 66: 365–378, 1989.

Galdeano CM, Perdigon G. Role of viability of probiotic strains in their persistence in the gut and mucosal immune stimulation. J Appl Microbiol 97: 673–681, 2004.

Gueniche A, Benyacoub J, Buetler T, Smola H, Blum S. Supplementation with oral probiotic bacteria maintains cutaneous immune homeostasis after UV exposure. Eur J Dermatol 16: 511–517, 2006.

Gueniche A, Benyacoub J, Breton L, Bastien P, Bureau-Franz I, Blum S, Leclaire J. A combination of *Lactobacillus paracasei* CNCM I-2116 and *Bifidobacterium lactis* CNCM I-3446 probiotic strains decreases skin reactivity. J Invest Dermatol 102: S17, 2007a.

Gueniche A, Benyacoub J, Bureau-Franz I, Castiel I, Blum S, Leclaire J. Probiotic strain *Lactobacillus paracasei* CNCM I-2116 decreases skin reactivity. J Invest Dermatol 103: S18, 2007b.

Gueniche A, Bastien P, Breton L, Castiel I. A novel ingredient for reactive skin. J Invest Dermatol 149: S25, 2007c.

Guslandi M, Mecí G, Sorghi M, Testoni PA. *Saccharomyces boulardii* in maintenance treatment of Crohn's disease. Dig Dis Sci 45: 1462–1464, 2000.

Haller D, Bode C, Hammes WP, Pfeifer AM, Schiffrin EJ, Blum S. Non-pathogenic bacteria elicit a differential cytokine response by intestinal epithelial cell/leucocyte co-cultures. Gut 47: 79–87, 2000.

Hosipovitch G. Evaluating subjective irritation and sensitive skin. Cosmetics Toiletries 414: 41–42, 1999.

Idson B. In vivo measurement of TEWL. J Soc Cosmet Chem 29: 573–580, 1978.

Isolauri E. Probiotics in the prevention and treatment of allergic disease. Pediatr Allergy Immunol 12: 56S–59S, 2001.

Isolauri E, Arvola T, Sutas Y, Moilanen E, Salminen S. Probiotics in the management of atopic eczema. Clin Exp Allergy 30: 1604–1610, 2000.

Isolauri E, Sutas Y, Kankaanpaa P, Arvilommi H, Salminen S. Probiotics: effects on immunity. Am J Clin Nutr 73: 444S–450S, 2001.

Jourdain R, de Lacharrière O, Bastien P, Maibach HI. Ethnic variations in self-perceived sensitive skin: epidemiological study. Contact Dermatitis 46: 162–169, 2002.

Kaila M, Isolauri E, Soppi E, Virtanen E, Laine E, Laine S, Arvilommi H. Enhancement of the circulating antibody secreting cell response in human diarrhea by a human *Lactobacillus* strain. Pediatr Res 32: 141–144, 1992.

Kalliomäki M, Salminen S, Arvilommi H, Kero P, Koskinen P, Isolauri E. Probiotics in primary prevention of atopic disease: a randomised placebo-controlled trial. Lancet 357: 1057–1059, 2001.

Kalliomäki M, Salminen S, Poussa T, Arvilommi H, Isolauri E. Probiotics and prevention of atopic disease: 4-year follow-up of a randomised placebo-controlled trial. Lancet 361: 1869–1871, 2003.

Kalliomäki M, Salminen S, Poussa T, Isolauri E. Probiotic during the first 7 years of life: a cumulative risk reduction of eczema in a randomized, placebo-controlled trial. J Allergy Clin Immunol 119: 1019–1021, 2007.

Krutmann J, Ahrens C, Roza L, Arlett CF. The role of DNA damage and repair in ultraviolet B radiation–induced immunomodulation: relevance for human photocarcinogenesis. Photochem Photobiol 63: 394–396, 1996.

Kukkonen K, Savilahti E, Haahtela T, Juntunen-Backman K, Korpela R, Poussa T, Tuure T, Kuitunen M. Probiotics and prebiotic galacto-oligosaccharides in the prevention of allergy diseases: a randomized, double-blind, placebo-controlled trial. J Allergy Clin Immunol 119: 192–198, 2007.

Lammers KM, Brigidi P, Vitali B, Gionchetti P, Rizello F, Caramelli E, Matteuzzi D, Campieri M. Immunomodulatory effects of probiotics bacteria DNA: IL-1 and IL-10 response in human peripheral blood mononuclear cells. FEMS Immunol Med Microbiol 38: 165–172, 2003.

Langhendries JP, Detry J, Van Hees J, Lamboray JM, Darimont J, Mozin MJ, Secretin MC, Senterre J. Effect of fermented infant formula containing viable bifidobacteria on the fecal flora composition and pH of healthy full-term infants. J Pediatr Gastroenterol Nutr 21: 177–181, 1995.

Link-Amster H, Rochat F, Saudan KY, Mignot O, Aeschlimann JM. Modulation of a specific humoral immune response and changes in intestinal flora mediated through fermented milk intake. FEMS Immunol Med Microbiol 10: 55–63, 1994.

Macfarlane GT, McBain AJ. The human colonic microbiota. In: *Colonic Microbiota, Nutrition and Health*. Gibson GR, (Ed.), Kluwer Academic Publishers, The Netherlands, pp. 1–25, 1999.

Marriott M, Holmes J, Peters L, Cooper K, Rowson M, Basketter A. The complex problem of sensitive skin. Contact Dermatitis 53: 93–99, 2005.

Matsuzaki T, Yamazaki R, Hashimoto S, Yokokura T. The effect of oral feeding of *Lactobacillus casei strain Shirota* on immunoglobulin E production in mice. J Dairy Sci 81: 48–53, 1998.

Meydani SN, Ha WK. Immunologic effects of yogurt. Am J Clin Nutr 71: 861–872, 2000.

Midolo PD, Lambert JR, Hull R, Luo F, Grayson ML. In vitro inhibition of *Helicobacter pylori* NCTC11637 by organic acids and lactic acid bacteria. J Appl Bacteriol 79: 475–479, 1995.

Miettinen M, Vuopio-Varkila J, Varkila K. Production of human tumor necrosis factor, interleukin 6 and interleukin 10 is induced by lactic acid bacteria. Infect Immun 64: 5403–5405, 1996.

Misery L, Myon E, Martin N, Verriere F, Nocera T, Taieb C. Peaux sensibles en France: approche épidémiologique. Ann Dermatol Venereol 132: 425–429, 2005.

Mohan R, Koebnick C, Schildt J, Schmidt S, Mueller M, Possner M, Radke M, Blaut M. Effect of *Bifidobacterium lactis* Bb12 supplementation on intestinal microbiota of preterm infants. J Clin Microbiol 44: 4025–4031, 2006.

Nickoloff BJ, Naidu Y. Perturbation of epidermal barrier function correlates with initiation of cytokine cascade in human skin. J Am Acad Dermatol 30: 535–546, 1994.

Nilsson GE. Measurement of water exchange through skin. Med Biol Eng Comput 15: 209–218, 1977.

Ohta M, Hikima R, Ogawa T. Physiological characteristics of sensitive skin classified by stinging test. J Cosmet Sci Jpn 23: 163–167, 2000.

Ouwehand AC, Batsman A, Salminen S. Probiotics for the skin: a new area of potential application? Lett Appl Microbiol 36: 327–331, 2003.

Ouwehand AC, Salminen S, Isolauri E. Probiotics: an overview of beneficial effects. Antonie Van Leeuwenhoek 82: 279–289, 2002.

Peguet Navarro J, Dezutter-Dambuyant C, Buetler T, Leclaire J, Smola H, Blum S, Bastien P, Breton L, Gueniche A. Oral Skin Probiotic™ facilitate early recovery of cutaneous immune homeostasis after UV exposure in humans. J Invest Dermatol 128: S75, 2006.

Perdigón G, de Macias ME, Alvarez S, Oliver G, Ruiz Holgado AP. Systemic augmentation of the immune response in mice by feeding fermented milks with *Lactobacillus casei* and *Lactobacillus acidophilus*. Immunology 63: 17–23, 1988.

Pinnagoda J, Tupker RA, Agner T, Serup J. Guidelines for TEWL measurement. Contact Dermatitis 22: 164–178, 1990.

Pochapin M. The effects of probiotics on *Clostridium difficile* diarrhea. Am J Gastroenterol 95: S11–S13, 2000.

Pohjavuori E, Viljanen M, Korpela R, Kuitunen M, Tiittanen M, Vaarala O, Savilahti E. *Lactobacillus GG* effect in increasing IFN-gamma production in infant with cow's milk allergy. J Allergy Clin Immunol 114: 131–136, 2004.

Pons-Guiraud A. Sensitive skin: a complex and multifactorial syndrome. J Cosmet Dermatol 3: 145–148, 2004.

Primavera G, Berardesca E. Sensitive skin: mechanisms and diagnosis. Int J Cosmet Sci 27: 1–10, 2005.

Rautava S, Isolauri E. The development of gut immune responses and gut microbiota: effects of probiotics in prevention and treatment of allergic disease. Curr Issues Intest Microbiol 3: 15–22, 2002.

Roberfroid MB, Bornet F, Bouley C, Cummings JH. Colonic microflora: nutrition and health. Summary and conclusions of an International Life Sciences Institute (ILSI) [Europe] workshop held in Barcelona, Spain. Nutr Rev 53: 127–130, 1995.

Rodriguez KL, Caputo LRG, Carvalho JCT, Evangelista J, Schneedorf JM. Antimicrobial and healing activity of kefir and kefiran extract. Int J Antimicrob Agents 25: 404–408, 2005.

Rogiers V, The EEMCO Group. EEMCO guidance for the assessment of transepidermal water loss in cosmetic sciences. Skin Pharmacol Appl Skin Physiol 14: 117–128, 2001.

Rolfe RD. The role of probiotic cultures in the control of gastrointestinal health. J Nutr 130: 396S–402S, 2000.

Salminen SJ, Gueimonde M, Isolauri E. Probiotics that modify disease risk. J Nutr 135: 1294–1298, 2005.

Sarker SA, Sultana S, Fuchs GJ, Alam NH, Azim T, Brussow H, Hammarstrom L. *Lactobacillus paracasei* ST11 has no effect on rotavirus but ameliorates the outcome of nonrotavirus diarrhea in children from Bangladesh. Pediatrics 116: 221–228, 2005.

Sawada J, Morita H, Tanaka A, Salminen S, He F, Matsuda H. Ingestion of heat-treated *Lactobacillus rhamnosus GG* prevents development of atopic dermatitis in NC/Nga mice. Clin Exp Allergy 37: 296–303, 2007.

Scharffetter-Kochanek K, Brenneisen P, Wenk J, Herrmann G, Ma W, Kuhr L, Meewes C, Wlaschek M. Photoaging of the skin from phenotype to mechanisms. Exp Gerontol 35: 307–316, 2000.

Schiffrin EJ, Rochat F, Link-Amster H, Aeschlimann JM, Donnet-Hughes A. Immunomodulation of human blood cells following the ingestion of lactic acid bacteria. J Dairy Sci 78: 491–497, 1995.

Schmeltz M, Petersen LJ. Neurogenic inflammation in human and rodent skin. News Physiol Sci 16: 33–37, 2001.

Seidenari S, Francomano M, Mantovani L. Baseline biophysical parameters in subjects with sensitive skin. Contact Dermatitis 38: 311–316, 1998.

Seite S, Zucchi H, Moyal D, Tison S, Compan D, Christiaens F, Gueniche A, Fourtanier A. Alterations in human epidermal Langerhans cells by ultraviolet radiation: quantitative and morphological study. Br J Dermatol 148: 291–299, 2003.

Steinhoff M, Stander S, Seelinger S. Modern aspect of cutaneous neurogenic inflammation. Arch Dermatol 139: 1479–1488, 2003.

Szajewska H, Mrukowicz JZ. Probiotics in the treatment and prevention of acute infectious diarrhea in infants and children: a systematic review of published randomized, double-blind, placebo-controlled trials. J Pediatr Gastroenterol Nutr 33: S17–S25, 2001.

Tejada-Simon MV, Pestka JJ. Proinflammatory cytokine and nitric oxide induction in murine macrophages by cell wall and cytoplasmic extracts of lactic acid bacteria. J Food Prot 62: 1435–1444, 1999.

Thurin JT, Baumann N. Stress, Pathologie et Immunité. Flammarion (Coll Medecine-Sciences) (ed), Paris, 2003.

Uhlig HH, Powrie F. Dendritic cells and the intestinal bacterial flora: a role for localized mucosal immune responses. J Clin Invest 112: 648–651, 2003.

Ullrich SE. Mechanisms underlying UV-induced immune suppression. Mutat Res 571: 185–205, 2005.

Ullrich SE. Photoimmune suppression and photocarcinogenesis. Front Biosci 7: 684–703, 2002.

Ullrich SE. The role of epidermal cytokines in the generation of cutaneous immune reactions and ultra-violet radiation-induced immune suppression. Photochem Photobiol 62: 389, 1995.

Viljanen M, Pohjavuori E, Haahtela T, Korpela R, Kuitunen M, Sarnesto A, Vaarala O, Savilahti E. Induction of inflammation as a possible mechanism of probiotic effect in atopic eczema-dermatitis syndrome. J Allergy Clin Immunol 115: 1254–1259, 2005.

Vink AA, Faith M, Strickland FM, Bucana C, Cox PA, Roza L, Yarosh DB, Kripke M. Localization of DNA damage and its role in altered antigen-presenting function in ultra-violet-irradiated mice. J Exp Med 183: 1491–1500, 1996.

Von der Weid T, Bulliard C, Schiffrin EJ. Induction by a lactic acid bacterium of a population of CD4(+) T cells with low proliferative capacity that produce transforming growth factor beta and interleukin-10. Clin Diagn Lab Immunol 8: 695–701, 2001.

Watson JL, McKay DM. The immunophysiological impact of bacterial CpG DNA on the gut. Clin Chim Acta 364: 1–11, 2006.

Wertz PW, Michniak BB. Epidermal lipid metabolism and barrier function of stratum corneum. In: *Biochemical Modulation of Skin Reactions*. Kydonieus AF, Wille JJ (Eds.), CRC Press, Boca Raton, Florida, US, pp. 35–43, 2000.

Willis CM, Shaw S, de Lacharrière O, Baverel M, Reiche L, Jourdain R, Bastien P, Wilkinson JD. Sensitive skin: an epidemiological study. Br J Dermatol 145: 258–263, 2001.

Woods GM, Malley RC, Muller HK. The skin immune system and the challenge of tumour immunosurveillance. Eur J Dermatol 15: 63–69, 2005.

Yokota T, Matsumoto M, Sakamati T. Classification of sensitive skin and development of a treatment system appropriate for each group. IFSCC Mag 6: 303–307, 2003.

Zegarska B, Lelinska A, Tyrakowski T. Clinical and experimental aspects of cutaneous neurogenic inflammation. Pharmacol Rep 58: 13–21, 2006.

23

The Beauty of Soy for Skin, Hair, and Nails

Robert M. Blair, PhD, and Aaron Tabor, MD

Physicians Pharmaceuticals Inc., Kernersville, NC, USA

Aaron Tabor and Robert M. Blair (eds.), *Nutritional Cosmetics: Beauty from Within*, 441–468, © 2009 William Andrew Inc.

23.1 Introduction

Dietary soy consumption has been shown to have beneficial effects on several aspects of human health. Soy consumption has been reported to modestly improve plasma lipid profiles, improve bone health, reduce menopausal symptoms, enhance cognitive function, and potentially reduce the risk of breast and prostate cancers. The health benefits of dietary soy have been attributed to its isoflavones as well as to the biological actions of its constituent proteins. These potential health benefits of soy consumption have been extensively reviewed elsewhere [1,2] and will not be discussed in this chapter.

The amount of soy to consume in order to achieve appreciable health benefits has long been a topic of debate. Initial estimates were based on Asian population intakes because the incidences of breast and prostate cancer have been historically lower in Asian populations compared to the United States and Europe [3,4]; however, determining soy protein and isoflavone intake was problematic. Recently, it has been determined that typical soy intake by older Japanese adults is approximately 6–11 g soy protein and 25–50 mg soy isoflavones, though it is uncertain if this amount provides maximum health benefits [5].

The potential health benefits of soy have led to its inclusion in an ever-growing number of cosmetic products. The vast majority of these products are designed for topical application; however, it is becoming clearer that proper nutrition and a variety of dietary ingredients impact dermatological health. This chapter will discuss some basic information on soy and its biological actions and will elucidate the evidence regarding the potential benefits of soy and soy isoflavones for dermatological health. Additionally, we will discuss our own studies designed to explore the potential benefits of dietary soy for skin, hair, and nail appearance.

23.2 Biology of Soy

23.2.1 Nutritional Components of Soy

Soy is rich in macronutrients like protein, fat, and carbohydrates, and it contains a variety of micronutrients such as calcium, iron, zinc, riboflavin, and folate. These will be discussed briefly below; however, more detailed reviews are available [6,7].

23.2.1.1 Protein

Soybeans are best known as a rich source of nonanimal protein. The nutritional quality of soy protein has been extensively studied and reviewed [8,9]. Metabolic studies of nitrogen balance have been used to assess protein quality [10–12]. When nitrogen balance, digestibility, and net protein utilization were examined, no differences were found between beef and soy proteins [11]. Additionally, these investigators found no differences between these protein sources in the amount of nitrogen intake needed to maintain nitrogen equilibrium. Other studies have reported that both soy protein concentrate and isolated soy protein are capable of maintaining nitrogen balance [10,12].

A more recent study examined ileal digestibility and utilization of a soy protein isolate in adult human subjects [13]. The results of this study revealed that soy protein digestibility was 91% and that soy protein had a biological value of 86% and a postprandial net utilization of 78%, yielding a protein digestibility corrected amino acid score (PDCAAS) of 1.0. These values were similar to milk protein values (95% digestibility, 89% biological value, and 85% postprandial net utilization) reported in a separate study using the same experimental design [14].

The amino acid pattern of soy protein provides adequate levels of each indispensable amino acid for normal growth and development. The high digestibility of properly processed soy protein and the bioavailability of its amino acids and nitrogen content make soy protein a high quality protein. Based on PDCAAS calculations, soy protein achieves a score of 1.0, the highest score possible and on par with other high quality proteins like egg white and milk proteins. Therefore, the addition of soy to the diet is a viable way to meet all of one's protein nutritional requirements.

23.2.1.2 Fat

Soybeans typically contain more dietary fat than other legumes; however, the fats in soybeans are of the healthy varieties. According to the USDA National Nutrient Database for Standard Reference, Release 20 [15], raw, mature soybeans contain approximately 20 g of fat in a 100 g portion. Of the fat present in soybeans, approximately 15% is saturated, 24% is monounsaturated, and 61% is polyunsaturated; therefore, approximately 85% of the fat in soy is of the healthy, unsaturated kind. The predominant unsaturated fats found in soy include linoleic acid and alpha-linolenic acid, two essential fatty acids. Together with the monounsaturated oleic acid, these fatty acids make up nearly all of the unsaturated fat in soybeans. Alpha-linolenic acid is an essential omega-3 fatty acid that is metabolized in the body to form eicosapentaenoic acid and docosahexaenoic acid, two fatty acids with numerous reported health benefits [16,17].

23.2.1.3 Carbohydrates

Carbohydrates in soy, though present in only low levels, consist primarily of fiber and the oligosaccharides raffinose and stachyose [6,7]. The presence of the oligosaccharides can lead to flatulence in human beings due to their lack of alpha-galactosidase, the enzyme necessary for oligosaccharide digestion. However, the low level of carbohydrates in soybeans and their poor digestibility is responsible for the low glycemic index of soybeans and many soy products [18]. Furthermore, research has suggested that the oligosaccharides in soybeans may support intestinal health by acting as a probiotic and stimulating the growth of beneficial bacteria [19].

23.2.1.4 Vitamins and Minerals

Soybeans are a quality source of several vitamins and minerals including folate, calcium, zinc, and iron [6,7]. Folate, an essential B-vitamin, is present in raw, green soybeans at a level of 165 μg/100 g of soybeans, according to the USDA National Nutrient Database. A 100 g portion of raw soybeans also contains 197 mg calcium (~20% of the RDA), 3.55 mg iron (~20% of the RDA for women and ~40% of the RDA for men), and 0.99 mg zinc (~10% of the RDA). The bioavailability of these minerals from soy is an area of increasing research interest. Recent studies indicate that calcium-fortified soymilk is an excellent source of calcium [20] and that soybean ferritin, a source of iron, is readily bioavailable [21]. Additionally, zinc absorption has been reported to be similar between meat and soy proteins [22]. In contrast, it has been reported that non-heme iron and zinc are more poorly absorbed from soy protein than from beef protein [23] and that there is a decline in calcium bioavailability when meat protein is replaced by isoflavone-free soy protein [24].

23.2.1.5 Saponins

Saponins are glycoside compounds with a triterpenoid or steroid structure attached to water-soluble mono- or oligosaccharides found in a wide variety of plants [25]. The saponins present in soy, called soyasaponins, are found in a variety of concentrations ranging from 0.6–6.5% dry weight depending on soy variety, growing conditions, and degree of maturity [26]. Soyasaponins have been classified into two main groups, Group A and Group B, based on their aglycone structure [27,28]. Though soyasaponins are generally considered to have a low bioavailability, they have been shown to have a multitude of biological actions and potential health benefits [29,30].

23.2.2 Soy Isoflavones

Though soy has been shown to have a variety of health benefits related to both the protein and isoflavone components of the soybean, it is generally thought that most of the beneficial effects of soy are due to its isoflavones.

23.2.2.1 Structure

Isoflavones belong to the flavonoid class of phytochemicals, which also includes flavonols, flavones, flavonones, anthocyanins, and proanthocyanins. Isoflavones are structurally characterized by two benzene rings linked to a central heterocyclic pyrane ring and are produced in a specific pathway of flavonoid biosynthesis [31].

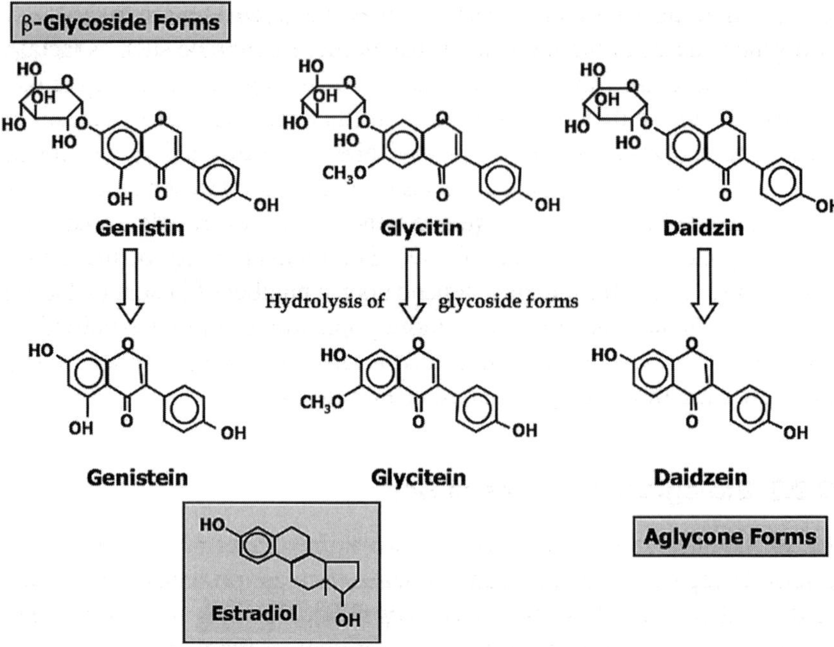

Figure 23.1 Structures of the β-glycoside and aglycone forms of soy isoflavones. Estradiol is shown in the inset to demonstrate the structural similarities between estradiol and the soy isoflavones.

The three primary isoflavones present in soy are found in either glycoside or aglycone forms. The glycoside forms include the β-glycosides (genistin, daidzin, and glycitin) as well as the acetyl- and malonyl-glycosides [32,33]. The glycoside forms are the predominant isoflavone form in the soybean; however, it is the aglycone forms (genistein, daidzein, and glycitein) that are the best studied and have reported health benefits. The structures of the β-glycoside and aglycone isoflavones and their structural similarity to estradiol are depicted in Fig. 23.1.

23.2.2.2 Metabolism

It is generally accepted that the glycoside forms of the soy isoflavones are poorly absorbed intact [34]. Therefore, after consumption, the soy isoflavone glycosides must undergo metabolism in the intestinal tract prior to absorption. In the small intestine, the isoflavone glycosides are hydrolyzed by β-glucosidases of the intestinal brush border membrane such as lactase phlorizin hydrolase [35]. After conversion to the aglycone form, the soy isoflavones may either be absorbed or further metabolized by intestinal bacteria to such metabolites as p-ethyl phenol, O-desmethylangolensin, and equol [36–38]. Equol production is of particular interest because it has been implicated in a variety of health benefits; however, only about 30% of the population produces equol [39]. The bioavailability of the isoflavones is highly variable and may depend upon a number of factors including form of soy intake, intestinal microflora populations, and inter-individual differences [2]. The metabolism, absorption, and bioavailability of soy isoflavones are thoroughly reviewed elsewhere [40,41].

23.2.3 Biological Activities of Soy

The isoflavones in soy have been shown to have a variety of biological actions. In regard to human health benefits, they are probably best known for the ability to bind estrogen receptors (ER). However, soy isoflavones also have a number of non-ER-mediated biological functions.

23.2.3.1 Estrogen Receptor Binding

The isoflavones present in soy have been shown to bind and activate both ERα and ERβ [42,43], though with much less potency than estradiol. Binding of the soy isoflavones to ERβ is generally more potent than binding to ERα [42,43]. It has been suggested that the soy isoflavones may act more like a selective estrogen receptor modulator than an estrogen because genistein binds to the ER more like raloxifene than estradiol [44]. Additionally, it has been suggested that isoflavones may act as estrogen antagonists in a high estrogen environment, but act as estrogen agonists in a low estrogen environment [45]. Many of the purported health benefits of soy, especially in relation to menopausal symptoms, osteoporosis, and cardiovascular disease, have been suggested to be due to the isoflavones' ability to bind to ER.

23.2.3.2 Non-ER-Binding Activities

Soy and soy isoflavones can affect biological systems in a variety of ways other than through ER binding. It has long been known that the isoflavone genistein is a potent tyrosine kinase inhibitor [46] and thus may inhibit cell proliferation and differentiation. Interestingly, the tyrosine kinase effects of genistein have been postulated to be irrelevant to genistein's potential health benefits due to the doses required to observe these effects [47].

It has also been reported that soy and soy isoflavones have antioxidant functions, which may be partly responsible for the health benefits of soy. Soy isoflavones have been reported to reduce lipid peroxidation [48,49], scavenge oxidative free radicals [50], reduce oxidative damage to DNA [51], and upregulate the expression of antioxidant genes [52]. In contrast, it has been suggested that soy has minimal antioxidant functions [53].

Soy isoflavones have also been reported to inhibit some steroidogenic enzymes. Early studies suggested that soy isoflavones weakly inhibit aromatase [54,55], the enzyme responsible for the conversion of androgens to estrogens. Additional studies have demonstrated that soy isoflavones inhibit other enzymes in the steroidogenic pathway including 3β- and 17β-hydroxysteroid dehydrogenase [56,57] and 5α-reductase [58]. These enzyme-inhibitory actions may have beneficial effects in regard to cancer risk.

23.3 Soy for Skin Care

23.3.1 *In Vitro* Evidence

The majority of *in vitro* research to date on the potential benefits of soy for skin health has focused on the soy isoflavones and their effects on skin aging and photocarcinogenesis. Both photocarcinogenesis and extrinsic skin aging have been attributed to the damaging effects of solar ultraviolet (UV) radiation. Some early research demonstrated that 10 μM genistein treatment of UVA-exposed human epidermoid carcinoma cells suppressed the production of 8-hydroxy-2′-deoxyguanosine (8-OHdG), a form of oxidative DNA damage [59]. The ability of genistein to reduce oxidative DNA damage in this system was suggested to be due to its ability to scavenge free oxygen radicals.

Recent research has confirmed the antioxidant and DNA-protective effects of soy isoflavones. Treatment of human epidermal carcinoma A431 cells with high doses of genistein (>100 μM) has been shown to block UV irradiation–induced apoptosis [60]. The protective effect of gensitein in this study appeared to be due to a reduction in intracellular oxidative stress because genistein pretreatment suppressed the UV irradiation–induced increase in intracellular reactive oxygen species. Using an *in vitro* full thickness, three-dimensional human reconstituted skin model (EpiDermFT™), Moore et al. [61] demonstrated that pretreatment with genistein (10 μM, 20 μM, and 50 μM) resulted in a dose-dependent reduction in UVB-induced DNA damage as assessed by pyrimidine dimer formation and a preservation of the skin's histological architecture. Furthermore, these authors reported that genistein treatment suppressed UVB-induced reduction in proliferating cell nuclear antigen (PCNA), suggesting that genistein dose-dependently maintained the skin cells' ability to proliferate and repair in the face of UVB radiation. In addition to its antioxidant properties, it has been reported that genistein has anti-inflammatory properties in normal human keratinocytes [62].

Soy isoflavones have also been reported to support skin health by additional mechanisms, which may or may not be related to their antioxidant properties. Kim et al. [63] reported that treatment of human fibroblasts with a soy isoflavone extract consisting of predominantly genistein and daidzein at a dose of 10 ppm significantly reduced UV-induced matrix metalloproteinase-1 secretion. Similar, but non-significant, results were observed with 1 ppm isoflavone treatment. Both purified genistein and varying soy extracts have been shown to have positive effects on primary human dermal fibroblasts [64]. In their study, Sudel et al. [64] reported that *in vitro* treatment of human fibroblasts with purified genistein (100 ng/ml) resulted in increased collagen synthesis. Additionally, treatment with two different soy extracts, one containing approximately 20% isoflavones and one containing about 11% isoflavones plus 14% soy saponins, also enhanced fibroblast collagen synthesis. Sudel et al. [64] further reported that treatment of dermal fibroblasts with the soy extract containing both isoflavones and saponins increased hyaluronan and sulfated glycosaminoglycan synthesis, suggesting that not only the soy isoflavones may have benefits for skin health.

Other studies have also suggested that the benefits of soy for skin health do not appear to be due solely to the presence and activity of the isoflavones. Application of a soy peptide (Phytokine®) to an *in vitro* skin equivalent model resulted in a significant increase in hyaluronic acid and a tendency for increased collagen production [65]. Treatment of human skin organ cultures with a soy extract (4–40 μg/ml) reduced retinoid-induced epidermal hyperplasia by up to 41% [66]. Additionally, the soy extract (40 μg/ml) was reported to enhance type I procollagen production by 50% by dermal fibroblasts in this same study.

Overall, these data suggest that various components of soy, including isoflavones, saponins, and peptides, may have a variety of beneficial effects on skin health. The mechanisms of these effects appear to be varied and may include antioxidant potential, suppression of inflammatory processes, production of extracellular matrix (ECM) components, and suppression of enzymes involved in ECM breakdown.

23.3.2 *In Vivo* Evidence: Topical Application

In addition to *in vitro* studies on the potential protective effects of soy against photoaging and photocarcinogenesis, a number of studies have been conducted to determine the effects of the topical application of soy components. The majority of these studies have focused on the potential protective effects of soy and soy isoflavones against photocarcinogenesis. Wei et al. [67] demonstrated that topical application of genistein inhibits both the initiation and promotion of skin tumorigenesis. In this study, genistein applied topically to mice at a dose of 10 μmol/day for 1 week prior to 7,12-dimethylbenz[a]anthracene (DMBA) treatment reduced tumor incidence and multiplicity by 20% and 50%, respectively. Also, genistein dose-dependently inhibited 12-*O*-tetradecanoyl phorbol-13-acetate (TPA) tumor promotion by up to 75% in mice [67]. In hairless mice, UVB radiation–induced initiation and promotion of photocarcinogenesis were inhibited by topical administration of genistein (1 μmol and 5 μmol), resulting in the reduction of tumor incidence and multiplicity [68]. Daily topical applications of equol, a metabolite of the soy isoflavone daidzein, reduced the proportion of tumors progressing from benign papillomas to malignant squamous cell carcinomas by 33–58% and reduced the size of these carcinomas by 71–82% [69]. This study also demonstrated that the topical application of equol (10 μmol) lengthened tumor latency and reduced tumor multiplicity.

The mechanism(s) responsible for the protective actions of soy on photocarcinogenesis appear to by manifold. Wei et al. [67] reported that genistein treatment dose-dependently reduced DMBA-induced DNA adduct formation and TPA-induced hydrogen peroxide formation. This research group further reported that topical genistein application (10 μmol) suppressed the UVB-induced expression of the proto-oncogenes *c-fos* and *c-jun*, likely through inhibition of UVB-induced epidermal growth factor receptor phosphorylation [70]. Genistein might also protect against skin cancer by suppressing UVB-induced formation of pyrimidine dimers and 8-hydroxy-deoxyguanosine (8-OHdG), inhibiting UVB-induced suppression of PCNA expression, reducing UVB-induced phosphorylation of MAP kinase, and blocking the damages (cutaneous ulceration, apoptosis, cleavage of poly[ADP-ribose] polymerase) caused by treatment with psoralen plus UVA [68,71]. Other studies have shown that equol may protect against solar-simulated UV radiation (SSUV)-induced photocarcinogenesis by reducing contact hypersensitivity, inflammatory edema, DNA damage, and epidermal hyperplasia [72–74]. Suppression of SSUV-induced epidermal hyperplasia and contact hypersensitivity appears to be mediated by the upregulation of metallothionein expression [74].

In addition to potential protection against skin photocarcinogenesis, soy and its isoflavones have been reported to have beneficial effects on photoaging. Wei et al. [68] demonstrated that topical application of 5 μmol genistein to mouse skin 60 minutes prior to exposure to UVB radiation blocked UVB-induced sunburn. Furthermore, these authors reported that application of genistein both before and after chronic exposure to UVB radiation reduced photodamage. In human subjects, topical application of 5 μmol genistein suppressed UVB-induced erythema. This effect was best observed when genistein was applied before UVB exposure because both erythema and discomfort were alleviated, whereas application post-UVB exposure only improved discomfort [68]. These benefits do not appear to be limited to just genistein. Topical application of a 2% soy extract has been shown to significantly increase the number of dermal papillae in human skin, suggesting a rejuvenation of the dermal-epidermal junction [64]. Additionally, topical application of equol has also been shown to reduce SSUV-induced epidermal hyperplasia, mast cell numbers, elastosis, collagen degradation, and glycosaminoglycan deposition in mice [75]. More recently, a standardized soy extract containing nondenatured soybean trypsin inhibitor (STI) and nondenatured Bowman-Birk inhibitor (BBI), two soy peptides with protease activities, has been shown to improve skin texture and tone, reduce fine lines, and improve overall appearance as early as 2 weeks after topical application to human facial skin [76].

23.3.3 *In Vivo* Evidence: Dietary Consumption

Despite the mounting *in vitro* and topical evidence elucidating the potential benefits of soy for skin care, little research has been done to examine the dietary benefits of soy for skin health. In a study by Kim et al. [63], mice were orally administered 500 mg of a soy extract/kg of body weight per day for 4 weeks and were exposed to UV radiation 3 times per week. Administration of the soy extract significantly improved parameters of skin roughness and nonsignificantly reduced transepidermal water loss, fine wrinkles, and UV radiation–induced thickening of the epidermis [63]. A separate study demonstrated that the addition of genistein to the drinking water of mice for 27 weeks starting 2 weeks prior to chronic UVB exposure resulted in a reduction in skin photocarcinogenesis [68].

23.4 Soy for Hair Care

In addition to the benefits of soy and/or soy isoflavones for skin care, recent research has suggested that soy may be beneficial for hair health as well. McElwee et al. [77] demonstrated in a mouse model that dietary consumption of soy oil reduced the incidence of alopecia areata in a dose-response fashion. In this study, the incidence of alopecia areata in soy oil–treated mice was 86%, 39%, and 18% in mice receiving 1%, 5%, and 20% soy oil, respectively. These authors also reported that only 4 of 10 mice injected with genistein developed alopecia areata compared to 9 of 10 control mice [77]. The effect of soy on hair loss does not appear to be restricted to only the isoflavones. Tsuruki et al. [78] reported on the isolation of an immunostimulating peptide, "soymetide-4," from a trypsin digest of soy protein, and demonstrated that this peptide inhibited chemotherapy-induced alopecia in neonatal rats. It has been suggested that the effects of soymetide-4 may be via prostaglandin E2 suppression of hair follicle apoptosis [79]. Harada et al. [80] demonstrated the benefits of coadministration of soy isoflavones with capsaicin for hair health in both mice and human subjects. In mice, subcutaneous treatment with isoflavones and capsaicin for 4 weeks promoted hair regrowth to a significantly greater extent than did capsaicin alone, with both treatments enhancing hair growth more than the study control. Oral administration of capsaicin plus soy isoflavones for 5 months promoted hair growth in 64.5% of human subjects with alopecia, compared to only 11.8% of subjects consuming the placebo treatment. It was suggested that these benefits were likely due to increases in IGF-1, because dermal and serum IGF-1 levels were increased in mice and human subjects, respectively, after treatment with capsaicin plus isoflavones [80].

Soy has also been reported to have some hair-related cosmetic benefits. In mice, topical application of STI and BBI as well as application of soymilk containing these protease inhibitors delayed hair growth and reduced hair shaft length, hair follicle size, hair shaft thickness, hair bulb diameter, and hair pigmentation [81]. These investigators also demonstrated that topical application of soymilk containing STI and BBI to facial and leg skin reduced the length and thickness of hair shafts and reduced hair growth in human subjects [81].

23.5 Effect of Dietary Soy on the Skin, Hair, and Nails of Postmenopausal Women

The studies discussed above have demonstrated the potentially beneficial effects of soy and/or soy isoflavones on hair loss, sunburn, photoaging, and photocarcinogenesis. However, there have been no studies examining the effects of dietary soy and soy isoflavones on dermatologic conditions in postmenopausal women, a population with an increasing interest in the use of soy as an alternative to hormone therapy. Therefore, we conducted a study to investigate the benefits of dietary soy for postmenopausal skin, hair, and nail appearance.

23.5.1 Design and Methods

We hypothesized that supplementation of soy products to the diet of postmenopausal women would (1) improve the health and appearance of the skin by normalizing skin pigmentation and reducing wrinkles, (2) improve the health and appearance of hair by enhancing sheen and ease of combing, and (3) improve the health and appearance of nails by enhancing shine and flexibility.

For this pilot study, we treated postmenopausal women (50–65 years of age with only mild to moderate photoaging; n = 40) with one soy shake (Soy; 20 g soy protein with 160 mg total isoflavones; n = 20) daily or no dietary intervention (Control; n = 20) for 6 months. The subjects were required to make three visits to the board-certified dermatologist assisting with the study, at baseline, at 3 months, and at 6 months. During each visit, subjects were asked to complete a questionnaire in order to provide self-reported improvements. Following completion of the questionnaire, subjects were examined by the dermatologist and staff to assess the health and appearance of their skin, hair, and nails. Statistical analysis was conducted using the Mann-Whitney nonparametric two-tailed paired test. Significance was defined as $P \leq 0.05$.

Four measurements were conducted to assess the health and appearance of the skin. Transepidermal water loss was measured with an evaporimeter, skin hydration was tested with a corneometer, and facial skin pigmentation and fine wrinkles were photographically recorded for later assessment. Facial skin parameters of roughness, wrinkling, flaking, and discoloration were rated on a scale of 0–4, with 0 = none, 1 = slight, 2 = mild, 3 = moderate, and 4 = severe.

Hair was visually assessed for sheen and tested for ease of combing to determine the potential beneficial effects of soy on hair health and appearance. To properly compare treatment differences and improvements over time, the women were provided a list of acceptable shampoos for their use during the study. They were required to choose one and use it for the duration of the study period. Hair appearance parameters of roughness, dullness, lack of manageability, and scalp flaking were rated on a scale of 0–4, with 0 = none, 1 = slight, 2 = mild, 3 = moderate, and 4 = severe.

To measure changes in the health and appearance of fingernails, the nails were visually appraised for shine, splitting, and flexibility. Fingernail parameters of roughness, ridging, flaking, and splitting were rated on a scale of 0–4, with 0 = none, 1 = slight, 2 = mild, 3 = moderate, and 4 = severe.

Study participants provided self-assessments of their facial skin (roughness, wrinkling, brown spots, and itching/burning), hair (roughness, breakage, splitting, and combing difficulty), and fingernails (roughness, splitting, and breakage) using a scale of 0–4, with 0 = none, 1 = minimal, 2 = mild, 3 = moderate, and 4 = severe.

23.5.2 Results and Discussion

23.5.2.1 Facial Skin Appearance

Facial skin appearance was assessed for changes in roughness, wrinkling, flaking, discoloration, and overall appearance. The results of this study revealed a number of investigator-assessed physical benefits from use of the product for the 6-month study period (Fig. 23.2). After 3 months, improvements from baseline were significantly greater in the Soy group compared to the Control group for facial skin flaking (P = 0.028), discoloration (P = 0.048), and overall appearance (P = 0.05). Similarly, improvements from baseline after 6 months were significantly greater in the Soy group compared to the Control group for facial wrinkling (P = 0.004), discoloration (P = 0.016), and overall appearance (P = 0.0001).

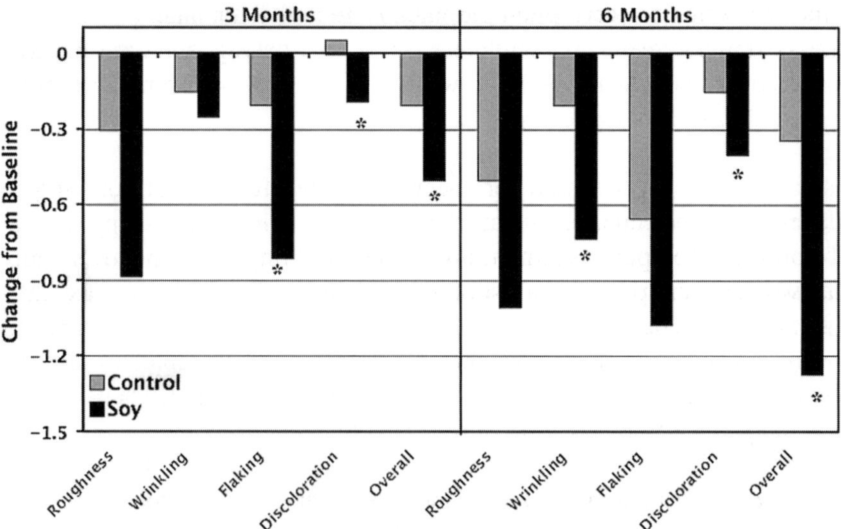

Figure 23.2 Change from baseline in facial skin appearance of postmenopausal women as determined by dermatologist assessment after 3 months and 6 months of dietary soy consumption. The asterisk indicates significant differences between groups for change from baseline (P < 0.05).

23.5.2.2 Fingernail Appearance

The nails of the subjects using the soy supplement also demonstrated investigator-assessed improvements (Fig. 23.3). Though no changes were apparent after 3 months, improvement was noted after 6 months of soy consumption. Compared to the Control group, improvements from baseline after 6 months were observed for nail roughness (P = 0.017), ridging (P = 0.006), flaking (P = 0.049), splitting (P = 0.007), and overall appearance (P = 0.008) in the Soy group.

23.5.2.3 Hair Appearance

Hair appearance was improved at both the 3- and 6-month time points as determined by dermatologist evaluations (Fig. 23.4). Compared to the Control group, improvements from baseline values were significantly greater in the Soy group for hair roughness (P = 0.041), manageability (P = 0.018), and overall appearance (P = 0.016) after 3 months of dietary

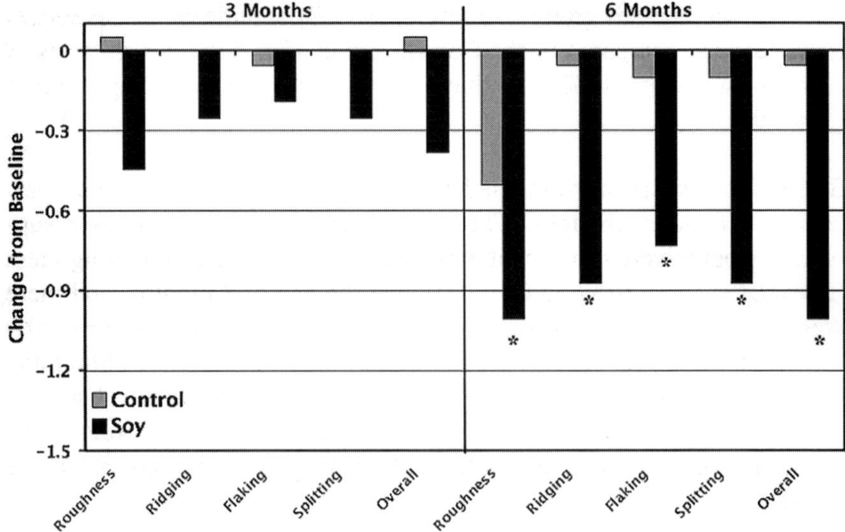

Figure 23.3 Change from baseline in fingernail appearance of postmenopausal women as determined by dermatologist assessment after 3 months and 6 months of dietary soy consumption. The asterisk indicates significant differences between groups for change from baseline ($P < 0.05$).

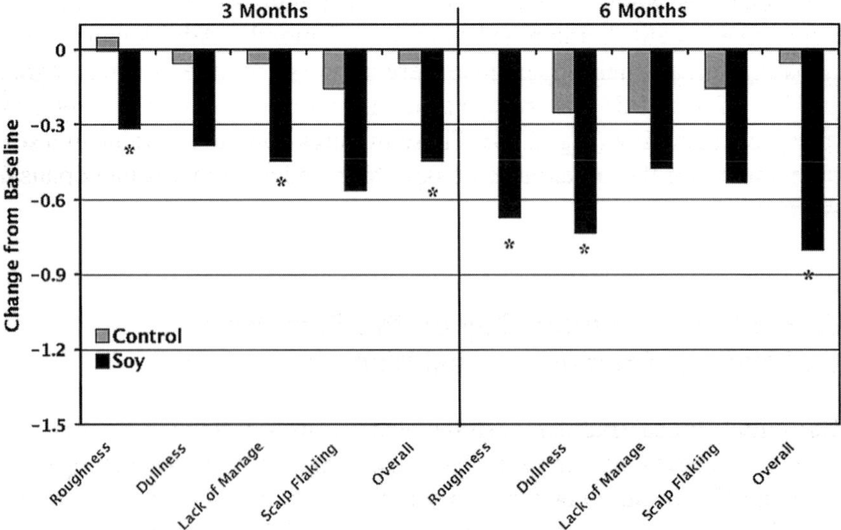

Figure 23.4 Change from baseline in hair appearance of postmenopausal women as determined by dermatologist assessment after 3 months and 6 months of dietary soy consumption. The asterisk indicates significant differences between groups for change from baseline ($P < 0.05$).

intervention. The Soy group also saw statistically significant improvements from baseline compared to the Control group after 6 months in terms of hair roughness ($P = 0.004$), dullness ($P = 0.048$), and overall appearance ($P = 0.005$).

Similar to investigator-assessed benefits, the study subjects also perceived beneficial changes. The women consuming a daily soy supplement perceived significant improvements in facial skin roughness ($P = 0.013$) and wrinkling ($P = 0.014$) at the end of 3 months, and there was a continued perception of improvement in facial roughness ($P = 0.009$) after 6 months. Self-perception of changes in the appearance of hair and nails after 3 months of consuming the soy supplement were not observed. However, after 6 months of soy supplementation, there was some self-perceived improvement noted in hair roughness, but no difference in nails.

Changes in skin hydration and transepidermal water loss were not significantly different between treatment groups after 6 months.

The results of this study demonstrate that daily soy consumption may improve dermatologic health. Dermatologist-assessed improvements in skin, hair, and nails were evident after 6 months of soy consumption, with some benefits being noted as early as 3 months. Additionally, overall skin, nail, and hair appearance were improved to at least some degree in 93%, 67%, and 53% of study participants, respectively, after 6 months of soy consumption (Fig. 23.5). These findings point to the value of a soy supplement for the appearance of skin, hair, and nails in postmenopausal women.

23.6 Effect of Dietary Soy on the Skin, Hair, and Nails of Premenopausal Women

Our initial clinical trial demonstrated that dietary soy supplementation has beneficial effects on the skin, hair, and fingernails of postmenopausal women [82]. To our knowledge, our initial study was the first to examine the effects of dietary soy on dermatologic conditions in postmenopausal women. To date there have been no studies on the effects of dietary soy consumption on dermatologic health in premenopausal women, so we conducted a second study to investigate the potential benefits of dietary soy for premenopausal skin, hair, and nail appearance.

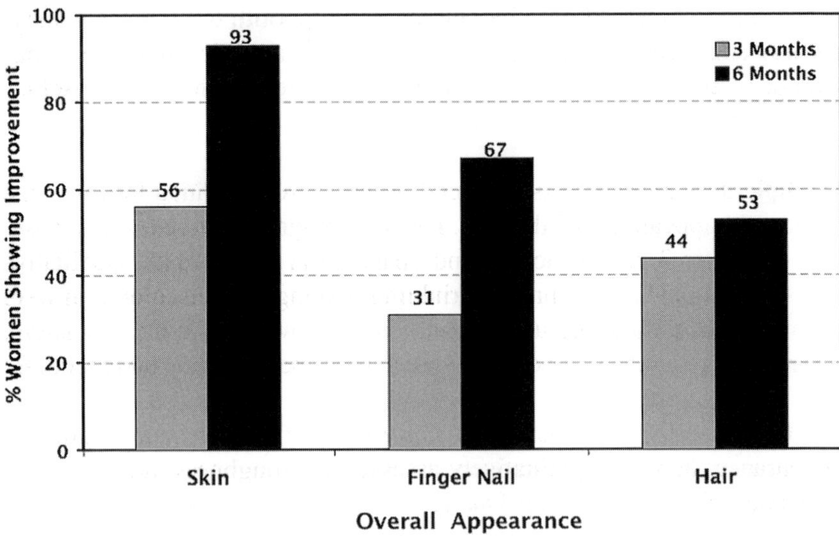

Figure 23.5 Percent of postmenopausal women showing improvement in overall skin, fingernail, and hair appearance after consumption of soy daily for 3 months and 6 months.

23.6.1 Design and Methods

We hypothesized that supplementation of dietary soy to overweight premenopausal women as part of a soy-based meal replacement diet plan would (1) improve the health and appearance of the skin by normalizing skin pigmentation and reducing wrinkles, (2) improve the health and appearance of hair by enhancing sheen and ease of combing, and (3) improve the health and appearance of nails by enhancing shine, flexibility, and strength.

For this study, we treated overweight or obese (BMI = 30–40 kg/m^2) premenopausal women (25–45 years of age) with only mild to moderate photoaging (n = 40) with soy protein (Soy; ~20 g soy protein with 160 mg total isoflavones; n = 20) or milk protein (Milk; ~20 g of protein with 0 mg isoflavones; n = 20) for 6 months. The subjects were asked to come into the office for a total of seven visits, at baseline and at monthly intervals for 6 months. At baseline, month 3, and month 6, the subjects were asked to complete a questionnaire in order to provide self-reported improvements related to hair, skin, and nail health. Following completion of the questionnaire, the investigator and the dermatology clinic staff examined the subjects to assess the health and appearance of their skin, hair, and nails at baseline, month 3, and month 6. Statistical analysis was conducted using Mann-Whitney nonparametric, two-tailed tests. Comparisons were made between treatment groups and from baseline. Significance was defined as $P \leq 0.05$.

Multiple dermatologist-assessed end points were examined to assess the health and appearance of the skin. Facial photographs were taken at baseline and after 3 and 6 months and analyzed at a later date. Qualitative assessments of skin roughness, wrinkling, flaking, and discoloration were performed and classified using a scale of 0–4, with 0 = none, 1 = slight, 2 = mild, 3 = moderate, and 4 = severe at baseline, month 3, and month 6.

To determine the potential beneficial effects of soy on hair health and appearance, hair was qualitatively assessed for roughness, dullness, lack of manageability, and scalp flaking and classified using a scale of 0–4, with 0 = none, 1 = slight, 2 = mild, 3 = moderate, and 4 = severe at baseline, month 3, and month 6. To properly compare treatment differences and improvements over time, the women were provided a list of acceptable shampoos for their use during the study. They were required to choose one and use it for the duration of the study period.

Changes in the health and appearance of fingernails were visually appraised for roughness, ridging, flaking, and splitting and rated using a scale of 0–4, with 0 = none, 1 = slight, 2 = mild, 3 = moderate, and 4 = severe at baseline, month 3, and month 6.

Study volunteers were asked to provide self-assessments of their facial skin (roughness, wrinkling, presence of brown spots, and itching/burning), hair (roughness, breakage, split ends, and combing difficulty), and fingernails (roughness, splitting, and breakage) using a scale of 0–4, with 0 = none, 1 = minimal, 2 = mild, 3 = moderate, and 4 = severe.

23.6.2 Results and Discussion

When comparing change from baseline between treatment groups, a statistically significant difference in skin wrinkling was observed at 3 months, with the Soy group exhibiting a greater decrease in wrinkling than the Milk group ($P = 0.04$). No other treatment differences were observed at either 3 months or 6 months. In contrast, numerous benefits were observed from baseline when the data was examined within individual treatments groups.

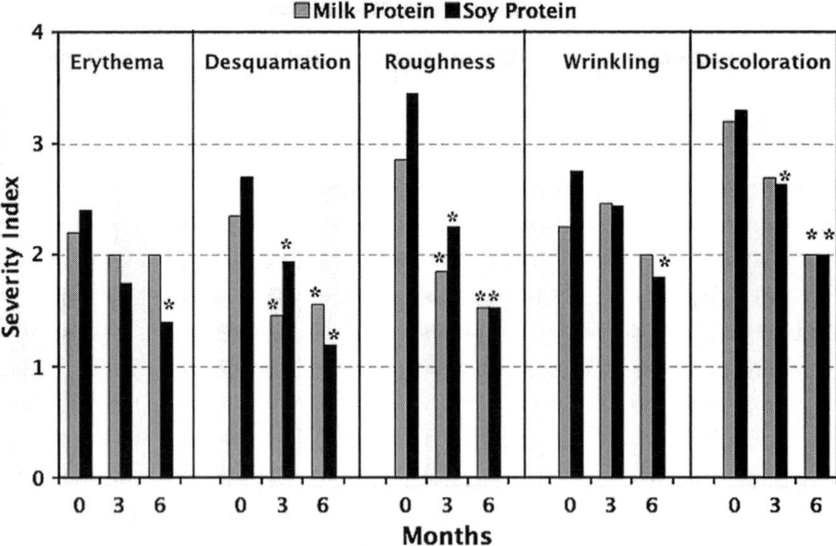

Figure 23.6 Severity index for parameters of facial skin appearance of premenopausal women as determined by dermatologist assessment after 3 months and 6 months of dietary soy consumption. The asterisks indicate significant differences from baseline (P < 0.05).

23.6.2.1 Facial Skin Appearance

After 3 months of dietary supplementation with either soy or milk protein, skin roughness and desquamation were significantly improved in both treatment groups (P < 0.05). However, skin discoloration was only significantly improved in the soy protein group (P = 0.005) at 3 months. Skin desquamation, roughness, and discoloration were each significantly (P < 0.05) improved by dietary supplementation with both soy and milk protein shakes after 6 months (Fig. 23.6). In contrast, only soy protein supplementation significantly (P < 0.05) improved skin wrinkling and erythema after 6 months compared to baseline. These data suggest that both soy and milk proteins have multiple benefits for facial skin appearance; however, only dietary soy provided benefits for erythema and wrinkling in the current study.

23.6.2.2 Fingernail Appearance

Parameters of fingernail appearance (roughness, ridging, flaking, and splitting) were all significantly (P < 0.01) improved at 3 months compared

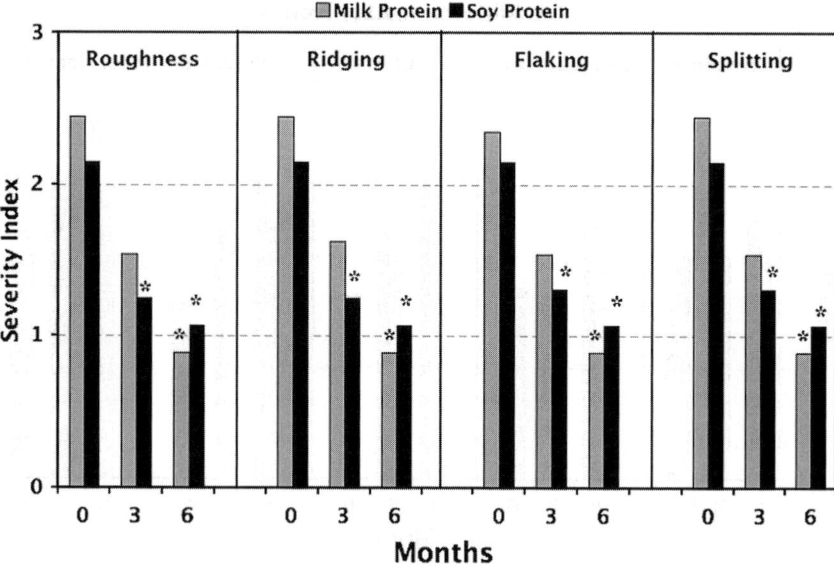

Figure 23.7 Severity index for parameters of fingernail appearance of premenopausal women as determined by dermatologist assessment after 3 months and 6 months of dietary soy consumption. The asterisks indicate significant differences from baseline ($P < 0.05$).

to baseline in the soy protein group. Though improvements were observed in the milk protein group at 3 months, these did not reach statistical significance. In contrast, both the soy protein and milk protein groups exhibited significant ($P < 0.01$) improvements in each parameter of nail appearance after 6 months compared to baseline (Fig. 23.7). These data suggest that whereas dietary supplementation with both soy protein and milk protein can support improved fingernail health, improvements may appear earlier with soy protein supplementation.

23.6.2.3 Hair Appearance

Few statistically significant effects were observed in regard to hair appearance in premenopausal women (Fig. 23.8). Though modest benefits in hair roughness, dullness, manageability, and scalp flaking were observed from baseline in both treatment groups, only roughness and manageability scores were significantly ($P < 0.05$) improved, and these only in the soy protein group after 6 months of dietary supplementation. These data

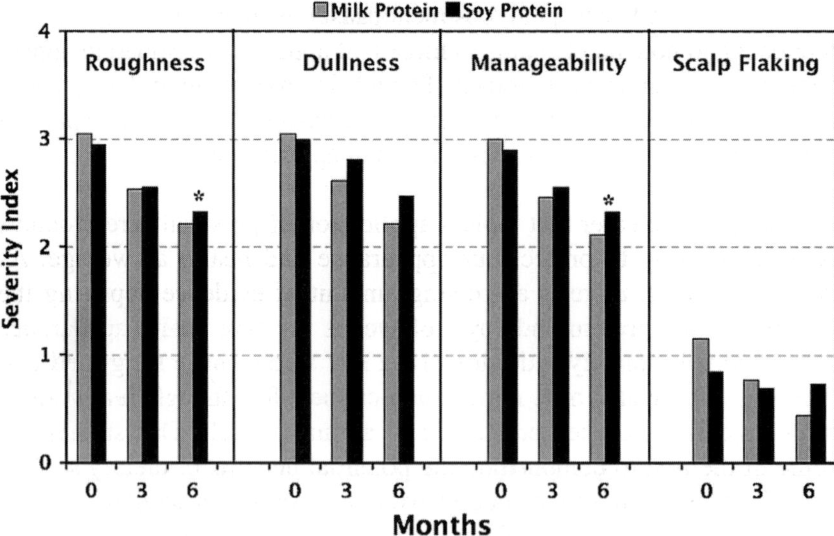

Figure 23.8 Severity index for parameters of hair appearance of premenopausal women as determined by dermatologist assessment after 3 months and 6 months of dietary soy consumption. The asterisks indicate significant differences from baseline ($P < 0.05$).

suggest that whereas both dietary milk and soy protein may have beneficial effects on hair appearance in premenopausal women, the benefits appear to be modest at best.

The results of this study demonstrate that both dietary soy and milk proteins may have multiple beneficial effects on the appearance of skin, hair, and nails in premenopausal women. While both dietary supplements appeared to support the appearance of healthy skin, hair, and nails, dietary soy supplementation seemed to support more individual parameters of dermatologic appearance and seemed to provide earlier benefits in some instances.

23.7 Summary and Conclusion

Our studies demonstrated that dietary soy consumption for 3–6 months improved several parameters of skin health in both pre- and postmenopausal women. It was also observed that fingernail appearance was improved in both pre- and postmenopausal women; however, it was interesting to note that benefits to fingernail appearance seemed to appear sooner in premenopausal women. Though improvements in hair appearance and manageability were evident, these changes seemed more modest than the benefits observed for skin and nail appearance.

It is becoming clearer that topical application of personal care products is only one way to protect our appearance and health as we age. As described above, there is a growing amount of evidence depicting the benefits of soy protein and soy isoflavones for skin health appearance when applied topically. Additionally, a few studies have suggested that dietary soy may also have dermatological benefits, though nearly all of these studies have been conducted in animal models. Our studies are some of the first to demonstrate the potential benefits of dietary soy for skin, hair, and nail appearance in women. Overall, these data indicate that dietary soy (protein and/or isoflavones) consumption can improve the appearance of the skin, hair, and nails and that dietary consumption may complement topical application of soy-containing products to provide a full inside-outside approach to personal care.

References

1. Omoni AO, Aluko RE. Soybean foods and their benefits: potential mechanisms of action. Nutr Rev 2005; 63:272–283.
2. Cassidy A, Albertazzi P, Nielsen IL, Hall W, Williamson G, Tetens I, Atkins S, Cross H, Manios Y, Wolk A, Steiner C, Branca F. Critical review of health effects of soyabean phyto-oestrogens in post-menopausal women. Proc Nutr Soc 2006; 65:76–92.
3. Parkin DM. Cancers of the breast, endometrium and ovary: geographic correlations. Eur J Cancer Clin Oncol 1989; 25:1917–1925.
4. Pisani P, Parkin DM, Bray F, Ferlay J. Estimates of the worldwide mortality from 25 cancers in 1990. Int J Cancer 1999; 83:18–29.
5. Messina M, Nagata C, Wu AH. Estimated Asian adult soy protein and isoflavone intakes. Nutr Cancer 2006; 55:1–12.
6. Messina MJ. Legumes and soybeans: overview of their nutritional profiles and health effects. Am J Clin Nutr 1999; 70 (3 Suppl):439S–450S.
7. Choi MS, Rhee KC. Production and processing of soybeans and nutrition and safety of isoflavone and other soy products for human health. J Med Food 2006; 9:1–10.
8. Erdman Jr. JW, Fordyce EJ. Soy products and the human diet. Am J Clin Nutr 1989; 49:725–737.
9. Young VR. Soy protein in relation to human protein and amino acid nutrition. J Am Diet Assoc 1991; 91:828–835.

10. Istfan N, Murray E, Janghorbani M, Young VR. An evaluation of the nutritional value of a soy protein concentrate in young adult men using the short-term N-balance method. J Nutr 1983; 113:2516–2523.
11. Wayler A, Queiroz E, Scrimshaw NS, Steinke FH, Rand WM, Young VR. Nitrogen balance studies in young men to assess the protein quality of an isolated soy protein in relation to meat proteins. J Nutr 1983; 113: 2485–2491.
12. Young VR, Wayler A, Garza C, Steinke FH, Murray E, Rand WM, Scrimshaw NS. A long-term metabolic balance study in young men to assess the nutritional quality of an isolated soy protein and beef proteins. Am J Clin Nutr 1984; 39:8–15.
13. Mariotti F, Mahe S, Benamouzig R, Luengo C, Dare S, Gaudichon C, Tome D. Nutritional value of [15N]-soy protein isolate assessed from ileal digestibility and postprandial protein utilization in humans. J Nutr 1999; 129:1992–1997.
14. Gaudichon C, Mahe S, Benamouzig R, Luengo C, Fouillet H, Dare S, Van Oycke M, Ferriere F, Rautureau J, Tome D. Net postprandial utilization of [15N]-labeled milk protein nitrogen is influenced by diet composition in humans. J Nutr 1999; 129:890–895.
15. U.S. Department of Agriculture, Agricultural Research Service. 2007. USDA Nutrient Database for Standard Reference, Release 20. Nutrient Data Laboratory Home Page, http://www.ars.usda.gov/nutrientdata. Note: Release numbers change as new versions are released.
16. Moyad MA. An introduction to dietary/supplemental omega-3 fatty acids for general health and prevention: Part I. Urol Oncol 2005; 23:28–35.
17. Moyad MA. An introduction to dietary/supplemental omega-3 fatty acids for general health and prevention: Part II. Urol Oncol 2005; 23:36–48.
18. Foster-Powell K, Holt SHA, Brand-Miller JC. International table of glycemic index and glycemic load values: 2002. Am J Clin Nutr 2002; 76:5–56.
19. Rycroft CE, Jones MR, Gibson GR, Rastall RA. A comparative in vitro evaluation of the fermentation properties of prebiotic oligosaccharides. J Appl Microbiol 2001; 91:878–887.
20. Zhao Y, Martin BR, Weaver CM. Calcium bioavailability of calcium carbonate fortified soymilk is equivalent to cow's milk in young women. J Nutr 2005; 135:2379–2382.
21. Lönnerdal B, Bryant A, Liu X, Theil EC. Iron absorption from soybean ferritin in nonanemic women. Am J Clin Nutr 2006; 83:103–107.
22. Sandstrom B, Kivisto B, Cederblad A. Absorption of zinc from soy protein meals in humans. J Nutr 1987; 117:321–327.
23. Etcheverry P, Hawthorne KM, Liang LK, Abrams SA, Griffin IJ. Effect of beef and soy proteins on the absorption of non-heme iron and inorganic zinc in children. J Am Coll Nutr 2006; 25:34–40.
24. Kerstetter JE, Wall DE, O'Brien KO, Caseria DM, Insogna KL. Meat and soy protein affect calcium homeostasis in healthy women. J Nutr 2006; 136:1890–1895.
25. Price KR, Johnson IT, Fenwick GR. The chemistry and biological significance of saponins in food and feeding stuffs. CRC Crit Rev Food Sci Nutr 1987; 26:27–135.

26. Rupasinghe HPV, Jackson CJC, Poysa V, Berardo CD, Bewley JD, Jenkinson J. Soyasapogenol A and B distribution in soybean (Glycine max L. Merr.) in relation to seed physiology, genetic variability, and growing location. J Agric Food Chem 2003; 51:5888–5894.
27. Shiraiwa M, Kudo S, Shimoyamada M, Harada K, Okubo K. Composition and structure of "group A saponin" in soybean seed. Agric Biol Chem 1991a; 55:315–322.
28. Shiraiwa M, Harada K, Okubo K. Composition and structure of "group B saponin" in soy seed. Agric Biol Chem 1991b; 55:911–917.
29. Francis G, Kerem Z, Makkar HPS, Becker K. The biological action of saponins in animal systems: a review. Br J Nutr 2002; 88:587–605.
30. Shi J, Arunasalam K, Yeung D, Kakuda Y, Mittal G, Jiang Y. Saponins from edible legumes: chemistry, processing, and health benefits. J Med Food 2004; 7:67–78.
31. Dixon RA. Phytoestrogens. Annu Rev Plant Biol 2004; 55:225–261.
32. Wang HJ, Murphy PA. Isoflavone content in commercial soybean foods. J Agric Food Chem 1994; 42:1666–1673.
33. Wang HJ, Murphy PA. Isoflavone composition of American and Japanese soybeans in Iowa: effects of variety, crop year, and location. J Agric Food Chem 1994; 42:1674–1677.
34. Setchell KD, Brown NM, Zimmer-Nechemias L, Brashear WT, Wolfe BE, Kirschner AS, Heubi JE. Evidence for lack of absorption of soy isoflavone glycosides in humans, supporting the crucial role of intestinal metabolism for bioavailability. Am J Clin Nutr 2002a; 76:447–453.
35. Day AJ, Canada FJ, Diaz JC, Kroon PA, McLauchlan R, Faulds CB, Plumb GW, Morgan MR, Williamson G. Dietary flavonoid and isoflavone glycosides are hydrolysed by the lactase site of lacatase phlorizin hydrolase. FEBS Lett 2000; 468:166–170.
36. Axelson M, Sjovall J, Gustafsson BE, Setchell KD. Soya—a dietary source of the non-steroidal oestrogen equol in man and animals. J Endocrinol 1984; 102:49–56.
37. Setchell KD, Borriello SP, Hulme P, Kirk DN, Axelson M. Nonsteroidal estrogens of dietary origin: possible roles in hormone-dependent disease. Am J Clin Nutr 1984; 40:569–578.
38. Xu X, Harris KS, Wang HJ, Murphy PA, Hendrich S. Bioavailability of soybean isoflavones depends upon gut microflora in women. J Nutr 1995; 125:2307–2315.
39. Setchell KD, Brown NM, Lydeking-Olsen E. The clinical importance of the metabolite equol—a clue to the effectiveness of soy and its isoflavones. J Nutr 2002b; 132:3577–3584.
40. Rowland I, Faughnan M, Hoey L, Wahala K, Williamson G, Cassidy A. Bioavailability of phyto-oestrogens. Br J Nutr 2003; 89 (Suppl 1):S45–S58.
41. Yuan JP, Wang JH, Liu X. Metabolism of dietary soy isoflavones to equol by human intestinal microflora—implications for health. Mol Nutr Food Res 2007; 51:765–781.
42. Kuiper GGJM, Lemmen JG, Carlsson B, Corton JC, Safe SH, van der Saag PT, van der Burg B, Gustafsson JA. Interaction of estrogenic chemicals and phytoestrogens with estrogen receptor β. Endocrinology 1998; 139:4252–4263.

43. Morito K, Hirose T, Kinjo J, Hirakawa T, Okawa M, Nohara T, Ogawa S, Inoue S, Muramatsu M, Masamune Y. Interaction of phytoestrogens with estrogen receptors α and β. Biol Pharm Bull 2001; 24:351–356.
44. Setchell KDR. Soy isoflavones—benefits and risks from nature's selective estrogen receptor modulators. J Am Coll Nutr 2001; 20:354S–362S.
45. Hwang CS, Kwak HS, Lim HJ, Lee SH, Kang YS, Choe TB, Hur HG, Han KO. Isoflavone metabolites and their in vitro dual functions: they can act as estrogenic agonist or antagonist depending on the estrogen concentration. J Steroid Biochem Mol Biol 2006; 101:246–253.
46. Akiyama T, Ishida J, Nakagawa S, Ogawara H, Watanabe S, Itoh N, Shibuya M, Fukami Y. Genistein, a specific inhibitor of tyrosine-specific protein kinases. J Biol Chem 1987; 262:5592–5595.
47. McCarty MF. Isoflavones made simple—genistein's agonist activity for the beta-type estrogen receptor mediates their health benefits. Med Hypotheses 2006; 66:1093–1114.
48. Wiseman H, O'Reilly JD, Adlercreutz H, Mallet AI, Bowey EA, Rowland IA, Sanders TAB. Isoflavone phytoestrogens consumed in soy decrease F2-isoprostane concentrations and increase resistance of low-density lipoprotein to oxidation in humans. Am J Clin Nutr 2000; 72:395–400.
49. Bazzoli DL, Hill S, DiSilvestro RA. Soy protein antioxidant actions in active, young adult women. Nutr Res 2002; 22:807–815.
50. Rufer CE, Kulling SE. Antioxidant activity of isoflavones and their major metabolites using different in vitro assays. J Agric Food Chem 2006; 54:2926–2931.
51. Wu HJ, Chan WH. Genistein protects methylglyoxal-induced oxidative DNA damage and cell injury in human mononuclear cells. Toxicol In Vitro 2007; 21:335–342.
52. Borras C, Gambini J, Gomez-Cabrera MC, Sastre J, Pallardo FV, Mann GE, Vina J. Genistein, a soy isoflavone, up-regulates expression of antioxidant genes: involvement of estrogen receptors, ERK1/2, and NFκB. FASEB J 2006; 20:E1476–E1481.
53. Vega-Lopez S, Yeum KJ, Lecker JL, Ausman LM, Johnsono EJ, Devaraj S, Jialal I, Lichtenstein AH. Plasma antioxidant capacity in response to diets high in soy or animal protein with or without isoflavones. Am J Clin Nutr 2005; 81:43–49.
54. Adlercreutz H, Bannwart C, Wahala K, Makela T, Brunow G, Hase T, Arosemena PJ, Kellis Jr. JT, Vickery LE. Inhibition of human aromatase by mammalian lignans and isoflavonoid phytoestrogens. J Steroid Biochem Mol Biol 1993; 44:147–153.
55. Pelissero C, Lenczowski MJP, Chinzi D, Davail-Cuisset B, Sumpter JP, Fostier A. Effects of flavonoids on aromatase activity, an in vitro study. J Steroid Biochem Mol Biol 1996; 57:215–223.
56. Makela S, Poutanen M, Lehtimaki J, Kostian ML, Santii R, Vihko R. Estrogen-specific 17 beta-hydroxysteroid oxidoreductase type 1 (E.C. 1.1.1.62) as a possible target for the action of phytoestrogens. Proc Soc Exp Biol Med 1995; 208:51–59.
57. Le Bail JC, Champavier Y, Chulia AJ, Habrioux G. Effects of phytoestrogens on aromatase, 3β- and 17β-hydroxysteroid dehydrogenase activities and human breast cancer cells. Life Sci 2000; 66:1281–1291.

58. Evans BA, Griffiths K, Morton MS. Inhibition of 5 alpha-reductase in genital skin fibroblasts and prostate tissue by dietary lignans and isoflavonoids. J Endocrinol 1995; 147:295–302.
59. Liu Z, Lu Y, Lebwohl M, Wei H. PUVA (8-methoxy-psoralen plus ultraviolet A) induces the formation of 8-hydroxy-2′-deoxyguanosine and DNA fragmentation in calf thymus DNA and human epidermoid carcinoma cells. Free Radic Biol Med 1999; 27:127–133.
60. Chan WH, Yu JS. Inhibition of UV irradiation–induced oxidative stress and apoptotic biochemical changes in human epidermal carcinoma A431 cells by genistein. J Cell Biochem 2000; 78:73–84.
61. Moore JO, Wang Y, Stebbins WG, Gao D, Zhou X, Phelps R, Lebwohl M, Wei H. Photoprotective effect of isoflavone genistein on ultraviolet B induced pyrimidine dimer formation and PCNA expression in human reconstituted skin and its implications in dermatology and prevention of cutaneous carcinogenesis. Carcinogenesis 2006; 27:1627–1635.
62. Trompezinski S, Denis A, Schmitt D, Viac J. Comparative effects of polyphenols from green tea (EGCG) and soybean (genistein) on VEGF and IL-8 release from normal human keratinocytes stimulated with the proinflammatory cytokine TNFα. Arch Dermatol Res 2003; 295:112–116.
63. Kim SY, Kim SJ, Lee JY, Kim WG, Park WS, Sim YC, Lee SJ. Protective effects of dietary soy isoflavones against UV-induced skin-aging in hairless mouse model. J Am Coll Nutr 2004; 23:157–162.
64. Sudel KM, Venzke K, Mielke H, Breitenbach U, Mundt C, Jaspers S, Koop U, Sauermann K, Knuβmann-Hartig E, Moll I, Gercken G, Young AR, Stab F, Wenck H, Gallinat S. Novel aspects of intrinsic and extrinsic aging of human skin: beneficial effects of soy extract. Photochem Photobiol 2005; 81:581–587.
65. Andre-Frei V, Perrier E, Augustin C, Damour O, Bordat P, Schumann K, Forster Th, Waldmann-Laue M. A comparison of biological activities of a new soya biopeptide studied in an in vitro skin equivalent model and human volunteers. Int J Cosmet Sci 1999; 21:299–311.
66. Varani J, Kelley EA, Perone P, Lateef H. Retinoin-induced epidermal hyperplasia in human skin organ culture: inhibition with soy extract and soy isoflavones. Exp Mol Pathol 2004; 77:176–183.
67. Wei H, Bowen R, Zhang X, Lebwohl M. Isoflavone genistein inhibits the initiation and promotion of two-stage skin carcinogenesis in mice. Carcinogenesis 1998; 19:1509–1514.
68. Wei H, Saladi R, Lu Y, Wang Y, Palep SR, Moore J, Phelps R, Shyong E, Lebwohl MG. Isoflavone genistein: photoprotection and clinical implications in dermatology. J Nutr 2003; 133:3811S–3819S.
69. Widyarini S, Husband AJ, Reeve VE. Protective effect of the isoflavonoid equol against hairless mouse skin carcinogenesis induced by UV radiation alone or with a chemical co-carcinogen. Photochem Photobiol 2005; 81:32–37.
70. Wang Y, Yaping E, Zhang X, Lebwohl M, DeLeo V, Wei H. Inhibition of ultraviolet B (UVB)-induced c-fos and c-jun expression in vivo by a tyrosine kinase inhibitor, genistein. Carcinogenesis 1998; 19:649–654.
71. Shyong EQ, Lu Y, Lazinsky A, Saladi RN, Phelps RG, Austin LM, Lebwohl M, Wei H. Effects of the isoflavone 4′,5,7-trihydroxyisoflavone (genistein) on psoralen plus ultraviolet A radiation (PUVA)-induced photodamage. Carcinogenesis 2002; 23:317–321.

72. Widyarini S, Spinks N, Husband AJ, Reeve VE. Isoflavonoid compounds from red clover (*Trifolium pratense*) protect from inflammation and immune suppression induced by UV radiation. Photochem Photobiol 2001; 74:465–470.
73. Widyarini S. Protective effect of the isoflavone equol against DNA damage induced by ultraviolet radiation to hairless mouse skin. J Vet Sci 2006; 7: 217–223.
74. Widyarini S, Allanson M, Gallagher NL, Pedley J, Boyle GM, Parsons PG, Whiteman DC, Walker C, Reeve VE. Isoflavonoid photoprotection in mouse and human skin is dependent on metallothionein. J Invest Dermatol 2006; 126:198–204.
75. Reeve VE, Widyarini S, Domanski D, Chew E, Barnes K. Protection against photoaging in the hairless mouse by the isoflavone equol. Photochem Photobiol 2005; 81:1548–1553.
76. Leyden JJ, Nebus J, Wallo W. Efficacy of a soy moisturizer in photoaging: a double-blind, vehicle-controlled, 12-week study. J Drugs Dermatol 2007; 6:917–922.
77. McElwee KJ, Niiyama S, Freyschmidt-Paul P, Wenzel E, Kissling S, Sundberg JP, Hoffmann R. Dietary soy oil content and soy-derived phytoestrogen genistein increase resistance to alopecia areata onset in C3H/HeJ mice. Exp Dermatol 2003; 12:30–36.
78. Tsuruki T, Takahata K, Yoshikawa M. A soy-derived immunostimulating peptide inhibits etoposide-induced alopecia in neonatal rats. J Invest Dermatol 2004; 122:848–850.
79. Tsuruki T, Takahata K, Yoshikawa M. Anti-alopecia mechanisms of soymetide-4, an immunostimulating peptide derived from soy β-conglycinin. Peptides 2005; 26:707–711.
80. Harada N, Okajima K, Arai M, Kurihara H, Nakagata N. Administration of capsaicin and isoflavone promotes hair growth by increasing insulin-like growth factor-1 production in mice and in humans with alopecia. Growth Horm IGF Res 2007; 17:408–415.
81. Seiberg M, Liu JC, Babiarz L, Sharlow E, Shapiro S. Soymilk reduces hair growth and hair follicle dimensions. Exp Dermatol 2001; 10:405–413.
82. Draelos Z, Blair R, Tabor A. Oral soy supplementation and dermatology. Cosmetic Dermatology 2007; 20:202–204.

PART 8
NATURAL PROTECTION FROM PHOTOCARCINOGENESIS

24

Green Tea and Skin Cancer: Immunological Modulation and DNA Repair

Suchitra Katiyar[1], MPH, Craig A. Elmets[1], MD, and Santosh K. Katiyar[1,2], PhD

[1]*Department of Dermatology, University of Alabama at Birmingham, Birmingham, AL, USA*
[2]*Birmingham VA Medical Center, Birmingham, AL, USA*

Aaron Tabor and Robert M. Blair (eds.), *Nutritional Cosmetics: Beauty from Within*, 469–499,

24.1 Introduction

Tea is consumed worldwide as a popular beverage because of its characteristic aroma, flavor, and health benefits [1]. Of the total commercial tea production worldwide, about 78% is consumed in the form of black tea, mainly in western countries and some Asian countries, and 20% is consumed in the form of green tea. Green tea is primarily consumed in some Asian countries, like Japan, China, Korea, and parts of India; a few countries in North Africa; and the Middle East [2,3]. Approximately 2% of tea is consumed in the form of oolong tea, a partially fermented tea, in some parts of Southeastern China [1,2]. The basic steps of manufacturing different tea varieties are more or less similar except to protect and develop their aroma during the fermentation process, which also controls the oxidation status of the individual catechin/epicatechin molecules and their derivatives present in fresh tea leaves [1]. The characteristic aroma and health benefits of green tea are associated with the presence of catechin/epicatechin derivatives, which are commonly called "polyphenols." The major polyphenolic constituents present in green tea are (–)-epicatechin, (–)-epigallocatechin, (–)-epicatechin-3-gallate, and (–)-epigallocatechin-3 -gallate (EGCG) [1–3]. In addition to a small amount of catechins, black tea contains thearubigins and theaflavins, which are the polymerized forms of catechin monomers and are the major components formed during enzymatic oxidation and fermentation processes [1,3].

24.2 Solar Ultraviolet (UV) Radiation

Solar UV radiation makes up about 5% of the electromagnetic spectrum that reaches the earth's surface. Three UV spectral regions have been designated based on their biological effects. Terrestrial UV radiation consists of 5–6% UVB and 90–95% UVA. Negligible amounts of UVC reach the earth's surface due to the filtering capacity of the ozone layer. Thus, the solar UV spectrum is mainly divided into three main categories based on their wavelengths and adverse biological effects [4,5]: long-wave UVA (320–400 nm), mid-wave UVB (290–320 nm), and short-wave UVC (200–290 nm). Briefly, these are summarized below.

24.2.1 Long-Wave UVA

UVA comprises the largest spectrum of solar UV radiation (90–95%) and is considered as the "aging ray." UVA penetrates deeper into the epidermis and dermis of the skin and it has recently been shown that extensive UVA exposure can lead to benign tumor formation as well as malignant cancers [6,7]. Its exposure induces the generation of singlet oxygen and hydroxyl free radicals, which can cause damage to cellular macromolecules, like proteins, lipids, and DNA [8]. In contrast to UVC or UVB, UVA is hardly able to excite the DNA molecule directly and produces only a small number of cyclobutane pyrimidine dimers in the skin. Therefore it is presumed that much of the mutagenic and carcinogenic action of UVA radiation appears to be mediated through reactive oxygen species [9,10]. However, it is still a matter of debate. It has been suggested that bipyrimidine photoproducts rather than oxidative lesions are the main type of DNA damage involved in the genotoxic effect of solar UVA radiation [11]. UVA is a significant source of oxidative stress in human skin and causes photoaging in the form of skin sagging rather than wrinkling [12] and can suppress some immunological functions [13].

24.2.2 Mid-Wave UVB

UVB radiation constitutes approximately 5% of the total solar UV radiation and is mainly responsible for a variety of skin diseases including nonmelanoma and melanoma skin cancers. UVB radiation can penetrate inside the epidermis of skin and can induce both direct and indirect adverse biological effects including induction of oxidative stress, DNA damage, cancer, and premature aging of the skin [4,14]. Excessive exposure of UVB radiation decreases the levels of antioxidant defense enzymes in the skin, impairing their ability to protect from harmful effects [4,14].

24.2.3 Short-Wave UVC

UVC radiation is largely absorbed by the atmospheric ozone layer and does not reach the surface of the earth. These wavelengths have enormous energy and are mutagenic in nature.

24.3 Characteristics of UV Radiation

Solar UV radiation, particularly UVB (290–320 nm), possesses suppressive effects on the immune system [15] and can act as a tumor initiator [16], tumor promoter [17], and cocarcinogen [18,19]. Exposure of the skin to UVB radiation induces a variety of biological effects, including inflammation, sunburn cell formation, immunologic alterations, induction of oxidative stress, and DNA damage, which play important roles in the generation and maintenance of UV-induced neoplasms [20–22]. Although skin possesses an elaborate defense system consisting of enzymatic and nonenzymatic components to protect the skin from adverse biological effects of UV radiation, excessive exposure to UV radiation overwhelms and depletes the cutaneous defense system and its ability, leading to the development of various skin disorders and/or diseases including the risk of skin cancers [17,22–24].

UVB radiation has multiple effects on the immune system [25,26]. There is ample clinical and experimental evidence to suggest that immune factors contribute to the pathogenesis of solar UV–induced skin cancer in mice and probably in humans as well [27,28]. Chronically immunosuppressed patients living in regions of intense sun exposure experience an exceptionally high rate of skin cancer [29]. This observation is consistent with the hypothesis that immune surveillance is an important mechanism designed to prevent the generation and maintenance of neoplastic cells. Further, the incidence of skin cancers, especially squamous cell carcinoma, is also increased among organ transplant recipients [30–32]. The increased frequencies of squamous cell carcinomas, especially in transplant patients, are presumably attributable to a long-term immunosuppressive therapy [33], however nonimmune mechanisms may also play a role [34]. These studies provide evidence in support of the concept that UV-induced immune suppression promotes the development of skin tumors.

24.4 Skin

The skin is the largest organ of the body and comprises a surface area of approximately 1.5–2.0 m^2. It protects the internal organs of the body by acting as an effective barrier against the detrimental effects of environmental and xenobiotic agents. Thus the major role of the skin is to provide a protective covering at this crucial interface between inside and outside. Morphologically, skin is a composite of a variety of cell types and organellar bodies, each of which has a particular function. Achieving this goal has resulted in the evolution of a complex structure involving several different layers, each with particular properties. The major layers include the epidermis, the dermis, and the hypodermis (Fig. 24.1). The epidermis is a stratifying layer of epithelial cells that overlies the connective tissue layer, the dermis. The dermis is divided into papillary dermis and reticular dermis. The epidermis and dermis are supported by an internal layer of adipose tissue, called the hypodermis. Skin cancer is mainly associated with the epidermal layer and its cell types. The major cell type of the epidermis is the keratinocytes. They comprise > 90% of the cells of the epidermal layer. Other cell types are: Langerhans cells, melanoma cells, and $\gamma\delta$ T cells. In laboratory animals like the mouse, the epidermis is about 2–3 cell layers thick (Fig. 24.1B), whereas in human skin, the epidermis is quite thick and about 8–15 cell layers thick (Fig. 24.1A). Among many environmental and xenobiotic factors, the exposure of the skin to solar UV radiation is the key factor in the initiation of several skin disorders, such as wrinkling, scaling, dryness, mottled pigment abnormalities consisting of hypopigmentation and hyperpigmentation, and skin cancer [4,35,36]. Statistical analysis indicates that the average annual UV dose that an average American typically receives in a year is about 2,500–3,300 mJ/cm^2. Further, an average female is exposed to an average 2,200 mJ/cm^2 and males 2,800 mJ/cm^2 each year, with an additional exposure of about 800 mJ/cm^2 of solar UVB radiation during a conservative vacation period [37,38].

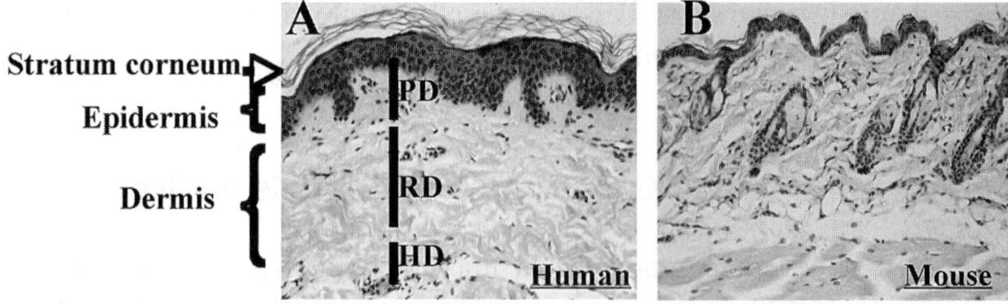

Figure 24.1 Structure of the normal skin. The morphology of the human and mouse skin is shown in Panel A and Panel B, respectively. The epidermis in human skin is thicker than that of mouse skin. The stratum corneum is a layer of dead cells and hydrophobic in nature. PD: papillary dermis; RD: reticular dermis; HD: hypodermis.

24.5 Phytochemicals and Photoprotection

There has been considerable interest among the human population for the use of naturally occurring botanicals/phytochemicals for the prevention of UV-induced photodamaging effects, including photoaging of the skin and nonmelanoma and melanoma skin cancers. Botanical supplements, specifically dietary botanicals, possessing anti-inflammatory, immunomodulatory, and antioxidant properties are among the most promising group of phytochemicals that can be exploited as ideal chemopreventive agents for a variety of skin disorders in general and skin cancer in particular. Recent advances in our understanding at the cellular and molecular levels of carcinogenesis have led to the development of promising strategies for the prevention of cancer or "chemoprevention." Chemoprevention is a means of cancer control by the use of specific natural or synthetic chemical substances that can suppress, retard, or reverse the process of carcinogenesis and other skin diseases. Thus, chemoprevention offers a realistic promise or strategy for controlling the risk of skin diseases. Further, the chemopreventive approach appears to have practical implications in reducing skin cancer risk because, unlike the carcinogenic or polluted environmental factors that are difficult to control, individuals can modify their dietary habits and lifestyle in combination with careful use of skin care products to prevent photodamaging effects of solar UV radiation. Studies have shown the chemopreventive potential of several naturally occurring phytochemicals, such as green tea polyphenols (GTPs), silymarin, retinoids, and grapeseed proanthocyanidins, etc., against UV radiation–induced inflammation, oxidative stress, and photocarcinogenesis [2,5,39,40]. Here, we will only summarize and discuss the photoprotective potential of green tea polyphenols and their mechanisms of action.

24.6 Green Tea and Green Tea Polyphenols (GTPs)

Commercial beverage–grade tea is manufactured from the leaves and buds of the plant *Camellia sinensis* and is commercially available in three forms: green, black, and oolong tea [1–3]. The basic steps of manufacturing the various forms of teas are similar except in the development of their aroma and in the fermentation process, which is dependent on the oxidation states of catechins present in tea leaves. The term *green tea* refers to the product manufactured from the fresh leaves of the tea. The production of green tea is characterized by an initial heating process, which kills the enzyme polyphenol oxidase, which is responsible for the conversion of the flavanols in the leaf into the dark polyphenolic compounds that are found in black tea.

The catechins present in green tea are commonly called polyphenols and are flavanols in nature. The major catechins found in green tea are: (–)-epicatechin, (–)-epigallocatechin, (–)-epicatechin-3-gallate, and (–)-epigallocatechin-3-gallate (EGCG). The chemical structures of these catechins or polyphenols are shown in Fig. 24.2. These polyphenols are antioxidant and anti-inflammatory in nature and have been shown to possess anticarcinogenic activity in several *in vitro* and *in vivo* systems [5,41]. Of these major polyphenols, EGCG is the major and most effective chemopreventive agent and has been extensively studied in several disease models including its use in skin photoprotective activity. It is noteworthy to mention that during the black tea manufacturing process the catechin derivatives are oxidized, polymerized, and converted into a less well-defined group of compounds known as thearubigens. The thearubigen fraction is a mixture of substances with a molecular weight distribution of 1,000–40,000 and accounts for about 15% of the dry weight solids of black tea [1]. The experimental evidence from several *in vitro* and *in vivo* studies indicate that the green tea polyphenols (GTPs) are better chemopreventive in nature than those present in black tea. Hence, extensive studies were conducted with GTPs. Here we describe the photoprotective potential, in particular antiphotocarcinogenic potential, of green tea polyphenols and their mechanisms of action, specifically the recent developments on the photoimmunology and DNA repair mechanisms in prevention of photocarcinogenesis by GTPs or EGCG.

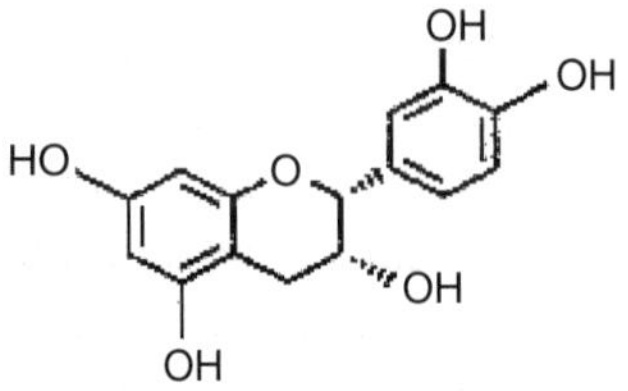

(-)-Epicatechin

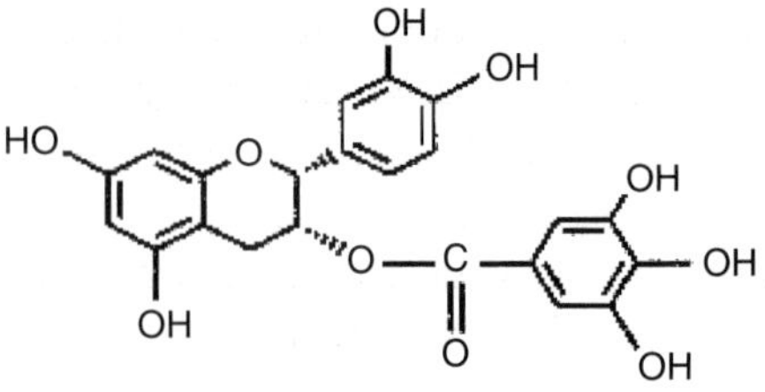

(-)-Epicatechin-3-gallate

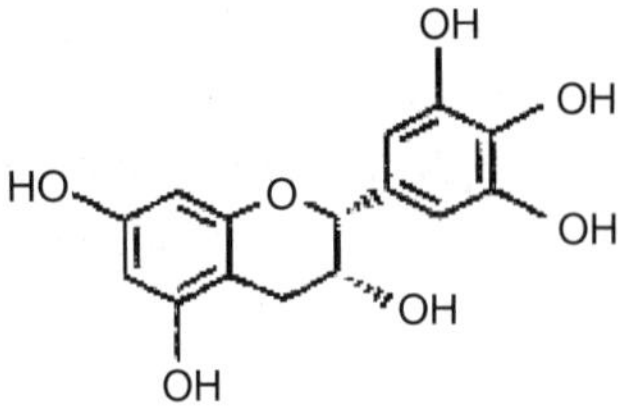

(-)-Epigallocatechin

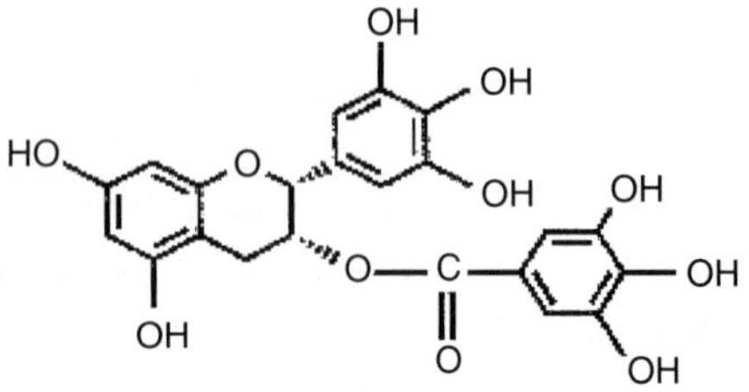

(-)-Epigallocatechin-3-gallate

Figure 24.2 Chemical structures of major epicatechin derivatives found in green tea. These are also called "polyphenols."

24.7 Bioavailability of Green Tea Catechins or Polyphenols

In order to understand the health benefits of green tea polyphenols, it is also equally important to examine and understand the bioavailability and metabolism of individual epicatechin derivatives, such as (–)-epicatechin (EC), (–)-epicatechin-gallate (ECG), (–)-epigallocatechin (EGC), and EGCG. For this purpose, it is necessary to evaluate their biological activity within the target tissues [42]. Following oral administration of tea catechins to rats, the four principal catechins (EC, ECG, EGC, and EGCG) have been identified in the portal vein, indicating that tea catechins are absorbed in the intestine [43]. In rats given GTPs (0.6%, w/v) in their drinking water over a period of 28 days, plasma concentrations of EGCG were lower than those of EGC or EC. When the same GTP preparation was given to mice, plasma levels of EGCG were much higher than those of EGC and EC. So, there appear to be species differences in the bioavailability of EGCG compared to other catechins [44]. The levels of catechins in human plasma reach their peak at 2–4 hours after ingestion [45]. A study conducted in humans compared the pharmacokinetics of equimolar doses of pure EGC, ECG, and EGCG in 10 healthy volunteers. Average peak plasma concentrations after a single dose of 1.5 mmol were 5.0 μmol/l for EGC, 3.1 μmol/l for ECG, and 1.3 μmol/l for EGCG. After 24 hours, plasma EGC and EGCG returned to baseline, but plasma ECG remained elevated [46]. In humans, ECG has been found to be more highly methylated than EGC and EGCG, and EGCG has been found to be less conjugated than EGC and EC [47]. Kim et al. [44] observed that when rats were given GTPs (0.6%, w/v) in their drinking water over a period of 28 days, substantial amounts of EGC and EC were found in the esophagus, large intestine, kidney, bladder, lung, and prostate. EGC and EC concentrations were relatively lower in the liver, spleen, heart, and thyroid. EGCG levels were higher in the esophagus and large intestine, but lower in other organs, likely due to poor systemic absorption of EGCG. The studies conducted in rats indicated that EGCG is mainly excreted through the bile, whereas EGC and EC are excreted through both urine and bile. Lu et al. [48] reported that after oral administration of green tea to rats, about 14% of EGC, 31% of EC, and <1% of EGCG appeared in the blood; however, the bioavailability of EGCG was higher in mice. But the biological activities of the catechin metabolites still need to be investigated. Inter-individual variations in the bioavailability of GTPs can be substantial and may be due, in part, to differences in colonic microflora and genetic polymorphisms among the enzymes involved in the metabolism of polyphenols [49]. The effect of green tea drinking may also differ by genotype [50]. There appear to be species differences in the bioavailability of EGCG compared to other tea catechins. Further, the addition of milk to tea does not interfere with catechin absorption [51–53], but milk may affect the antioxidant potential of tea, depending upon milk fat content, milk volume added, and the method used to assess this parameter [51–54]. It is suggested that determination of the actual bioavailability of metabolites in tissues may be much more important than knowledge of their plasma concentration, but data are still very scarce even in animals. Hence, the metabolism and bioavailability of individual tea catechins and the pharmacokinetics of their metabolites require further attention.

24.8 Prevention of Photocarcinogenesis

Extensive studies have been conducted in *in vitro* and *in vivo* models to assess the antiphotocarcinogenic effects of green tea. It has been found that oral administration of GTPs (a mixture of polyphenolic components) in drinking water to mice results in significant protection against UV-induced skin carcinogenesis in terms of tumor incidence, tumor multiplicity, and tumor growth or size compared to those mice that were not given

GTPs in drinking water [3,5,41,55]. The mice that were given a crude water extract of green tea as a sole source of drinking water had developed fewer tumors compared to those mice that were not given a water extract of green tea [56,57]. Topical application of EGCG or administration of GTPs in drinking water also induced partial regression or inhibition of tumor growth of established skin papillomas in mice [58]. Oral administration of green tea as a sole source of drinking water resulted in a decreased number of skin tumor types, such as papillomas, keratoacanthomas, and squamous cell carcinomas induced by chronic exposure of mice to UVB radiation. The authors' laboratory developed a hydrophilic ointment–based formulation for the topical treatment of green tea polyphenols [59]. Treatment of EGCG in this topical formulation has resulted in exceptionally high protection against photocarcinogenesis in the SKH-1 hairless mouse model in terms of tumor incidence (60% inhibition), tumor multiplicity (86% inhibition), and tumor growth in terms of total tumor volume per group (95% inhibition) [59]. These results indicated that the use of EGCG with topical formulation might increase the penetration or absorption capacity inside the skin layers and the presence of a higher concentration may be responsible for the higher photoprotection. The antioxidative effects or photoprotective effects of GTPs or EGCG by topical treatment were greater than in oral administration, which may be due to the presence of a higher content of these polyphenols in topical application [60]. This information also suggests that nutrient supply to the skin through topical treatment might be better than oral administration.

It has been shown that green tea can protect against both UVA and UVB radiation–induced skin cancer in mice [61]. Experimental studies conducted in *in vitro* and *in vivo* models indicate that GTPs or EGCG prevents photocarcinogenesis following several mechanisms that involve multiple molecular targets [5,14]. Importantly, long-term oral administration or topical application of GTPs or EGCG did not show signs of visible toxicity in laboratory animals.

24.9 Modulation of Immune System

24.9.1 Prevention of UVB-Induced Immune Suppression: CHS Model

The immunosuppressive effects of solar UV radiation are well established, having been demonstrated most clearly on the inhibition of contact hypersensitivity (CHS) response induced by the contact allergens. CHS response is considered as a prototypic T-cell mediated immune response [62,63]. Many of the adverse effects of solar UV radiation on human health, including exacerbation of infectious diseases, premature aging of the skin, and induction of skin cancer, are mediated at least in part by the ability of UV radiation to induce immune suppression [64–66]. As UV-induced immunosuppression is considered to be a risk factor for the induction of skin cancer [67,68], prevention of UV-induced immunosuppression represents a potential strategy for the management of skin cancer. Topical treatment of C3H/HeN mice with GTPs resulted in significant protection against local and systemic models of CHS where 2,4-dinitrofluorobenzene was used as a contact sensitizer [69]. The chemopreventive effect of GTPs on UVB-induced suppression of CHS response was dependent on the doses of GTPs administered. Similar effects were also noted when water extract of green tea was given to mice as the sole source of drinking water. Among the four major epicatechin derivatives present in GTPs, EGCG was found to be the most effective polyphenol affording protection against UVB-induced suppression of CHS. The mechanisms involved in UV-induced immune suppression differ greatly. There are studies defining the role of immunoregulatory cytokine interleukin (IL)-12 in the induction and elicitation of CHS. CHS has been considered to be a Th1-mediated immune response [70]. Epidermal Langerhans cells, which are critical antigen presenting cells in the induction phase of CHS [71], have been described as an additional source for IL-12 production. Exposure of the skin to UV radiation induces inflammatory leukocyte infiltration in the skin, and most prominent are the activated macrophages and neutrophils. These infiltrating leukocytes (also called CD11b$^+$ cell subset) have been shown to have a role in UV-induced immune suppression, and are the major source of reactive oxygen species generation at the UV-exposed site, as demonstrated in Fig. 24.3. We were interested in determining whether topical treatment of EGCG blocks UVB-induced infiltration of CD11b$^+$ cells and result in downregulation of IL-10 or upregulation of IL-12 in skin and/or draining lymph nodes in mice, thereby preventing UV-induced immune suppression. The prevention of UVB-induced immunosuppression by EGCG treatment was found to be associated with a reduction in the number of infiltrating CD11b$^+$ cells in UVB-irradiated skin [72]. Hammerberg et al. [73,74] demonstrated that blocking of infiltrating leukocytes using anti-CD11b$^+$ antibody or treatment with soluble complement receptor type I inhibited UV-induced immune suppression and tolerance induction in C3H/HeN mice. Therefore, it seems that inhibition of UV-induced immunosuppression by EGCG is mediated, at least in part, through the inhibition of infiltrating CD11b$^+$ cells in the UVB-irradiated mouse skin.

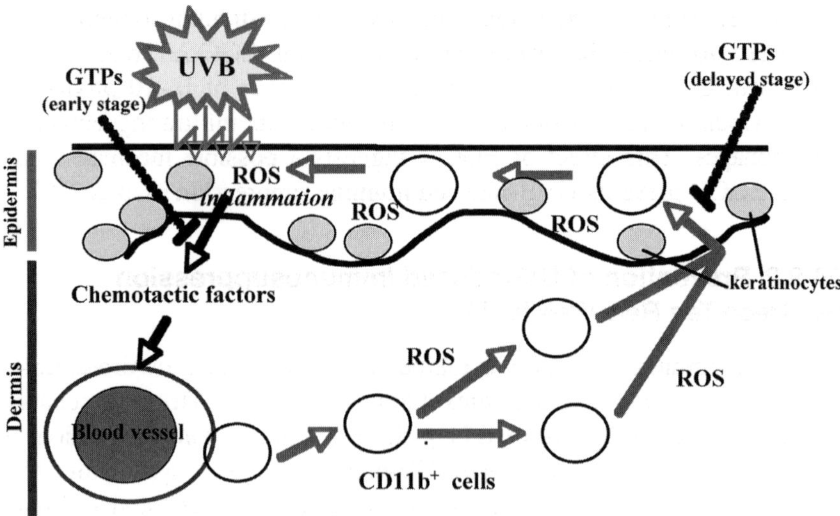

Figure 24.3 The schematic diagram depicts the mechanism of UV-induced generation of oxidative stress in the skin. UV exposure induces inflammatory responses, chemotactic factors and reactive oxygen species (ROS) at early stage or early time points. Inflammatory responses and chemotactic factors stimulate the infiltration of leukocytes at UV-irradiated skin sites. Infiltrating leukocytes, particularly activated macrophages and neutrophils ($CD11b^+$ cells), are the major source of ROS. The generation of ROS through $CD11b^+$ cells is termed as the effect at delayed stage or delayed time points. Treatment of skin with green tea polyphenols, topical or dietary, may prevent UV-induced ROS generation at early time points, and also at delayed time points. Inhibition of UV-induced oxidative stress by green tea could be an effective strategy to prevent the risk of photocarcinogenesis. UV-induced infiltrating $CD11b^+$ cells also have a role in suppression of the immune system.

Some other molecular targets also have a role in altering the immune system in UV-exposed skin. The cytokine IL-10 possesses immunosuppressive activity and inhibits the antigen presenting capacity of antigen presenting cells [75,76] in *in vitro* and *in vivo* systems. In UVB-irradiated skin, IL-10 is primarily secreted by activated macrophages [77,78], thereby downregulating CHS responses. It has been shown that intraperitoneal administration of IL-10 to mice inhibits delayed type hypersensitivity responses [79], whereas intraperitoneal injection of anti-IL-10 antibody prevents UV-induced tolerance induction in mice [80]. In accordance with these observations, the treatment of mouse skin with EGCG resulted in a decreased level of IL-10 in UV-irradiated skin as well as in draining lymph nodes compared to mice that were not treated with EGCG. Treatment of mice with EGCG significantly reduces the number of IL-10 producing cells, which is accompanied with a reduction in infiltrating activated macrophages. This effect of EGCG suggests a possible mechanism by which EGCG prevents UVB-induced immune suppression in mice [72].

24.9.2 Prevention of UV-Induced Immunosuppression by Green Tea Requires IL-12

IL-12 is an immunoregulatory cytokine that regulates the growth and functions of T-cells [81] and augments the development of Th1 type cells by stimulating the production of IFN-γ [82–84]. Intraperitoneal injection of recombinant IL-12 in mice prevents UV-induced immune suppression [85]. These studies imply that a cytokine imbalance between Th1 and Th2 may be responsible for the development of UVB-induced immunosuppression. Topical treatment of EGCG prior to UVB exposure resulted in increased levels of IL-12 in the skin and draining lymph nodes compared to non-EGCG treated but UVB-exposed mice [72]. Increased levels of IL-12 may contribute in stimulating the immune responses, and therefore it appears that EGCG treatment is capable of promoting the development of a Th1 immune response through the induction of IL-12 and may be one of the mechanisms responsible for inhibiting UVB-induced immune suppression in mice. To further confirm the role of IL-12 in EGCG-mediated prevention of UV-induced immunosuppression, IL-12 knockout (KO) mice were used. Recently, Meeran et al. [86] have reported that topical treatment of EGCG prevented UV-induced suppression of CHS in wild-type mice as shown by a significant enhancement of the CHS response (ear swelling) to 2,4-dinitrofluorobenzene, a contact allergen. In contrast, UV-exposed IL-12 KO mice remained unresponsive to 2,4-dinitrofluorobenzene despite the application of EGCG on the mouse skin, indicating that the immunopreventive effect of EGCG on UV-induced suppression of CHS requires IL-12 or is mediated through IL-12. To further confirm that prevention of UV-induced suppression of CHS by EGCG requires IL-12, wild-type mice were treated i.p. with anti-IL-12 monoclonal antibody. In EGCG-treated mice, the i.p. injection of anti-IL-12 antibody significantly blocked the preventive effect of EGCG on UV-induced suppression of CHS. These studies provide convincing evidence that prevention of UV-induced suppression of CHS by EGCG is mediated, at least in part, through IL-12. They suggest that green tea polyphenols have the ability to modify the immunological responses in *in vivo* system, which can result in beneficial effects.

24.9.3 Prevention of UV-Induced Immunosuppression by Green Tea Is Mediated through IL-12-Dependent DNA Repair

Ultraviolet radiation-induced DNA damage, particularly in the form of cyclobutane pyrimidine dimers (CPDs), is an important molecular trigger for UV-induced immunosuppression [87], and reduction in UV-induced CPDs through application of DNA repair enzymes can prevent UV-induced immunosuppression [88,89]. IL-12 has the ability to remove or repair UV-induced CPDs [90]. Schwarz et al. [91] reported that the prevention of UV radiation–induced immunosuppression by IL-12 is dependent on DNA repair and acts through the induction of a nucleotide excision repair mechanism. Based on these observations, our laboratory conducted experiments to examine whether IL-12 contributes to the ability of EGCG to prevent UV-induced immunosuppression by stimulating repair of UV-induced DNA damage [86]. The effect of EGCG was determined on UV-induced CPDs in wild-type mice and compared with IL-12 KO mice. Twenty-four hours after UV irradiation it was observed that the numbers of CPD-positive cells were significantly lower in the EGCG-treated wild-type mice than the wild-type mice that were not treated with EGCG but were exposed to UVB radiation. In contrast, the UVB-induced DNA damage in the IL-12 KO mice that had been treated with EGCG did not differ from that in the IL-12 KO mice that had not been treated with EGCG [86]. This information suggests that EGCG-induced IL-12 may contribute to the repair of UV-damaged DNA, and that the difference in DNA repair between wild-type and IL-12 KO may be due to the absence of IL-12 in IL-12 KO mice. These studies indicate that endogenous stimulation of IL-12 by EGCG can be used as a strategy for the prevention of UV-induced immunosuppression in humans.

UV-induced DNA damage is an important molecular trigger for the migration of antigen presenting cells (i.e., epidermal Langerhans cells) from the skin to the draining lymph nodes. DNA damage in antigen presenting cells impairs their capacity to present Ag, which in turn results in a lack of sensitization [92]. CPD-containing antigen presenting cells have been found in the draining lymph nodes of UV-exposed mice [93]. These antigen presenting cells were identified to be of epidermal origin and exhibited an impaired Ag presentation capacity. Because the treatment of EGCG induces IL-12 in mice [72], and IL-12 has the ability to induce DNA repair [90], the effect of EGCG on the migration of CPD-positive cells from the UV-exposed skin to the draining lymph nodes was studied.

Immunohistochemical analysis of CPD-positive cells in lymph nodes after 36 hours of UV irradiation showed significant numbers of CPD-positive cells in the draining lymph nodes in both UV-exposed wild-type and IL-12 KO mice; however, the numbers of CPD-positive cells in the draining lymph nodes of the UV-exposed IL-12 KO mice were significantly higher than in the draining lymph nodes of their wild-type mice. The decreased numbers of CPD-positive cells in the draining lymph nodes of UV-exposed wild-type mice compared to IL-12 KO mice may be attributable to the presence of endogenous IL-12 in the wild-type mice at levels that are capable of partial removal of the damaged DNA or partial repair of the damaged DNA in the migrating cells. Treatment with EGCG resulted in a significant reduction in the numbers of CPD-positive cells in the draining lymph nodes of UV-exposed wild-type mice compared to UV-exposed wild-type mice that did not receive EGCG. In contrast, there was no significant difference in the number of CPD-positive cells in the draining lymph nodes between EGCG-treated and non-EGCG-treated UV-exposed IL-12 KO mice. This observation further supports the evidence that the reduction in the numbers of CPD-positive cells in the draining lymph nodes of wild-type mice after EGCG treatment may be due to EGCG-stimulated IL-12-mediated repair of CPD-positive cells. The major molecular targets of green tea that are discussed in this chapter are summarized in Fig. 24.4.

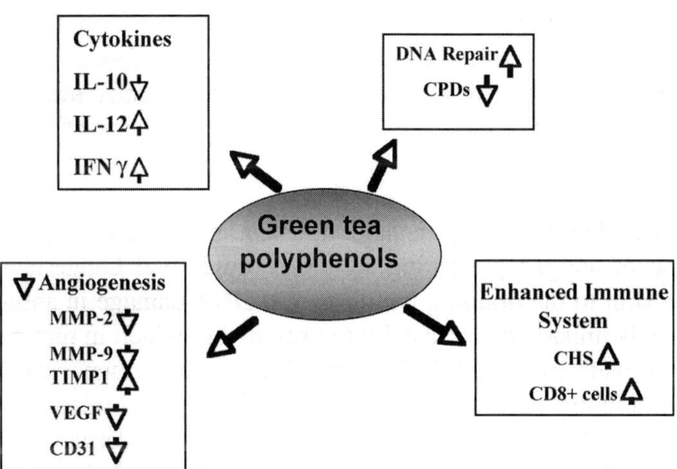

Figure 24.4 Summary of major molecular targets of GTPs. ↑: Upregulation; ↓: downregulation. CPDs: cyclobutane pyrimidine dimers; CHS: contact hypersensitivity.

24.10 Prevention of Photocarcinogenesis by Green Tea through IL-12-Dependent DNA Repair

Administration of GTPs in drinking water or topical application of EGCG inhibits photocarcinogenesis in mice [55,59]. Because treatment of EGCG prevents UV-induced immunosuppression through stimulation of IL-12, and IL-12 has antitumor activity and the ability to repair UVB-induced DNA damage, the effect of EGCG on photocarcinogenesis in IL-12 KO mouse model was studied. Meeran et al. [94] have reported that topical treatment with EGCG significantly inhibits photocarcinogenesis in terms of tumor incidence and tumor multiplicity in wild-type (C3H/HeN) mice but does not prevent it in IL-12 KO mice, indicating that the prevention of UVB-induced skin cancer by EGCG requires IL-12 cytokine. Because UVB-induced DNA damage, particularly the formation of CPDs, has a role in initiation of photocarcinogenesis, this study further demonstrated that treatment with EGCG more rapidly removed or repaired UVB-induced CPDs in wild-type mice than in IL-12 KO mice. Further treatment of EGCG-treated IL-12 KO mice with recombinant IL-12 rapidly removes or repairs UVB-induced CPD-positive cells. This information supports the evidence that EGCG promotes the removal or repair of damaged DNA in UVB-exposed skin through a mechanism that requires IL-12.

Epidermal sunburn cell formation in UV-exposed skin is primarily a consequence of DNA damage [94,95]. Sunburn cells are keratinocytes undergoing apoptosis after they have been exposed to a physiologic UVB dose that has severely and irreversibly damaged their DNA. Using IL-12 KO and their wild-type counterparts as a tool, Meeran et al. [94] reported that treatment with EGCG results in a significant reduction in the numbers of sunburn cells after UVB irradiation in a time-dependent manner in wild-type mice. In contrast to wild-type mice, the treatment of IL-12 KO mice with EGCG did not result in the rapid removal of sunburn cells after UVB exposure. This difference in the repair kinetics of sunburn cells in the IL-12 KO and their wild-type counterparts may be attributed to the endogenous stimulation of IL-12 in the wild-type mice after EGCG treatment.

24.11 Repair of UV-Induced DNA Damage by Green Tea Is Mediated through Nucleotide Excision Repair (NER) Mechanism

The next important question arises as to how the UVB-induced DNA damage was repaired by IL-12 in EGCG-treated mice. To elucidate the DNA repair mechanism by EGCG in UV-exposed skin, Meeran et al. [94] used the fibroblasts from the patients suffering from xeroderma pigmentosum complementation group A (*XPA*-deficient) and from normal healthy persons. A distinct property of IL-12 was utilized in these studies. IL-12 has been shown to have the ability to repair UVB-induced CPDs in healthy cells but not in cells from patients suffering from *XPA* disease [86,90,94]. The *XPA* gene is an essential component of nucleotide excision repair (NER), thus, cells with a mutated *XPA* gene completely lack functional NER. To examine whether the NER mechanism is required for the EGCG-induced IL-12-mediated repair of UVB-induced CPDs, NER-deficient fibroblasts from *XPA* patients and NER-proficient fibroblasts from healthy persons were exposed to UVB with or without prior treatment with EGCG. When the cells were analyzed for CPDs immediately after UVB exposure, no differences were observed in the cells treated with or without EGCG in terms of the number of CPD-positive cells. This finding implies that EGCG does not prevent immediate formation of CPDs after UVB exposure and excludes a UVB radiation filtering effect. However, when the cells were analyzed after 24 hours of UVB irradiation, the numbers of CPD-positive cells were significantly decreased in the NER-proficient cells but did not significantly remove or repair CPDs in NER-deficient cells from *XPA* patients, suggesting that EGCG might accelerate the repair of UVB-induced CPDs through the NER mechanism. Some studies have reported that UVB-caused DNA damage can be reduced by the application of exogenous DNA-repair enzymes. The bacterial DNA-repair enzyme T4 endonuclease V (T4N5) can be delivered into cells by liposomes [96]. Topically applied T4N5 liposomes increase the removal of CPDs and reduce the incidence of skin cancer [97,98]. In accordance with the accelerated removal of DNA lesions by T4N5, sunburn cell formation is also reduced [99]. In contrast to the external application of DNA-repair enzymes, the studies from Meeran et al. [94] indicate that EGCG affects the cell's own NER system through the endogenous stimulation of IL-12.

24.12 Stimulation of Cytotoxic T Cells (CD8+) in Skin Tumors

To explore further the chemopreventive mechanisms of green tea against photocarcinogenesis, other molecular targets were also investigated. IL-12 has been shown to stimulate the production of IFNγ and stimulate the development of cytotoxic T cells (e.g., $CD8^+$ T cells), which are tumoricidal in nature and thus may result in inhibition or regression of the tumors. Mantena et al. [55] have reported that administration of GTPs in drinking water inhibits photocarcinogenesis in mice, and this chemopreventive activity of GTPs is mediated, at least in part, through the recruitment of cytotoxic T cells in the tumor microenvironment. This study indicates another major pathway by which GTPs can inhibit skin tumor growth in UV-exposed skin. $CD8^+$ T cells are the effector cells in the cytotoxic response of the host to UV-induced skin tumor cells; they play an important role in the protection against tumor immunity, at least for skin tumors induced by chronic UV irradiation [100]. The ability of oral administration of GTPs to enhance the infiltration or recruitment of $CD8^+$ T cells in the tumor microenvironment may act to enhance the immunosurveillance that is mediated by these cells, thereby decreasing the growth as well as incidence of tumors. Similar observations were noted when mice were topically treated with EGCG in hydrophilic formulation and subjected to a photocarcinogenesis protocol [101]. However, the chemopreventive effect of EGCG applied topically was greater than GTPs given in drinking water. Presumably, this difference may be due to the higher concentration of EGCG available in the skin with topical application compared with the administration of GTPs in drinking water. However, the chemopreventive effect of orally administered GTPs in mice was substantial. It can be as important as EGCG because GTPs are affordable, less costly than pure EGCG, and easily obtained from green tea beverages.

24.13 Inhibition of Angiogenesis in Skin Tumors

Inhibition of tumor angiogenesis is an important mechanism by which tumor growth can be inhibited or regressed by chemopreventive agents. Tumor angiogenesis is the proliferation of a network of blood vessels that penetrates into cancerous growths, and which supply extra nutrients and oxygen for the developing tumors. It has been identified that angiogenic factors, such as matrix metalloproteinases (MMP) and vascular endothelial growth factors (VEGF), are important regulators of tumor growth, both at the primary tumor site and at distant metastasis [102,103]. Hence, MMP and their regulatory pathways are considered promising targets for anticancer drugs or chemopreventive agents. In an effort to define the possible mechanism of inhibition of tumor growth in the photocarcinogenesis model by GTPs, Mantena et al. [55] reported that administration of GTPs in drinking water inhibits the protein expression as well as the activity of both MMP-2 and MMP-9 in tumors compared to those mice that were not given GTPs in drinking water. The inhibition of MMPs was accompanied by elevated expression of their natural inhibitor, TIMP1. The administration of GTPs also inhibited the expression of vascular endothelial cell antigens, such as CD31 and VEGF, in the UVB-induced tumors (Fig. 24.4). The increased expression of these proteins play an important role in tumor growth, invasion, and metastasis due to their promotion of the new vasculature formation that supports the growth of the tumor. Thus, administration of GTPs in drinking water has a significant antiangiogenic effect and has the potential to reduce the growth, or cause regression, of the tumors through this mechanism. Similar effects were also observed when SKH-1 hairless mice were topically treated with EGCG [101]. Garbisa et al. [104] have reported that green tea inhibits the tumor invasion potential of cancer cells.

24.14 UV-Induced Immune Suppression in Humans

The majority of information on UV-induced immune suppression is derived from studies of experimental animal models. There is, however, evidence that UV radiation also suppresses the generation of contact hypersensitivity response to contact allergens in humans. In humans, the induction of delayed-type hypersensitivity [105,106] and contact hypersensitivity [105,107] is suppressed after a single, or short-term, exposure to UV radiation. In addition, immunosuppressive cytokines, such as IL-10, are produced in human skin in response to UV exposure, although the identity of the cell that releases IL-10 may be different in humans and mice [108]. Exposure of human skin to UV radiation suppresses human antigen–presenting cell function, with the doses of UV irradiation that induce immune suppression in humans and mice being comparable [105]. Many of the mechanisms involved in UV-induced immune suppression in humans are similar to those described in experimental animals. UV-induced DNA damage triggers immunoregulatory cytokine production and triggers immune suppression. Wolf et al. [109] observed that exposure of human skin to solar simulated light resulted in the upregulation of IL-10 and TNF-α in the epidermis. Application of T4N5-containing liposomes to the skin of UV-irradiated individuals immediately after UV exposure inhibits cytokine production, and Stege et al. [89] found that repair of CPDs in human skin using photolyase prevented UV-induced suppression of the contact hypersensitivity response. Moreover, sensitivity to sunburn appears to be associated with susceptibility to UV radiation–induced suppression of cutaneous cell-mediated immunity in humans [107]. Therefore, it is suggested that the regular use of chemopreventive agents, such as green tea polyphenols, might be useful for the prevention of UV-induced immunosuppression in humans.

24.15 Role of IL-12 in Human Cutaneous Malignancies: Outcome of Clinical Trials

IL-12 is recognized as a master regulator of adaptive type 1, cell-mediated immunity, the critical pathway involved in protection against neoplasia. This is supported by the analysis of numerous animal [110,111] and human clinical studies that attribute improved clinical outcomes [112] and mechanisms of IL-12-based therapy [113] to strong type 1 responses *in situ*. The present overview of literature suggests that IL-12 deficiency enhances immune suppression, reduces UVB-induced damaged-DNA repair ability, and enhances tumor cell proliferation in mice, which leads to an enhanced risk of photocarcinogenesis. To examine the potent chemotherapeutic effect of IL-12 against cutaneous malignancies in humans, some clinical trials have been conducted. In the human melanoma model, treatment with DNA coding for IL-12 induced regression of tumors in all cases, with complete disappearance of the tumor in two out of five animals. An antivascular effect of IL-12 treatment was evident on histological examination, with endothelial thickening and abrupt changes in vessel diameters [114]. This study suggests that intratumoral plasmid DNA coding for IL-12 holds some promise as a new therapeutic tool for accessible melanoma lesions and should be tested in clinical trial. Rook et al. [115] conducted a phase I dose escalation trial with recombinant human IL-12 (rhIL-12), where patients suffering from cutaneous T-cell lymphoma were given rhIL-12 twice weekly subcutaneously or intralesionally for up to 24 weeks. It was observed that subcutaneous dosing resulted in complete responses in two of five plaques and partial responses in two of five plaques, and one of two Sezary syndrome patients. Intralesional dosing resulted in individual tumor regression in two of two patients. Adverse effects of rhIL-12 on this regimen were minor and limited and included low-grade fever and headache. The results from this phase I trial suggest that rhIL-12 augments antitumor cytotoxic T-cell responses and may represent a potent and well-tolerated therapeutic agent for cutaneous T-cell lymphoma. Further phase I and phase II trials conducted by the same group with rhIL-12 against cytotoxic T-cell lymphoma revealed an overall response rate approaching 50%. These results suggest that rhIL-12 induces lesion regression by augmenting antitumor cytotoxic T cell response [116]. Another clinical trial was conducted in melanoma patients. In this study rhIL-12 was administered subcutaneously and intravenously. IL-12 therapy was well tolerated. Clinical responses included a complete response in a subject with small-volume subcutaneous disease and a partial response in a subject with hepatic metastases. The efficacy of IL-12 was also tested in patients with malignant melanoma. In this study, the plasmid DNA encoding human IL-12 was injected into lesions of nine patients. The therapy was well tolerated. Three of nine patients experienced a clinical response: two stable diseases and one complete remission [117]. All patients but one experienced a transient response at the intratumoral injection site. These results show that intratumoral injection of DNA encoding IL-12 produced some beneficial clinical effect. Further clinical studies could be developed by exploiting the anti-angiogenic program promoted by IL-12 in lymphocytes and neoplastic cells. To this end, IL-12-based therapy might be combined with other anti-angiogenic chemopreventive agents that also stimulate endogenous IL-12 production in mouse models, such as green tea polyphenols [59,73,81], grapeseed proanthocyanidins [118], and silymarin from milk thistle [119].

Acknowledgments

The work reported from the authors' laboratory was supported from the funds from the National Institutes of Health (CA104428, CA105368, ES11421, AT002536) and a VA Merit Review Award. Grateful thanks are also due to our former and current colleagues and postdoctoral fellows for their outstanding contributions.

References

1. Hara Y. Fermentation of Tea. In: Y. Hara (Ed.), Green Tea, Health Benefits and Applications. Marcel Dekker, New York, 2001, pp. 16–21.
2. Katiyar SK and Elmets CA. Green tea polyphenolic antioxidants and skin photoprotection. *Int. J. Oncol.* 18:1307–1313, 2001.
3. Katiyar SK and Mukhtar H. Tea in chemoprevention of cancer: epidemiologic and experimental studies (review). *Int. J. Oncol.* 8:221–238, 1996.
4. deGruijl FR and van der Leun JC. Estimate of the wavelength dependency of ultraviolet carcinogenesis in humans and its relevance to the risk assessment of stratospheric ozone depletion. *Health Phys.* 67:319–325, 1994.
5. Baliga MS and Katiyar SK. Chemoprevention of photocarcinogenesis by selected dietary botanicals. *Photochem. Photobiol. Sci.* 5:243–253, 2006.
6. Bachelor MA and Bowden GT. UVA-mediated activation of signaling pathways involved in skin tumor promotion and progression. *Semin. Cancer Biol.* 14:131–138, 2004.
7. Wang SQ, Setlow R, Berwick M, Polsky D, Marghoob AA, Kopf AW and Bart RS. Ultraviolet A and melanoma: a review. *J. Am. Acad. Dermatol.* 44:837–846, 2001.
8. DiGiovanni J. Multistage carcinogenesis in mouse skin. *Pharmacol. Ther.* 54:63–128, 1992.

9. de Gruijl FR. Photocarcinogenesis: UVA vs UVB. Singlet oxygen, UVA, and ozone. *Methods Enzymol.* 319:359–366, 2000.
10. Runger TM. Role of UVA in the pathogenesis of melanoma and non-melanoma skin cancer. A short review. *Photodermatol. Photoimmunol. Photomed.* 15:212–216, 1999.
11. Douki T, Reynaud-Angelin A, Cadet J and Sage E. Bipyrimidine photoproducts rather than oxidative lesions are the main type of DNA damage involved in the genotoxic effect of solar UVA radiation. *Biochemistry* 42: 9221–9226, 2003.
12. Krutmann J. The role of UVA rays in skin aging. *Eur. J. Dermatol.* 11: 170–171, 2001.
13. Ullrich SE. Potential for immunotoxicity due to environmental exposure to ultraviolet radiation. *Hum. Exp. Toxicol.* 14:89–91, 1995.
14. Katiyar S, Elmets CA and Katiyar SK. Green tea and skin cancer: photoimmunology, angiogenesis and DNA repair. *J. Nutr. Biochem.* 18:287–296, 2007.
15. Kripke ML. Photoimmunology. *Photochem. Photobiol.* 52:919–924, 1990.
16. Kligman LH, Akin FJ and Kligman AM. Sunscreens prevent ultraviolet photocarcinogenesis. *J. Am. Acad. Dermatol.* 3:30–35, 1980.
17. Katiyar SK, Korman NJ, Mukhtar H and Agarwal R. Protective effects of silymarin against photocarcinogenesis in a mouse skin model. *J. Natl. Cancer Inst.* 89:556–566, 1997.
18. Donawho CK and Kripke ML. Evidence that the local effect of ultraviolet radiation on the growth of murine melanomas is immunologically mediated. *Cancer Res.* 51:4176–4181, 1991.
19. Ziegler A, Jonason AS, Leffell DJ, Simon JA, Sharma HW, Kimmelman J, Remington L, Jacks T and Brash DE. Sunburn and p^{53} in the onset of skin cancer. *Nature* 372:773–776, 1994.
20. Taylor CR, Stern RS, Leyden JJ and Gilchrest BA. Photoaging/photodamage and photoprotection. *J. Am. Acad. Dermatol.* 22:1–15, 1990.
21. Hruza LL and Pentland AP. Mechanisms of UV-induced inflammation. *J. Invest. Dermatol.* 100:35S–41S, 1993.
22. Katiyar SK and Mukhtar H. Green tea polyphenol (-)-epigallocatechin-3-gallate treatment to mouse skin prevents UVB-induced infiltration of leukocytes, depletion of antigen presenting cells and oxidative stress. *J. Leukoc. Biol.* 69:719–726, 2001.
23. Katiyar SK, Afaq F, Perez A and Mukhtar H. Green tea polyphenol (-)-epigallocatechin-3-gallate treatment of human skin inhibits ultraviolet radiation–induced oxidative stress. *Carcinogenesis* 22:287–294, 2001.
24. Mittal A, Elmets CA and Katiyar SK. Dietary feeding of proanthocyanidins from grape seeds prevents photocarcinogenesis in SKH-1 hairless mice: relationship to decreased fat and lipid peroxidation. *Carcinogenesis* 24: 1379–1388, 2003.
25. Meunier L, Raison-Peyron N and Meynadier J. UV-induced immunosuppression and skin cancers. *Rev. Med. Interne.* 19:247–254, 1998.
26. Parrish JA. Photoimmunology. *Adv. Exp. Med. Biol.* 160:91–108, 1983.
27. Urbach F. Incidences of nonmelanoma skin cancer. *Dermatol. Clin.* 9: 751–755, 1991.

28. Yoshikawa T, Rae V, Bruins-Slot W, vand-den-Berg JW, Taylor JR and Streilein JW. Susceptibility to effects of UVB radiation on induction of contact hypersensitivity as a risk factor for skin cancer in humans. *J. Invest. Dermatol.* 95:530–536, 1990.
29. Kinlen L, Sheil A, Peta J and Doll R. Collaborative United Kingdom–Australia study of cancer in patients treated with immunosuppressive drugs. *Br. Med. J.* 2:1461–1466, 1979.
30. Cowen EW and Billingsley EM. Awareness of skin cancer by kidney transplant patients. *J. Am. Acad. Dermatol.* 40:697–701, 1999.
31. Otley CC and Pittelkow MR. Skin cancer in liver transplant recipients. *Liver Transpl.* 6:253–262, 2000.
32. Fortina AB, Caforio AL, Piaserico S, Alaibac M, Tona F, Feltrin G, Livi U and Peserico A. Skin cancer in heart transplant recipients: frequency and risk factor analysis. *J. Heart Lung Transplant* 19:249–255, 2000.
33. DiGiovanna JJ. Posttransplantation skin cancer: scope of the problem, management and role for systemic retinoid chemoprevention. *Transplant Proc.* 30:2771–2775, 1998.
34. Hojo M, Morimoto T, Maluccio M, Asano T, Morimoto K, Lagman M, Shimbo T and Suthanthiran M. Cyclosporin induces cancer progression by a cell-autonomous mechanism. *Nature (Lond.)* 397:530–534, 1999.
35. Ichihashi M, Ueda M, Budiyanto A, Bito T, Oka M, Fukunaga M, Tsuru K and Horikawa T. UV-induced skin damage. *Toxicology* 189:21–39, 2003.
36. Mukhtar H and Elmets CA. Photocarcinogenesis: mechanisms, models and human health implications. *Photochem. Photobiol.* 63:355–447, 1996.
37. Godar DE, Wengraitis SP, Shreffler J and Sliney DH. UV doses of Americans. *Photochem. Photobiol.* 73:621–629, 2001.
38. Godar DE. UV doses of American children and adolescents. *Photochem. Photobiol.* 74:787–793, 2001.
39. Surh Y-J. Molecular mechanisms of chemopreventive effects of selected dietary and medicinal phenolic substances. *Mutat. Res.* 428:305–327, 1999.
40. Pinnell SR. Cutaneous photodamage, oxidative stress, and topical antioxidant protection. *J. Am. Acad. Dermatol.* 48:1–19, 2003.
41. Katiyar SK. Skin photoprotection by green tea: antioxidant and immunomodulatory effects. *Curr. Drug Targets Immune, Endocr. Metabol. Disord.* 3:234–242, 2003.
42. Manach C, Scalbert A, Morand C, Rémésy C and Jiménez L. Polyphenols: food sources and bioavailability. *Am. J. Clin. Nutr.* 79:727–747, 2004.
43. Okushio K, Matsumoto N, Kohri T, Suzuki M, Nanjo F and Hara Y. Absorption of tea catechins into rat portal vein. *Biol. Pharm. Bull.* 19:326–329, 1996.
44. Kim S, Lee MJ and Hong J. Plasma and tissue levels of tea catechins in rats and mice during chronic consumption of green tea polyphenols. *Nutr. Cancer* 37:41–48, 2000.
45. Yang CS, Chen L, Lee MJ, Balentine D, Kuo M and Schantz SP. Blood and urine levels of tea catechins after ingestion of different amounts of green tea by human volunteers. *Cancer Epidemiol. Biomarkers Prev.* 7:351–354, 1998.
46. Higdon JV and Frei B. Tea catechins and polyphenols: health effects, metabolism, and antioxidant functions. *Crit. Rev. Food Sci. Nutr.* 43:89–143, 2003.

47. Chow HH, Cai Y and Alberts DS. Phase I pharmacokinetic study of tea polyphenols following single-dose administration of epigallocatechin gallate and polyphenon E. *Cancer Epidemiol. Biomarkers Prev.* 10:53–58, 2001.
48. Lu H, Meng XF, Lee MJ, Li C, Maliakal P and Yang CS. Bioavailability and biological activity of tea polyphenols. *Food Factors in Health Promotion and Disease Prevention Symposium Series* 851:9–15, 2003.
49. Scalabert A and Williamson G. Dietary intake and bioavailability of polyphenols. *J. Nutr.* 130:2073S–2085S, 2000.
50. Loktionov A, Bingham S, Vorster H, Jerling J, Runswick S and Cummings J. Apolipoprotein E genotype modulates the effect of black tea drinking on blood lipids and blood coagulation factors: a pilot study. *Br. J. Nutr.* 79: 133–139, 1998.
51. Leenen R, Roodenburg A, Tijburg L and Wiseman S. A single dose of tea with or without milk increases plasma antioxidant activity in humans. *Eur. J. Clin. Nutr.* 54:87–92, 2000.
52. Van het Hof K, Kivits G, Weststrate J and Tijburg L. Bioavailability of catechins from tea: the effect of milk. *Eur. J. Clin. Nutr.* 52:356–359, 1998.
53. Hollman PC, Van Het Hof K, Tijburg L and Katan MB. Addition of milk does not affect the absorption of flavonols from tea in man. *Free Radic. Res.* 34:297–300, 2001.
54. Langley-Evans S. Consumption of black tea elicits an increase in plasma antioxidant potential in humans. *Int. J. Food Sci. Nutr.* 51:309–315, 2000.
55. Mantena SK, Meeran SM, Elmets CA and Katiyar SK. Orally administered green tea polyphenols prevent ultraviolet radiation–induced skin cancer in mice through activation of cytotoxic T cells and inhibition of angiogenesis in tumors. *J. Nutr.* 135:2871–2877, 2005.
56. Wang ZY, Huang MT, Ferraro T, Wong CQ, Lou YR, Iatropoulos M, Yang CS and Conney AH. Inhibitory effect of green tea in the drinking water on tumorigenesis by ultraviolet light and 12-O-tetradecanoylphorbol-13-acetate in the skin of SKH-1 mice. *Cancer Res.* 52:1162–1170, 1992.
57. Wang ZY, Agarwal R, Bickers DR and Mukhtar H. Protection against ultraviolet B radiation–induced photocarcinogenesis in hairless mice by green tea polyphenols. *Carcinogenesis* 12:1527–1530, 1991.
58. Wang ZY, Huang MT, Ho CT, Chang R, Ma W, Ferraro T, Reuhl KR, Yang CS and Conney AH. Inhibitory effect of green tea on the growth of established skin papillomas in mice. *Cancer Res.* 52:6657–6665, 1992.
59. Mittal A, Piyathilake C, Hara Y and Katiyar SK. Exceptionally high protection of photocarcinogenesis by topical application of (-)-epigallocatechin-3-gallate in hydrophilic cream in SKH-1 hairless mouse model: relationship to inhibition of UVB-induced global DNA hypomethylation. *Neoplasia* 5:555–565, 2003.
60. Vayalil PK, Elmets CA and Katiyar SK. Treatment of green tea polyphenols in hydrophilic cream prevents UVB-induced oxidation of lipids and proteins, depletion of antioxidant enzymes and phosphorylation of MAPK proteins in SKH-1 hairless mouse skin. *Carcinogenesis* 24:927–936, 2003.
61. Record IR and Dreosti IE. Protection by tea against UV-A + B-induced skin cancers in hairless mice. *Nutr. Cancer* 32:71–75, 1998.

62. Toews GB, Bergstresser PR, Streilein JW and Sullivan S. Epidermal Langerhans cell density determines whether contact hypersensitivity or unresponsiveness follows skin painting with DNFB. *J. Immunol.* 124:445–453, 1980.
63. Cooper KD, Oberhelman L, Hamilton TA, Baadsgaard O, Terhune M, LeVee G, Anderson T and Koren H. UV exposure reduces immunization rates and promotes tolerance to epicutaneous antigens in humans: relationship to dose, CD1a-DR+ epidermal macrophage induction, and Langerhans cell depletion. *Proc. Natl. Acad. Sci. U S A.* 89:8497–8501, 1992.
64. Chapman RS, Cooper KD, De Fabo EC, Frederick JE, Gelatt KN, Hammond SP, Hersey P, Koren HS, Ley RD and Noonan F. Solar ultraviolet radiation and the risk of infectious disease. *Photochem. Photobiol.* 61:223–247, 1995.
65. Fisher GJ, Datta SC, Talwar HS, Wang ZQ, Varani J, Kang S and Voorhees JJ. Molecular basis of sun-induced premature skin ageing and retinoid antagonism. *Nature* 379:335–339, 1996.
66. de Gruijl FR, Sterenborg HJ, Forbes PD, Davies RE, Cole C, Kelfkens G, van Weelden H, Slaper H and van der Leun JC. Wavelength dependence of skin cancer induction by ultraviolet irradiation of albino hairless mice. *Cancer Res.* 53:53–60, 1993.
67. Meunier L, Raison-Peyron N and Meynadier J. UV-induced immunosuppression and skin cancers. *Rev. Med. Interne.* 19:247–254, 1998.
68. Yoshikawa T, Rae V, Bruins-Slot W, vand-den-Berg JW, Taylor JR and Streilein JW. Susceptibility to effects of UVB radiation on induction of contact hypersensitivity as a risk factor for skin cancer in humans. *J. Invest. Dermatol.* 95:530–536, 1990.
69. Katiyar SK, Elmets CA, Agarwal R and Mukhtar H. Protection against ultraviolet-B radiation-induced local and systemic suppression of contact hypersensitivity and edema responses in C3H/HeN mice by green tea polyphenols. *Photochem. Photobiol.* 62:855–861, 1995.
70. Hauser C. Cultured epidermal Langerhans cells activate effector T cells for contact sensitivity. *J. Invest. Dermatol.* 95:436–440, 1990.
71. Toews GB, Bergstresser PR, Streilein JW and Sullivan S. Epidermal Langerhans cell density determines whether contact hypersensitivity or unresponsiveness follows skin painting with DNFB. *J. Immunol.* 124: 445–453, 1980.
72. Katiyar SK, Challa A, McCormick TS, Cooper KD and Mukhtar H. Prevention of UVB-induced immunosuppression in mice by green tea polyphenol (-)-epigallocatechin-3-gallate may be associated with alterations in IL-10 and IL-12 production. *Carcinogenesis* 20:2117–2124, 1999.
73. Hammerberg C, Duraiswamy N and Cooper KD. Reversal of immunosuppression inducible through ultraviolet-exposed skin by *in vivo* anti-CD11b treatment. *J. Immunol.* 157:5254–5261, 1996.
74. Hammerberg C, Katiyar SK, Carroll MC and Cooper KD. Activated complement component 3 (C3) is required for ultraviolet induction of immunosuppression and antigenic tolerance. *J. Exp. Med.* 187:1133–1138, 1998.
75. Fiorentino DF, Zlotnik A, Vieira P, Mosmann TR, Howard M, Moore KW and O'Garra A. IL-10 acts on the antigen-presenting cell to inhibit cytokine production by Th1 cells. *J. Immunol.* 146:3444–3451, 1991.

76. de Waal Malefyt R, Haanen J, Spits H, Roncarolo MG, te Velde A, Figdor C, Johnson K, Kastelein R, Yssel H and de Vries JE. Interleukin 10 (IL-10) and viral IL-10 strongly reduce antigen-specific human T cell proliferation by diminishing the antigen-presenting capacity of monocytes via downregulation of class II major histocompatibility complex expression. *J. Exp. Med.* 174:915–924, 1991.
77. Howard M and O'Garra A. Biological properties of interleukin 10. *Immunol. Today* 13:198–200, 1992.
78. Fiorentino DF, Zlotnik A, Mosmann TR, Howard M and O'Garra A. IL-10 inhibits cytokine production by activated macrophages. *J. Immunol.* 147:3815–3822, 1991.
79. Schwarz A, Grabbe S, Riemann H, Aragane Y, Simon M, Manon S, Andrade S, Luger TA, Zlotnik A and Schwarz T. *In vivo* effects of interleukin-10 on contact hypersensitivity and delayed-type hypersensitivity reactions. *J. Invest. Dermatol.* 103:211–216, 1994.
80. Niizeki H and Streilein JW. Hapten-specific tolerance induced by acute, low-dose ultraviolet B radiation of skin is mediated via interleukin-10. *J. Invest. Dermatol.* 109:25–30, 1997.
81. Kobayashi M, Fitz L, Ryan M, Hewick RM, Clark SC, Chan S, Loudon R, Sherman F, Perussia B and Trinchieri G. Identification and purification of natural killer cell stimulatory factor (NKSF), a cytokine with multiple biologic effects on human lymphocytes. *J. Exp. Med.* 170:827–845, 1989.
82. Manetti R, Parronchi P, Giudizi MG, Piccinni P, Maggi E, Trinchieri G and Romagnani S. Natural killer cell stimulatory factor (interleukin-12 [IL-12]) induces T helper type 1 (Th1)-specific immune responses and inhibits the development of IL-4-producing Th cells. *J. Exp. Med.* 177:1199–1204, 1993.
83. Hsieh C, Macatonia SE, Tripp CS, Wolf SF, O'Garra A and Murphy KM. Development of Th1 $CD4^+$ T cells through IL-12 produced by Listeria-induced macrophages. *Science* 260:547–549, 1993.
84. Scott P. IL-12: initiation cytokine for cell-mediated immunity. *Science* 260:496–497, 1993.
85. Schmitt DA, Owen-Schaub L and Ullrich SE. Effect of IL-12 on immune suppression and suppressor cell induction by ultraviolet radiation. *J. Immunol.* 154:5114–5120, 1995.
86. Meeran SM, Mantena SK and Katiyar SK. Prevention of ultraviolet radiation-induced immunosuppression by (-)-epigallocatechin-3-gallate in mice is mediated through interleukin 12-dependent DNA repair. *Clin. Cancer Res.* 12:2272–2280, 2006.
87. Applegate LA, Ley RD, Alcalay J and Kripke ML. Identification of the molecular target for the suppression of contact hypersensitivity by ultraviolet radiation. *J. Exp. Med.* 170:1117–1131, 1989.
88. Kripke ML, Cox PA, Alas LG and Yarosh DB. Pyrimidine dimers in DNA initiated systemic immunosuppression in UV-irradiated mice. *Proc. Natl. Acad. Sci. U S A.* 89:7516–7520, 1992.
89. Stege H, Roza L, Vink AA, Grewe M, Ruzicka T, Grether-Beck S and Krutmann J. Enzyme plus light therapy to repair DNA damage in ultraviolet-B-irradiated human skin. *Proc. Natl. Acad. Sci. U S A.* 97:1790–1795, 2000.

90. Schwarz A, Stander S, Berneburg M, Bohm M, Kulms D, van Steeg H, Grosse-Heitmeyer K, Krutmann J and Schwarz T. Interleukin-12 suppresses ultraviolet radiation–induced apoptosis by inducing DNA repair. *Nat. Cell Biol.* 4:26–31, 2002.
91. Schwarz A, Maeda A, Kernebeck K, van Steeg H, Beissert S and Schwarz T. Prevention of UV radiation–induced immunosuppression by IL-12 is dependent on DNA repair. *J. Exp. Med.* 201:173–179, 2005.
92. Vink AA, Moodycliffe AM, Shreedhar V, Ullrich SE, Roza L, Yarosh DB and Kripke ML. The inhibition of antigen-presenting activity of dendritic cells resulting from UV irradiation of murine skin is restored by *in vitro* photorepair of cyclobutane pyrimidine dimers. *Proc. Natl. Acad. Sci. U S A.* 94:5255–5260, 1997.
93. Vink AA, Strickland FM, Bucana C, Cox PA, Roza L, Yarosh DB and Kripke ML. Localization of DNA damage and its role in altered antigen-presenting cell function in ultraviolet-irradiated mice. *J. Exp. Med.* 183: 1491–1500, 1996.
94. Meeran SM, Mantena SK, Elmets CA and Katiyar SK. (-)-Epigallocatechin-3-gallate prevents photocarcinogenesis in mice through interleukin-12-dependent DNA repair. *Cancer Res.* 66:5512–5520, 2006.
95. Meeran SM, Mantena SK, Meleth S, Elmets CA and Katiyar SK. Interleukin-12-deficient mice are at greater risk of ultraviolet radiation–induced skin tumors and malignant transformation of papillomas to carcinomas. *Mol. Cancer Ther.* 5:825–832, 2006.
96. Yarosh D, Bucana C, Cox P, Alas L, Kibitel J and Kripke ML. Localization of liposomes containing a DNA repair enzyme in murine skin. *J. Invest. Dermatol.* 103:461–468, 1994.
97. Yarosh D, Alas LG, Yee V, Oberyszyn A, Kibitel JT, Mitchell D, Rosenstein R, Spinowitz A and Citron M. Pyrimidine dimer removal enhanced by DNA repair liposomes reduces the incidence of UV skin cancer in mice. *Cancer Res.* 52:4227–4231, 1992.
98. Yarosh D, Klein J, O'Connor A, Hawk J, Rafal E and Wolf P. Effect of topically applied T4 endonuclease V in liposomes on skin cancer in xeroderma pigmentosum: a randomised study. Xeroderma Pigmentosum Study Group. *Lancet* 357:926–929, 2001.
99. Wolf P, Cox P, Yarosh DB and Kripke ML. Sunscreens and T4N5 liposomes differ in their ability to protect against ultraviolet-induced sunburn cell formation, alterations of dendritic epidermal cells, and local suppression of contact hypersensitivity. *J. Invest. Dermatol.* 104:287–292, 1995.
100. de Gruijl FR. Ultraviolet radiation and tumor immunity. *Methods* 28: 122–129, 2002.
101. Mantena SK, Roy AM and Katiyar SK. Epigallocatechin-3-gallate inhibits photocarcinogenesis through inhibition of angiogenic factors and activation of $CD8^+$ T cells in tumors. *Photochem. Photobiol.* 81:1174–1179, 2005.
102. Yu AE, Hewitt RE, Connor EW and Stetler-Stevenson WG. Matrix metalloproteinases. Novel targets for directed cancer therapy. *Drugs Aging.* 11:229–244, 1997.
103. John A and Tuszynski G. The role of matrix metalloproteinases in tumor angiogenesis and tumor metastasis. *Pathol. Oncol. Res.* 7:14–23, 2001.

104. Garbisa S, Biggin S, Cavallarin N, Sartor L, Benelli R and Albini A. Tumor invasion: molecular shears blunted by green tea. *Nat. Med.* 5:1216, 1999.
105. Ullrich SE. Mechanisms underlying UV-induced immune suppression. *Mutat. Res.* 571:185–205, 2005.
106. Moyal D, Courbiere C, Le Corre Y, de Lacharriere O and Hourseau C. Immunosuppression induced by chronic solar simulated irradiation in humans and its prevention by sunscreens. *Eur. J. Dermatol.* 7:223–225, 1997.
107. Damian DL, Halliday GM and Barnetson RS. Broad-spectrum sunscreens provide greater protection against ultraviolet-radiation-induced suppression of contact hypersensitivity to a recall antigen in humans. *J. Invest. Dermatol.* 109:146–151, 1997.
108. Kang K, Hammerberg C, Meunier L and Cooper KD. CD11b+ macrophages that infiltrate human epidermis after *in vivo* ultraviolet exposure potently produce IL-10 and represent the major secretory source of epidermal IL-10 protein. *J. Immunol.* 153:5256–5264, 1994.
109. Wolf P, Maier H, Mullegger RR, Chadwick CA, Hofmann-Wellenhof R, Soyer HP, Hofer A, Smolle J, Horn M, Cerroni L, Yarosh D, Klein J, Bucana C, Dunner Jr K, Potten CS, Honigsmann H, Kerl H and Kripke ML. Topical treatment with liposomes containing T4 endonuclease V protects human skin *in vivo* from ultraviolet-induced upregulation of interleukin-10 and tumor necrosis factor-alpha. *J. Invest. Dermatol.* 114:149–156, 2000.
110. Hung K, Hayashi R, Lafond-Walker A, Lowenstein C, Pardoll D and Levitsky H. The central role of CD4(+) T cells in the antitumor immune response. *J. Exp. Med.* 188:2357–2368, 1998.
111. Tatsumi T, Huang J, Gooding WE, Gambotto A, Robbins PD, Vujanovic NL, Alber SM, Watkins SC, Okada H and Storkus WJ. Intratumoral delivery of dendritic cells engineered to secrete both interleukin (IL)-12 and IL-18 effectively treats local and distant disease in association with broadly reactive Tc1-type immunity. *Cancer Res.* 63:6378–6386, 2003.
112. Galon J, Costes A, Sanchez-Cabo F, Kirilovsky A, Mlecnik B, Lagorce-Pagès C, Tosolini M, Camus M, Berger A, Wind P, Zinzindohoué F, Bruneval P, Cugnenc PH, Trajanoski Z, Fridman WH and Pagès F. Type, density, and location of immune cells within human colorectal tumors predict clinical outcome. *Science* 313(5795):1960–1964, 2006.
113. van Herpen CM, Looman M, Zonneveld M, Scharenborg N, de Wilde PC, van de Locht L, Merkx MA, Adema GJ and de Mulder PH. Intratumoral administration of recombinant human interleukin 12 in head and neck squamous cell carcinoma patients elicits a T-helper 1 profile in the locoregional lymph nodes. *Clin. Cancer Res.* 10:2626–2635, 2004.
114. Heinzerling L, Dummer R, Pavlovic J, Schultz J, Burg G and Moelling K. Tumor regression of human and murine melanoma after intratumoral injection of IL-12-encoding plasmid DNA in mice. *Exp. Dermatol.* 11:232–240, 2002.
115. Rook AH, Wood GS, Yoo EK, Elenitsas R, Kao DM, Sherman ML, Witmer WK, Rockwell KA, Shane RB, Lessin SR and Vonderheid EC. Interleukin-12 therapy of cutaneous T-cell lymphoma induces lesion regression and cytotoxic T-cell responses. *Blood* 94:902–908, 1999.

116. Rook AH, Zaki MH, Wysocka M, Wood GS, Duvic M, Showe LC, Foss F, Shapiro M, Kuzel TM, Olsen EA, Vonderheid EC, Laliberte R and Sherman ML. The role for interleukin-12 therapy of cutaneous T cell lymphoma. *Ann. N Y Acad. Sci.* 941:177–184, 2001.
117. Heinzerling L, Burg G, Dummer R, Maier T, Oberholzer PA, Schultz J, Elzaouk L, Pavlovic J and Moelling K. Intratumoral injection of DNA encoding human interleukin 12 into patients with metastatic melanoma: clinical efficacy. *Hum. Gene Ther.* 16:35–48, 2005.
118. Sharma SD and Katiyar SK. Dietary grape-seed proanthocyanidin inhibition of ultraviolet B–induced immune suppression is associated with induction of IL-12. *Carcinogenesis* 27:95–102, 2006.
119. Meeran SM, Katiyar S, Elmets CA and Katiyar SK. Silymarin inhibits UV radiation–induced immunosuppression through augmentation of interleukin-12 in mice. *Mol. Cancer Ther.* 5:1660–1668, 2006.

25

Silibinin in Skin Health: Efficacy and Mechanism of Action

Manjinder Kaur*, PhD, Gagan Deep*, PhD, and Rajesh Agarwal, PhD

Department of Pharmaceutical Sciences, School of Pharmacy, University of Colorado Denver, and University of Colorado Cancer Center, Denver, CO, USA

*These authors contributed equally to this work.

Aaron Tabor and Robert M. Blair (eds.), *Nutritional Cosmetics: Beauty from Within*, 501–528,

25.1 An Overview of Silibinin

25.1.1 Silibinin Source, Distribution, and Properties

Silibinin (Fig. 25.1), a flavolignan, is one of the active ingredients of milk thistle extract, which is rich in a complex mixture of flavonoids such as silydianin, silychristin, isosilybin, and dehydrosilybin [1]. The milk thistle plant (*Silybum marianum* L.), belonging to family Compositae, is native to Europe and Asia, but it has been naturalized in North and South America. Today, this plant is cultivated in Europe, China, and Argentina for commercial preparation of milk thistle extract [2]. Fruits and seeds of this plant are the major source for silibinin, or milk thistle extract in general. Recent research has revealed that silibinin (also often referred to as silybin) is actually a mixture of two diastereomers, silybin A and silybin B, present in equal proportion [3]. It constitutes 40% of the total flavonolignans present in silymarin, an active extract of milk thistle. Silibinin exhibits antioxidant properties and is a scavenger of hydroxyl radicals [4,5]. It inhibits lipid peroxidation by acting as a chain-breaking antioxidant [6,7]. Silibinin partitions into the hydrophobic region of the membranes to

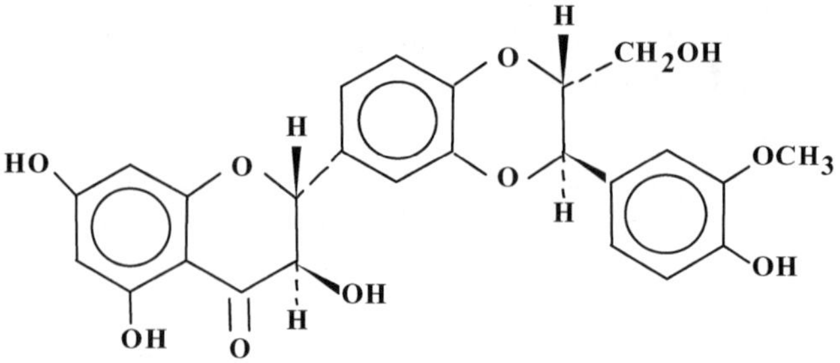

Figure 25.1 Chemical structure of silibinin.

some extent; however, it does not change the biophysical properties of phospholipid bilayers [8]. Silibinin possesses strong antibacterial activity, with selective potency against gram-positive bacteria, where it inhibits both RNA and DNA synthesis. The overall potency of silibinin is more than silymarin [9]. Multidrug resistance through overexpression of multidrug resistance transporters, such as P-glycoprotein (P-gp, ABCB1) and multidrug resistance associated protein (MRP1, ABCC1), is often the cause of failure of chemotherapy of various cancers. It has been shown that silibinin strongly inhibits the efflux of BCPCF (2′, 7′-bis-[3-carboxypropyl]-5-[and-6]-carboxyfluorescein), used as a fluorescent probe to measure MRP1 activity, from erythrocytes [10].

25.1.2 Traditional Use of Silibinin in Human Health

Silibinin, in the form of milk thistle extract, has been used traditionally to treat various human liver conditions for centuries as an herbal medication. Ancient herbalists and physicians have described milk thistle as a plant with hepatoprotective effects due to its excellent capacity in carrying off the bile, as well as in removing obstructions of the liver and spleen [11]. Today, it is sold throughout Europe as well as in the United States for "wellness of the liver." In recent years, silibinin and silymarin have been used for the treatment of liver conditions such as cirrhosis, chronic hepatitis, alcoholic liver disease, and liver toxicity due to environmental toxins [12]. Use of milk thistle extract in the treatment of liver conditions is more traditional; however, in recent years, its hepatoprotective activity has been supported by numerous studies conducted in experimental models of various liver conditions [13]. Protective effects against toxicity associated with various chemicals, drugs, and natural toxins such as carbon tetrachloride, cyclosporin, acetaminophen, ethanol, microcystin-LR, and *Amanita phylloides* have been shown under experimental conditions [14–17]. Other studies have shown the beneficial effects of silibinin and/or silymarin in conditions such as alcoholic liver disease, T-cell-dependent hepatitis, experimental models of choleostasis, diet-induced hypercholesterolemia, cisplatin-induced nephrotoxicity, atherosclerosis, colitis, etc. [7,13,18–21]. The beneficial effects of silibinin in these conditions have been attributed either to its antioxidant capacity (because it prevents oxidation of LDL for its antiatherosclerotic effects), or to its immunomodulatory, antiinflammatory (selectively inhibits leukotriene production by Kupffer cells), or antifibrolytic capacity [7,13,18–21].

A recent study has shown that silybin-vitamin E-phospholipid complex improves insulin resistance and liver damage in patients with nonalcoholic fatty liver disease [22].

25.1.3 Cancer Chemopreventive Efficacy of Silibinin

In the past two decades, in addition to hepatoprotective effects, anticancer efficacy of silibinin has been demonstrated in numerous studies conducted in cell culture and animal models of cancer of varying phenotypes. Silibinin has been shown to inhibit the proliferation of human colon cancer cells via the inhibition of cell cycle progression by modulating the levels of cyclins and cyclin-dependent kinases [23,24]. Various *in vitro* and *in vivo* studies have also demonstrated the anticancer efficacy of silibinin against human prostate carcinoma cells [25–27]. Silibinin has been shown to downregulate PSA secretion as well as to inhibit telomerase activity in human prostate carcinoma LNCaP cells, thereby exerting its anticancer effects [28,29]. Silibinin has also been reported to inhibit the growth and progression of lung carcinogenesis [30]. Further, synergistic anticancer effects of silibinin have been observed with conventional chemotherapeutic drugs such as doxorubicin, cisplatin, and carboplatin against human breast carcinoma cells [31]. However, there are conflicting reports for the cancer preventive efficacy of silymarin and silibinin against mammary carcinogenesis [32–34]. Silymarin was reported to enhance 1-methyl-1-nitrosourea (MNU)-induced mammary carcinogenesis as well as to increase the mammary tumor incidences in MMTV-neu/HER2 transgenic mice [32]. On the contrary, silibinin treatment was shown to strongly inhibit development of mammary tumors as well as lung metastasis in HER-2/neu transgenic mice [34]. In contrast, another study reported that silibinin treatment has no effect on breast cancer development in the C3(1) SV40 T, t antigen transgenic multiple mammary adenocarcinoma (T Ag) mouse [33]. These differences could be related to differences in the models used in these studies as well as to the fact that silymarin has other flavonolignans in addition to silibinin, which might possess weak estrogenic activity. In the case of gliomas, silibinin has been reported to sensitize tumor necrosis factor-related apoptosis-inducing ligand (TRAIL)-resistant glioma cells to TRAIL-mediated apoptosis [35]. Anticancer effects of silibinin were also observed in renal cell carcinoma, where oral administration of silibinin was found to suppress the growth of local and metastatic tumors in the xenograft model of renal cell carcinoma by increasing the plasma levels of IGFBP-3, a binding protein for insulinlike growth factor 1 (IGF-1) [36]. Silibinin has also been reported to exert antimetastatic effects through inhibition of urokinase-plasminogen activator and matrix metalloproteinase-9 and -2 expression involved in invasion and metastasis of cancerous cells [37,38]. In addition, silibinin has been shown to inhibit the invasion and motility of oral cancer cells [39]. Lastly, silibinin exerts protective effects against UV- and chemical carcinogen/tumor promoter-induced skin carcinogenesis through multiple mechanisms detailed in the following sections.

25.2 Silibinin and Skin Cancer

25.2.1 Skin Cancer Incidence and Etiological Factors

According to the estimates of the American Cancer Society, there will be 62,480 new cases of melanoma diagnosed in the United States in 2007, and 8,420 deaths may occur due to this cancer [40]. In the case of nonmelanoma skin cancers, including basal cell carcinoma and squamous cell carcinoma (SCC), it is estimated that around one million new cases will be diagnosed during the current year [40]. One of the major risk factors for skin cancer is ultraviolet (UV) radiation from sunlight, though exposure to various chemical pollutants and viral infections also contributes to skin cancer incidence [40,41]. Solar radiation represents the spectrum of electromagnetic radiation with wavelengths ranging from UV (200–400 nm), visible (400–700 nm), and infrared (700 nm–1 mm). In the case of skin cancer, exposure to the UV range of solar radiation is the major etiological factor. UV light in itself can be divided into three ranges: UVC (200–280 nm), UVB (280–315 nm), and UVA (315–400 nm). All three ranges of UV are capable of causing skin damage, though UVC is efficiently blocked in the outer ozone layer of the atmosphere [42]. However, UVB and UVA reach the earth's surface and thus cause damage to skin. Overall, exposure to solar UV radiation, irrespective of UVA and/or UVB, increases the risk of developing skin cancer or accelerates skin aging [41].

As mentioned above, skin cancer is broadly classified into two types: nonmelanoma and melanoma skin cancers. Most of the skin cancers fall into the former category, whereas in the latter case, there is an involvement of melanocytes only, a type of skin cell that is responsible for giving pigmentation/color to the skin. Depending upon the type of cells involved, nonmelanoma skin cancers are further classified into two types: basal cell and squamous cell carcinomas. Nonmelanoma skin cancers rarely spread to other parts of the body; however, melanoma skin cancers can spread/metastasize to other parts of the body. Actinic keratoses, as a result of chronic sunlight exposure, can also develop into nonmelanoma skin cancer, specifically SCC [43]. Apart from these, certain viral warts and Bowen's disease can develop into SCC, whereas clinical atypical moles can be precursors for melanoma [44]. Most of the mortality is due to melanoma skin cancer because of its ability to metastasize, whereas more than 90% of the cases of nonmelanoma skin cancer can be cured [45].

In today's world, with a depleting ozone level as a consequence of increased use of chlorofluorocarbons and other ozone-depleting substances, exposure to the UV range of sunlight has become a leading risk factor for developing skin cancer. DNA damage and generation of reactive oxygen species are the two key events involved in pathogenesis of UV-induced skin cancer, though to some extent immunosuppression due to UV exposure and other factors is also involved in skin cancer etiology [46]. For reactive oxygen species generation, the electromagnetic energy of UV light is absorbed by chromophores in the skin such as DNA, porphyrins, urocanic acid, and aromatic amino acids. These chromophores, after absorbing energy, become energized and react with oxygen molecules present in the intracellular milieu, and thus initiate the formation of reactive oxygen species [47]. In UV-induced DNA damage, cyclobutane pyrimidine dimers (CPDs) and 6-4 photoproducts (6-4 pp) are formed in the DNA strands upon exposure to UV, and these modifications in DNA lead to mutations [48,49]. As far as mutagenic potential is concerned, CPDs are far more mutagenic than 6-4 pps, because the latter are more efficiently repaired [49,50]. Both these photoproducts usually occur in runs of tandem pyrimidine residues often termed as "hot spots" for UV-induced mutations [49,50]. These mutations, such as C→T and CC→TT in response to UV exposure, are also usually referred to as UV-signature mutations [50]. However, within the skin, there exists a system of nucleotide excision repair (NER), which can normally take care of most of these deleterious mutations. There are two NER pathways: global genome repair, which repairs DNA lesions throughout the genome, and transcription coupled repair, which preferentially repairs UV-induced CPDs in the transcribing strand of DNA [51]. Consequently, human patients with the rare hereditary disorder Xenoderma pigmentosum have increased risk of developing skin cancers, and demonstrate increased sensitivity to sunlight due to a deficiency in the nucleotide excision repair. The other two hereditary human diseases associated with defective NER are Cockayne's disease, where patients exhibits hypersensitivity to cytotoxic as well as mutagenic effects of UV, and trichothiodystrophy, where human subjects have a defect in transcription coupled repair [46]. In addition, there also exists a base-excision repair system to repair the damage to DNA by reactive oxygen species such as 8-oxoguanine [52].

Regarding the susceptibility of developing cancer apart from the hereditary disorders listed above, findings of a recent study have shown a direct correlation between sunlight exposure, eye color, red hair, and frequency and age of getting sunburns to the development of basal cell carcinoma [53]. For preventing the development of skin cancer, efforts have been made to educate the population to minimize exposure to sunlight especially during midday hours, wear protective clothing when outdoors, and use sunscreens abundantly.

25.2.2 Chemopreventive Agents in Skin Health

Although the above-listed measures are strongly recommended to prevent photoaging and skin cancer, these measures often are inadequate because of limited compliance. Another effective alternate is the use of chemoprevention strategy, where prevention of skin cancer and other skin conditions is sought by the use of various pharmacological agents to either inhibit or reverse this process. These agents can be either of natural origin or synthetic in nature, and include vitamins, dietary factors/natural agents, nonsteriodal anti-inflammatory drugs, and other topical agents. Chemopreventive efficacy of these agents has been proven in various cell culture and preclinical animal models, and a few agents such as green tea polyphenols, curcumin, oral vitamin A, and difluoromethylornithine have reached the clinical trial stage in humans [54–57]. Here, we have briefly summarized a few of the most relevant studies conducted with natural agents in various models. In the ornithine decarboxylase/Ras double transgenic mouse model, where mice spontaneously develop skin tumors, administration of the green tea polyphenol (–)-epigallocatechin-3-gallate (EGCG) through drinking water significantly decreased tumor number as well as tumor burden [58]. It was observed that elevated levels of polyamines in tumor tissues as well as transformed cells sensitized them to EGCG-induced apoptosis as compared to normal cells/tissues [58]. EGCG has also been shown to inhibit photocarcinogenesis by inhibiting angiogenic factors (CD31 and VEGF, vascular endothelial growth factor) and inducing antitumor CD8+ T cells in skin tumors [59]. Antitumor-promoting activity of procyanidin B5-3′-gallate isolated from grapeseed extract was observed in 7,12-dimethylbenz(a) anthracene (DMBA)-initiated and 12-O-tetradecanoylphorbol 13-acetate (TPA) promoted a SENCAR mouse skin two-stage carcinogenesis model in terms of reduction in tumor incidence, multiplicity, and volume [60].

Sandalwood oil, traditionally used for the treatment of inflammatory and eruptive skin diseases, and its active constituent alpha-santalol, inhibited the skin tumor development and also exhibited anticancer efficacy against human epidermoid carcinoma A431 cell line [61–64]. Dwivedi and coworkers also found that pomegranate seed oil significantly decreases tumor incidence and multiplicity in the DMBA-initiated and TPA-promoted CD1 mouse skin tumor model [65]. These studies strongly suggest that phytochemicals play an important role in skin health and chemoprevention of skin cancer. Next, we elaborately discuss the important role for silibinin in managing skin health along with chemoprevention of skin cancer.

25.2.3 Silibinin and Skin Carcinogenesis Models

The SKH-1 hairless mouse serves as an excellent model to conduct skin photocarcinogenesis studies because it closely mimics an erythema-edema response, as seen in human skin exposed to UV light; the skin tumors developed in these mice are histologically similar to those seen in humans. The development of tumors in these mice depends on the dose, time, and wavelength of the UV radiation employed and has been well characterized [66]. Although haired mice after shaving could be potentially used for conducting such studies, these mice develop fibrosarcomas and tumors on eyes upon chronic exposure to UV [67]. Therefore, the SKH-1 hairless mouse is the most preferred model to study photocarcinogenesis. Our studies have clearly shown that silibinin treatment strongly inhibits UVB-induced photocarcinogenesis in SKH-1 hairless mice [68–74]. We observed that either topical application (pre-/post-UVB exposure) or dietary feeding of silibinin in SKH-1 hairless mice affords strong protection against photocarcinogenesis via inhibitory effects on DNA synthesis, cell cycle progression, and induction of apoptosis [72]. In another study, silibinin was found to inhibit UV-induced sunburn cell formation when applied topically pre- or post-UV exposure, thereby suggesting that mechanisms other than sunscreen effect might be involved in its protective effects [73]. At a clinically relevant UV dose, topical or dietary feeding of silibinin delayed the appearance of tumors and inhibited tumor multiplicity as well as tumor volume [71].

Another well-characterized *in vivo* model to study skin carcinogenesis is multistage chemical-induced carcinogenesis in mouse. In this model, animals are topically exposed on the dorsal side with a single subthreshold dose of a chemical carcinogen, such as DMBA, for tumor initiation. This is followed by repetitive application of a tumor promoter such as TPA.

Benign squamous papillomas generally develop within 10 weeks, and most of these papillomas contain Ha-ras mutations [75]. A small percentage of these papillomas eventually develop to malignant SCCs [75]. Our studies have clearly shown that silymarin has strong antitumor effect in chemical-induced skin carcinogenesis [76–78]. Silymarin treatment was reported to retard the DMBA-initiated and TPA-, benzoyl peroxide–, or okadaic acid–promoted skin carcinogenesis in SENCAR mice [76–79].

The JB6 mouse epidermal cell model is another important *in vitro* model to study skin carcinogenesis. This cell line is selectively sensitive to second-stage tumor promoters, thus is an excellent model to study tumor promotion in the two-stage process. As described by Colburn et al., JB6 cells serve as an excellent *in vitro* model to study tumor promotion as well as progression because this cell line is available in two variant forms: tumor promoter sensitive and an insensitive cell line. In this cell line, preneoplastic cells formed in response to promoters continue to exhibit anchorage dependence and are nontumorigenic even after repeated subculturing [80]. However, irreversible anchorage independence as well as tumorigenesis can also be achieved by exposure to promoters, and finally they can also be selected for promoter resistance. In the case of studies employing promoter sensitive JB6 cell system, our group found that silibinin prevents tumor promotion by inhibiting UVB- and epidermal growth factor–induced mitogenic and cell survival signaling involving both AP-1 and NF-κB [81].

The other two cell culture models that are extensively used in skin chemoprevention mechanistic studies are: A431, a human epidermoid carcinoma cell line, and human immortalized keratinocyte HaCaT cell line. Both these models (i.e., A431 and HaCaT cells) have been extensively used to evaluate the effect of silibinin on UV-induced damages [74,82]. These aspects will be dealt with in detail later in this chapter.

25.3 Mechanisms of Silibinin Action

25.3.1 Anti-Inflammatory Action of Silibinin

Inflammation is a critical component of tumor progression in skin carcinogenesis. Chronic inflammation has been shown to trigger various pro-oncogenic events, which include the release of cytokines, growth factors, chemotactic polypeptides, and prostaglandins, and these factors plays an important role in chemical- or UV-induced skin carcinogenesis [83].

Various studies focusing on the mediators of inflammation in cutaneous carcinogenic pathways have revealed the key roles for prostaglandins, cycloxygenase-2, tumor necrosis factor-α (TNF-1), AP-1, nuclear factor-kB (NF-κB), signal transducer and activator of transcription (STAT3), and others [83]. Further, several clinical conditions associated with inflammation appear to predispose the patient to increased susceptibility for skin cancer [83]. Therefore, phytochemicals that can target inflammation and associated molecular events could be promising in terms of skin protection from various damaging agents. Various studies have shown the strong anti-inflammatory action of silibinin.

Increased metabolism of arachidonic acid via lipoxygenase and cyclooxygenase (COX) pathways into hydroxyeicosatetraenoic acid and prostaglandin metabolites has been related with skin inflammation and tumor promotion [77]. Silymarin treatment has been reported to strongly inhibit the tumor promoter–induced arachidonic acid metabolism, accompanied by inhibition of lipoxygenase and cyclooxygnease activities in mouse skin [77]. Recently, we reported that chronic exposure to a physiological dose of UVB (30 mJ/cm^2) strongly increased the COX2 levels in the skin and skin tumors. Pre- or post-topical treatment or dietary feeding of silibinin was reported to strongly inhibit UV-induced COX2 levels [71].

Skin inflammation caused by UV radiation and chemical promoters is also marked by infiltration and accumulation of neutrophils. Silymarin treatment was shown to inhibit the tumor promoter–induced myeloperoxidase activity (MPO, marker for neutrophils) [77]. Silymarin was also shown to inhibit the tumor promoter-induced edema and increased levels of inflammatory cytokines TNFα and IL-1 in mouse epidermis [77]. There are numerous reports suggesting the critical role of iNOS (inducible nitric oxide synthase) and its product NO (nitric oxide) in the induction of inflammation [84]. Pre- or post-topical treatment or dietary feeding of silibinin has been shown to strongly inhibit the UV-induced iNOS expression [71]. Silymarin has also been shown to significantly inhibit the expression of iNOS as well as production of NO in UV-treated skin [85].

Various transcriptional factors (AP1, NF-κB, STATs, etc.) are involved in regulating the expression of various inflammatory cytokines [83]. Silibinin treatment was reported to strongly inhibit the UVB-induced activation of STAT3 and NF-κB in skin and skin tumors in SKH-1 hairless mice [71]. Further, silibinin treatment has also been shown to inhibit the activation of AP1 [81]. These studies suggest that silibinin targets

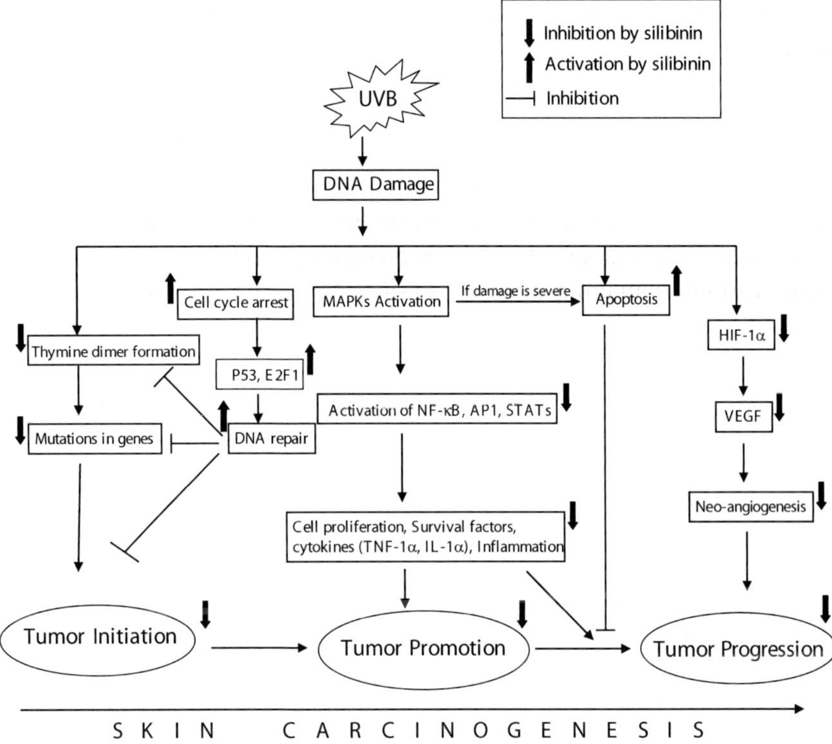

Figure 25.2 Proposed model for silibinin action against UVB-induced skin damage and carcinogenesis.

the production of inflammatory cytokines via inhibiting the activation of these transcription factors. Overall, these studies suggest a strong anti-inflammatory action of silibinin, which might be responsible for its skin-protective actions (Fig. 25.2).

25.3.2 Antiproliferative Action of Silibinin

Unlimited replicative potential is the essential characteristic of cancer cells. The basic principle of cancer chemotherapy or intervention is to inhibit the rate of cancer cell multiplication. Though most of the chemotherapeutic drugs have limitations of also affecting the normal cells adversely, cancer-chemopreventive phytochemical silibinin has been shown to be selective in action and kills primarily transformed or malignant cells [72]. As discussed earlier, silibinin treatment inhibits the growth and progression of carcinogenesis in various chemical and photocarcinogenesis models. These anticancer effects of silibinin have been correlated with its strong antiproliferative action. In uninvolved skin/normal skin, UVB (180 mJ/cm^2) irradiation for 25 weeks was shown to increase bromodeoxyuridine incorporation (marker for S-phase proliferating cells), which was inhibited strongly by dietary or topical application (pre- or post-UV) of silibinin [72]. Silibinin treatment was also reported to inhibit the UV-induced PCNA expression (marker for cell proliferation) in uninvolved skin and skin tumors, though it did not have any effect on these proliferation markers in unirradiated skin [72]. *In vitro* studies have also shown the antiproliferative efficacy of silibinin. It was shown that silibinin treatment inhibits the growth of human epidermoid carcinoma A431 cells [86]. These results were biologically significant because the growth-inhibitory effect of silibinin was at physiologically achievable doses [86].

A deregulated cell cycle has been considered as a hallmark of cancer cells and is one of the reasons for their unlimited replicative potential [87]. Cell cycle regulation involves various factors, namely: cyclins, cyclin-dependent kinases (CDKs), cyclin-dependent kinase inhibitors (CDKIs), and cell division cycle 25 (Cdc25) phosphatases. G1-S transition in the cell cycle is regulated by CDK4 and CDK6 (in association with D-type cyclins) and CDK2 (in association with cyclin E and cyclin A) [27]. The G2M transition in the cell cycle is positively regulated by Cdc2 kinase in association with cyclin B [27]. These activated CDKs phosphorylate the retinoblastoma family of proteins to release E2F transcription factors, which then regulate the expression of various genes required for cell cycle transition [27]. However, CDKIs (Kip1/p27, Cip1/p21, and INK4 family) are known to regulate the activity of CDKs [27]. Cdc25 phosphatases also regulate the cell cycle transition by removing the inhibitory phosphorylation present on CDKs [27]. Chronic exposure to UV radiation has been shown to alter the expression of key cell cycle regulatory molecules: cyclins, CDKs, and CDKIs [70,72]. Silibinin treatment was reported to inhibit the UV-induced expression of Cdk2, Cdk4, cyclin A, cyclin E, and cyclin D1 in SKH1 hairless mice [72]. Similarly, silibinin treatment decreased the UV-induced expression of Cdc25c, Cdc2, and cyclin B1 [70], but upregulated the expression of CDK inhibitors (p21 and p27) [72]. P53 is a tumor suppressor gene and is known to regulate various cell cycle regulatory molecules [88]. It is known to activate in response to stress and it stalls the cell cycle machinery via upregulating p21 [88]. In this regard, silibinin treatment was shown to further increase the UV-induced p53-positive cells in skin and skin tumors [69,72,73]. Similarly, silibinin also enhanced the UV-induced increase in the p21 positive cells in skin and skin tumors [69,73].

Also, in *in vitro* studies we observed that UV exposure (100 mJ/cm^2) resulted in S-phase arrest and silibinin pretreatment shifted it to G1 or G2M arrest in JB6 cells [89]. All these studies suggest that the antiproliferative action of silibinin could be due to its modulatory influence on cell cycle regulatory molecules, resulting in cell cycle arrest (Fig. 25.2).

25.3.3 Pro- or Antiapoptotic Action of Silibinin

Apoptosis refers to programmed cell death and is initiated by various intrinsic and extrinsic signals. Apoptosis is regulated by many players, including bcl2 family members, caspases, survivin, and p53. DNA is the primary target of UVB-induced damage, causing either apoptotic death of damaged cells or fixation of mutations in various genes. Generally, lower doses of UVB cause DNA mutation, leading to tumor initiation, whereas higher doses result in irreparable DNA damage, causing apoptosis (sunburn) and eventually leads to cell elimination. Silibinin has been shown to have dual action against UV-induced photodamage. Silibinin protects normal human skin keratinocytes from sunburn or apoptosis when the damage is moderate, but silibinin acts differently, causing cell death when the UVB damage is severe (Fig. 25.2). In this regard, silibinin treatment was reported to further increase the apoptosis induced by UVB alone (180 mJ/cm^2 for 25 weeks) in skin tumors and uninvolved skin in SKH1 hairless mice [72]. Caspases are known as executors of apoptosis, and silibinin treatment was shown to increase the activation of caspase 3 compared to UVB alone in uninvolved skin and skin tumors [72]. Survivin is another important negative apoptotic regulator. Though it is absent in most of the normal tissues, it is highly expressed in cancers, making it a promising target for cancer therapy. The level of survivin was strongly increased in skin tumors (UVB-generated) and silibinin treatment strongly inhibited the expression levels of survivin in tumor samples, which might be responsible for the apoptotic action of silibinin [72]. In another study, silibinin inhibited apoptosis and sunburn cell formation by UV180 mJ/cm^2 exposure (once or for 5 days) in SKH hairless mice [68].

In vitro studies have also suggested the variable effect of silibinin on UV-induced apoptosis, depending upon the severity of the damage. Silibinin treatment was reported to induce apoptotic cell death in human epidermoid carcinoma A431 cells at physiological achievable doses [86]. In another experiment, silibinin treatment before UVB (5–100 mJ/cm^2) exposure resulted in increased apoptosis, whereas silibinin treatment after UVB exposure decreased the extent of apoptosis in A431 cells [82]. Silibinin pretreatment increased the activation of caspases compared to UVB alone, whereas post-treatment decreased the activation of caspases. Furthermore, silibinin treatment prior to or immediately after UVB exposure altered Bcl2, Bax, Bak, and cytochrome c levels in mitochondria and cytosol in favor of or against apoptosis, respectively [82]. Similarly, the dual efficacy of silibinin was also observed in HaCaT human immortalized keratinocytes, where it strongly prevented the apoptosis induced by lower doses of UVB (15 and 30 mJ/cm^2) while it enhanced the apoptosis induced by higher doses of UVB (120 mJ/cm^2) [74]. It was suggested that silibinin protects HaCaT cells from lower doses of UVB-induced apoptosis at least in part via inhibition of caspase activation, restoration of survivin expression, and upregulation of NF-κB [74]. Silibinin treatment also modulated the levels, interaction, and translocation of Bcl2 family members to target intracellular membranes in affording protection against UVB-induced apoptosis. Additionally, silibinin inhibited the heterodimerization of Bad with Bcl-xL, compared to UV alone. These studies suggest that at least part of the protective effect of silibinin against UVB-caused apoptosis involves the modulation of Bcl2 family members [74]. In another study we reported that silibinin treatment enhances UVB-induced (100 mJ/cm^2) apoptosis via upregulation of DNA-protein-kinase-dependent p53 activation in mouse epithelial JB6 cells [89]. Overall, these *in vivo* and *in vitro* studies reveal the dual efficacy of silibinin in protecting or enhancing UVB-caused apoptosis, and they also suggest that silibinin works as a UVB damage sensor to exert its biological action.

25.3.4 Silibinin Modulates Aberrant Mitogenic Signaling

The activation of mitogen-activated protein kinases (MAPKs) by tumor-promoting agents plays a major role in tumor promotion and malignant transformation [2,41]. For example, the UVB-mediated signaling events are primarily mediated through MAPKs, including extracellular signal-regulated kinase 1/2 (ERK1/2), c-jun-NH$_2$-kinase 1/2 (JNK1/2), and p38 kinase. Once activated, these serine/threonine kinases translocate to the nucleus and phosphorylate target transcription factors, including AP1. Activation of SAPK/JNK and MAPK/p38 has been reported mainly in response to various stress signals such as UV, heat, and osmotic shock, which then activates numerous transcription factors involved in induction of apoptosis and promotion of tumorigenesis [2]. ERK1/2 is also an important signaling molecule that is upregulated in response to a variety of stresses and is known to be essential for cell survival. In addition, ERK1/2 has been implicated to mediate cell transformation and tumor promotion caused by diverse tumor promoters, including UV and hydrogen peroxide [2,90]. Other than MAPKs, PI3K/Akt is the other crucial cell survival and antiapoptotic pathway that has been shown to play a role in tumor promotion and progression [91]. Both MAPK and Akt signaling converge at downstream transcription factor AP1 and NF-κB activation.

We have reported in *in vitro* studies that silibinin treatment significantly inhibited ERK1/2 activation in A431 cells. The inhibition of ERK1/2 was in part responsible for growth inhibition and death of A431 cells by silibinin [86]. But silibinin treatment activated the other MAPKs, namely JNK1/2 and p38, and this was suggested to be associated with increased apoptosis by silibinin [86]. Similarly, silibinin pre- and post-treatment inhibited the UV-induced (400 mJ) phosphorylation of ERK1/2 and Akt in JB6 cells [81]. In another study, pretreatment of silibinin was shown to increase the ERK1/2 phosphorylation as compared to UVB treatment in HaCaT cells [74]. Similarly, silibinin treatment was shown to enhance the UVB-caused phosphorylation and kinase activity of ERK1/2 and Akt [89]. As mentioned earlier, activation of AP-1 and NF-κB transcription factors plays a critical role in UVB-induced tumor promotion (Fig. 25.2). Increased activity of these transcription factors is directly linked to the activation of mitogenic and survival signaling for tumor promotion as well as progression responses. Silibinin treatment inhibited the UV-induced AP-1 and NF-κB activation in JB6 cells [81].

In vivo studies have also shown the variable effect of silibinin on MAPK signaling. Silibinin treatment was reported to increase the activation of ERK1/2, JNK1/2, and p38 in skin tumors generated by UVB exposure [72]. In contrast to the findings in tumors, silibinin treatment inhibited the phosphorylation of ERK1/2, JNK1/2, and p38 in uninvolved skin [72]. Further, a single UVB exposure (180 mJ/cm^2) or chronic exposure for 5 days resulted in ERK1/2, JNK1/2, MAPK/p38, and Akt phosphorylation in SKH-1 mouse skin, and silibinin treatment strongly inhibited the activation of these molecules [68]. Similarly, silibinin inhibited the UVB-induced mitogenic and survival signaling in chronically UVB-exposed skin [70]. Silibinin treatment inhibited the chronic UV-induced phosphorylation of ERK1/2, JNK1/2, and p38 kinase as well as Akt in both 15-week and 25-week treatment [70].

These studies suggest that silibinin activates or inhibits MAPKs depending upon the extent of damage caused by UV and exerts its biological response accordingly. Overall, the modulation of mitogenic signaling could be the important component of silibinin efficacy against UVB-induced tumorigenesis (Fig. 25.2).

25.3.5 Antiangiogenic Action of Silibinin

Angiogenesis is considered an important mediator of tumor progression. As a tumor expands, diffusion distance from the existing vascular supply increases, resulting in hypoxic conditions within the tumor. Therefore, sustained expansion of a tumor needs new blood vessel formation to provide rapidly proliferating tumor cells with an adequate supply of oxygen and metabolites [92]. Recently, we reported that UV exposure (30 mJ/cm^2/day for 25 weeks) resulted in increased angiogenesis in the skin and skin tumors, which was inhibited by silibinin treatment [71]. Silibinin treatment (pre- or post-topical, or dietary) significantly inhibited the PECAM-1 levels (an endothelial cell marker) in UVB-exposed skin and UVB-induced skin tumors [71]. But silibinin treatment did not alter the PECAM-1 level in age-matched unexposed skin. VEGF is a known mitogen for endothelial cells and is known to promote angiogenesis [93]. In our studies we reported that clinically relevant doses of UVB treatment increased the VEGF expression in both skin and skin tumors, whereas silibinin treatment significantly reduced the VEGF level in both skin and skin tumors [71] (Fig. 25.1). Hypoxia inducible factor-1α (HIF-1α) is a transcriptional factor and is known to regulate the expression of key angiogenic factors, including VEGF, and plays a crucial role in pathophysiologic angiogenesis [94]. Our findings showed that silibinin treatment significantly inhibited the UVB-induced HIF-1α levels in exposed skin and skin tumors [71] (Fig. 25.2). These results strongly suggest the strong antiangiogenic effects of silibinin and might be contributing to its overall efficacy against photocarcinogenesis (Fig. 25.2).

25.3.6 Silibinin Repairs UV-Induced DNA Damage

As mentioned earlier, the carcinogenic and mutagenic effects of UV radiation are attributed to the induction of DNA damage and errors in repair and replication [95,96]. In normal cells there are varieties of mechanisms that constantly monitor and repair most of the damage inflicted by UV light [95,96]. But severe DNA damage could make DNA repair machinery redundant, and mutations might occur. We have shown that silibinin treatment strongly inhibits the UVB-induced (180 mJ/cm^2) thymine dimer-positive cells in the skin of SKH1 hairless mice [69,73]. These studies suggest that silibinin might alleviate the UV-caused DNA damage by potentiating the DNA repair machinery.

Tumor suppressor gene p53 also plays an important role in DNA damage response by causing cell cycle arrest, providing additional time for DNA repair, or by inducing cell death by apoptosis when DNA damages are too severe to repair [88,97]. Silibinin treatment was shown to further increase the UV-induced p53-positive cells in the epidermis. Similarly, silibinin also enhances the UV-induced increase in the p21 positive cells. These results were accompanied with decreased apoptosis and sunburn cell formation [69,73]. These studies suggest that silibinin activates p53, which upregulates the expression of its downstream target p21 and induces cell cycle arrest, which provides cells with sufficient time for DNA damage repair.

Various studies have shown that UV exposure of mouse or human skin or epidermal keratinocytes results in depletion of antioxidant enzymes and strong oxidative stress [98]. This effect is mediated via generation of various reactive free radicals, including reactive oxygen species (ROS), which play an important role in posing an oxidative challenge via targeting crucial cellular machinery: lipids, proteins, DNA, etc. [98]. Both silymarin and silibinin have been shown to possess potent antioxidant potential, which might be responsible for the protective efficacy of these compounds. In this regard, silymarin has been shown to provide protection against BPO-induced depletion of SOD, catalase, and glutathione peroxidase (GPX) in SENCAR mouse skin [76]. The phase II enzymes glutathione S transferase (GST) and quinine reductase (QR) are known to play important roles in detoxification of carcinogens and their metabolites. Silibinin was shown to increase the activity of GST and QR in various tissues including skin [99]. Thus, the antioxidant capacity of silibinin could also indirectly contribute in lessening the extent of DNA damage via decreasing ROS production or by decreasing oxidative stress.

Transcriptional factor E2F1 has been shown to play an important role in DNA repair and suppression of apoptosis [100]. Silibinin treatment was shown to upregulate E2F1 in skin chronically exposed to UVB (180 mJ/cm^2, 2 days/week for 15 weeks and 25 weeks) in SKH1 hairless mice. In contrast, silibinin inhibited the chronic UV-induced expression of E2F1 in skin tumors, suggesting the dual nature of silibinin action in exposed skin and skin tumors [70]. This study showed that silibinin treatment could repair the UV-induced damage in skin while it promotes apoptosis in tumors. Overall, these studies suggested that silibinin treatment could protect the skin from photodamage by enhancing the cellular DNA repair mechanisms (Fig. 25.2).

25.4 Toxicity and Bioavailability of Silibinin

25.4.1 Nontoxic Nature of Silibinin

Because of its widespread use for liver health in the form of milk thistle extract, toxicity due to silibinin's consumption has never been a serious issue. This might be the reason for lack of studies/reports about its acute or chronic toxicity even in preclinical animal models. With the development of more bioavailable forms such as siliphos, silipide, and legalon, studies have been conducted in animals and humans to evaluate the health beneficial effects. Various traditional toxicological tests have proven the nontoxic nature of siliphos, and it was reported safe for rats and monkeys at oral doses up to 2,000 mg/kg/day, after 13-week subacute toxicity studies [101]. In 26-week chronic toxicity studies, oral doses up to 1,000 mg/kg/day were well tolerated in dogs and rats [101]. In another 26-week oral toxicity study, rats were fed a daily 2,000 mg/kg dose of siliphos, which is equivalent to 160 g daily for a 176-pound (80 kg) human [101]. In this study, body weight, liver weight, and enzyme indicators of liver damage (AST, ALT) remained within the normal healthy range of the untreated control rats, confirming the nontoxic nature of this compound [101]. Even in patients with liver conditions, treatment with 240 mg or 360 mg daily dose of siliphos for the period of 4 months did not cause any severe adverse effects except in 5.2% of total patients enrolled in the study, who reported nausea, heartburn, dyspepsia, or transient headaches [101].

In our *in vivo* studies we have reported that silymarin and silibinin are well tolerated and have no toxicity in terms of body weight gain, diet consumption, and animal behavior [99,102,103]. In our phase I clinical trials we have reported that 13 g/day of silibinin in the form of phytosome (siliphos), in three divided doses was well tolerated in patients with advanced prostate cancer [104]. The asymptomatic liver toxicity at higher doses was the most commonly seen adverse effect in this study [104].

Studies conducted with silibinin in the form of a new pharmacological complex (silybin-vitamin E-phospholipids, sold under the trade name as RealSIL-BI by Lorenzi Pharmaceuticals, Italy) in patients with nonalcoholic fatty liver disease with or without hepatitis C virus–related chronic hepatitis demonstrated some therapeutic effects, such as improvement in insulin resistance and hepatic fibrosis, and did not report any adverse effects [22]. Patients enrolled in this study were treated with 4 pieces/day of complex, where each piece contained 94 mg of silibinin, 194 mg of phosphotidylcholine, and 90 mg of vitamin E [22]. In another long term, 12-month, open-controlled study with silymarin in two well-matched groups of insulin-treated diabetics with alcoholic cirrhosis, administration of 600 mg silymarin per day along with standard therapy did not show any toxicity in patients, though it reduced insulin resistance, endogenous insulin overproduction, and the need for exogenous insulin [105]. Overall, reports of adverse effects of silibinin or silymarin are rare; however, cases of nausea, epigastric discomfort, arthralgia, pruritus, headache, urticaria, and mild laxative effect have been reported [106].

25.4.2 Bioavailability of Silibinin

Silibinin, being insoluble in water, demonstrates low bioavailability. However, bioavailability of silibinin has been extensively studied in both animal models (preclinical) as well as, more recently, in humans. Pharmacokinetic analysis of silymarin, where silibinin is a major active constituent, given to healthy volunteers has revealed that these flavonoids are rapidly metabolized to their conjugates such as glucuronides, and can be detected in the human plasma [107]. In one of our earliest studies, peak levels of free silibinin after oral administration of silibinin (50 mg/kg dose) to SENCAR mice were achieved at 0.5 hour postadministration in liver, lung, stomach, and pancreas, whereas, in the case of skin and prostate, peak levels of silibinin were achieved 1 hour postadministration [99]. In another study conducted in rats, bioavailability of silibinin (silybin) phospholipid complex and silybin-N-methylglucamine was studied after oral administration. In this study, it was observed that the mean plasma concentration time curve for silybin in both the forms was first order [108]. In another study in humans, administration of a single dose of silibinin-phosphatidylcholine complex (Silipide, 80 mg of silibinin) to 12 healthy human volunteers revealed that free silibinin concentration in plasma reached a peak of 141 ± 32 ng/ml in 2.4 hours postadministration and half-life in plasma was found to be about 2 hours. It was also observed that conjugated silibinin metabolites had slower elimination as compared to free silibinin [109].

In recent years, studies have been conducted with different formulations of silibinin to achieve enhanced bioavailability. In one such study, oral administration of silibinin, in formulation with phosphatidylcholine

(Silipide) at the doses of 360 mg, 720 mg, and 1440 mg to colorectal cancer patients for 7 days did not produce any adverse effects, and dosages used were termed as safe [110]. At these dosage levels, various metabolites of silibinin such as silibinin mono- and di-glucuronide, silibinin monosulfate, and silibinin glucuronide sulfate were detectable in the plasma of the study subjects [110]. The free levels of silibinin were also achieved in liver and colorectal tissue [110]. Bioavailability of silybin phospholipid complex was greater than silybin-N-methylglucamine due to improved lipophilicity of this complex [108]. In our phase I clinical trial we reported that the half-life of plasma silibinin was short, ranging from 1.79–4.99 hours and peak plasma levels of silibinin were in excess of 100 μM [104]. Silibinin was quickly conjugated and excreted into the urine, with a mean silibinin level of 6.4 μM in the urine [104]. From the findings of the above-mentioned studies, it is clear that bioavailability of silibinin at physiologically relevant concentrations can be achieved by employing different formulations.

25.5 Conclusions and Future Perspectives

Recent statistics for skin cancer with almost one million new cases to be diagnosed in a year is a strong indication that more effective measures are required to control this malignancy than just educating the population to minimize the exposure to sunlight. In this direction, prevention/intervention of this malignancy by pharmacological agents of synthetic or natural origin hold promise. Based on numerous studies, it seems that silibinin can be effectively used in improving skin health in general, and reduction of skin cancer risk in particular. More interestingly, silibinin has shown efficacy against UV-induced skin damage when applied topically, as well as when supplemented in the diet in various preclinical models. Thus, it has a strong potential to be developed as a chemopreventive agent against skin cancer and photoaging and is already being marketed in improved bioavailable forms. Clinical studies conducted with these formulations have not reported any serious adverse effects, thus giving added advantage to this agent's being developed as a future cure for skin ailments.

Acknowledgements

Original studies are supported in part by the NCI RO1 grants CA64514, CA102514, CA104286, CA112304, CA113876, and CA116636. Authors Manjinder Kaur and Gagan Deep contributed equally to this work.

References

1. Hikino H, Kiso Y, Wagner H, Fiebig M. Antihepatotoxic actions of flavonolignans from *Silybum marianum* fruits. Planta Med 1984;50:248–250.
2. Singh RP, Agarwal R. Mechanisms and preclinical efficacy of silibinin in preventing skin cancer. Eur J Cancer 2005;41:1969–1979.
3. Kroll DJ, Shaw HS, Oberlies NH. Milk thistle nomenclature: why it matters in cancer research and pharmacokinetic studies. Integr Cancer Ther 2007; 6:110–119.
4. Gyorgy I, Antus S, Blazovics A, Foldiak G. Substituent effects in the free radical reactions of silybin: radiation-induced oxidation of the flavonoid at neutral pH. Int J Radiat Biol 1992;61:603–609.
5. Basaga H, Poli G, Tekkaya C, Aras I. Free radical scavenging and antioxidative properties of 'silibin' complexes on microsomal lipid peroxidation. Cell Biochem Funct 1997;15:27–33.
6. Valenzuela A, Garrido A. Biochemical bases of the pharmacological action of the flavonoid silymarin and of its structural isomer silibinin. Biol Res 1994;27:105–112.
7. Saller R, Meier R, Brignoli R. The use of silymarin in the treatment of liver diseases. Drugs 2001;61:2035–2063.
8. Wesolowska O, Lania-Pietrzak B, Kuzdzal M, Stanczak K, Mosiadz D, Dobryszycki P, Ozyhar A, Komorowska M, Hendrich AB, Michalak K. Influence of silybin on biophysical properties of phospholipid bilayers. Acta Pharmacol Sin 2007;28:296–306.
9. Lee DG, Kim HK, Park Y, Park SC, Woo ER, Jeong HG, Hahm KS. Gram-positive bacteria specific properties of silybin derived from *Silybum marianum*. Arch Pharm Res 2003;26:597–600.
10. Lania-Pietrzak B, Michalak K, Hendrich AB, Mosiadz D, Grynkiewicz G, Motohashi N, Shirataki Y. Modulation of MRP1 protein transport by plant, and synthetically modified flavonoids. Life Sci 2005;77:1879–1891.
11. Luper S. A review of plants used in the treatment of liver disease: part 1. Altern Med Rev 1998;3:410–421.
12. Kren V, Walterova D. Silybin and silymarin: new effects and applications. Biomed Pap Med Fac Univ Palacky Olomouc Czech Repub 2005;149:29–41.
13. Schumann J, Prockl J, Kiemer AK, Vollmar AM, Bang R, Tiegs G. Silibinin protects mice from T cell–dependent liver injury. J Hepatol 2003;39:333–340.
14. Jayaraj R, Deb U, Bhaskar AS, Prasad GB, Rao PV. Hepatoprotective efficacy of certain flavonoids against microcystin induced toxicity in mice. Environ Toxicol 2007;22:472–479.
15. Vogel G, Tuchweber B, Trost W, Mengs U. Protection by silibinin against *Amanita phalloides* intoxication in beagles. Toxicol Appl Pharmacol 1984; 73:355–362.
16. Mereish KA, Bunner DL, Ragland DR, Creasia DA. Protection against microcystin-LR-induced hepatotoxicity by silymarin: biochemistry, histopathology, and lethality. Pharm Res 1991;8:273–277.
17. Zima T, Kamenikova L, Janebova M, Buchar E, Crkovska J, Tesar V. The effect of silibinin on experimental cyclosporine nephrotoxicity. Ren Fail 1998;20:471–479.

18. Crocenzi FA, Roma MG. Silymarin as a new hepatoprotective agent in experimental cholestasis: new possibilities for an ancient medication. Curr Med Chem 2006;13:1055–1074.
19. Gaedeke J, Fels LM, Bokemeyer C, Mengs U, Stolte H, Lentzen H. Cisplatin nephrotoxicity and protection by silibinin. Nephrol Dial Transplant 1996; 11:55–62.
20. Cruz T, Galvez J, Crespo E, Ocete MA, Zarzuelo A. Effects of silymarin on the acute stage of the trinitrobenzenesulphonic acid model of rat colitis. Planta Med 2001;67:94–96.
21. Locher R, Suter PM, Weyhenmeyer R, Vetter W. Inhibitory action of silibinin on low density lipoprotein oxidation. Arzneimittelforschung 1998;48:236–239.
22. Federico A, Trappoliere M, Tuccillo C, de Sio I, Di Leva A, Del Vecchio Blanco C, Loguercio C. A new silybin-vitamin E-phospholipid complex improves insulin resistance and liver damage in patients with non-alcoholic fatty liver disease: preliminary observations. Gut 2006;55:901–902.
23. Agarwal C, Singh RP, Dhanalakshmi S, Tyagi AK, Tecklenburg M, Sclafani RA, Agarwal R. Silibinin upregulates the expression of cyclin-dependent kinase inhibitors and causes cell cycle arrest and apoptosis in human colon carcinoma HT-29 cells. Oncogene 2003;22:8271–8282.
24. Hogan FS, Krishnegowda NK, Mikhailova M, Kahlenberg MS. Flavonoid, silibinin, inhibits proliferation and promotes cell-cycle arrest of human colon cancer. J Surg Res 2007;143:58–65.
25. Singh RP, Deep G, Blouin MJ, Pollak MN, Agarwal R. Silibinin suppresses *in vivo* growth of human prostate carcinoma PC-3 tumor xenograft. Carcinogenesis 2007;28:2567–2574.
26. Raina K, Blouin MJ, Singh RP, Majeed N, Deep G, Varghese L, Glode LM, Greenberg NM, Hwang D, Cohen P, Pollak MN, Agarwal R. Dietary feeding of silibinin inhibits prostate tumor growth and progression in transgenic adenocarcinoma of the mouse prostate model. Cancer Res 2007;67:11083–11091.
27. Deep G, Singh RP, Agarwal C, Kroll DJ, Agarwal R. Silymarin and silibinin cause G1 and G2-M cell cycle arrest via distinct circuitries in human prostate cancer PC3 cells: a comparison of flavanone silibinin with flavanolignan mixture silymarin. Oncogene 2006;25:1053–1069.
28. Thelen P, Wuttke W, Jarry H, Grzmil M, Ringert RH. Inhibition of telomerase activity and secretion of prostate specific antigen by silibinin in prostate cancer cells. J Urol 2004;171:1934–1938.
29. Zi X, Agarwal R. Silibinin decreases prostate-specific antigen with cell growth inhibition via G1 arrest, leading to differentiation of prostate carcinoma cells: implications for prostate cancer intervention. Proc Natl Acad Sci U S A 1999;96:7490–7495.
30. Singh RP, Deep G, Chittezhath M, Kaur M, Dwyer-Nield LD, Malkinson AM, Agarwal R. Effect of silibinin on the growth and progression of primary lung tumors in mice. J Natl Cancer Inst 2006;98:846–855.
31. Tyagi AK, Agarwal C, Chan DC, Agarwal R. Synergistic anti-cancer effects of silibinin with conventional cytotoxic agents doxorubicin, cisplatin and carboplatin against human breast carcinoma MCF-7 and MDA-MB468 cells. Oncol Rep 2004;11:493–499.

32. Malewicz B, Wang Z, Jiang C, Guo J, Cleary MP, Grande JP, Lu J. Enhancement of mammary carcinogenesis in two rodent models by silymarin dietary supplements. Carcinogenesis 2006;27:1739–1747.
33. Verschoyle RD, Brown K, Steward WP, Gescher AJ. Consumption of silibinin, a flavonolignan from milk thistle, and mammary cancer development in the C3(1) SV40 T, t antigen transgenic multiple mammary adenocarcinoma (TAg) mouse. Cancer Chemother Pharmacol 2008;62:369–372.
34. Provinciali M, Papalini F, Orlando F, Pierpaoli S, Donnini A, Morazzoni P, Riva A, Smorlesi A. Effect of the silybin-phosphatidylcholine complex (IdB 1016) on the development of mammary tumors in HER-2/neu transgenic mice. Cancer Res 2007;67:2022–2029.
35. Son YG, Kim EH, Kim JY, Kim SU, Kwon TK, Yoon AR, Yun CO, Choi KS. Silibinin sensitizes human glioma cells to TRAIL-mediated apoptosis via DR5 up-regulation and down-regulation of c-FLIP and survivin. Cancer Res 2007;67:8274–8284.
36. Cheung CW, Vesey DA, Nicol DL, Johnson DW. Silibinin inhibits renal cell carcinoma via mechanisms that are independent of insulin-like growth factor-binding protein 3. BJU Int 2007;99:454–460.
37. Chu SC, Chiou HL, Chen PN, Yang SF, Hsieh YS. Silibinin inhibits the invasion of human lung cancer cells via decreased productions of urokinase-plasminogen activator and matrix metalloproteinase-2. Mol Carcinog 2004;40:143–149.
38. Lee SO, Jeong YJ, Im HG, Kim CH, Chang YC, Lee IS. Silibinin suppresses PMA-induced MMP-9 expression by blocking the AP-1 activation via MAPK signaling pathways in MCF-7 human breast carcinoma cells. Biochem Biophys Res Commun 2007;354:165–171.
39. Chen PN, Hsieh YS, Chiang CL, Chiou HL, Yang SF, Chu SC. Silibinin inhibits invasion of oral cancer cells by suppressing the MAPK pathway. J Dent Res 2006;85:220–225.
40. Jemal A, Siegel R, Ward E, Hao Y, Xu J, Murray T, Thun MJ. Cancer statistics. CA Cancer J Clin 2008;58:71–96.
41. Deep G, Agarwal R. Chemopreventive efficacy of silymarin in skin and prostate cancer. Integr Cancer Ther 2007;6:130–145.
42. Hussein MR. Ultraviolet radiation and skin cancer: molecular mechanisms. J Cutan Pathol 2005;32:191–205.
43. McIntyre WJ, Downs MR, Bedwell SA. Treatment options for actinic keratoses. Am Fam Physician 2007;76:667–671.
44. Sober AJ, Burstein JM. Precursors to skin cancer. Cancer 1995;75:645–650.
45. de Gruijl FR. Skin cancer and solar UV radiation. Eur J Cancer 1999; 35:2003–2009.
46. Ichihashi M, Ueda M, Budiyanto A, Bito T, Oka M, Fukunaga M, Tsuru K, Horikawa T. UV-induced skin damage. Toxicology 2003;189:21–39.
47. Xu Y, Shao Y, Voorhees JJ, Fisher GJ. Oxidative inhibition of receptor-type protein-tyrosine phosphatase kappa by ultraviolet irradiation activates epidermal growth factor receptor in human keratinocytes. J Biol Chem 2006; 281:27389–27397.
48. Mitchell DL. The induction and repair of lesions produced by the photolysis of (6-4) photoproducts in normal and UV-hypersensitive human cells. Mutat Res 1988;194:227–237.

49. Clingen PH, McCarthy PJ, Davies RJ. Enzymatic incision of UV-irradiated DNA at sites of purine photoproducts. Biochem Soc Trans 1992;20:67S.
50. Sarasin A. The molecular pathways of ultraviolet-induced carcinogenesis. Mutat Res 1999;428:5–10.
51. van der Wees C, Jansen J, Vrieling H, van der Laarse A, Van Zeeland A, Mullenders L. Nucleotide excision repair in differentiated cells. Mutat Res 2007;614:16–23.
52. Moriwaki S, Takahashi Y. Photoaging and DNA repair. J Dermatol Sci 2008;50:169–176.
53. Pelucchi C, Di Landro A, Naldi L, La Vecchia C. Risk factors for histological types and anatomic sites of cutaneous basal-cell carcinoma: an Italian case-control study. J Invest Dermatol 2007;127:935–944.
54. Stratton SP, Dorr RT, Alberts DS. The state-of-the-art in chemoprevention of skin cancer. Eur J Cancer 2000;36:1292–1297.
55. Linden KG, Carpenter PM, McLaren CE, Barr RJ, Hite P, Sun JD, Li KT, Viner JL, Meyskens FL. Chemoprevention of nonmelanoma skin cancer: experience with a polyphenol from green tea. Recent Results Cancer Res 2003;163:165–171; discussion 264–266.
56. Einspahr JG, Bowden GT, Alberts DS. Skin cancer chemoprevention: strategies to save our skin. Recent Results Cancer Res 2003;163:151–164; discussion 264–266.
57. Carbone PP, Douglas JA, Larson PO, Verma AK, Blair IA, Pomplun M, Tutsch KD. Phase I chemoprevention study of piroxicam and alpha-difluoromethylornithine. Cancer Epidemiol Biomarkers Prev 1998;7:907–912.
58. Paul B, Hayes CS, Kim A, Athar M, Gilmour SK. Elevated polyamines lead to selective induction of apoptosis and inhibition of tumorigenesis by (-)-epigallocatechin-3-gallate (EGCG) in ODC/Ras transgenic mice. Carcinogenesis 2005;26:119–124.
59. Mantena SK, Roy AM, Katiyar SK. Epigallocatechin-3-gallate inhibits photocarcinogenesis through inhibition of angiogenic factors and activation of CD8+ T cells in tumors. Photochem Photobiol 2005;81:1174–1179.
60. Zhao J, Wang J, Chen Y, Agarwal R. Anti-tumor-promoting activity of a polyphenolic fraction isolated from grape seeds in the mouse skin two-stage initiation-promotion protocol and identification of procyanidin B5-3′-gallate as the most effective antioxidant constituent. Carcinogenesis 1999;20:1737–1745.
61. Dwivedi C, Zhang Y. Sandalwood oil prevents skin tumour development in CD1 mice. Eur J Cancer Prev 1999;8:449–455.
62. Dwivedi C, Guan X, Harmsen WL, Voss AL, Goetz-Parten DE, Koopman EM, Johnson KM, Valluri HB, Matthees DP. Chemopreventive effects of alpha-santalol on skin tumor development in CD-1 and SENCAR mice. Cancer Epidemiol Biomarkers Prev 2003;12:151–156.
63. Bommareddy A, Hora J, Cornish B, Dwivedi C. Chemoprevention by alpha-santalol on UVB radiation–induced skin tumor development in mice. Anticancer Res 2007;27:2185–2188.
64. Kaur M, Agarwal C, Singh RP, Guan X, Dwivedi C, Agarwal R. Skin cancer chemopreventive agent, {alpha}-santalol, induces apoptotic death of human epidermoid carcinoma A431 cells via caspase activation together with

dissipation of mitochondrial membrane potential and cytochrome c release. Carcinogenesis 2005;26:369–380.

65. Hora JJ, Maydew ER, Lansky EP, Dwivedi C. Chemopreventive effects of pomegranate seed oil on skin tumor development in CD1 mice. J Med Food 2003;6:157–161.
66. de Gruijl FR, Forbes PD. UV-induced skin cancer in a hairless mouse model. Bioessays 1995;17:651–660.
67. van Kranen HJ, de Gruijl FR, de Vries A, Sontag Y, Wester PW, Senden HC, Rozemuller E, van Kreijl CF. Frequent p53 alterations but low incidence of ras mutations in UV-B-induced skin tumors of hairless mice. Carcinogenesis 1995;16:1141–1147.
68. Gu M, Dhanalakshmi S, Mohan S, Singh RP, Agarwal R. Silibinin inhibits ultraviolet B radiation–induced mitogenic and survival signaling, and associated biological responses in SKH-1 mouse skin. Carcinogenesis 2005; 26:1404–1413.
69. Gu M, Dhanalakshmi S, Singh RP, Agarwal R. Dietary feeding of silibinin prevents early biomarkers of UVB radiation–induced carcinogenesis in SKH-1 hairless mouse epidermis. Cancer Epidemiol Biomarkers Prev 2005;14: 1344–1349.
70. Gu M, Singh RP, Dhanalakshmi S, Mohan S, Agarwal R. Differential effect of silibinin on E2F transcription factors and associated biological events in chronically UVB-exposed skin versus tumors in SKH-1 hairless mice. Mol Cancer Ther 2006;5:2121–2129.
71. Gu M, Singh RP, Dhanalakshmi S, Agarwal C, Agarwal R. Silibinin inhibits inflammatory and angiogenic attributes in photocarcinogenesis in SKH-1 hairless mice. Cancer Res 2007;67:3483–3491.
72. Mallikarjuna G, Dhanalakshmi S, Singh RP, Agarwal C, Agarwal R. Silibinin protects against photocarcinogenesis via modulation of cell cycle regulators, mitogen-activated protein kinases, and Akt signaling. Cancer Res 2004;64:6349–6356.
73. Dhanalakshmi S, Mallikarjuna GU, Singh RP, Agarwal R. Silibinin prevents ultraviolet radiation–caused skin damages in SKH-1 hairless mice via a decrease in thymine dimer positive cells and an up-regulation of p53-p21/Cip1 in epidermis. Carcinogenesis 2004;25:1459–1465.
74. Dhanalakshmi S, Mallikarjuna GU, Singh RP, Agarwal R. Dual efficacy of silibinin in protecting or enhancing ultraviolet B radiation–caused apoptosis in HaCaT human immortalized keratinocytes. Carcinogenesis 2004;25:99–106.
75. Owens DM, Wei S, Smart RC. A multihit, multistage model of chemical carcinogenesis. Carcinogenesis 1999;20:1837–1844.
76. Zhao J, Lahiri-Chatterjee M, Sharma Y, Agarwal R. Inhibitory effect of a flavonoid antioxidant silymarin on benzoyl peroxide–induced tumor promotion, oxidative stress and inflammatory responses in SENCAR mouse skin. Carcinogenesis 2000;21:811–816.
77. Zhao J, Sharma Y, Agarwal R. Significant inhibition by the flavonoid antioxidant silymarin against 12-O-tetradecanoylphorbol 13-acetate-caused modulation of antioxidant and inflammatory enzymes, and cyclooxygenase 2 and interleukin-1alpha expression in SENCAR mouse epidermis: implications in the prevention of stage I tumor promotion. Mol Carcinog 1999;26:321–333.

78. Lahiri-Chatterjee M, Katiyar SK, Mohan RR, Agarwal R. A flavonoid antioxidant, silymarin, affords exceptionally high protection against tumor promotion in the SENCAR mouse skin tumorigenesis model. Cancer Res 1999;59:622–632.
79. Zi X, Mukhtar H, Agarwal R. Novel cancer chemopreventive effects of a flavonoid antioxidant silymarin: inhibition of mRNA expression of an endogenous tumor promoter TNF alpha. Biochem Biophys Res Commun 1997;239:334–339.
80. Colburn NH, Wendel E, Srinivas L. Responses of preneoplastic epidermal cells to tumor promoters and growth factors: use of promoter-resistant variants for mechanism studies. J Cell Biochem 1982;18:261–270.
81. Singh RP, Dhanalakshmi S, Mohan S, Agarwal C, Agarwal R. Silibinin inhibits UVB- and epidermal growth factor–induced mitogenic and cell survival signaling involving activator protein-1 and nuclear factor-kappaB in mouse epidermal JB6 cells. Mol Cancer Ther 2006;5:1145–1153.
82. Mohan S, Dhanalakshmi S, Mallikarjuna GU, Singh RP, Agarwal R. Silibinin modulates UVB-induced apoptosis via mitochondrial proteins, caspases activation, and mitogen-activated protein kinase signaling in human epidermoid carcinoma A431 cells. Biochem Biophys Res Commun 2004;320:183–189.
83. Nickoloff BJ, Ben-Neriah Y, Pikarsky E. Inflammation and cancer: is the link as simple as we think? J Invest Dermatol 2005;124:x–xiv.
84. Korhonen R, Lahti A, Kankaanranta H, Moilanen E. Nitric oxide production and signaling in inflammation. Curr Drug Targets Inflamm Allergy 2005; 4:471–479.
85. Katiyar SK. Treatment of silymarin, a plant flavonoid, prevents ultraviolet light–induced immune suppression and oxidative stress in mouse skin. Int J Oncol 2002;21:1213–1222.
86. Singh RP, Tyagi AK, Zhao J, Agarwal R. Silymarin inhibits growth and causes regression of established skin tumors in SENCAR mice via modulation of mitogen-activated protein kinases and induction of apoptosis. Carcinogenesis 2002;23:499–510.
87. Hanahan D, Weinberg RA. The hallmarks of cancer. Cell 2000;100: 57–70.
88. Ponten F, Berne B, Ren ZP, Nister M, Ponten J. Ultraviolet light induces expression of p53 and p21 in human skin: effect of sunscreen and constitutive p21 expression in skin appendages. J Invest Dermatol 1995;105: 402–406.
89. Dhanalakshmi S, Agarwal C, Singh RP, Agarwal R. Silibinin up-regulates DNA-protein kinase-dependent p53 activation to enhance UVB-induced apoptosis in mouse epithelial JB6 cells. J Biol Chem 2005;280:20375–20383.
90. Chang H, Oehrl W, Elsner P, Thiele JJ. The role of H_2O_2 as a mediator of UVB-induced apoptosis in keratinocytes. Free Radic Res 2003;37: 655–663.
91. Testa JR, Bellacosa A. AKT plays a central role in tumorigenesis. Proc Natl Acad Sci U S A 2001;98:10983–10985.
92. Liao D, Johnson RS. Hypoxia: a key regulator of angiogenesis in cancer. Cancer Metastasis Rev 2007;26:281–290.

93. Detmar M. Tumor angiogenesis. J Investig Dermatol Symp Proc 2000; 5:20–23.
94. Koga K, Nabeshima K, Nishimura N, Shishime M, Nakayama J, Iwasaki H. Microvessel density and HIF-1alpha expression correlate with malignant potential in fibrohistiocytic tumors. Eur J Dermatol 2005;15:465–469.
95. Matsumura Y, Ananthaswamy HN. Short-term and long-term cellular and molecular events following UV irradiation of skin: implications for molecular medicine. Expert Rev Mol Med 2002;4:1–22.
96. Matsumura Y, Ananthaswamy HN. Toxic effects of ultraviolet radiation on the skin. Toxicol Appl Pharmacol 2004;195:298–308.
97. Ouhtit A, Gorny A, Muller HK, Hill LL, Owen-Schaub L, Ananthaswamy HN. Loss of Fas-ligand expression in mouse keratinocytes during UV carcinogenesis. Am J Pathol 2000;157:1975–1981.
98. Pinnell SR. Cutaneous photodamage, oxidative stress, and topical antioxidant protection. J Am Acad Dermatol 2003;48:1–19; quiz 20–2.
99. Zhao J, Agarwal R. Tissue distribution of silibinin, the major active constituent of silymarin, in mice and its association with enhancement of phase II enzymes: implications in cancer chemoprevention. Carcinogenesis 1999;20:2101–2108.
100. Berton TR, Mitchell DL, Guo R, Johnson DG. Regulation of epidermal apoptosis and DNA repair by E2F1 in response to ultraviolet B radiation. Oncogene 2005;24:2449–2460.
101. Kidd P, Head K. A review of the bioavailability and clinical efficacy of milk thistle phytosome: a silybin-phosphatidylcholine complex (Siliphos). Altern Med Rev 2005;10:193–203.
102. Singh RP, Dhanalakshmi S, Tyagi AK, Chan DC, Agarwal C, Agarwal R. Dietary feeding of silibinin inhibits advance human prostate carcinoma growth in athymic nude mice and increases plasma insulin-like growth factor–binding protein-3 levels. Cancer Res 2002;62:3063–3069.
103. Singh RP, Mallikarjuna GU, Sharma G, Dhanalakshmi S, Tyagi AK, Chan DC, Agarwal C, Agarwal R. Oral silibinin inhibits lung tumor growth in athymic nude mice and forms a novel chemocombination with doxorubicin targeting nuclear factor kappaB-mediated inducible chemoresistance. Clin Cancer Res 2004;10:8641–8647.
104. Flaig TW, Gustafson DL, Su LJ, Zirrolli JA, Crighton F, Harrison GS, Pierson AS, Agarwal R, Glode LM. A phase I and pharmacokinetic study of silybin-phytosome in prostate cancer patients. Invest New Drugs 2007;25: 139–146.
105. Velussi M, Cernigoi AM, De Monte A, Dapas F, Caffau C, Zilli M. Long-term (12 months) treatment with an anti-oxidant drug (silymarin) is effective on hyperinsulinemia, exogenous insulin need and malondialdehyde levels in cirrhotic diabetic patients. J Hepatol 1997;26:871–879.
106. Wellington K, Jarvis B. Silymarin: a review of its clinical properties in the management of hepatic disorders. BioDrugs 2001;15:465–489.
107. Wen Z, Dumas TE, Schrieber SJ, Hawke RL, Fried MW, Smith PC. Pharmacokinetics and metabolic profile of free, conjugated, and total silymarin flavonolignans in human plasma after oral administration of milk thistle extract. Drug Metab Dispos 2008;36:65–72.

108. Yanyu X, Yunmei S, Zhipeng C, Qineng P. The preparation of silybin-phospholipid complex and the study on its pharmacokinetics in rats. Int J Pharm 2006;307:77–82.

109. Gatti G, Perucca E. Plasma concentrations of free and conjugated silybin after oral intake of a silybin-phosphatidylcholine complex (silipide) in healthy volunteers. Int J Clin Pharmacol Ther 1994;32:614–617.

110. Hoh C, Boocock D, Marczylo T, Singh R, Berry DP, Dennison AR, Hemingway D, Miller A, West K, Euden S, Garcea G, Farmer PB, Steward WP, Gescher AJ. Pilot study of oral silibinin, a putative chemopreventive agent, in colorectal cancer patients: silibinin levels in plasma, colorectum, and liver and their pharmacodynamic consequences. Clin Cancer Res 2006;12:2944–2950.

Index